Chromatographic Analysis
of the Environment

Chromatographic Analysis of the Environment

EDITED BY

Robert L. Grob, Ph.D., F.A.I.C.

Professor of Analytical Chemistry
Villanova University
Villanova, Pennsylvania

MARCEL DEKKER, INC.   New York and Basel

To My Wife and Children

CONTRIBUTORS

OSMAN M. ALY, Department of Environmental Engineering, Campbell Soup Company, Campbell Place, New Jersey

DANIEL F. BENDER, Environmental Protection Agency, Cincinnati, Ohio

ROBERT S. BRAMAN, Department of Chemistry, University of South Florida, Tampa, Florida

THOMAS G. BUNTING,* Chase Brass and Copper Company, Kennecott Copper Corporation, Cleveland, Ohio

GORDON CHESTERS, Water Resources Center, Department of Soil Science and Water Chemistry Program, University of Wisconsin-Madison, Madison, Wisconsin

RENATO C. DELL'ACQUA, Division of Laboratories and Research, New York State Health Department, Albany, New York

WALTER C. ELBERT, Water Science Branch, National Field Investigations Center — Cincinnati, Office of Enforcement and General Counsil, U.S. Environmental Protection Agency, Cincinnati, Ohio

SAMUEL D. FAUST, Department of Environmental Science, Rutgers — The State University, New Brunswick, New Jersey

D. A. GRAETZ,** Water Resources Center, Department of Soil Science and Water Chemistry Program, University of Wisconsin-Madison, Madison, Wisconsin

ROBERT L. GROB, Department of Chemistry, Villanova University, Villanova, Pennsylvania

CHARLES E. HAMILTON, Waste Control, The Dow Chemical Company, Midland, Michigan

---

*Present Address: Ross Laboratories, Division of Abbott Laboratories, Columbus, Ohio

**Present Address: Soil Department, Institute of Food and Agricultural Sciences, University of Florida, Gainesville, Florida

JOSEPH V. HUNTER, Department of Environmental Science, Cook College, Rutgers University, New Brunswick, New Jersey

BARBARA S. JACOBSON, Ross Laboratories, Columbus, Ohio

KINGSLEY KAY, Department of Environmental Medicine, Mt. Sinai School of Medicine, New York, New York

EUGENE J. McGONIGLE, Merck Sharp and Dohme Research Laboratories, Pharmaceutical Research and Development, West Point, Pennsylvania

E. J. SOWINSKI, Western Electric Company, Inc., Allentown, Pennsylvania

IRWIN H. SUFFET, Department of Chemistry, Environmental Engineering and Science, Drexel University, Philadelphia, Pennsylvania

WAYNE THORNBURG, Del Monte Corporation Research Center, Walnut Creek, California

HAROLD F. WALTON, Department of Chemistry, University of Colorado, Boulder, Colorado

PREFACE

A monograph of this type, in order to cover the particular subject, must be complex. Environmental studies, by nature, are complex and a condensed survey-type book would serve little purpose to those engaged in the various environmental services. As a meaningful and working definition of environment I have chosen "the gross amount of surrounding entities and/or conditions that influence the well-being or preservation of mankind."

The need for a comprehensive series of methods for determining the various toxic substances in the atmosphere, water, waste effluents, and soil is generally recognized. This book has been written to provide a comprehensive work of chromatographic techniques that are of practical value to consultants, engineers, chemists, and students. It is hoped that it can end or lessen the need for tedious searches through the mass of scattered literature. Every effort has been made to present the material as simply and clearly as possible. Information contained herein may prove to be valuable in establishing the criteria for choosing one technique in lieu of another for a specific sample type and analysis. It certainly should be of value in teaching because it provides a primary framework for existing knowledge.

The ultimate goal in the scientific understanding of the environment as well as the development of scientific technology for its control is the benefit of mankind.

Some of our present-day drugs and pesticides are hazardous; their presence causes need for solicitude. Tobacco smoke, automobile exhaust fumes, drinking water, industrial effluents, and bad air may easily be included as of concern. Perhaps the two notorious examples of substances that hypothetically menace the health of mankind are pesticides and radioactive isotopes.

Chemists, because of their training and special skills, must play an important role in the understanding of the air, water, and land environment as well as in the prevention of their degradation. Due to the ecological interrelationships of the environment there needs to be more interdependence of research between chemists and other scientific disciplines concerned with man's environment. Our responsibility of monitoring and controlling toxic materials would be very simple if all such contaminants were radioactive — a radiation detector would suffice.

The term "environment" not only includes the vast out of doors (macro-climates) but the enclosed environment where man toils (localized or specific microclimates). Industry must have the capability of detecting and preventing atmospheric contaminants in these immediate surroundings. Toxic materials are a necessity for many industries; methods of detection and prevention means must be available so that concentrations are maintained at minimum levels. Thus, it is not amiss to emphasize that the most important single step toward prevention of environmental illness is the control of foreign bodies in our air, soil, and water supplies.

Frequently, speed, accuracy, and specificity of analyses have been greatly enhanced by the development and/or modification of procedures. Equally significant are the discoveries and utilization of powerful and improved methods of separation, indispensable in the investigation of complex systems where even the most refined of available methods are inadequate for distinguishing and measuring with sufficient precision.

In a field as extensive and as rapidly expanding as chromatography the practicing chemist is obliged to maintain close scrutiny over the current literature. He or she must be sufficiently grounded in the basic principles of chemistry to enable him or her to carry out tasks in the most intelligent, efficient, and scientific manner. It is hoped that the contributors to this monograph have aided in this understanding.

To have a self-contained reference source this volume commences with a plenary chapter on the theory and practice of chromatography. This enables minimization and repetition from chapter to chapter on such similar subjects as general theory, experimental techniques, and commonly used chromatographic systems. Thus, the book should be useful to the chemist (novice or expert in chromatography), practicing engineer, technician, or teacher.

In organizing the chapters of this monograph it has been decided to cover the main chromatographic techniques (gas, liquid, paper, thin-layer, and ion exchange) for each of the four environmental areas, i.e., air, water, soil, and waste. Only in the areas of air pollution and waste chemistry has it been decided not to cover all techniques. In these two areas, the extent to which ion exchange has been utilized does not warrant a chapter.

In preparing a work of this nature it is impossible to succeed without the support and cooperation of many people. I am grateful to many people, who in some way have made the monograph a reality. Special thanks would have to go to Dr. Calvin Calmon for aid in securing a contributor; to Dr. Harold F. Walton, Dr. Eugene J. McGonigle, and Mrs. Barbara S. Jacobson for their willingness to come to the aid of a fellow scientist; and to Dr. Thomas G. Bunting who read and commented on most of the manuscripts. Last, but by no means least, my thanks go to my family for their understanding during the long and quiet evenings.

Robert L. Grob

CONTENTS

PART I

PLENARY SECTION

Chapter 1

THEORY AND PRACTICE OF CHROMATOGRAPHY

Thomas G. Bunting*

Chase Brass and Copper Company
Kennecott Copper Corporation
Cleveland, Ohio

*Present Address: Ross Laboratories
                   Division of Abbott Laboratories
                   Columbus, Ohio

3

I. INTRODUCTION

A Russian botanist, Mikhail Tswett [1], used the word "chromatography" to describe his separation of plant pigments, which was effected by passing an extract of the pigments through a column packed with calcium carbonate. The result was a series of colored zones on the column and, thus, the name "chromatography" from the Greek <u>chromatus</u> and <u>graphein</u>, meaning "color" and "to write." Since Tswett, a wide variety of independent techniques that have little or nothing to do with color have come to be called chromatography. At least one, paper chromatography, was recognized centuries before Tswett, while most evolved from his work [2].

"Chromatography" now refers to any of a diverse group of techniques that effect a separation through a distribution of sample between two immiscible phases. Further qualification is necessary to distinguish chromatography from other separation techniques, such as extraction. The stipulation is thus added that one phase be stationary while the second phase be mobile and percolate through the first phase. Practically, the mobile phase is gas or liquid, while the stationary phase is a liquid or a solid. The separation of the components, or solutes, of a sample results from differences in their rates of adsorption, solution, or reaction with the mobile and stationary phases. The nature of the mobile and stationary phases, the type of inter-action between the two phases and the solute, and the physical arrangement of the stationary phase must be considered in distinguishing the many types of chromatography. The physical states of the mobile and stationary phases give rise to four basic types of chromatography: gas-liquid chromatography, GLC; gas-solid chromatography, GSC; liquid-liquid chromatography, LLC; and liquid-solid chromatography, LSC.

Of the four basic types of chromatography, the two gas systems are independent and not subject to subdivisions. Thus, the terms GLC and GSC adequately identify the technique involved. The liquid systems are not so easily described and are not mutually exclusive, which can lead to consider-able confusion. Liquid-solid chromatography may include column chroma-tography, thin-layer chromatography, and ion-exchange chromatography. "Liquid-liquid chromatography" generally is reserved for the liquid analog of gas-liquid chromatography, but other forms are possible. Finally, paper chromatography appears to be a combination of liquid-liquid and liquid-solid chromatography. Unfortunately, the profusion of nomenclature does not end here. Table 1, which omits some specialized systems, notably gel perme-ation and electrophoresis, describes 11 frequently encountered designations. The significance of Table 1 is that the common terms are insufficient to adequately describe a chromatographic system. Ideally, one should specify the physical states of the two phases, e.g., liquid-solid, the nature of the separation, e.g., adsorption, and the configuration of the system, e.g., columnar.

TABLE 1

Types of Chromatography

Adsorption	Any system where the solutes are resolved by selective adsorption on a solid stationary phase; generally refers to columnar liquid-solid chromatography but may be used for gas-solid, thin-layer, and paper
Column	Any technique in which the stationary phase is contained in a column: usually designates a liquid-solid adsorption system but may include ion exchange and, rarely, gas-liquid and gas-solid systems
Gas (GC)	Gas-liquid or gas-solid chromatography
Gas-liquid (GLC)	The system utilizing a gaseous mobile phase and a liquid stationary phase which is supported either by fine particles packed in a tube or by the walls of the tube itself
Gas-solid (GSC)	A gaseous mobile phase and a solid adsorbent stationary phase contained in a column
Ion exchange	Any system in which the stationary phase is an ion-exchange resin: configuration is usually columnar, but may be thin-layer or paper
Liquid (LC)	Technically refers to any system with a liquid mobile phase but is commonly used for columnar liquid-solid adsorption chromatography

TABLE 1 (Continued)

Liquid-liquid (LLC)	Mobile and stationary phases are liquid: generally refers to the analog of the gas-liquid system but may include ion-exchange and paper chromatography
Liquid-solid (LSC)	Liquid mobile phase; solid stationary phase: name used infrequently but technically includes columnar adsorption, ion-exchange, thin-layer, and paper chromatography
Paper	Paper strip or sheet is the stationary phase; cellulose or shredded paper in a column is generally treated as columnar liquid-solid adsorption chromatography
Thin-layer	Literally, any system in which the stationary phase is in the form of a thin layer: usually refers to liquid-solid adsorption but may be used for ion exchange in the form of a thin layer

TABLE 2

Chromatographic Systems

System	Mobile phase	Stationary phase	Configuration	Separation
Gas	Gas	Liquid	Column	Partition
	Gas	Solid	Column	Adsorption
Liquid	Liquid	Liquid	Column	Partition
	Liquid	Solid	Column	Adsorption
Paper	Liquid	Paper	Sheet or strip	Partition or adsorption
Thin-layer	Liquid	Solid	Thin film	Adsorption
Ion exchange	Liquid	Solid	Column	Ionic replacement reactions

For the purposes of environmental analysis, the classification of chromatographic systems according to Table 2 is expedient. The development of chromatographic theory and practice has been sufficiently nonstructured as to allow reclassification of the 11 systems of Table 1 into the five broad types in Table 2. This organization is convenient for a general consideration of chromatographic theory as well as for a specific treatment of chromatographic practice as applied to environmental analysis.

II. THEORY

A. General

1. Analysis Type

In a chromatographic system a solute is distributed between a stationary phase and a mobile phase that flows through the stationary phase. One may classify the analysis by the nature, not the physical state, of the mobile phase or eluent. If sample is used as the mobile phase, the technique is frontal analysis. Displacement analysis involves the use of an eluent that has a greater affinity for the stationary phase than does the solute. The use of a mobile phase that simply transports the solute through the system constitutes elution chromatography.

Frontal analysis, the simplest form of chromatography, requires the continuous addition of sample to the system. The results obtained from this system are depicted in Fig. 1. The chromatogram is interpreted in the following manner. If the components of the sample have different affinities for the stationary phase, a number of zones will be formed. The leading edge of a zone is called the "front"; thus the name, "frontal analysis." The first zone contains only the least retained solute A since it is moving fastest. The second zone will contain solutes A and B, the third zone, A and B and C, etc. The amount of A decreases as B appears and both A and B decrease as C appears because displacement analysis is involved in

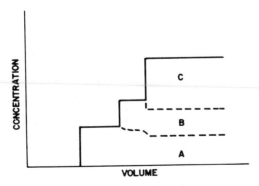

Figure 1. Frontal analysis.

all zones but A. Eventually, since sample is added continuously, the effluent from the system becomes identical in composition to the sample entering the system. In frontal analysis only one component, A, can be obtained in pure form, but the number of components in a sample can be determined since the steps in the chromatogram equal the number of components in the sample.

In displacement analysis, the mobile phase contains a substance that has a greater affinity for the stationary phase than does the solute. A single portion of sample is added and the components are moved through the system by virtue of their displacement in the stationary phase by the eluent. It should be noted that the displacer will replace the most strongly retained component C, which in turn will replace B, which will displace the least retained solute A. A chromatogram similar to Fig. 2 results. Each component can be recovered in pure form, but not quantitatively since the zones ideally have an interface and actually some overlap. One can, however, quantitatively measure each component with this technique. If equilibrium exists, a given amount of stationary phase can retain only a given amount of solute. As the solute concentration increases, then, the lengths, not the heights, of the zones must increase. Consequently, the volume of mobile phase required to elute a solute is proportional to the concentration of the solute.

Elution analysis involves the use of a mobile phase, which ideally serves only to transport the solutes through the system. Separation occurs because the various solutes have different affinities for the stationary phase and thus different speeds through the system. A typical elution chromatogram is presented in Fig. 3. Conceivably, elution analysis can completely resolve even the most complex mixtures into their individual components, which can then be analyzed by conventional techniques. The great separating

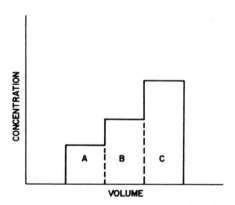

Figure 2. Displacement analysis.

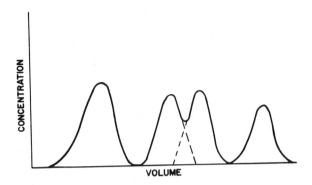

Figure 3. Elution analysis.

power of elution chromatography accounts for its being the most common form of chromatography, almost to the exclusion of the other forms. Consequently, unless otherwise noted, the remainder of this chapter applies specifically to elution chromatography.

2. The Chromatographic Process

A representation of the chromatographic process for the elution of a single solute is presented in Fig. 4. The sample is introduced to the system in what ideally would be an infinitely small volume and is immediately divided between the mobile and stationary phases through some physical or chemical process, e.g., adsorption, partition, or ion exchange. The unit volume Δx is described by:

$$\Delta x = \Delta V_m + \Delta V_s \tag{1}$$

where ΔV_m and ΔV_s are the unit volumes of the mobile and stationary phases, respectively. If C_m and C_s are the respective concentrations of solute in the mobile and stationary phases, then an equilibrium distribution coefficient K can be defined as the ratio of the concentration of the solute in the stationary phase to the concentration of solute in the mobile phase, i.e.,

$$K = C_s / C_m \tag{2}$$

This gives rise to the situation depicted in Fig. 4a. Now as the mobile phase progresses through the system, C_m is carried to the next unit volume and is redistributed between the two phases according to Eq. (2).

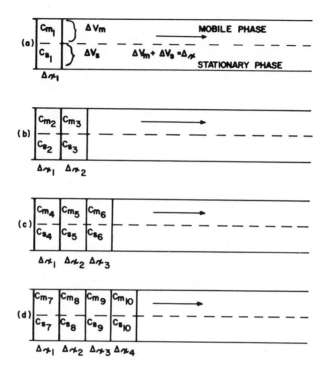

Figure 4. Schematic representation of solute distribution.

Simultaneously, C_S has remained at the first unit volume and has also redistributed according to Eq. (2). This leads to Fig. 4b, where:

$$K = C_{s_2}/C_{m_2} = C_{s_3}/C_{m_3} \tag{3}$$

The next step in the process leads to Fig. 4c, where C_{m_2} has moved to ΔV_2 and C_{m_3} has moved to ΔV_3. C_{m_2} plus C_{s_3} redistribute as does C_{s_2} and also C_{m_3}. This phenomenon of solute movement–solute distribution continues until the sample has moved through the system. It should be noted that the process, conveniently depicted here as discontinuous, is actually continuous.

Consideration of Eq. (2) and Fig. 4 leads to the observation that the relative rate of movement of a given solute is inversely proportional to the distribution coefficient K. For large values of K most of the solute is in the stationary phase and thus immobile. Conversely, if K is small, the solute moves rapidly through the system. Now, the various constituents of a multicomponent sample will behave as the single solute described in Fig. 4.

Assuming a lack of solute-solute interaction, each sample component equilibrates between the two phases according to its distribution coefficient and moves through the system at a rate that is inversely proportional to the distribution coefficient. This is the basis of all chromatographic separations. Any two materials with different distribution coefficients will travel with different velocities and consequently will eventually be separated.

Figure 4 presents the chromatographic process as a series of extractions analogous to Craig's countercurrent distribution system [3]. In extraction systems, the effectiveness of the separation is described by a separation factor β which is defined as:

$$\beta = K_1/K_2 \tag{4}$$

where K_1 and K_2 are the distribution coefficients of the two solutes. Two species can only be effectively separated by an extraction, then, if β is very small or very large, i.e., K_1 is very different than K_2. The tremendous contribution of chromatography stems from its ability to resolve a pair of solutes for which β approaches unity, i.e., where K_1 is very similar to K_2. This resolving power of chromatography is due to its ability to perform what amounts to hundreds and even thousands of batch extractions rapidly and simultaneously.

3. Plate Theory

The plate concept of chromatography is inherited from distillation theory, where originally the more efficient distillation columns actually contained discrete plates. Martin and Synge [4] considered a chromatographic system to be composed of a number of imaginary or theoretical plates. Now the number of theoretical plates N that a column appears to contain is a poor measure of chromatographic performance since the number of plates is independent of the system dimensions. Consequently, the height equivalent of a theoretical plate, defined as the thickness of a layer necessary to permit the solution coming from the layer to have been in equilibrium with the average concentration of the solute in the mobile phase in that layer, presents a better measure of chromatographic effectiveness. Then the efficiency of the system is described by:

$$H = HETP = L/N \tag{5}$$

where H or HETP is the height equivalent of a theoretical plate, L is the length of the system in the direction of mobile phase velocity, and N is the number of plates.

Now, if a quantity of solute C in an M-ml volume is placed in the system as in Fig. 4, the fractions Y and Z of the solute in the stationary and mobile phases respectively are:

$$Y = C_s \Delta V_s / (C_m \Delta V_m + C_s \Delta V_s) \tag{6a}$$

and

$$Z = C_m \Delta V_m / (C_m \Delta V_m + C_s \Delta V_s) \tag{6b}$$

The distribution of solute in Fig. 4a then has fraction Z in ΔV_m volume of the mobile phase and fraction Y in ΔV_s volume of the stationary phase. If a volume ΔV_m is now added to the column, an amount ZM ml of solute is moved to plate 2, where it again equilibrates to give YZM ml in the stationary phase and Z^2M ml in the mobile phase, while plate 1 reequilibrates to give YZM ml in the mobile phase and YM^2 ml in the stationary phase. Continuation of this process by successive additions of ΔV_m results in the solute distribution described in Table 3. This is the binominal distribution defined by $(Y + Z)^R$, where R is the number of ΔV_m volumes added. The quantity of solute in any plate is then given by:

$$Q_{N+1} = MR! \, (Y)^{R-N} (Z)^N / N! \, (R - N)! \tag{7}$$

where Q_{N+1} is the quantity of solute in plate number (N + 1).

If plate number (N + 1) contains more solute than any other plate when R mobile phase volumes have been added, it follows that (N + 1) contains a greater quantity of solute when R has been added than it does when either (R − 1) or (R + 1) volumes have been added. Then Eq. (7) yields:

$$\frac{M(R + 1)! \, Y^{(R+1-N)} Z^N}{N! \, (R + 1 - N)!} < \frac{MR! \, Y^{R-N} Z^N}{N! \, (R - N)!} > \frac{M(R - 1)! \, Y^{(R-1-N)} Z^N}{N! \, (R - 1 - N)!} \tag{8}$$

The solution of this inequality gives:

$$N \cong RZ \tag{9}$$

and when R is very large the following approximations are valid:

$$Y^{R-N} \cong Y^R \cong (1 - Z)^R \cong e^{-RZ} \tag{10}$$

and

$$R! / (R - N)! \cong R^N \tag{11}$$

TABLE 3

Solute Distribution in Chromatographic Systems

R, No. of ΔV_m's	Plate numbers				
	1	2	3	4	5
0	M				
1	MY	MZ			
2	MY^2	2MYZ	MZ^2		
3	MY^3	$3MY^2Z$	$3MYZ^2$	MZ^3	
4	MY^4	$4MY^3Z$	$6MY^2Z^2$	$4MYZ^3$	MZ^4

The use of Eqs. (10) and (11) with Sterling's approximation permits Eq. (7) to be rewritten as:

$$Q_{N+1} = M(RZ)^N e^{(N-RZ)}/(2\pi N)^{1/2} N^N \tag{12}$$

However, Eq. (9) indicates that for the concentration maximum $N = RZ$, and thus:

$$Q_{max} = M/(2\pi N)^{1/2} \tag{13}$$

The maximum rate of escape S of solute from the system is:

$$S = NQ_{max}/t \tag{14}$$

since the rate of movement of solute through a system of N plates in time t is N/t. Equations (13) and (14) yield:

$$N = 2\pi(St/M)^2 \tag{15}$$

and since S is related to the maximum peak height while M is related in the same way to peak area,

$$N = 2\pi(\text{height} \times \text{time/area})^2 \tag{16}$$

N can be experimentally determined from a chromatogram.

The plate theory of chromatography, as introduced by Martin and Synge [4] and utilized by James and Martin [5], depends on the following assumptions: (1) that the distribution coefficient α is constant, (2) that diffusion in the direction of flow is negligible, (3) that the rate of equilibration of solute between the two phases is fast compared with the velocity of the mobile phase, and (4) that the system be considered as discontinuous, consisting of a number of volume elements with one equilibration occurring in each. In general, the first assumption is usually valid, while the others are not particularly defensible. However, while development of various rate theories based on continuous flow models for gas chromatography systems by Glueckauf [6], van Deemter et al. [7], and Purnell [8] is more rigorous than the plate model, the mathematically simpler plate theory provides essentially the same result. The reader may also be interested in the relatively simple conservation of mass approach developed by Rony [9].

Alternate expressions for the number of theoretical plates include the Glueckauf equation:

$$N = 8(t/\beta)^2 \tag{17}$$

where t is the elution time (or volume) and β is the peak width at $1/e$ of the peak height and is equal to $2\,\sigma\sqrt{2}$ for a gaussian distribution; and the van Deemter equation:

$$N = (4t/w)^2 \tag{18}$$

where w is the distance between the points of intersection of the tangents to the inflection points with the baseline. The experimental measurements of N described by Eqs. (16)-(18) are explained graphically in Fig. 5.

4. Selectivity

While the plate theory and the various rate theories of chromatography describe the distribution of a solute in the system, selectivity is a consideration of the relative distributions of two or more solutes in a chromatographic process. The ability of chromatography to separate two solutes depends on the selectivity of the process and the degree to which the system can thermodynamically distinguish between the solutes. Selectivity, then, is simply a measure of the relative retentions or relative mobilities of two solutes. The relative retention difference is defined in Fig. 6 as the difference between zone centers, ΔV, where the zone or chromatographic peak center is at the peak maximum for a gaussian-shaped curve.

As seen above, a solute is retained, i.e., has a long retention time or a low mobility, in proportion to the size of its distribution coefficient. The magnitude of the distribution coefficient is determined by the physicochemical nature of the solute and of the mobile and stationary phases. Consequently, it is only necessary to select a mobile and stationary phase combination that

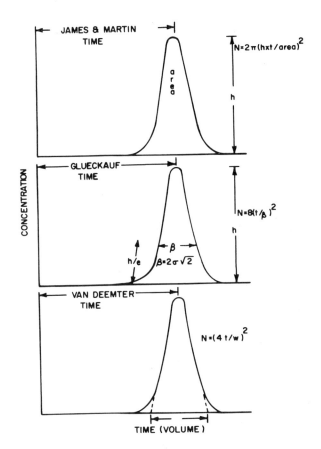

Figure 5. Experimental measurement of N, the number of theoretical plates.

provides at least slightly different distribution coefficients and separation becomes possible. Therefore, the primary requirement for any chromatographic system demands that the solutes have different distribution coefficients, i.e.,

$$K_1 \neq K_2 \tag{19}$$

Besides depending on the nature of the two phases, however, selectivity is also a function of the relative amounts of the two phases. Equation (2) defined the distribution coefficient as:

$$K = C_s/C_m \tag{20}$$

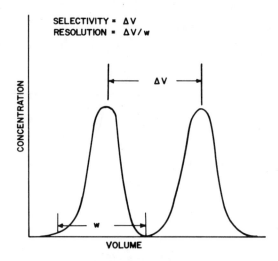

Figure 6. Graphical definition of selectivity.

Now, since K is reasonably constant, a change in the amount of one of the phases, normally the stationary phase, results in corresponding changes in solute distribution to preserve Eq. (2). The net effect is a change in the elution time due to the change in the amount of solute in the mobile phase.

Equation (6) defined the fraction of solute in the mobile phase Z as:

$$Z = C_m \Delta V_m / (C_m \Delta V_m + C_s \Delta V_s) \tag{6}$$

Substitution of $C_s = KC_m$ [cf. Eq. (1)] leads to:

$$Z = \Delta V_m / (\Delta V_m + K \Delta V_s) \tag{20}$$

Now according to Eq. (9), $N = RZ$ when the peak maximum is in the $(N + 1)$ plate. Then if N is the total number of plates in the system, plate $(N + 1)$ is the collection or detection system, and $R \Delta V_m$ is the total volume of eluent required to move the solute peak through the system. Therefore, the retention or elution volume V is:

$$V = R \Delta V_m = N \Delta V_m / Z \tag{21}$$

and from Eq. (20),

$$V = N \Delta V_m + KN \Delta V_s \tag{22}$$

However, $N \Delta V_m$ is the total mobile phase volume of the system V_m (void or dead volume) and $N \Delta V_s$ is the total volume of stationary phase V_s. Thus,

$$V = V_m + KV_s \tag{23}$$

For a pair of solutes, then, chromatographic selectivity can be measured by the difference in their zone centers (Fig. 6), which is the difference in their retention volumes ΔV,

$$\Delta V = V_2 - V_1 = (V_m + K_2 V_s) - (V_m + K_1 V_s) = \Delta K V_s \tag{24}$$

Equation (24) demonstrates that selectivity is proportional to differences in distribution coefficients and to the quantity of stationary phase. Once the two phases are chosen, ΔK at a given temperature is fixed and zone center separation can be improved only by changing V_s. This stationary phase volume can be varied by changing the size of the chromatographic system, e.g., length, or by increasing the amount of stationary phase per unit volume. Therefore, ΔV can be increased by increasing the surface area of adsorbents, by increasing the liquid content of partition systems, and by increasing the charge density of ion-exchange resins.

Changing the length of the chromatographic system is the most common manner of increasing selectivity since doubling the length doubles V_s and therefore ΔV. It should be noted, however, that while increasing the system length gives a corresponding increase in zone center separation, a similar increase in zone separation or resolution with an increase in length cannot be inferred.

5. Resolution

The resolution of a chromatographic system is a measure of the completeness of separation of two solutes. While selectivity is a description of zone center separation, resolution is a measure of zone separation. Resolution, R, is a function of zone center separation and zone width as shown in Fig. 6 and Eq. (25):

$$R = \Delta V / [(w_1 + w_2)/2] \tag{25}$$

where ΔV is the retention or elution volume difference and w is the baseline band width. For adjacent peaks of similar size, R is essentially given by:

$$R = \Delta V / w \tag{26}$$

When Eqs. (24) and (26) are combined,

$$R = \frac{\Delta K V_s}{w} \tag{27}$$

If a solute retained the infinitely small volume it was assumed to have had on entering the system, the chromatogram would appear as a series of very sharp peaks and zone center separation would essentially represent true resolution. Unfortunately, as a solute migrates through the system, a number of factors conspire to cause the solute zone to broaden. Figure 7 illustrates that resolution depends on both band center separation and band width and that one can increase separation by decreasing the zone width as well as by increasing zone center separation. In Fig. 7a, solutes A and B

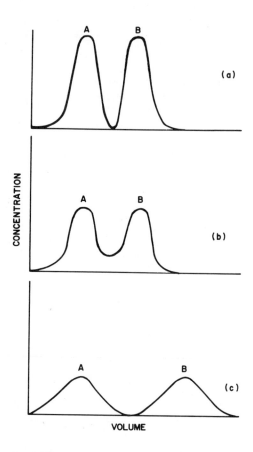

Figure 7. Difference between selectivity and resolution.

have a zone center separation equal to the zone center separation in Fig. 7b, yet because of increased band widths A and B are not resolved completely in Fig. 7b. The band widths in Fig. 7c are greater than those in Fig. 7b but since the band center separation in Fig. 7c is much larger than that in Fig. 7b complete separation still occurs in Fig. 7c.

Thus, resolution is a function of both selectivity and band width. Equation (24) indicates that selectivity, and therefore resolution, is a function of differences in distribution coefficients and the quantity of stationary phase. However, ΔV is proportional to V_s and therefore to L, the system length. From Eq. (27), then, R is also a function of L. However, R is not directly proportional to L because of the denominator in Eq. (27). When Eq. (18) is rewritten, it is found that:

$$w = 4t/\sqrt{N} \tag{28}$$

However, elution time t is equivalent to elution volume V and therefore is proportional to L [cf. Eq. (23)], while $\sqrt{N}$ is proportional to $\sqrt{L}$ [cf. Eq. (5)]. Therefore,

$$w \propto \sqrt{L} \tag{29}$$

Now, from Eq. (27),

$$R \propto L \quad \text{and} \quad R \propto 1/w \tag{30}$$

Thus, resolution is proportional to the square root of the system length in the direction of mobile phase velocity,

$$R \propto \sqrt{L} \tag{31}$$

Consequently, zone center separation is directly proportional to the length of the system but zone separation is proportional to the square root of the system length.

6. Efficiency

Chromatographic efficiency is a measure of the ability of the system to prevent zone spreading. The height equivalent of a theoretical plate, H, defined by Eq. (5) as:

$$H = L/N \tag{5}$$

is the parameter used to characterize band broadening, which arises from a variety of sources. Classically (cf. Ref. 7), band broadening is attributed to three concurrent phenomena assumed to be independent: flow velocity inequalities, molecular diffusion, and nonequilibrium. The van Deemter equation in simplified form is:

$$H = A + B/\bar{v} + C\bar{v} \tag{32}$$

where A represents the flow velocity inequality, the so-called eddy diffusion contribution; $B/\bar{v}$ is the molecular diffusion term; and $C\bar{v}$ accounts for the nonequilibrium effect, which is also referred to as "resistance to mass transfer." A, B, and C are constants for a given system and $\bar{v}$ is the average mobile phase velocity. Figure 8 describes the variation of H with flow rate and indicates the classical contribution of the three phenomena involved in zone spreading.

Giddings [10] challenged the validity of the van Deemter equation by rejecting the concept of additive contributions for eddy diffusion and non-equilibrium and suggested that they be coupled in a nonadditive way as:

$$1/[\,(1/A) + (1/C\bar{v})\,]$$

at least for gas-liquid chromatography. That Giddings' theory is valid has apparently been demonstrated for a GLC system by Sie and Rijnders [11]. For gas chromatography in particular, a variety of changes in the basic

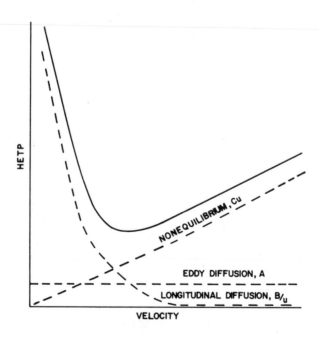

Figure 8. Plate height vs flow rate.

van Deemter equation have been recommended in the various rate theories of chromatography (cf. Ref. 8). Fortunately, for the purpose of performing a chromatographic analysis, the particular expression of the van Deemter equation is of little consequence. Practically, the experimental parameters that contribute to H in Eq. (32) affect H in the same way in the modified van Deemter expressions. Consequently, if a system is constructed so as to minimize H in Eq. (32), H is essentially minimized regardless of the relationship of the parameters involved. In addition, the variation of H with velocity (cf. Fig. 8) has the same general form for the several expressions of H. Therefore, this discussion is limited to the classical van Deemter approach, not as a rigorous treatment but because Eq. (32) apparently contains the significant factors affecting efficiency and because this equation indicates the proper approach to minimizing H and therefore maximizing resolution.

Classically, the chromatographic process can be considered as a random walk problem [12] in which each step is taken in a direction completely independent of the direction of the preceding step. A plot of the number of individuals at a given distance as a function of distance at the completion of a random walk results in a gaussian curve with the amount of spreading measured by the standard deviation σ, defined as:

$$\sigma = l\sqrt{n} \tag{33}$$

where l is the length of the step and n is the number of steps. Since the three contributions to plate height in Eq. (32) are random as well as independent phenomena, the total standard deviation is given by:

$$\sigma^2 = \sum_i \sigma_i^2 \tag{34}$$

where σ_i is the standard deviation of the individual process.

The first term in the van Deemter equation represents flow velocity inequalities. If the mobile-phase velocity of a chromatographic system is v, then the solute velocity is some fraction R of the mobile-phase velocity or Rv. Actually, the mobile-phase velocity is an average velocity $\bar{v}$ since the eluent has a great variety of velocities in the system. If the chromatographic medium is considered as consisting of a close-packed arrangement of spherical particles, mobile-phase flow velocity inequalities arise from three sources: capillary flow, multipath effect, and transcolumn flow. The voids around the particles through which the eluent travels are essentially capillaries where the flow rate is zero at the wall and twice the average at the center of the channel. The random particle size results in a variety of channel lengths and thus a range of velocities, since some molecules travel a longer, more tortuous path than others. Finally, the difference in packing structure between the wall and the center of the system causes unequal flow rates.

Now the length of a channel, or step, will average about one particle diameter d_p and the number of steps n will be essentially the length of the system L divided by the length of the step, i.e., $n = L/d_p$. Then, from Eq. (33),

$$\sigma_A = \sqrt{Ld_p} \tag{35}$$

where A indicates the flow velocity inequality term.

Molecular diffusion, or longitudinal diffusion, in the direction of flow contributes to zone spreading because the solute molecules spread in all directions from the zone center. The contribution of diffusion to zone spreading is given by Einstein's equation:

$$\sigma_B^2 = 2D_m t \tag{36}$$

where B identifies the term as the diffusion contribution, D_m is the diffusion coefficient for the solute in the mobile phase (diffusion in the stationary phase is generally much less significant and is usually ignored), and t is the time the solute spends in the mobile phase. If t is replaced by $L/\bar{v}$, Eq. (36) becomes:

$$\sigma_B^2 = 2D_m L/\bar{v} \tag{37}$$

The nonequilibrium contribution arises from the random transfer of solute between the mobile and stationary phases since a solute molecule in the mobile phase moves faster than the average solute velocity, while a molecule in the stationary phase moves slower than the average. After a rather complex derivation is followed, the nonequilibrium contribution is:

$$\sigma_C^2 = C\bar{v}d^2 L/D_s \tag{38}$$

where C is a constant dependent on the retention ratio R, d is the stationary phase thickness, and D_s is the diffusion coefficient of the solute in the stationary phase.

Then, from Eq. (34), the total zone spreading is given by:

$$\sigma^2 = Ld_p + 2D_m L/\bar{v} + Cd^2 L\bar{v}/D_s \tag{39}$$

and since:

$$H = \sigma^2/L \tag{40}$$

then:

$$H = d_p + 2D_m/\bar{v} + Cd^2\bar{v}/D_s \tag{41}$$

Actually, the classical values for A, B, and C in the van Deemter equation give a slightly different result than Eq. (41), i.e.,

$$H = 2\lambda d_p + 2\gamma D_m/\bar{v} + (C_s d^2\bar{v}/D_s + C_m d_p^2\bar{v}/D_m) \tag{42}$$

where λ is a constant characteristic of packing irregularity, γ is a correction for the obstruction of diffusion by the packing, and C_s and C_m are constants characteristic of the stationary- and mobile-phase resistance to mass transfer, respectively. Once a particular system is constructed, all terms except H and $\bar{v}$ are fixed and Eq. (42) reduces to Eq. (32).

The value of the van Deemter equation is in its indication of the requirements for the minimization of zone spreading and, therefore, for the maximization of resolution. From Eq. (42), eddy diffusion is minimized by avoidance of packing irregularity and by the use of small particles. In addition, it has been found that for column chromatography λ decreases with a decrease in the column diameter. The longitudinal diffusion term also recommends fine particle size and uniform packing to obstruct molecular diffusion in the mobile phase. The second term recommends a high flow rate to minimize H. The nonequilibrium term (in parentheses) recommends a thin stationary phase to reduce the distance a solute must migrate in the stationary phase before it reaches the surface and enters the mobile phase. Moreover, there is the requirement for a small particle to give closer packing and less mobile-phase volume through which a solute must migrate before it reaches the stationary phase. Finally, the mass transfer, or nonequilibrium, term demands a low flow rate to minimize H.

Obviously, Eq. (42) contains a number of contradictory requirements. The molecular diffusion term suggests a minimization of solute diffusion to prevent zone spreading, while the nonequilibrium expression depends on rapid diffusion to maintain equilibrium. A fine particle size is recommended by each term; yet if the particles are too fine, mobile-phase velocity is obstructed. The molecular diffusion term demands a high flow rate, while the mass transfer expression requires a low mobile-phase velocity. Finally, the thin film requirement [d in Eq. (42)] conflicts with the large stationary-phase volume recommended by Eq. (27).

In practice, then, H must be minimized by arranging the variables to obtain the best compromise. The particle size is usually as small as is consistent with a reasonable flow rate for the particular chromatographic system. Once the system is arranged, H can be varied only by changing the flow rate. Figure 8 indicates that H as a function of flow has a minimum and that molecular diffusion is the predominant contribution to H at low velocities, while nonequilibrium is dominant at high flow rates.

Finally, the effect of efficiency or H on resolution follows from Eqs. (27) and (40) and the recognition that for a gaussian distribution the peak width is essentially equal to four times the standard deviation, i.e.,

$$w \cong 4\sigma \tag{43}$$

Consequently, Eq. (27) can be rewritten as:

$$R = \Delta K V_s / 4\sqrt{HL} \tag{44}$$

7. Temperature Effects

A temperature variation in any chromatographic system results in a change in band center separation and in zone width. The specific effects of a temperature change are complex, involving changes in distribution coefficients, longitudinal diffusion, and mass transfer terms. Generally, an increase in temperature decreases retention time, decreases zone separation, and decreases zone spreading. Retention time decreases with increasing temperature due to the increased rate of transfer of solute to the mobile phase. Zone center separation is reduced at increased temperatures as a consequence of the shorter retention times. The effect of temperature on band width is opposite to the effect of eluent velocity on plate height. While an increase in velocity reduces longitudinal diffusion and increases nonequilibrium contributions to plate height [cf. Eq. (32)], an increase in temperature, because of the corresponding increase in solute mobility, results in increased molecular diffusion and decreased resistance to mass transfer. The net effect of a temperature change is usually improved resolution with decreased temperature, since the resulting increase in zone center separation is generally greater than the increase in band width. Consequently, time is sacrificed for resolution and the highest temperature that provides the required resolution is usually selected.

8. Summary

The purpose of a chromatographic system is the separation of the components of a sample. Equation (26) defines resolution as the zone center separation divided by zone width, i.e.,

$$R = \Delta V / w \tag{26}$$

while Eq. (44) presents resolution as a function of K, V_s, H, and L and provides the general requirements for chromatographic separation. The primary requirement is the selection of mobile and stationary phases that provide different distribution coefficients for the various solutes. Establishment of different K's makes separation possible. The other terms in Eq. (44) determine the degree of separation. The V_s term indicates that resolution is improved by an increase in the quantity of stationary phase, which can be effected by increasing the amount of stationary phase per unit length of the system or by increasing the length of the system. An increase in system length improves resolution by a factor of $\sqrt{L}$, since the numerator of Eq. (26) is proportional to L, while the denominator is proportional to $\sqrt{L}$. Increasing the amount of stationary phase per unit volume is actually detrimental to resolution because of the diffusion effect on H [cf. Eq. (42)]. The H term in Eq. (44) should be minimized by selection of a finely divided uniform stationary phase and an experimentally determined optimum mobile-phase velocity. Finally, resolution is generally improved by decreasing temperature, at the sacrifice of analysis time.

B. Gas Chromatography

Although gas-liquid chromatography is a relatively new technique, some 40 years younger than liquid chromatography, GLC is responsible for the development of most chromatographic theory. There have been at least three primary reasons for the rapid development of gas-liquid chromatography theory at the expense of other systems of chromatography: (1) Adsorption systems are very difficult to handle theoretically because the distribution coefficients primarily depend on the nature and extent of the available adsorbent surface, which is not readily reproducible; (2) the GLC system is relatively easy to treat since gas-solute and gas-stationary phase interactions are usually negligible; (3) and the great speed of GC has made it the method of choice with liquid systems generally used only as a last resort. Consequently, until the recent development of instruments to provide high-speed liquid chromatography, GLC has received the most attention and the most theoretical development.

Gas chromatographic (GC) theory is now sufficiently developed to provide a variety of physical and chemical information from GC data. The retention volume can be related to various thermodynamic properties (cf. Chapter 10 in Ref. 8) so that solution thermodynamic information can be obtained from gas chromatographic retention data [13]. Alternatively, retention behavior can be predicted if the thermodynamic properties are known. A mathematical expression for structural determinations from retention data was developed by Pierotti et al. [14], based on a correlation between activity coefficients and the structures of solute and stationary phase. Martire and Purnell [15] were able to determine the molecular weights of stationary phases from retention data. These and similar developments are beyond the scope of this work and are mentioned only to indicate the power of gas chromatography as an analytical tool. The interested reader should consult the original literature.

1. Compressibility and Retention

Although the gaseous mobile phase generally simplifies GLC theory, the use of a gas also adds a complication. At the pressures used in gas chromatography systems, the compressibility of the mobile phase is significant. This has been a phenomenon unique to GC, but the recent introduction of high-speed liquid systems designed to operate at pressures approaching 1000 atm require consideration of compressibility for these systems also, at least for the more exacting work. The following discussion of mobile phase compressibility and its effect on retention volume is essentially that deduced by James and Martin [5].

The quantity of carrier gas (mobile phase) moving through any point x in the column is defined by:

$$A_m v = kdP/dx = F_c P_o/P \tag{45}$$

where A_m is the mobile phase cross-section of the column, v is the carrier gas velocity at point x, k is a column constant related to the gas-phase viscosity and the packing density, F_c is the gas volumetric flow rate, and P and P_o are the gas pressures at x and at the column outlet, respectively. Integration of Eq. (45) and solution for F_c yields:

$$F_c = (P_o k/2x)[(P/P_o)^2 - 1] \tag{46}$$

Now, the velocity of the solute is given by vR_f, where R_f is the retardation factor defined by Consden et al. [16] as the rate of movement of solute divided by the rate of movement of the carrier gas. Then, the time dt required to move the solute a distance dx is:

$$dt = dx/vR_f \tag{47}$$

Therefore, when Eq. (45) is utilized, the retention time t_R is:

$$t_R = \int_0^L dx/vR_f = \int_0^L A_m Pdx/P_o R_f F_c = \int_{P_o}^{P_i} kA_m P^2 dP/P_o^2 R_f F_c^2 \tag{48}$$

where P_i is the column inlet pressure. Integration of Eq. (48) over the definite integral defines retention time as:

$$t_R = kA_m(P_i^3 - P_o^3)/3P_o^2 R_f F_c^2 \tag{49}$$

The retention volume V_R is defined as the volume of gas required to move the zone center to the end of the column. Therefore,

$$V_R = F_c t_R \tag{50}$$

and from Eq. (49), with rearrangement,

$$V_R = kA_m P_o [(P_i/P_o)^3 - 1]/3R_f F_c = kA_m P_o (a^3 - 1)/3R_f F_c \tag{51}$$

where a is the column pressure drop, i.e., P_i/P_o.

When Eq. (46) is used to eliminate F_c, and column length L is substituted for x, V_R becomes:

$$V_R = (2A_m L/3R_f) \cdot (a^3 - 1)/(a^2 - 1) = jA_m L/R_f \tag{52}$$

where j is the compressibility factor. However, $A_m L/R_f$ is V_R^o, the corrected retention volume, so that

$$V_R^o = jV_R \tag{53}$$

Thus, the limiting or corrected retention volume is the observed retention volume corrected for the compressibility of the mobile phase. It should be noted that V_R^o, not V_R, is the appropriate term for use as V in Eq. (23).

C. Liquid, Thin-Layer, and Paper Chromatography

As indicated in Sec. II, B, the development of the theory of solute retention for the various modes of liquid chromatography has lagged far behind work in the gas chromatography field. To a great extent, the advantages of gas over liquid chromatography, i.e., speed and efficiency, resulted in an immense popularity for GLC at the expense of liquid chromatography. In addition, most liquid chromatography systems involve adsorption, which presents two obstacles to theoretical development. First, the lack of an understanding of the relationship between chemical structure and adsorption phenomena requires an empirical approach at the expense of theoretical development. Second, from a practical standpoint, the lack of reproducible absorbents severely restricts further developments in liquid-solid chromatography.

For adsorption chromatography, exact reproduction of adsorbent characteristics is difficult even among portions of adsorbent from a single source. The lack of reproducible adsorbent properties stems from an inability to control the nature of active surfaces, primarily because of the large variation of surface properties caused by even trace amounts of impurities. The affinity of most adsorbents for water is a severe problem since the presence of any water alters the characteristics of the active surface. The consequence of this lack of reproducibility is an imperfect knowledge of the characteristics of the active adsorbent and, therefore, failure to elucidate the nature of the chromatographic adsorption process.

The situation for paper chromatography may be even more complex in view of the probable dual nature of paper chromatographic separation. Burma [17] demonstrated that the cellulose in paper does allow adsorption of the solute based on an argument for two types of water in paper: chemically bound water and free adsorbed water. A solute can be partitioned between the eluent and the adsorbed water in a true liquid-liquid partition system. However, the solute might also be retarded by the chemically bound water, which would constitute an adsorption process. Generally, both adsorption and partition mechanisms are involved in the separation, with the partition contribution being the dominant phenomenon. Additional complications stem from the complex chemical nature of cellulose, which contains aldehyde, carboxyl, hydroxyl, and ketone functions. For a detailed discussion of the complex nature of cellulose and the paper chromatography system, the reader should consult the review by Cassidy [18].

The liquid-liquid partition system has recently received great attention following the introduction of high-pressure liquid chromatography, which provides the speed and efficiency previously available only in GC. The resurgence of interest in liquid chromatography led to the development of a thermodynamic theory of solute retention for liquid-liquid chromatography by Locke and Martire [19]. This theory of solute retention is based on a liquid system analogous to GLC systems; i.e., the carrier gas is replaced by a high-pressure liquid mobile phase. The retention volume is related to thermodynamic properties of the solute and solvents, and the effects of temperature, pressure, and solvent molecular weights are deduced.

D.   Ion-Exchange Chromatography

Like most other forms of liquid chromatography, ion-exchange chromatography is generally an empirical affair as a consequence of the present inability to relate ion-exchange phenomena to the chemical and physical properties of the system. Some of the difficulty in theoretical progress is attributable to the nature of the ion-exchange resins, which are usually branched polymers containing a high density of ionic functional groups. Selectivity is often governed only by electrostatic forces, but van der Waals' forces can contribute significantly, especially in systems involving organic solutes. One must be concerned with interfacial phenomena, specifically, the mechanism of solute transfer between phases. Finally, the great majority of exchangeable ions are not on the surface of the resin but are internal, thus requiring the migration of ions within the resins.

Fortunately, resins generally behave as concentrated aqueous electrolyte solutions and can usually be treated as such. However, this is precisely the obstacle that most hinders development of retention theory. Both the resin and the eluent, at least in the regions about zone centers, must be treated as concentrated aqueous electrolytes, for which present solution theories fail to explain interionic forces. Until solution theory for concentrated electrolytes can be perfected, there is little prospect for further development of selectivity theory for ion-exchange chromatography.

Actually, perfectly rigorous thermodynamic treatments of retention have been developed by Glueckauf [20] and by Myers and Boyd [21]. The theory as developed by Myers and Boyd provides for the calculation of activity coefficients in the resin from measurements of activities in water. Because of the difficulty in determining ionic activities in concentrated aqueous solutions, however, utilization of the thermodynamic treatment awaits development of adequate solution theory. Various less rigorous equilibrium theories have been advanced to explain solute distribution. Two of these, the Donnan membrane equilibrium theory and the application of the mass-action law, frequently provide reasonable estimates of exchange behavior and are briefly discussed below.

The Donnan theory was advanced in 1911 [22, 23] to describe the distribution of electrolytes between two aqueous phases separated by a semipermeable membrane when one phase contained a nondiffusible ionic species. Now, since the organic portion of the resin is certainly a non-diffusible ionic species and, as noted above, the resin behaves as an aqueous electrolyte solution, it need only be assumed that the resin surfaces behaves as a semipermeable membrane to apply the Donnan theory to ion-exchange systems. If, then, two solutions are considered, one containing AB and the other AS, where S^- is the immobile portion of the resin, or stationary phase, the chemical potential μ of AB at equilibrium must be equal on both sides of the membrane, i.e.,

$$\mu AB_m = \mu AB_s \tag{54}$$

where m and s refer to the two phases. Now the potential of an electrolyte is the sum of the potentials of the ions. Thus,

$$\mu^o_{A^+} + RT \ln (a_{A^+})_m + \mu^o_{B^-} + RT \ln (a_{B^-})_m =$$
$$\mu^o_{A^+} + RT \ln (a_{A^+})_s + \mu^o_{B^-} + RT \ln (a_{B^-})_s \tag{55}$$

and,

$$(a_{A^+})_m (a_{B^-})_m = (a_{A^+})_s (a_{B^-})_s \tag{56}$$

where a represents the ionic activities, R is the gas constant, and T is the absolute temperature. If it is now assumed that the solutions are dilute, i.e., that concentrations are equivalent to activities, and that electroneutrality applies, then,

$$[B^-]_m^2 = [B^-]_s^2 + [B^-]_s [S^-]_s \tag{57}$$

and,

$$[A^+]_m [B^-]_m = [B^-]_m^2 = [A^+]_s [B^-]_s = [B^-]_s^2 + [S^-]_s [B^-]_s \tag{58}$$

If another diffusible ion C^+ is added to the system:

$$[A^+]_m [B^-]_m = [A^+]_s [B^-]_s \tag{59}$$

and,

$$[C^+]_m [B^-]_m = [C^+]_s [B^-]_s \tag{60}$$

Division of Eq. (59) by Eq. (60) yields:

$$[A^+]_m / [C^+]_m = [A^+]_s / [C^+]_s \tag{61}$$

and, in general, for ions of different charges,

$$[A^{\pm y}]_m^{1/y} / [C^{\pm z}]_m^{1/z} = [A^{\pm y}]_s^{1/y} / [C^{\pm z}]_s^{1/z} \tag{62}$$

where y and z are the respective valences.

As indicated above, the solutions in ion-exchange are generally concentrated and the errors resulting from the substitution of concentrations for activities are significant when the Donnan theory is used to predict distribution behavior. However, this theory does explain the inability of electrolytes to diffuse into resins of high fixed ionic concentrations from solutions of lower ionic concentrations [cf. $[B^-]$ in Eq. (57)]. This phenomenon provides the basis for ion-exclusion separations. In addition, Eq. (62) gives a qualitative description of the ionic distributions and predicts the effect of valence on distribution.

The application of the law of mass action to ion-exchange equilibria provides selectivity coefficients that indicate the relative retentions of various species. For the general reaction,

$$yM_1 S_z + zM_2^{+y} = zM_2 S_y + yM_1^{+z} \tag{63}$$

The equilibrium constant K_{eq} is given by:

$$K_{eq} = (a_{M_2}S_y)^z (a_{M_1} + z)^y / (a_{M_1}S_z)^y (a_{M_2} + y)^z \tag{64}$$

However, the use of mole fractions and activity coefficients gives an apparent equilibrium coefficient, defined as:

$$K'_{eq} = K_{eq}(\gamma_{M_1}S_z)^y / (\gamma_{M_2}S_y)^z =$$

$$(a_{M_1} + z)^y (x_{M_2}S_y)^z / (a_{M_2} + y)^z (x_{M_1}S_z)^y \tag{65}$$

where γ represents the activity coefficients in the resin, x represents the mole fractions in the resin, and K'_{eq} is the apparent equilibrium coefficient. There have been no successful attempts to evaluate the ratio of activity coefficients in the resin phase and, consequently, the true equilibrium constants K_{eq} cannot be evaluated. However, K'_{eq} can be determined since the terms on the right side of Eq. (65) can be experimentally measured. Unfortunately, K'_{eq} is not a constant; it generally varies with the mole fraction ratio because of resin phase nonideality and the variation of activity coefficient ratios with mole fraction ratios. However, K'_{eq} does give a qualitative or semiquantitative measure of the relative affinity of a resin for a series of ions and, therefore, is a measure of relative retention and separability. Hence the name "separability coefficient" is frequently applied to K'_{eq}.

III. PRACTICE

When faced with the analysis of a sample that requires a separation, the first determinant of the choice of an analytical method is the nature of the sample. If the sample is relatively simple, separation may often be effected by nonchromatographic techniques, e.g., extraction, precipitation, distillation. However, if the sample has many components or if the components of interest have similar chemical and physical properties, one of the chromatographic systems is generally necessary to accomplish the analysis. The selection of a particular chromatographic system is determined by a variety of factors, which include the nature of the sample, available equipment, type of information required, and the preference of the analyst.

The nature of the sample is the first consideration in the choice of a chromatographic method. Frequently, the sample alone determines the method. For example, inorganic gases are resolved only by gas chromatography, very light organic compounds are best handled by gas chromatography, and solutions of inorganic ions are usually resolved by ion-exchange chromatography. Generally, however, the sample precludes the use of one or more techniques while permitting the application of several others. For example, labile, nonionic materials are not susceptible to gas or ion-exchange chromatography but can be analyzed, in general, by any liquid system including thin-layer and paper. Other samples, e.g., mixtures of low molecular weight carboxylic acids, have been analyzed by all five systems of interest here. In general, volatile, nonlabile samples are subjected to gas chromatographic analysis; ionic species are resolved by ion-exchange chromatography; and nonionic, nonvolatile, or thermally unstable compounds are separated by some form of liquid chromatography, including paper and thin-layer.

Occasionally, the type of information required must be considered in choosing a chromatographic system. For qualitative analysis the various systems perform about equally well. For quantitative analysis the techniques that include a detector as an integral part of the system, i.e., gas, columnar liquid, and ion exchange, are generally preferable. If it is desired to collect relatively large quantities of pure components by preparative scale chromatography, paper and thin-layer are usually excluded because of their relative inability to accommodate large samples.

Except for those cases where the nature of the sample dictates the choice of method, e.g., inorganic gases by gas chromatography, the final determinant in the choice of method is the personal inclination of the investigator. The analyst's choice may be influenced by a variety of factors in addition to those mentioned above. Analysis time can be a significant consideration, especially if it is desired to develop a relatively routine method to be applied to a large number of similar samples. When speed is a valid consideration, the high-pressure gas and liquid systems are the preferred methods. Literature data will also influence the analyst's decision. If a similar analysis can be found in the literature, application of the described system is generally preferable to development of a different system because of the great saving of time. Finally, the ultimate criterion often reflects the prejudice of the investigator, who may opt for a system with which he or she is familiar regardless of the choice suggested by objective criteria. If not pursued to extremes, familiarity is a valid consideration in the choice of a chromatographic method.

A. Gas Chromatography

Gas chromatography, the most popular chromatographic system, has a number of advantages over other separation techniques. Gas chromatography is fast and extremely sensitive. For example, gas chromatographic analysis of a fraction of a microliter of a petroleum cut routinely provides qualitative and quantitative data for some hundred components in a couple of hours, often automatically since gas chromatography readily lends itself to automatic control operations. The system is versatile; it has been applied to

inorganic gases, metals via formation of volatile chelates, and most
organic compounds, including polymers, either directly or through volatile
derivatives or pyrolysis products. Further proof of versatility is provided
by the sample weight, which has varied from picograms to grams. The
extreme sensitivity of GC is due to the variety of detectors, which are
generally unavailable to other forms of chromatography. Finally, with the
selection of a nondestructive detector, resolved sample components are
readily recovered from the gaseous mobile phase.

However, gas chromatography has two severe limitations: volatility
and stability requirements. The components of interest in a sample must
be sufficiently volatile to allow migration through the system at a temperature
that does not induce thermal degradation, although, as noted above, lack of
volatility has sometimes been overcome by derivative formation and pyrolysis. In addition, the sample should be inert with respect to the system
except for the distribution process. Finally, gas chromatography is
expensive relative to most other chromatographic systems, ranging in
price from a few hundred to about $15,000, unless a mass spectrometer
detector or computerized data reduction system is added, in which case the
GC expense is nominal relative to the cost of the accessories.

1. Instrumentation

The minimum requirements for a gas chromatographic system are a pressurized gas supply and a column containing the stationary phase. Actually,
most experimental arrangements consist, at least, of the components shown
schematically in Fig. 9. The usual configuration is a pressurized gas
supply with a pressure regulator; a flow controller for maintenance of a
constant flow rate; a heated injection port, which is usually independent of
the column oven; a detector with its own temperature control system; and
a strip chart recorder. In addition, flowmeters are often included in front
of the column and/or after the detector. Frequently encountered accessories
include manometers, which are necessary for absolute but not for relative
retention measurements; temperature programming facilities to change

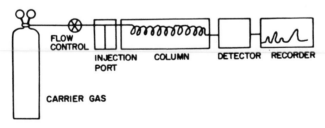

Figure 9. Schematic gas chromatography system.

column temperature as a function of time; automatic injection devices to perform repetitive analyses; trapping devices to collect solutes; and data handling equipment, e.g., disk and digital integrators and computer configurations. A list of manufacturers is provided by the annual Analytical Chemistry Laboratory Guide [24], which attests to the proliferation of gas chromatographic equipment by devoting two and one-quarter pages of the 1970-1971 issue to a list of names of suppliers of instruments, supplies, and accessories. The functions of the various components are considered in the following discussions.

2. Carrier Gas

To fulfill the primary purpose of the carrier gas, transport of the sample through the column, any gas is suitable. However, a number of restrictions must be placed on the selection of a carrier gas, and these reduce the available choices to a very few gases. The first restriction is inertness; the carrier gas must not react with the solute or with any components of the system. There are a host of additional restrictions that are consequences of the various detection systems. The detector restrictions is discussed in Section III, A, 7, on detectors, but the gases along with their compatible detectors are listed in Table 4. Less severe restrictions that may be imposed include availability, purity, expense, and safety. With the exception of hydrogen, the use of which may be considered relatively hazardous, the gases in Table 4 all meet the above requirements for carrier gases, and consequently, gas selection is usually dictated by the detector. Finally, with the exception of argon, which generally must be dried before use with the argon ionization detector, the common carrier gases are available in a sufficient state of purity that further purification is usually unnecessary.

TABLE 4

Carrier Gas-Detector Compatibilities

Thermal conductivity	He, H_2, N_2
Flame ionization	He, N_2
Electron capture	N_2, Ar + 10% CH_4
Helium	He
Cross-section	H_2, He + 3% CH_4
Gas density balance	N_2, CO_2, Ar
Alkali flame	He, N_2
Argon ionization	Ar

Ideally, an experimental design would include two manometers to determine inlet and outlet pressures and a flowmeter to measure carrier velocity. The manometers are usually omitted, however, since most analytical procedures record retention data relative to a standard solute measured under identical conditions. A gas velocity measurement device, a rotameter or flowmeter, is employed to determine the optimum carrier gas flow rate and, thereafter, to insure reproduction of experimental conditions. The optimum velocity is determined by the preparation of an HETP vs velocity plot (cf. Fig. 8) with the use of Eqs. (16), (17), or (18). Flow rates normally range from 50 to 100 ml/min for 1/4 in. o.d. columns, from 15 to 50 ml/min for 1/8 in. o.d. columns, and from 1 to 5 ml/min for capillary columns.

3. Sample Introduction

The gas chromatographic sample is introduced to the system through an injection port, which may be a separate heated chamber immediately in front of the column or may be the front of the column itself. Gaseous samples are best handled by on-column introduction, which eliminates band broadening due to the dead volume of the injection port. Liquids are usually introduced into the separate, heated, minimal-volume injection port to obtain flash vaporization and, thus, instant equilibration of the sample in the column. The injection port temperature, which is usually much higher than the column temperature, must be hot enough to vaporize the sample without loss of efficiency, yet cool enough to avoid sample degradation. The adequacy of the injection port temperature is readily established by increasing the temperature. If an increase improves efficiency or peak shape, the temperature has been too low. If, however, the peak area or shape or retention time changes markedly, the temperature may have been too high, causing decomposition.

The mechanics of sample introduction involve the use of either a syringe or a bypass loop, i.e., a gas-sampling valve. Liquids are introduced with syringes through self-sealing septums, while gases are injected with gas-tight syringes or gas-samplings valves. Syringes are commercially available in various sizes down to 1 μl, with a precision of about 2% relative. The bypass sample loops can provide a relative precision of approximately 0.5%.

The sample introduction are performed as follows. The syringe is held in both hands, one acting as a guide, the other supporting the syringe with the thumb on the plunger to prevent the carrier gas pressure from expelling the plunger. The needle is rapidly inserted as far as possible through the septum; the plunger is immediately and rapidly depressed, and after a delay of 1 sec, the needle is quickly and smoothly withdrawn. The injection technique is readily tested by reducing the sample size. If a smaller sample size produces an increased number of theoretical plates, i.e., improves efficiency, either the injection port temperature is too low, as discussed

TABLE 5

Typical Sample Volumes

Column	Gas	Liquid
Capillary, liquid film	0.1-10 μl	0.001-0.5 μl
1/8 in. o.d., 2-5% liquid	0.01-5 ml	0.01-5 μl
1/4 in. o.d., 5-20% liquid	0.1-100 ml	0.1-50 μl
Preparative, 5-30% liquid	0.01-10 liter	0.01-10 ml

above, or a poor injection technique has been used. A bypass loop is arranged so that either sample or carrier may be passed through the loop. The loop is swept with sample to remove air and any previous sample or carrier gas and then closed to trap the sample. When the valve is opened to the carrier gas, the sample is carried into the column. Commonly used sample volumes are shown in Table 5 as a function of column diameter and stationary-phase concentration. The smaller volumes in Table 5 are obtained with a sample splitter, which separates the carrier-sample mixture into two streams, the larger of which is vented while the smaller is directed to the column. Common splitters provide a vent to column ratio ranging from 10:1 to several thousand to one.

4. Column Preparation

The gas chromatographic column consists of a small diameter tube that contains the stationary phase. The stationary phase for gas-solid chromatography is a finely divided, surface-active solid capable of selective adsorption of the sample components. For gas-liquid chromatography, the nonvolatile liquid stationary phase is supported on a fine inert solid or, in the case of capillary columns, on the tubing wall.

Equation (42) presents one indirect and two direct restrictions that apply equally to adsorbents and supports. Efficiency increases both as the particle size decreases and as the particle size distribution range decreases because of reduction in the multipath effect. However, as particle size decreases, flow is impeded. Consequently, the practical requirement is a uniform particle size with the smallest diameter particle consistent with efficient flow rates at reasonable pressures. The as-received solids, supports, or adsorbents must be passed through sieves to insure uniformity, despite the nominal mesh range usually provided on the manufacturer's label. A mesh range of 20 is usually satisfactory, and, for example, the material that passes a No. 100 and is retained on a No. 120 U.S. standard sieve may be used. The optimum particle size depends on the particular experiment, but the usual ranges are between 100 and 200 for 1/16 in. columns, 100 and 160 for 1/8 in. columns, 60 and 100 for 1/4 in. columns, and 20 and 60 for preparative columns.

The solid adsorbents for GSC must be activated before use, usually by heating at 500° C for 4-8 hr. For GLC, the liquid stationary phase must be deposited on the support. The experimental difficulties encountered in preparing capillary columns are such as to highly recommend their purchase in prepared form. The stationary phase for a packed column is readily prepared as follows. The liquid phase is dissolved in a very volatile solvent of sufficient volume to thoroughly wet the solid support. The solution and the solid are then mixed and the stationary phase is deposited on the solid by evaporation of the solvent with gentle agitation but not stirring, which fractures most solids.

For a column to be packed, one end of the tube is loosely plugged with a small portion of glass wool. A funnel is then attached to the open end with rubber tubing and stationary phase is added to the funnel. The packing is transfered to the column by vibration. For metal columns, which are packed while straight, the vibration can be supplied by a mechanical or electrical vibrator, by tapping with a stick, or by gently bouncing the tube on the floor. For glass columns, which must usually be packed after the glass is coiled to fit the instrument, the vibration is generally provided by a hand vibrator while vacuum is applied to the plugged end of the column. Vibration is continued until packing can no longer be added to the column. The open end of the column is then plugged with glass wool and metal columns are coiled or bent to fit the chromatograph.

Before use, the gas chromatographic column must be conditioned to provide stable operation. The conditioning process removes volatile materials which, if not removed, can interfere with column performance. The column is conditioned in the chromatographic oven with a low carrier gas flow rate (5-10 ml/min) and with the exit end of the column disconnected to avoid detector contamination. Nearly all columns are conditioned by heating for several hours at a temperature at least 25° C above the temperature at which the column is to be operated but below the maximum temperature limit of the liquid phase. In addition, columns generally need to be reconditioned only after long periods of exposure to the atmosphere.

Notable exceptions to the standard conditioning procedure above are FFAP (free fatty acid phase), SE-30 (a methyl silicone polymer), and STAP (steroid analysis phase). FFAP is heated for 2 days at 100° C with no flow and then conditioned with flow at the operating temperature until stable. SE-30 is conditioned overnight at 250° C without flow and then at the operating temperature with flow until stable. STAP must be baked at 250°C overnight with no flow and then conditioned at 225° C with flow until stable.

5. Column Selection

The success of a gas chromatographic experiment depends primarily on the choice of a suitable column. Adsorbents for GSC are relatively easy to select due to the limited variety of satisfactory adsorbents. There is also only a limited number of supports available for GLC, but their selection can be complicated by the need to consider support compatibility with the nature and quantity of the liquid phase. The choice of a liquid phase can be exceedingly difficult, since there are no hard and fast rules for liquid selection, and since, in general, any material that is a liquid at the operating temperature can be used. Fortunately, the difficulty of column selection is somewhat alleviated by the need only of a satisfactory column as opposed to the optimum column for a given separation.

Prior knowledge about the sample is essentially a prerequisite for a successful chromatographic analysis. The difficulty of a completely unknown sample can sometimes be overcome through the use of multiple columns, but the analysis is greatly simplified by a knowledge of sample composition since column selection is dictated by the nature of the sample components. The minimum desirable sample information is the type of compounds in the sample. Column selection is further simplified by such additional information as suspected components, relative amounts, structure, thermal stability, and boiling range. Prior knowledge about the sample is then used in selecting a column with the aid of experience, or the literature, or trial and error.

Although a great variety of solids have been used as GSC adsorbents, only a few are used with any frequency as gas-solid chromatography is mostly applied to permanent gases and low molecular weight hydrocarbons. The common adsorbents are alumina, charcoal, molecular sieves (4A, 5A, and 13X), and silica gel. While the molecular sieves are relatively consistent, the adsorptive properties of alumina, charcoal, and silica gel are very dependent on their sources, methods of preparation, and treatments prior to chromatographic use. Generally, the molecular sieves, which are alkali metal aluminosilicates similar to natural clay, are used for the permanent gases, while silica gel, charcoal, and alumina are used for the permanent gases and hydrocarbons.

Hollis [25] introduced the use of porous polymer beads of styrene cross-linked with divinyl benzene. Porapak (registered trademark of Waters Associates) provides a sample distribution phenomenon that is different from that of conventional gas-solid or gas-liquid chromatography. In gas chromatography, the partition of solutes is from a gas to a thin film of liquid supported on a solid or from the gas to a solid adsorptive surface, while with Porapak, partition apparently occurs directly from the gas into the solid amorphous polymer. There are six types of Porapak, N, P, Q, R, S, and T. Type-Q material is the most versatile polymer and is non-polar, as is type P. The others are moderately polar materials that are modifications of the basic styrene-divinyl benzene polymer arising from the addition of various monomers during polymerization. Porapak columns are remarkable in their lack of adsorption of polar solutes. Consequently, they are normally used for the rapid separation of very polar materials. Particularly useful is their ability to rapidly separate water from polar and nonpolar organics while maintaining band symmetry; i.e., peak tailing is generally absent.

Only a very few materials are used for supports in GLC, but a proliferation of brands exist. Unfortunately, the various brand-name supports generally have somewhat different characteristics for the same reasons that adsorptive solids are different, i.e., source material, preparation, and treatment. Except for a limited use of glass and fluorocarbon, e.g., Teflon (registered trademark of E. I. duPont de Nemours & Co.) most GLC supports are prepared from diatomaceous earth, which is composed of the skeletons of diatoms. The diatomite is basically an amorphous hydrous silica that usually contains some metallic oxide impurities. The various diatomite supports may be used directly, after acid washing, which reduces surface adsorption; or after silylation, e.g., with hexamethyldisilazane or dimethyldichlorosilizane, which masks the surface hydroxyl groups, thereby minimizing adsorption of solutes by the support.

Liquid-phase selection is the most difficult aspect of gas chromatography, but fortunately much help is available. A set of requirements for liquid stationary phases provides considerable information about column selection. The stationary phase must be a good solvent for the solutes, since solutes with low solubility elute rapidly and are poorly resolved. The liquid must also be a good differential solvent for the solutes; i.e., the solutes should have different solubilities for separation to occur. Thermal stability is a requirement necessary to prevent extraneous peaks, detector contamination, and column deterioration, all of which also result from a too volatile solvent, i.e., one the vapor pressure at operating temperature of which is above about 0.1mm. The liquid phase should be inert toward solutes of interest to preserve the partition nature of GLC, to prevent column degradation, and to avoid changes in sample composition. Finally, the solute solubility requirement generally necessitates a stationary phase with a chemical structure similar to that of the solutes.

Now, when one is faced with a chromatographic separation for which no personal experience is available, one should take immediate recourse to the literature. An excellent starting point for a literature search is available in the form of a literature survey prepared periodically by Analabs, Inc. [26]. The Analabs compilation consists of three sections. The first section is divided into 22 compound classifications, each of which is subdivided into a variety of specific analyses for members of the general class. Data provided in this section includes the stationary phases and the original literature references. The second section is an alphabetical list of stationary phases and includes the composition of trade-name phases, maximum recommended temperature, a solvent for column preparation, appropriate uses, and literature references. The third section is simply a listing of stationary phases by the chemical nature of the phase.

A much more comprehensive reference source is available from the American Society for Testing and Materials (ASTM) [27]. The ASTM compilation presents the Tennessee Eastman system of chromatographic data retrieval [28] in either punched card or text form. The text is generally easier to use and much less expensive than the cards, but the cards represent a more complete informative file. The text consists of two sections, one tabulated on the basis of stationary phases, either liquid and support or active solid, and the other arranged according to solute molecular formula. The compiled data include: liquid phase and support, or active solid; reference solute; retention data; operating temperature; and literature references.

With the aid of literature similar to that described above, column selection is relatively simple provided that the solutes of interest are chemically similar. If, however, the components of the sample are of different chemical classes but have similar vapor pressures, column selection becomes more difficult, and the literature must be searched for each class to find a column that, it is hoped, has been used for the types of solutes of interest. If this approach fails, a trial and error approach must generally be used, although even here some help is available.

Ewell et al. [29] proposed a series of classifications for solutes and liquid phases based on hydrogen bonding, active hydrogen, and donor atoms. The following brief description of the application of Ewell's classifications to column selection is similar to the more detailed presentation by McNair and Bonelli [30]. In this system, solutes and liquid phases are divided into five classifications: (I) capable of hydrogen-bond network formation; (II) a donor atom (O, N, F) and active hydrogen present; (III) a donor atom available; (IV) active hydrogen available; and, (V) no hydrogen bonding capacity. Then, if a sample contains a number of solutes of the same class, a liquid phase of the same classification is selected since this generally provides greater solubility and better separation. If, however, the mixture is of solutes from two classes, e.g., II and IV, a liquid phase from either class can retain one group and generally the second group can be eluted rapidly. This may provide an adequate separation that can frequently be enhanced by proper selection of polarity of the liquid phase to obtain dipole interactions between solute and liquid.

Finally, some solute mixtures are not readily susceptible to analysis by a single column. The analysis can then be performed with the aid of sample modification, e.g., distillation, extraction, or derivative formation, or utilization of multiple columns. Multiple columns may be used independently with subsequent recombination of data, or two or more columns can be arranged in series to obtain data as if a single column has been employed. Combinations of individual columns are easier to select but data treatment is more complex. A series of multiple columns is much easier to operate and eliminates the need for column switching or multiple instruments and provides simplified data handling, but the selection of a series of columns is rather difficult and complex.

TABLE 6

Stationary Phases Applications

Class	Stationary phases
Acids	FFAP, SE-52, Apiezon L, SE-30
Alcohols	Porapak Q, DEGS, Carbowax 20M, FFAP
Aldehydes	Porapak Q, DC-550, Ucon 280X, Carbowax 20M
Amides	Versamid 900, Apiezon L
Amines	Porapak R, DC-550, THEED, Dowfax 9N9/KOH
Aromatics	TCEPE, Ucon LB 550-X, dibutyltetrachlorophthalate
Essential oils	Carbowax 20M, FFAP
Esters	Porapak Q, dinonyl phthalate, EGS, SE-30
Ethers	Carbowax 20M, β,β'-oxydipropionitrile
Glycols	Porapak Q
Halogens	SE-52, DC-550, triphenyl phosphate, Carbowax 20M
Ketones	Porapak Q, DC-550, Carbowax 20M, Lexan
Nitriles	Carbowax 400, TCEPE, XF-1150
Olefins	AgNO$_3$/ethylene glycol, tricresyl phosphate, squalane
Paraffins	Squalane, SE-30, Carbowax 400, tricresyl phosphate
Phenols	Dinonyl phthalate, XE-60
Steroids	STAP, QF-1, SE-30, XE-60, Epon 1001
Sugars	Carbowax 6000, PEG 4000, SE-52, DEGS, QF 1
Sulfur	Porapak Q, Apiezon M, Carbowax 1500
Water	Porapak Q

A very brief and incomplete list of liquid stationary phases that have been used for the analysis of various function groups is presented in Table 6.

6. Column Temperature

Both the magnitude and the control of column temperature are important in gas chromatography. The resolution and analysis times decrease with increasing temperature, and according to Giddings [31], the retention time is halved by a 30°C increase in temperature. For most solutes, the ratio of partition coefficients, and therefore resolution, increases with a decrease in temperature. Generally, a column is operated at the highest temperature that permits the desired separation. Accurate temperature control is necessary to obtain reproducible data.

In many cases, the solute retentions vary so widely that temperature programming is desirable. Temperature programming may involve a linear rate of temperature increase, or a stepwise increase. Generally, if a sample contains solutes with boiling points ranging from low to high, iso-thermal operation can provide either rapid elution and poor separation of the low boilers or very slow elution of the high boilers, with peaks sometimes indistinguishable from the baseline. A temperature program operation allows resolution of low boilers at a low temperature and subsequent elution of high boilers at an elevated temperature, so that solutes with a wide boiling point range can be conveniently separated in a single analysis.

7. Detectors

The wide variety of detectors available for gas chromatographic analysis is partly responsible for the rapid development of the technique. Table 4 lists most of the detectors that are frequently incorporated into gas chromatography systems and that may be labeled "GC detectors." Solutes of interest can also be collected and measured by any appropriate technique, e.g., titration, atomic absorption, nuclear magnetic resonance. In addition, the coupling of gas chromatographs with mass spectrometers and rapid-scan infrared spectrophotometers offers immense capabilities for the identification of unknoqn solutes. Some of the more common detectors are briefly described below. Greater detail may be obtained from refs. 8 and 30 or from data sheets available from the various manufacturers. The detectors described below are all differential detectors; i.e., the electrical signal generated by the detector represents the detector response to a change in the composition of its contents. The result is the familiar series of gaussian peaks, with each peak indicating the arrival of a solute at the detector.

The thermal conductivity (TC) detector [32], shown schematically in Fig. 10, is the most widely used gas chromatography detector because it is inexpensive, convenient, sensitive, and universal; i.e., it responds to any solute. The detector consists of two resistance wires in a Wheatstone-bridge circuit. One wire is placed in the effluent carrier gas stream, while the other wire, the reference, is confronted with pure carrier gas; both wires are then heated by the passage of a direct current. Now, the temperature of an object is determined by the ability of its surroundings to conduct heat, while the electrical resistance of an object is inversely proportional to its temperature. Consequently, if the Wheatstone bridge is balanced with pure carrier in each side, any solute entering the detector causes an imbalance and generates an electrical signal. The magnitude of this signal is determined

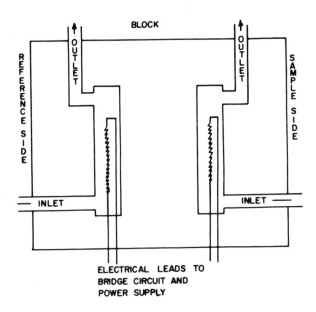

Fig. 10. Thermal conductivity detector.

by the nature and operating conditions of the detector and by the difference in thermal conductivities of the solute and carrier gas. The maximum temperatures and filament currents are determined by the composition of the resistor elements and are available from the manufacturers. The sensitivity of a thermal conductivity detector, on the order of 5×10^{-10} g/sec, can be increased by increasing filament current. It may also be increased by decreasing detector block temperature (but block temperature must be high enough to prevent solute condensation) and by selecting a carrier gas with a thermal conductivity very different from that of the sample. Consequently, helium is preferred for the TC detector because the thermal conductivity of helium is very much greater than that of any other material except hydrogen. The reader should be cautioned that current must never be applied to a TC detector in the absence of carrier gas flow, as the filaments rapidly burn when heated in air.

The gas density balance [33] is similar to the thermal conductivity detector in that the signal is generated by a change in resistance due to a differential heat loss. However, unlike the TC detector, which responds to a change in gas composition, the gas density detector responds to a change in gas velocity. With pure carrier gas emerging from the column the two resistance elements (Fig. 11) are balanced and no signal is generated. When the composition of the column effluent changes, the corresponding density changes result in a flow and, therefore, a resistance change. If the sample

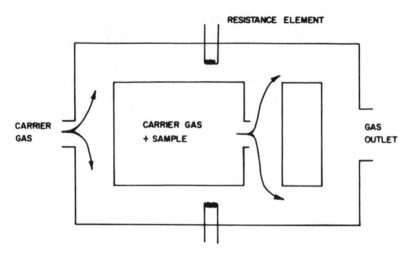

Figure 11. Gas density balance.

is more dense than the carrier, the effluent tends to flow down, restricting
the reference flow in the lower channel and thereby establishing a resistance
difference between the two sensing elements. Conversely, if the solute is
lighter than the carrier, column flow tends toward the upper path. Consequently, both positive and negative peaks are obtained as solute densities
range from light to heavy relative to the carrier density. The sensitivity of
the balance depends on the difference in density between the solute and eluent.
The reference flow must be at least 15 ml/min faster than column flow to
prevent diffusion of the sample to the sensing elements; since diffusion occurs
very readily in hydrogen and helium, they are generally not used with this
detector. The primary advantage of the gas density balance is the capability
of quantitative analysis without calibration, with the weight of solute given
by:

$$\text{Solute wt.} = XAM_s/(M_s - M_c) \tag{66}$$

where X is an instrument constant, A is the peak area, and M_s and M_c are
the molecular weights of the solute and carrier, respectively. The gas
density balance has a sensitivity comparable to that of the TC detector.

With the exception of the thermal conductivity and gas density balance
detectors, the detectors listed in Table 4 are all ionization detectors that
operate as indicated schematically in Fig. 12. The column effluent enters
a chamber that contains an ionization source and a current-sensing electrode
system. In some of the ionization chambers, the solutes are ionized by the
source to give an increase in current with sample concentration. In other

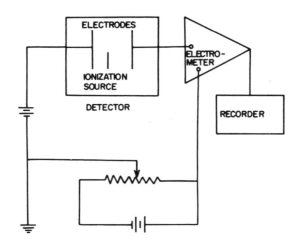

Figure 12. Ionization detection system.

ionization detectors, the carrier is ionized to produce a constant or "standing current," which is diminished by the presence of electron-absorbing solutes. The ionizing source is the distinguishing characteristic of the detectors described below.

The flame ionization detector (FID) is the most common ionization detector. The column effluent is mixed with hydrogen and burned in air or oxygen to produce a current when a solute is present. The mechanism of solute ionization in the flame is uncertain and apparently complex [34], but nearly all materials respond to the flame with the major exceptions of H_2; O_2; N_2; the inert gases; nitrogen oxides; hydrogen sulfide; the various combinations of carbon, oxygen, and sulfur; and water. The FID is very sensitive, capable of detecting 1×10^{-12} g/sec, with the detector response proportional to mass flow, i.e., to grams of solute per second, rather than to mass. Conversion from mass flow to mass is accomplished by multiplication of mass flow values by the baseline peak width expressed in seconds. Consequently, the flow rates of carrier, hydrogen, and air or oxygen must be well controlled and usually in the ratio of 1:1:10. In addition, there exists an optimum hydrogen flow rate for each instrument. This optimum hydrogen flow is determined from a plot of detector response vs hydrogen flow. The optimum flow rate for hydrogen is usually near 30 ml/min, but the best operating condition may be quite different since this is determined generally by the column flow rate and the 1:1:10 flow ratio.

The alkali flame detector (AFD) is essentially the FID described above with an alkali salt pellet, CsBr or Rb_2SO_4, placed on the burner jet. The AFD operates at about 130 ml/min of air compared to 300-400 ml/min for the FID. This relatively fuel-rich operation results in a selective sensitivity

to phosphorus compounds that is approximately 5000 times the detector
sensitivity to hydrocarbons. Moreover, with a flow modification, the
detector becomes selectively sensitive to nitrogen at a 50:1 ratio relative to
hydrocarbons [28]. The AFD is capable of detecting approximately 10^{-12} g
in the phosphorus mode and 10^{-10} g in the nitrogen mode.

The electron capture detector uses tritium or ^{63}Ni to ionize the carrier
gas and produce a constant current. The presence of solutes capable of
absorbing electrons causes a decrease in the standing current that is proportional to solute concentration. The ^{63}Ni allows a higher operating temperature
and, therefore, less detector contamination than is possible with a tritium
source. The detector is excellent for pesticide analysis because of its
extreme sensitivity (2×10^{-14} g/sec) to organic halides. Other compounds
that are susceptible to electron capture are organometallics and conjugated
carbonyls. The detector has essentially no response to most nonconjugated
organics except the halides, nitrites, and nitrates. The maximum operating
temperature is $225°$ C for tritium and $350°$C for nickel.

The cross-section detector [35] uses a 250 mCi tritium foil, as does the
electron capture (EC) detector. However, unlike the EC detector, the carrier
gas (H_2 or He) has a very low ionization cross-section and produces only a
very low background current. Any solute entering the detector has a higher
probability of β capture with electron release than does the carrier. Consequently, any solute entering the detector generates an increase in current
proportional to solute concentration. The detector is relatively insensitive,
with a detection limit of about 20 μg, with a sensitivity on the order of
2×10^{-9} g/sec. The detector does respond to all compounds, and detector
signal output has a larger linear range than any other detector. If helium is
the carrier, 3% methane must be added to quench metastable helium, which
would interfere with detector performance.

Metastable argon is produced by collisions between β radiation and argon
carrier gas in the argon ionization detector [36]. The metastable argon
has excess energy of about 11.7 eV relative to ground-state argon. Collisions
between the excited argon and solute molecules result in energy transfer
and subsequent ionization of solutes with ionization potentials below 11.7 eV.
In addition, a small percentage of argon atoms will be raised to states or
even ions more energetic than 11.7 eV. Solutes with ionization potentials
greater than 11.7 eV are thus also ionized, but with very poor sensitivity.
Ionization of solute molecules results in a current flow, which is fed to the
amplifier. The detector has a sensitivity on the order of 10^{-14} g/cm for
materials that meet the ionization potential requirement. The detector
responds very poorly or not at all to water, fixed gases, CO, CO_2, CH_4,
and C_2H_6, while giving a nearly constant response to most other materials.
The detector has a narrow range and is inoperable for concentrations of
solute in argon of 0.1% and above.

The helium detector also uses tritium β radiation, but with a 4000 V/cm
field gradient to promote the helium carrier molecules to a metastable state
of 19.8 eV. The metastable helium will ionize any solute. The sensitivity

is very good, on the order of 10^{-14} g/sec, but the detector is very sensitive to heat and flow stability. In addition, the extreme sensitivity requires ultrapure helium and generally must be used with bleed-free columns, i.e., with solid adsorbents and Porapak. Consequently, the detector is generally used only for the fixed gases.

The reader should carefully note that the detector characteristics presented above are guides, not absolute values. The various detector parameters, such as optimum flows, filament currents, or voltages, and temperatures, are functions of detector composition and design, which are determined by the manufacturer and consequently vary among instruments. In addition, the magnitude of a detector's response is a function of the solute, itself. Thus, sensitivity values found here or elsewhere should be considered only as guides since they vary both with the instrument and the solute. Finally, a variety of sensitivity expressions are currently used, and care must be exercised to ascertain that a comparison is actually between like quantities.

8. Qualitative Analysis

This section is concerned with solute identification from retention data only, although retention data alone are usually not sufficient to characterize a sample. Additional data are nearly always necessary and preferably concern the functionality and structures of the solutes. With no additional data, one could still identify the solutes, but only through the rather tedious process of obtaining retention data from two or more dissimilar columns. A variety of systems for the identification of solutes is briefly mentioned here, not because all are equally satisfactory but because all have considerable followings.

Under reproducible pressure and temperature conditions, retention volume or time is characteristic of a solute for a given column. Uncorrected retention volume, the volume from injection to peak maximum, can be used, but the data cannot be compared to that obtained on another column or instrument. The uncorrected retention volume is a function of the liquid phase (material and amount), the column (diameter, length, and temperature), the carrier gas (material, flow rate, and pressure drop), and the system dead volume. The adjusted retention volume is a function of the same variables as the uncorrected retention volume except that the adjusted volume is corrected for dead volume. The time from the maximum of a nonsorbed solute, usually air or solvent, to the solute peak maximum is the adjusted retention time. Corrected retention volume is the adjusted retention volume corrected for the pressure drop across the column [cf. Eqs. (52) and (53)].

One of the preferred methods of identification from retention data is the use of the relative retention R_{IS}, shown in Fig. 13 along with the uncorrected and adjusted retention volumes. The adjusted retention volumes are obtained

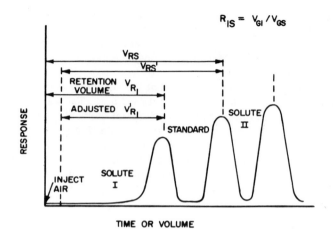

Figure 13. Retention volume or time measurements.

as indicated; then each is divided by the adjusted retention volume of any arbitrarily selected standard. Generally, the relative retention volume depends only on the liquid-phase material and the column temperature.

One of the best retention volume expressions for purposes of interlaboratory comparison is the specific retention volume. This was defined by Littlewood et al. [37] to express retention volume as a standard value at standard temperature and pressure per unit weight of stationary phase. The specific retention volume V_g is given by:

$$V_g = 273 \, V_N / T_c W_s \tag{67}$$

where V_N is the corrected retention volume, T_c is the column temperature, and W_s is the weight of stationary phase in grams.

An excellent but somewhat tedious method of data comparison is the retention index I of Kovats [38], which is defined as:

$$I = 100 \left[\frac{\log V_{N_x} - \log V_{N(n-C_z)}}{\log V_{N(n-C_{z+1})} - \log V_{N(n-C_z)}} \right] + 100z \tag{68}$$

where V_N is the net retention volume ($V_N = jW_R$), x refers to the unknown substance $n - C_z$ is a normal paraffin containing z even number of carbon atoms, and $n - C_z + 1$ is a normal paraffin with z + 1 carbons. When the Kovats retention index system was first proposed [39, 40] it was based on even number normal paraffins because it was felt that there would be an ascillation in the chromatographic properties of successive numbers of the complete n-paraffin series. This was found not to be the case and the equation was appropriately changed. In its initial form the equation had the form:

$$I = (200 \log \alpha_x / \log \alpha_{P_{Z+2}}) + 100z \tag{69}$$

where α_x is the relative retention of the unknown referred to some normal paraffin, P_z, containing an even number of carbon atoms; $\alpha_{P_{Z+2}}$ is the relative retention of the normal paraffin, P_{Z+2}, referred to P_z; and Z is the number of carbon atoms. Structural information can sometimes be obtained from the use of ΔI, where ΔI is the retention index on a polar column minus the retention index on a nonpolar column.

9. Quantitative Analysis

With the use of gas chromatography, quantitative measurements are relatively simple since each solute is presented to the measuring device in essentially pure form. The discussion here is concerned only with quantitative analysis as performed with the detectors described above. These all require some kind of calibration or standardization except the gas density balance. The resolved solutes can be collected and direct quantitization methods applied, but these are rather tedious and are usually unnecessary. The first portion of this discussion presents means of data collection, while the later portion is concerned with data treatment.

Data for quantitative analysis generally are obtained from a recorder, a recorder with disk integrator, or a digital integrator. Either peak height or peak area can be measured from a chromatogram, and height measurement and two of the possible area measurements are described in Fig. 14. Another common means of obtaining area measurements from a chromatogram involves the use of a planimeter, which when moved over the peak perimeter gives a digital area measurement. One may also obtain area measurements from a disk integrator (Disc Instrument Company) attached to a recorder that provides a "sawtooth" trace on the chromatogram. The number of pen traverses is proportional to peak area. By far the best means of peak integration is the use of a digital integrator, which converts electrometer voltage to a digital printout of peak area; usually, retention times are also printed. The order of preference for the data measurements listed above is digital integrator; disk integrator; others, primarily determined by operator preference.

Data having been obtained, they must somehow be related to the concentrations of solutes in the sample. The means of correlating data with concentration include calibration, area normalization, internal

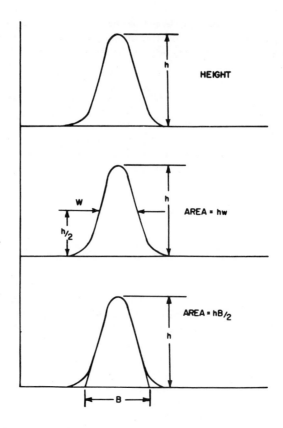

Figure 14. Peak measurements.

standardization, and standard addition. Except with direct calibration and
standard addition methods, the variation of detector response among different
compounds is significant and must be accounted for in data treatment.

The direct calibration method of quantitization is no different from that
employed in a great variety of analytical techniques. A plot of peak height
or area vs the amount of solute is prepared. Unknown concentrations are
then determined by comparison of peak areas or heights with the calibration
curve. This method is most useful for repetitive measurements of no more
than a few solutes, as calibration for a number of compounds is tedious and
time consuming. In addition, sample weight or volume must be accurately
measured, which can be disadvantageous in view of the sample sizes usually
encountered.

Area normalization is very easy but, strictly speaking, is applicable
only if all solutes are identifiable. Then, if all solutes in the sample give

similar detector responses within the desired error limits, simple area normalization provides solute concentrations in weight percent; i.e., for solute A, the area of A times 100, divided by the total area of all peaks, is equal to the weight percent of A. However, in the more normal case where detector response varies with the compound, correction (response) factors must be applied to the peak areas before normalization. Both flame ionization and thermal conductivity response factors have been tabulated for a large number of compounds by Dietz [41]. The relative response factors for FID are determined as follows: (1) some compound A (often benzene) is chosen as a standard; (2) a solution of known concentration of A, B, C, etc., is prepared; (3) the chromatogram is obtained and the areas are measured; (4) the ratio of area to weight is determined for each peak; and (5) the ratio for each peak is divided by the ratio for the standard peak to give the relative response factor. Area normalization for a sample is then applied after division of each peak by the appropriate FID response factor. It is not necessary that the original standard, e.g., benzene, be present; only that the relative responses of each solute be known. The data for a TC detector have historically been treated as follows [42]: (1) the relative response is determined as described for the FID and then converted to a mole basis; (2) the molecular weight is divided by the thermal response per mole relative to benzene, the molar response of which is assumed to be 100; (3) the weight factor thus obtained is multiplied by the peak area; and (4) the products obtained in (3) are normalized to give weight percent.

Internal standardization also proceeds in GC as in other analytical techniques. A given weight or volume of standard is mixed with a known amount of sample and the mixture is chromatographed. Then, if the standard and solute of interest have similar responses, the weight of the unknown can be calculated by multiplying the weight to area ratio of the known by the area of the unknown. If the relative responses are not similar, the appropriate correction factors must be applied to the measured areas. The internal standard method is best suited to nonrepetitive analyses of one or a few solutes when all solutes are not identified or all response factors are not known.

The standard addition method involves the use of two chromatograms, one from a known volume of sample and the other from the same volume of sample plus a known amount of the solute of interest. Response factors are no longer a concern, and weight of solute is determined by comparing the ratio of unknown area to unknown plus standard area. This method is best suited for nonroutine analysis of one or a few solutes the response factors of which are unknown.

B. Liquid Chromatography

Liquid chromatography, as discussed here, is concerned with those systems having a liquid mobile phase, a column configuration stationary phase, and a distribution process that involves adsorption or partition. Since the eluent is a liquid and temperature is usually not a significant factor, liquid systems have been very simple and inexpensive. However, the recent introduction of high-pressure, high-speed liquid chromatography has brought about an abrupt change in equipment sophistication and cost. The liquid systems can handle a relatively large sample, regardless of sample volatility or thermal stability. In addition to the analysis capabilities, the liquid systems are also able to easily and quickly concentrate very dilute solutions, although, except for the high-pressure systems, liquid chromatography is much slower than gas chromatography. Liquid chromatography also suffers from a paucity of detection systems.

1. Instrumentation

Liquid chromatography equipment may be as simple as that shown in Fig. 15, which is simply a glass tube with a porous support in the tapered end to support the stationary phase. A great variety of improvements can be made on this basic device [43] and these usually involve modifications of the reservoir to assure constant pressures and flow rate or modifications at the column exit, including vacuum devices for increasing column flow,

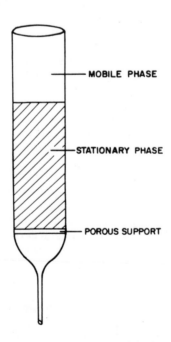

Figure 15. Simple liquid chromatography column.

automatic sample collection, and effluent monitors. However, the apparatus for high-speed liquid chromatography is similar to the gas chromatography apparatus of Fig. 9, with the exception that the gas cylinder is replaced by a liquid reservoir, a pump, and a flow regulation system to provide a constant high liquid velocity. A recent addition [44] is the use of a centrifuge as the driving force for eluents in very small columns.

2. Mobile Phase

The eluent in liquid adsorption chromatography is unique in that the only difference between solute and eluent is the quantities present; i.e., the adsorbent acts on both solute and solvent. The phenomenon is much less true in liquid-liquid systems because of the requirement that the two liquids be immiscible. Generally, the choice of solvent for adsorption systems is determined only by the nature of the solutes. If the solutes of interest have a relatively narrow range of distribution coefficients, a poorly adsorbed solvent should be selected to avoid interference of the adsorption phenomena by competition of the eluent for the active sites. If the solutes have a wide range of affinities for the adsorbent, however, gradient elution or stepwise elution should be used, in which the solvent adsorption is continuously or discontinuously increased by additions of more strongly adsorbed solvents. This represents a considerable saving of analysis time since the weakly adsorbed solutes are eluted rapidly by the poorly adsorbed eluent, and the strongly adsorbed solutes are eluted relatively fast by the strongly adsorbed mobile phase. Generally, adsorption affinity increases with increasing polarity. It should be noted that the above comments apply only to elution analysis; for displacement analysis the solvent must be more strongly adsorbed than any solute of interest. A very low solvent velocity is usually advantageous for the attainment of sorption equilibrium and because of minimization of column packing irregularities. These advantages normally outweight the disadvantage of diffusion at low flow rates. Consequently, for difficult separations, flow velocities of a few milliliters per hour may be desirable.

For liquid-liquid systems, the mobile-phase requirements are somehwat more severe than those for liquid-solid chromatography. The ultimate mobile-phase requirement is immiscibility with the stationary liquid. In addition, the eluent must be capable of solvating a significant quantity of solute but the solute should be much more readily soluble in the stationary phase to avoid rapid elution and poor resolution of solutes. Suitable mobile-stationary phase combinations for given solutes can readily be obtained from solvent extraction data. Flow rates are on the order of 1 ml/min.

3. Sample Introduction

For apparatus similar to that in Fig. 15, the sample is introduced onto the
top of the column in a very small volume of a solvent that is less polar than
the eluting solvent, as this avoids band spreading. Addition may be made
with a pipet or syringe but must be uniform over the surface. Care must
be taken to avoid disturbing the top of the column, as this leads to channeling.
A thin layer of fine sand on top of the column is a good preventive. Liquid-
liquid injection is essentially identical to that described in Sec. III, A, 3 for
gas chromatography. However, a separate heated injection port is not
required, as samples should be introduced at operating temperatures.

4. Column Preparation

Column preparation for high-speed liquid chromatography, both solid and
liquid, is exactly as described above for gas chromatography. For the
larger, open-end columns (cf. Fig. 15), liquid stationary phases are packed
as above but may also be packed by tamping with a rod. In addition, adsorbents may be packed in the same fashion, or they may be packed wet, in the
form of a slurry with the mobile phase, which usually gives a more homogeneous column. Generally, once a packed column has been wet by solvent,
care must be taken to keep it wet to prevent crack and channel formation.
Moreover, for high-speed liquid-liquid systems, a precolumn, which is
usually identical to the analytical column must be used. The precolumn
allows eluent equilibration with the stationary phase and thereby minimizes
bleed from the analytical column, which can occur even when the eluent has
been saturated with stationary phase. The precolumn is eliminated by the
use of Permaphase packings in which the stationary liquid is permanently
bonded to the support [45].

5. Column Selection

Generally, column selection in liquid chromatography is much easier than in
gas chromatography. As in GC, only a limited number of adsorbents are
used with any frequency. Unlike in GC, however, liquid-liquid choices can
be made very easily from an observation of solvent extraction behavior.
Table 7 presents a partial list of the more common adsorbents, while Table
8 lists the order of adsorption for common solutes as well as solvents (more
complete listings are available from refs. 2 and 46). There is no generally
satisfactory method for adsorbent selection. The usual procedure is an
experience-guided trial and error variation of eluent-adsorbent polarity
combinations as indicated by solute polarities. Thin-layer techniques can
also be used to rapidly estimate the utility of a given system, which is then
optimized for column work.

TABLE 7

Common Adsorbents in Order of Increasing
Adsorption Power

Sucrose

Cellulose

Starch

Calcium carbonate

Silica

Charcoal

Magnesia

Alumina

Fuller's earths

TABLE 8

Functional Groups in Order of Increasing
Adsorption Affinity

Paraffins

Olefins

Aromatics

Halogens

Carbonyls

Hydroxyl, amino, thio, and nitro

Water

Organic acids

6. Temperature

Temperature is much less critical in liquid chromatography than in gas
chromatography. As indicated in Sec. III, B, 6, no separate injection part
is used as solutes are introduced at column temperature. Column temperature must be well controlled to obtain reproducible data, but for most work

this usually requires only a good insulation system since most liquid work is performed at room temperature. For certain continuous detection systems, especially differential refractometers, accurate temperature control is mandatory.

7. Detectors

Whereas nearly all gas chromatography detection involves continuous effluent monitoring, a lack of suitable detectors has hindered liquid chromatography monitoring. A considerable portion of adsorption experiments still utilize noncontinuous monitoring. One of the more tedious procedures involves fraction collection; i.e., small increments of effluent (one to a few milliliters) are collected consecutively, with subsequent measurement by any appropriate analytical technique. Another common detection procedure involves the physical removal of column packing from the column after resolution but before elution. The solutes are then located on the solid by visual observation for colored solutes, with fluorescent light for susceptible solutes; or by staining, i.e., spraying or brushing of the column with reagents that give colored products with the solutes. When quantitative analysis is required, the sections of the column that contain solutes are extracted with an appropriate solvent, in a soxhlet apparatus if necessary, and the separated solutes are analyzed by conventional techniques.

The most universal continuous detector for liquid chromatography is the differential refractometer. This detector is universal only to the extent that one can select an eluent the refractive index of which differs from that of the solutes. Pure eluent flows through one cell of the detector while column effluent flows through the other cell. When a solute with a refractive index different from that of the solvent enters the detector, an imbalance occurs, which is transmitted to the recorder. The differential refractometers are relatively insensitive to trace and minor components, and detector stability is greatly affected by small temperature and flow fluctuations, thus requiring very good flow and temperature controls.

One of the most sensitive detectors available utilizes the absorption of electromagnetic energy, usually in the ultraviolet region, by the solutes. These detectors are very sensitive, even to trace components. However, they are generally fixed wavelength units, often 254 and 280 nm, which severely limits their universal application. A beam of ultraviolet (UV) light passes through the nonabsorbing eluent and impinges on a photomultiplier. When a susceptible solute enters the light path, energy is absorbed and the photomultiplier senses a decrease in light intensity. The UV detector is somewhat less sensitive to temperature and flow irregularities than is the differential refractometer.

A flame ionization detector is also available but suffers some severe limitations. The eluent is fed to a treadmill-like wire that passes through a heater to evaporate the solvent, since most solvents are ionized in the flame, with water and carbon disulfide the only notable exceptions. The dried sample is then carried directly to the flame on the wire, volatilized and carried to the burner, or pyrolyzed with the pyrolysis products directed to the flame. Ionization occurs in the flame, causing a current to flow in the detector circuit. The detector is very sensitive but is suitable only for those systems where the eluent boils at a very much lower or very much higher temperature than all of the solutes. In addition, the column bleed problem is a severe one for liquid-liquid systems unless permanent liquid phases are employed.

A variety of very selective detectors has been used to monitor selected solutes, but their application is very limited. Most of these involve electrical measurements to which many materials are not susceptible. Among those reported in the literature are electrolytic conductivity cells, dielectric constant measurements, and polarography cells. Radioactivity and atomic absorption measurements have also been employed to monitor effluents. Finally, detectors have been developed that operate by monitoring heats of reaction between eluent, solute, and stationary phase.

8. Qualitative Analysis

As indicated in Section III, B, 7, qualitative analysis can be performed by any convenient technique after solute isolation. Qualitative analysis from retention data is very similar to that employed in gas chromatography. Uncorrected retention volume or time is characteristic of the solute but, because of the dead volume effect, cannot be used for comparisons between columns or instruments. Adjusted retention volume, retention volume minus the retention volume of a nonsorbed solute, is corrected for instrument dead volume but is still relatively difficult to reproduce. The relative retention, i.e., adjusted retention divided by the retention of some standard, is the preferred solute characteristic for liquid chromatography as well as gas chromatography. Column pressure drop corrections are unnecessary even for high-pressure systems in all but the most exacting work.

If, instead of eluting the solutes, chromatographic development is stopped after solute resolution but before elution, retention measurements for identification purposes take the form of R_f data. The R_f value is character-istic of a solute for a given system and is defined as the ratio of the distance traveled by the leading edge of the solute zone to the distance traveled by the leading edge of the eluent. The R_f value for a given solute and system is a function only of temperature, solute concentration, and mobile-phase velocity. Since R_f is a function of the amount of solute, and therefore band width, it is essential that bands be sharp and narrow to obtain reproducible data.

9. Quantitative Analysis

Again, isolated solutes may be measured by any suitable technique.
Analysis from chromatographic data is as described above for gas chromatography, i.e., by calibration, area normalization, correction factors,
internal standardization, or standard additions. For correction factor
analysis, the preferred method is that described for the flame ionization
detector in gas chromatography.

C. Thin-Layer Chromatography

Originally, "thin-layer chromatography" referred specifically to liquid-solid
chromatography performed on a thin film in an analogy to paper chromatography. Now, however, thin-layer chromatography is also employed in
systems where the separation is based on liquid-liquid partition or ion-
exchange phenomena. The primary advantage of thin-layer chromatography
is the ability to rapidly analyze minute quantities of very similar materials.
The improvements in plate coating methods, and especially the availability
of prepared plates, have virtually eliminated the major difficulties formerly
encountered in thin-layer chromatography, i.e., fragile and nonuniform
coatings that were extremely difficult to reproduce. And, like the column
forms of liquid chromatography above, thin-layer is applicable to all
compounds regardless of volatility, while thermal stability is usually not a
problem since most analyses are performed at room temperature. However,
in both column and thin-layer adsorption chromatography, adsorbents may
catalyze solute degradation even at room temperature.

1. Instrumentation

One of the advantages of thin-layer chromatography is the lack of a need for
expensive equipment. The only necessary equipment is the development
chamber, shown schematically in Fig. 16. In its simplest form, the chamber
is a closeable container that is larger than the plates to be developed. More
sophisticated chambers include "sandwich," i.e., very small volume construction; provisions for an air-tight seal; and a provision for maintaining
an inert atmosphere. Provisions for temperature control are usually
unnecessary unless room temperature fluctuations are severe. The development chamber contains all the components of the chromatographic system,
i.e., eluent, sample introduction, and separating medium. No on-stream
detectors or recorder are used.

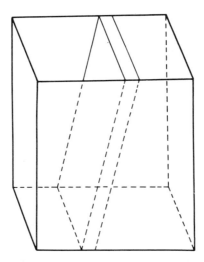

Figure 16. TLC development chamber.

2. Mobile Phase

The solvents used in TLC are the same as those used in liquid-solid chromatography (cf. Table 8). As in column chromatography, moreover, solute
movement is increased with increasing solvent polarity. The solvent is
applied in the form of a pool at the bottom of the developing chamber. The
solvent must come into contact with the adsorbent layer but should be no
more than 5 mm above the plate edge. The solvent may also be made to
descend or to travel horizontally on the plate, but these techniques are
rarely employed. The solvent travels through the system by capillary
action; thus, solvent velocity is determined by the nature and packing
structure of the adsorbent. In TLC development, the saturation of the
chamber with solvent vapor has a significant effect on solute migration.
Specifically, with undersaturation, solvent and therefore solutes move faster
at the plate edges than in the center. This phenomena results from increased
evaporation at the plate edges in an effort to saturate the chamber on the
uncoated side of the plate. The saturation problem can be overcome by
lining the developing chamber with a filter paper that is saturated with the
developing solvent to provide uniform chamber saturation.

As in liquid chromatography, elution may be stepwise, with two or more
solvents, or may be with a continuous solvent gradient. For stepwise
development, the plate is usually dried at 130° C for 30 min between each

solvent application. One technique that is available only to thin-layer and paper chromatography is two-dimensional chromatography in which the chromatoplate is developed with a given solvent; the plate is then rotated 90° and developed with another solvent of different polarity. A variety of special development techniques is described by Stahl [47].

3. Sample Introduction

Sample application is performed by spotting or streaking the thin-layer plate with the sample. Analytical plates are usually spotted, while preparative (large scale) plates are streaked, although either sample application can be used for analytical or preparative work. Spotting is effected by literally placing a drop of sample on the plate, usually with a micropipet or microliter syringe. Sample volume is usually in the range of 1-10 μl. If dilute solutions of a sample are to be applied, the technique of successive spotting must be used, where a number of drops are placed on the same spot but only after each previous spot is dry. Sample streaking is usually accomplished manually or automatically by movement of a microliter syringe across the surface with a constant pressure on the syringe plunger. For streaking, the automative device usually provides a much neater sample application and, therefore, much neater elution bands. The sample should normally be applied about 25 mm from the edge of the plate. A number of samples can often be separated simultaneously on the same plate. Multiple samples usually must be applied by spotting, and the spots should be separated from each other by at least 15 mm. Finally, after sample application, the plate is usually scored with a sharp pencil about 10 cm from the point of sample application. The scoring prevents solvent travel to the end of the plate by interrupting capillary flow.

4. Plate Preparation

Thin-layer plates may be prepared by pouring, dipping, spraying, or spreading. However, the best source of uniform plates may be the prepared plates that are available from various manufacturers and suppliers. The requirements for the supporting plate include uniform thickness; inertness with respect to solvent, solute, stationary phase, and identification reagents and procedures; and sufficient strength to allow vertical development. Common plate materials include glass, aluminum, and plastic, while typical, but noncritical, dimensions are 5 × 10 to 20 × 20 cm, with most accessories based on a 20 × 20 cm size.

The adsorbent is applied in the form of a suspension or a slurry in a suitable solvent. For the pouring procedure, the suspension is dumped in one motion onto the plate center and the plate is shaken and tilted to spread the adsorbent. If two plates, held together face to face, are immersed in a

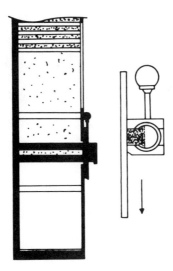

Figure 17. Schematic drawing of TLC stationary phase applicator (reproduced through the courtesy of Brinkmann Instruments, Inc., Westbury, New
York).

suspension of the adsorbent, one face of each plate becomes coated with a
thin layer of the stationary phase. With considerable practice, a skilled
operator can prepare plates by spreading the suspension with a commercial
sprayer. Generally, the best procedure for the preparation of uniform films
is the use of a spreader, such as that shown schematically in Fig. 17 and
in operation in Fig. 18. The device features an inner cylindrical chamber
the open side of which can be turned up for filling or down for dispensing
the slurry. The layer thickness is continuously adjustable from zero to
2000 μm. A series of plates 20 cm wide and up to a total length of 100 cm
on a mounting board can be prepared in a single operation as shown in
Fig. 18. Normal layer thickness, after drying, is in the range of 150 μm
for analytical and 2 mm for preparative systems.

For any of the above techniques, the adsorbent may be mixed with usually
5-15% of a suitable binder, e.g., calcium sulfate or starch, to aid in
adhesion of the layer. After they are coated, the plates must be dried,
often in a stream or warm air for 30 min, then in an oven at 150° C for
several hours. The drying conditions may vary with the nature of adsorbent,
binder, and solvent. The adsorbent may also be mixed with a variety of
materials to aid in spot location and measurement as discussed in Sec.
III,C,6. After they are dry, the plates should be stored in a dessicator
until use.

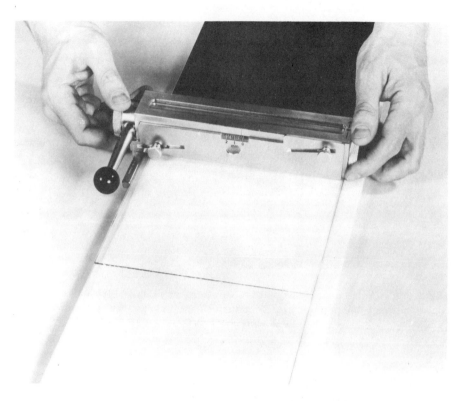

Figure 18. Operation of TLC stationary phase applicator (reproduced through the courtesy of Brinkmann Instruments, Inc., Westbury, New York).

5. Plate Selection

Only a limited number of adsorbents are available. Generally, these are the same as those listed in Table 7 for liquid chromatography. In practice, only a few, alumina, silica, cellulose, and a polyamide, are used with any frequency. Consequently, most systems for a given separation are developed by a selection of solvent polarity rather than by any significant choice of adsorbent. Admittedly, some samples are better handled on one adsorbent than another. However, there appears to be no suitable mechanism for adsorbent selection, and experience, trial and error, or the literature must be invoked [47-49]. Some simplifying generalizations can be made, however, and include: (1) silica gel is the most versatile adsorbent, suitable for nearly all solutes with a proper solvent choice; (2) alumina is very strongly adsorbing and is generally restricted to nonpolar solutes; (3) microcrystalline cellulose is used to convert a paper chromatography experiment to a thin-layer system; and (4) polyamides are very useful for the resolution of solutes that contain hydroxyl bonded to an aromatic ring, i. e., phenolic functions.

6. Detection

Since TLC development is always interrupted after solute separation but before complete elution, detectors similar to those for gas and liquid chromatography are not available. Consequently, detection efforts in TLC are directed toward the location or visualization of solutes, most of which are colorless. There are three widely applied visualization techniques: fluorescence, spray reagent, and charring.

Fluorescent visualization may be direct or indirect and is accomplished with the aid of a UV lamp that emits radiation usually at 254 nm (short wavelength) and 350 nm (long wavelength). Direct fluorescent visualization is possible only for those few solutes that are capable of fluorescing. Indirect fluorescence is much more applicable and involves the use of a fluorescent material with the adsorbent. The entire plate then fluoresces under UV light, and solutes that quench or block the fluorescence appear as dark spots on a fluorescent background.

Chemical reagents, usually called "spray reagents" because the proper method of application is spraying with an atomizer or from an aerosol can, locate solutes by forming colored reaction products with one or more types of functional groups. With the proper choice of reagent, which must be applied uniformly by spraying with both a vertical and a horizontal motion, any solute can be located. However, most reagents are fairly specific and their use for a variety of solute types in the same sample can be quite tedious. Most spray reagent visualization procedures require some color development, usually heat, after application. A few of the more common reagents with their applications are listed in Table 9. A fairly complete list of 266 reagents, methods of preparation used, and treatments after application is available in ref. 47.

Solutes may also be located on a TLC plate by either chemical or thermal charring. Thermal charring, i. e., carbonization by dry pyrolysis, is infrequently used because of the difficulty due to heat radiation from the white adsorbent. Chemical charring is usually accomplished either by treatment with concentrated sulfuric acid or with a saturated solution of potassium dichromate in 70% sulfuric acid. Both charring reagents are usually aided by heat treatment at about 100° C for 1/2 hr.

TABLE 9

TLC Spray Reagents

Reagent	Use
Aniline phthalate	Reducing sugars
Bromocresol Green	Acids and bases
Bromothymol Blue	Lipids
Dichlorofluoresceine	Lipids
4-Dimethylaminobenzaldehyde	Amino sugars, indols, alkaloids, urea, and others
Ninhydrin	Amino acids, amines
Phosphomolybdic acid	Reducible compounds
Rhodamine B	General fluorescent indicator

7. Qualitative Analysis

Specific reagents may be used to identify solutes on a TLC plate, but their application is practiced only when the types of compounds present are known in advance. Identification from retention data involves the use of R_f values and especially relative R_f values R_{st}. For thin-layer chromatography, R_f is defined as the distance traveled by the spot center divided by the distance traveled by the solvent front. The relative value R_{st} is defined as the solute R_f divided by a standard R_f or, simply, solute distance divided by reference distance. The R_{st} value must be determined from a system where the solute and sample are simultaneously chromatographed on the same plate.

The R_f values, although characteristic of a solute in a given system, are very difficult to reproduce since they are affected by adsorbent activation and surface area, by layer thickness, by chamber saturation, and by humidity and temperature. The R_{st} values are much more useful, as are relative retention values in the other systems of chromatography. Often, however, even documentation of R_{st} values is insufficient for purposes of TLC comparisons, and it is necessary to record details of spot shapes and areas and the shape of the solvent front. These data may be recorded in writing or by a schematic drawing, but for important samples the best procedure is the actual preservation of the plate or a reproduction of the plate. The plates can often be preserved with a coat of lacquer. The plates can be reproduced by a variety of techniques, including photography and xerography.

8. Quantitative Analysis

The quantitative measurement of TLC solutes may be performed on the plate or after removal of the solute from the plate. The latter is generally preferred for the analysis of a limited number of solutes, while the former is advantageous for the determination of a large number of compounds. The solute, along with the adsorbent, can be scraped from the plate, separated by extraction, and measured by any appropriate technique. On-plate analysis may involve visual observation, fluorescent measurements, spot area determination, reflectance measurements, or light transmission evaluation.

The quantitative analysis of solutes by visual comparison of spots is, at best, semiquantitative. This method is best performed by alternating sample and standard spots on the thin-layer plate. The same amount of sample is applied in each spot, while the amount of standard is progressively increased. After development, the spot sizes and color intensities for sample and standard are compared and the solute concentration is estimated.

Fluorescent measurements can be applied to compounds that can be reacted to form a fluorescent product as well as to those that naturally fluoresce. Since the fluorescent energy of the spot is not significantly affected by spot shape, solute concentrations can be determined from a calibration curve of solute weight vs fluorescent intensity. Commercial instruments designed for TLC measurements are available.

The spot area measurement is a rather tedious technique that involves comparisons of spot sizes. The area measurements are not linear with concentration, but the plot of the square root of the surface area vs concentration is usually linear. If the spots are sharp and distinct, measurement may be accomplished by any of the following: visual comparison to standard sizes; manual or reproductive copying followed by weighing of the spot; and evaluation by tracing the perimeter with a planimeter.

Quantitization by transmission measurements is one of the better techniques. This procedure depends on the absorbance of light by a colored or charred spot. A plot of the light absorption vs the amount of solute is generally linear. The commercial TLC densitometers generally operate on this principal. The surface reflection method is basically the same phenomenon except that reflected rather than transmitted light is measured. Commercial equipment is also available for reflectance measurements on TLC plates.

D. Paper Chromatography

The technique of paper chromatography is very similar in practice, if not in theory, to the thin-layer chromatography technique described in Sec. III, C. Just as thin layer now utilizes partition and ion-exchange stationary phases, paper has been impregnated with solid adsorbents, liquids, and ion-exchange materials. The advantage of paper chromatography lies in its great ability to resolve minute amounts of material without regard for their thermal stability. The technique is also easy, rapid, and inexpensive. Major disadvantages include the uncertain nature of the distribution mechanism and the empirical nature of paper chromatography experimentation.

1. Instrumentation

The apparatus for paper chromatography is essentially identical to that used for thin-layer chromatography and can be readily made, in simplest form, from a jar or jug by adding a wire or clip to the lid to support the paper strip. Simple devices for the performance of both ascending and descending paper chromatography are shown in Fig. 19. Usually, temperature control is not a significant concern unless large fluctuations occur. A useful addition to the development tank is a provision for placing the paper in the system, but not in the solvent. After the solvent vapor-liquid equilibrium

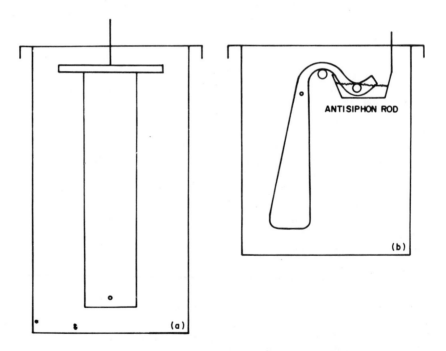

Figure 19. Ascending (a) and descending (b) paper chromatography apparatus.

has been established, the paper is placed in the solvent without opening the chamber. This can be easily accomplished with a variable position paper support, the position of which can be adjusted from without the chamber. As in TLC, the development chamber contains the entire system, as no detectors or recorders are used.

2. Mobile Phase

The eluent for paper chromatography is determined by the type of investigation, i.e., on the separating mechanism and, specifically, on the nature of the stationary phase. If the paper has been impregnated with an organic liquid, the distribution mechanism is predominatly, at least, liquid-liquid partition, and solvent choice is determined from extraction data (cf. Sec. III, B). For adsorbent-impregnated papers, liquid-solid adsorption dominates, and solvent selection is similar to that for liquid solid or TLC systems; i.e., the more polar the solvent, the higher the R_f value. If the paper has ion-exchange properties, the solvent is usually water or a mixture of water with polar organics. For a pure paper system, the solvent may be any liquid that can move through the paper but is frequently a water-organic mixture. A guide to solvent performance is available in the form of solute solubilities. Generally, the higher the solubility of a solute in a solvent, the greater the solute mobility, R_f, in that solvent. Finally, a relatively universal solvent for paper systems is the upper phase formed from a 1:4:5 mixture of acetic acid, n-butanol, and water, respectively.

For paper development, a variety of carrier flow directions are available. Figure 19 illustrates two of the types of paper chromatography as determined by the direction of eluent movement. Figure 19a depicts the system for ascending chromatography in which about 5 mm of the paper is immersed in the solvent, which moves up the paper by capillary action. Ascending chromatography is limited in resolving power by the inability of the solvent to flow more than about 8-9 in. because of retardation by gravity. Descending chromatography, Fig. 19b, involves a slightly more complex experimental arrangement in which the top of the paper dips into a solvent tank, with the solvent flowing over an antisiphon bar and then down the paper. In this system, flow is aided rather than hindered by gravity. This is the method of choice for solutes with low R_f values that are difficult to separate, since the solvent flow length is unrestricted and may even be allowed to flow past the end of the paper. Paper chromatography can also be performed in a horizontal position, in either a linear or a radial direction. For the linear horizontal system, solvent is applied by a wick or buret to one end of a paper strip, while for the radial system, sample and then solvent are applied at the center of the paper. The only advantages of the horizontal versions are the relatively small space required and, therefore, the ease of transfer of the entire system to an oven, incubator, or refrigerator for temperature control purposes. Two-dimensional chromatography is identical to that for

TLC; development with a solvent proceeds in one direction; then, after the paper is dried, another solvent is applied at a right angle to the direction of flow of the first solvent.

3. Sample Application

Sample application in paper chromatography is identical to that described above for thin-layer chromatography; streaking is less frequently used, however, even though it generally provides more narrow, better defined solute zones. Spotting may be done with a syringe, a pipet, or a small loop of platinum wire. The spot or streak should be about 15 mm from the bottom for ascending, 15 mm below the antisiphon bar for descending, 15 mm from the edge for linear horizontal, and in or near the center for circular or radial chromatography. For the radial system, care must be taken to assure that the spot center is also the center of solvent application. Spot or streak locations may be marked with a lead pencil to indicate start- ing points for those samples that leave no residue at the origin.

4. Paper

Commercial paper used for chromatography consists of bundles of cellulose fibers held together by hydrogen bonds. A small amount of water is chemi- cally bound to the cellulose fibers, while a larger amount of water is loosely held in the fiber interstices. The specific roles of the cellulose and the two types of water in the separation process have not been well defined. The picture is further complicated by the presence of hydroxyl, adelhyde, ketone, and carboxyl groups in the cellulose and by the presence of impurities in the paper, e.g., lignin and metal salts that remain from the paper pro- cessing system [18]. Paper for chromatographic work is available in a great variety of sizes and shapes, but only one type of paper has been used with considerable frequency. By far the most frequently used paper has been Whatman No. 1 or its equivalent, and this is still the preferred paper for most applications.

5. Detection

The methods of detection or spot location for paper chromatography are similar to those for thin-layer chromatography except that the nature of the paper complicated a number of detection methods. Charring is not practical for paper systems as the paper itself is readily charred. When spot test reagents are selected from Table 9 or elsewhere, care must be exercised to avoid interference from the cellulose functional groups, i.e., hydroxyl, carboxyl, aldehyde, and ketone, and from the impurities, e.g., lignin and metal salts.

6. Qualitative and Quantitative Analysis

The procedures for the identification and quantitative measurement of solutes are exactly as described above for thin-layer chromatography, and the respective sections above should be consulted.

E. Ion-Exchange Chromatography

Ion-exchange chromatography, as the name implies, is used for the chromatographic resolution of mixtures of ions. Moreover, while ion-exchange systems can be used to effect a separation of organic ions, e.g., amino acids or carboxylic acids, these separations can usually be performed equally well by one of the other chromatographic techniques. Some mixtures of ionic species, however, can only be satisfactorily resolved by ion-exchange chromatography. The most notable of these separations include the resolution of mixtures of the rare earths and mixtures of the transuranium elements. Primary difficulties of ion exchange include a somewhat unsatisfactory theoretical basis and a failure to elute pure materials; i.e., the solutes of interest are resolved from other solutes in the sample but on elution are contained in large volumes of eluent, which often contain high concentrations of such foreign ions as buffers and chelating agents.

1. Instrumentation

The more common variety of ion-exchange chromatography, the column configuration, may be performed at atmospheric pressure or in a high-speed, high-pressure variation. Equipment for the column operation is identical to that described above for liquid chromatography. Similarly, thin-layer ion-exchange equipment is as described for TLC, while impregnated paper systems use the paper chromatography chambers described above.

2. Mobile Phase

The proper selection of a mobile phase for ion-exchange chromatography may be more difficult than the eluent choice for any other chromatographic system. The eluent is nearly always water; however, a buffer must usually be added for pH control, while complexing agents are often used to enhance separation or to selectively elute or retain components. Consequently, a thorough knowledge of ionic equilibria in aqueous solutions is essential to the selection of elution solvents. In view of the complexities involved in mobile-phase selection, much laboratory time and effort can be saved by recourse to ion-exchange literature [50, 51]. However, experimental conditions can be arrived at by employing a series of batch equilibrations

between the exchange media, the solute, and solutions of various pH values
and chelate concentrations. Once this procedure has been repeated for
each solute, the conditions of analysis can be established.

3. Sample Introduction

Just as instrumentation for ion-exchange systems is analogous to that of the
chromatographic systems above with the same stationary phase configurations, sample application for columnar ion exchange is identical to that for
liquid chromatography. Moreover, ion exchange on thin films and paper
involves the same sample techniques as described for TLC and paper
chromatography, respectively. For sample introduction procedures, the
proper selections above should be consulted.

4. Column Preparation

Ion-exchange resins on thin layers are prepared as described under TLC
above, while impregnated paper systems are usually purchased. Ion-
exchange columns must be packed wet, in slurry form, because of the
tendency of the beads to swell when wet. The resins should be sieved to
insure a uniform particle size which, for analytical separations, is usually
from 1 to 75 μm, i.e., 200–400 mesh and higher. The slurry is prepared
from the eluent, or first eluent if more than one is to be used, and is poured
into the column, which already contains a small amount of eluent the presence
of which helps prevent air entrapment with subsequent channeling. Immediately after the slurry is added, the excess eluent is forced by pressure or
drawn by vacuum through the column to aid in rapid settling and avoidance of
classification by particle size, which also causes channeling. After packing,
the column should be kept wet at all times, as rewetting after drying may
cause sufficient swelling to burst the columns.

5. Exchanger Selection

Selection of ion-exchange material is relatively easy because of the limited
number of exchangers available. Most ion-exchange materials have been
organic resins of polystyrene crosslinked with divinylbenzene. The resins
incorporate various functional groups, strong or weak acids or bases, to
provide the ion-exchange capacities. Inorganic ion exchangers are also
available but to date represent only a small contribution to experimental
work. For the organic resins, the basic consideration is resin type, e.g.,
strong base, which determines the types of ions that can be separated. A

TABLE 10

Resin Properties Summary[a]

RESIN TYPE	ACTIVE GROUP (ϕ Benzene Ring)	SELECTIVITY (X8 resin)	ORDER OF SELECTIVITY For Monovalent ions . . . For Divalent ions.	STABILITY		
				Thermal.	[1] Solvent (alcohols, hydrocarbons, etc.) [2] Oxidation............	Reduction.
AG® 1 Strongly Basic Anion Exchanger	ϕ-CH$_2$N$^+$(CH$_3$)$_3$	Cl$^-$/OH$^- \cong$ 25	I>phenolate>HSO$_4$>ClO$_3$> NO$_3$>Br>CN>HSO$_3$>NO$_2$> Cl>HCO$_3$>IO$_3$>H$_2$COO>Ac> OH>F	OH$^-$ form - fair up to 50°C Cl$^-$ and other forms good up to 150°C	[1] Very good [2] Slow solution in hot 15% HNO$_3$ or conc. H$_2$O$_2$	Good[b]
AG 2 Strongly Basic Anion Exchanger	ϕ-CH$_2$N$^+$(CH$_3$)$_2$ C$_2$H$_4$OH	Cl$^-$/OH$^- \cong$ 1.5	phenolate>I>HSO$_4$>ClO$_3$>NO$_3$> Br>CN>HSO$_3$>NO$_2$>Cl> OH>IO$_3$>H$_2$COO>Ac>F	OH$^-$form - Fair up to 30°C Cl$^-$ form good up to 150°C	[1] Very good [2] Slow solution in hot 15% HNO$_3$ or conc. H$_2$O$_2$	Good[b]
Bio-Rex® 9 Strongly Basic Anion Exchanger	NH$^+$	· · ·	· · ·	Good up to 38°C	[1] · · · [2] Very good	· · ·
AG 21K Strongly Basic Anion Exchanger	ϕ-CH$_2$N$^+$(CH$_3$)$_3$	Cl$^-$/OH$^- \cong$ 15	I>NO$_3$>Br>Cl>Ac>OH>F	OH$^-$ form fair to 50° C Cl$^-$ & others good to 150°C	[1] Very good [2] Slow solution in hot 15% HNO$_3$ or conc. H$_2$O$_2$	Good[c]
Bio-Rex 5 Intermediate Base Anion Exchanger	R-N$^+$(CH$_3$)$_2$ (C$_2$H$_5$OH) & R-N$^+$(CH$_3$)$_2$	· · ·	· · ·	Good up to 60°C	[1] Good [2] Good	Excellent
AG 3 Weakly Basic Anion Exchanger	Polyamine	· · ·	ϕSO$_3$H>HCit>CrO$_3$>H$_2$SO$_4$> tartaric>oxalic>H$_3$PO$_4$> H$_3$AsO$_4$>HNO$_3$>HI>HBr> HCl>HF>HCOO>HAc>H$_2$CO$_3$	Extensive information not available. Tentatively limited to 65°C	[1] Very Good [2] Good	Unknown
AG 11A8 Strongly Basic Anion plus Weakly Acidic Cation (Zwitterion)	ϕ-CH$_2$N$^+$(CH$_3$)$_3$ & R-CH-COO$^-$	· · ·	· · ·	Tentatively good to 100°C	[1] Good [2] Slow solution in hot 15% HNO$_3$ or conc. H$_2$O$_2$	Good
Chelex® 100 Weakly Acid Cation Chelating Resin	ϕ-CH$_2$N⟨CH$_2$-COO CH$_2$-COO	Cu^{++}/Na $\gg$ 100	H>Li>Na>K Cu>Pb>Fe^{+3}>Al^{+3}>Cr^{+3}>Ni> Zn>Ag^{+1}>Co>Cd>Fe>Mn> Ba>Ca>Na^{+1}	Good up to 75°	[1] Good [2] Breakdown in strong oxidizing agents	Unknown
Bio-Rex 70 Weakly Acidic Carboxylic Cation Exchanger	R-COO$^-$	· · ·	H $\gg$ Ag>K>Na>Li H $\gg$ Fe>Ba>Sr>Ca>Mg	Good up to 100°C	[1] Good [2] Good	Good
Bio-Rex 63 Intermediate Acid Cation Exchanger	ϕ-PO$_3^-$	· · ·	H $\gg$ Li>Na>K	Good up to 100°C	[1] Very good [2] Very good	Very good
Bio-Rex 40 Strongly Acidic Cation Exchanger	R-CH$_2$SO$_3^-$	· · ·	Cs>Rb>K>Na>H>Li	Good up to 40°C	[1] Good [2] Breakdown in strong oxidizing agents	Good
AG 50 or 50W Strongly Acidic Cation Exchanger	ϕ-SO$_3^-$	Na$^+$/H$^+ \cong$ 1.2	Ag>Rb>Cs>K>NH$_4$>Na> H>Li Zn>Cu>Ni>Co	Good up to 150°C	[1] Very good [2] Slow solution in hot 15% HNO$_3$	Very good

[a] Reprinted from Ref. 54, p. 12, courtesy of Bio Rad Laboratories.

[b] AG 1 and AG 2 will break down in the presence of sulfur-containing reducing agents.

[c] Apply in acid solution only.

TABLE 11

Resin Trade-name Index[a,b]

Type and Exchange Group	Bio-Rad Analytical Grade Ion Exchange Resins	Dow Chem. Company "Dowex"	Diamond-Shamrock "Duolite"	Rohm & Haas Co. "Amberlite"	Permutit Company (England)	Permutit Company (U.S.A.)	Nalco Chemical Co. "Nalcite"
CATION EXCHANGE RESINS							
Strong Acidic, Phenolic type $R\text{-}CH_2SO_3^-\ H^+$	Bio Rex 40		C-3		Zeocarb 215		
Strong Acidic, Polystyrene type $\phi\text{-}SO_3^-\ H^+$	AG 50W-X1	50-X1					
	AG 50W-X2	50-X2					
	AG 50W-X4	50-X4		IR-112	Zeocarb 225 (X4)		
	AG 50W-X5	50-X5	C-25D				
	AG 50W-X8	50-X8	C-20	IR-120	Zeocarb 225	Permutit Q	HCR
	AG 50W-X10	50-X10	C-20X10	IR-122		Q-100	HGR
	AG 50W-X12	50-X12	C-20X12	IR-124		Q-110	HDR
	AG 50W-X16	60-16				Q-130	
Intermediate Acid, polystyrene type $\phi\text{-}PO_3^-\ (Na^+)_2$	Bio Rex 63		ES-63				X-219
Weakly Acidic, acrylic type $R\text{-}COO^-Na^+$	Bio Rex 70		CC-3	IRC-50 IRC-84	Zeocarb 226	Q-210	
Weakly Acidic chelating resin, polystyrene type $\phi\text{-}CH_2N\diagup{}^{CH_2COO^-H^+}_{\diagdown CH_2COO^-H^+}$	Chelex 100	A-1					
ANION EXCHANGE RESINS							
Strongly basic, polystyrene type $\phi\text{-}CH_2N^+(CH_3)_3Cl$	AG 1-X1	1-X1			DeAcidite	S-100	
	AG 1-X2	1-X2			FF (lightly		
	AG 1-X4	1-X4	A-101D	IRA-401	crosslinked)		
	AG 1-X8	1-X8		IRA-400	DeAcidite FF		
	AG 1-X10	1-X10					SBR
	AG 21K	21K					SBR-P
$\phi\text{-}CH_2N^+(CH_3)_2$ $(C_2H_4OH)\ Cl^-$	AG 2-X4	2-4	A-102D			S-200	
	AG 2-X8	2-X8		IRA-410			
$\diagup{}^{}_{}NH^+$	AG 2-X10						SAR
	Bio-Rex 9					S-180	
Intermediate base, epoxypolyamine $R\text{-}N^+(CH_3)_3Cl^-$ and $R\text{-}N^+(CH_3)_2(C_2H_4OH)\ Cl^-$	Bio Rex 5		A-30 A-30B		F	S-310 S-380	
Weakly basic, polystyrene or phenolic polyamine $R\text{-}N^+H(R)_2Cl^-$ $R\text{-}N^+H(R)_2Cl^-$	AG 3-X4A	3-X4	A-2 A-6 A-7 A-4t	IR-45 IR-4B IRA-68	G	S-300 S-350	WBR
MIXED BED RESINS $\phi\text{-}SO_3^-H^+$ & $\phi\text{-}CH_2N^+(CH_3)_3OH^-$	AG 501-X8		GPM-331 G	MB-1	Bio Demineralit	M-100	
$\phi\text{-}SO_3H^+$ & $\phi\text{-}CH_2N^+(CH_3)_3OH^-$ indicator dye	AG 501-X8 (D)				Indicator Bio-Demineralit		
$\phi\text{-}SO_3^-H^+$ & $\phi\text{-}CH_2N^+(CH_3)_2$ $(C_2H_4OH)OH^-$	Supplied in Reactor Grade only		GPM-331A	MB-3		M-103	

[a]Reprinted from Ref. 54, p. 17, courtesy of Bio Rad Laboratories.

[b]The table lists manufacturers and distributors throughout the world and the trade names and numerical designations of their commercial grade ion-exchange resins. Resins with the same active groups made by different manufacturers may in many cases be used interchangeably, although they are not necessarily identical in composition or physical properties. Some resins have been discontinued by a particular manufacturer and are listed here for reference purposes.

second important consideration is the degree, or amount, of crosslinkage, which has a variety of effects on exchanger performance [52]. Effects of increased crosslinkage include decreased tendency of resin to swell with change in ionic form; decreased permeability, favoring retention of ions of smaller hydrated radii; measured selectivity because of increased density of ionic sites; and showing of equilibration because of more difficult ion diffusion in the resin. The properties of the more common ion-exchange resins are presented in Table 10, where selectivity [53] is defined as the affinity of a resin for one ion relative to another. Table 11 presents a resin trade-name index for ease in the determination of interchangeable resins.

6. Detection

Continuous monitoring systems that have any degree of universal applicability are not available. A great variety of ionic solution properties can be monitored except that the ionic complexity of the eluent usually renders this approach fruitless. The effluent can be monitored by UV or visible light, but this is of little value for detecting a variety of solutes since these usually absorb at different wavelengths. The same difficulty is encountered in atomic absorption monitoring. The difficulty for light adsorption monitoring cannot easily be overcome by the use of nonchromatic radiation because of sensitivity and precision problems inherent in the use of nonchromatic radiation. Consequently, one is frequently reduced to collecting successive small fractions for subsequent analysis by any suitable technique. Fortunately, automatic sample collection devices are available to relieve the operator of considerable work.

7. Qualitative and Quantitative Analysis

Generally, ion-exchange chromatography is the least useful chromatographic technique for the purposes of qualitative or quantitative analysis. The need to elute the sample from the system combined with the lack of continuous detection essentially eliminates the use of retention data for identification. In any event, moreover, identification is not the usual function of ion exchange, as most ions, at least most inorganic ions, are better and more easily identified by other techniques. Similarly, quantitative analysis for most materials is better performed otherwise. The primary functions of ion-exchange chromatography are, or should be, separation and/or concentration. If two or more ionic species in a sample prevent the measurement of one or more of these due to interference, ion exchange will allow separation of the interfering species in order that they may be measured individually. Likewise, if a sample is too dilute for conventional analysis, the species of interest may often be sufficiently concentrated by an ion-exchange procedure to permit analysis.

Finally, it should be noted that the above discussion generally applies only to column forms of ion-exchange chromatography. When the ion-exchange medium is a thin layer or a paper, the sample is not eluted from the system and the detection, qualitative, and quantitative procedures described above under thin-layer and paper chromatography apply directly to the ion-exchange systems.

REFERENCES

1. M. Tswett, Ber. Deut. Botan. Ges., 24, 384 (1906).

2. For a brief history with references see L. Zechmeister, in Chromatography (E. Heftmann, ed.), Reinhold, New York, 1961, Chapter 1.

3. L. C. Craig, J. Biol. Chem., 155, 519 (1944).

4. A. J. P. Martin and R. L. M. Synge, Biochem. J., 35, 1358 (1941).

5. A. T. James and A. J. P. Martin, Biochem. J., 50, 679 (1952).

6. E. Glueckauf, Trans. Faraday Soc., 51, 34 (1955).

7. J. J. van Deemter, F. J. Zuiderweg, and A. Klinkenberg, Chem. Eng. Sci., 5, 271 (1956).

8. H. Purnell, Gas Chromatography, Wiley, New York, 1962, provides a detailed discussion and critique of plate theory, Chapter 7, and a review of the various rate theories, Chapter 8.

9. P. R. Rony, Separation Sci., 3, 425 (1968); 4, 413 and 447 (1969).

10. J. C. Giddings, Anal. Chem., 34, 1186 (1962).

11. S. T. Sie and G. W. A. Rijnders, Anal. Chim. Acta, 38, 3 (1967).

12. W. Feller, An Introduction to Probability Theory and its Application, 3rd ed., Wiley, New York, 1968.

13. J. H. Purnell, Endeavor, 23, 142 (1964).

14. G. J. Pierotti, C. H. Deal, E. L. Derr, and P. E. Porter, J. Amer. Chem. Soc., 78, 2989 (1956).

15. D. E. Martire and J. H. Purnell, Trans. Faraday Soc., 62, 610 (1966).

16. R. Consden, A. H. Gordon, and A. J. P. Martin, Biochem. J., 38, 224 (1944).

17. D. P. Burma, Anal. Chem., 25, 549 (1953).

18. H. G. Cassidy, Anal. Chem., 24, 1415 (1952).

19. D. C. Locke and D. E. Martire, Anal. Chem., 39, 921 (1967).

20. E. Glueckauf, Proc. Roy. Soc., A214, 207 (1952).

21. G. E. Meyers and G. E. Boyd, J. Phys. Chem., 60, 521 (1956).

22. F. G. Donnan, Z. Electrochem., 17, 572 (1911).

23. F. G. Donnan, Z. Phys. Chem., A168, 369 (1934).

24. The Annual Analytical Chemistry Laboratory Guide provides an excellent list of sources for any type of chromatographic equipment.

25. O. L. Hollis, Anal. Chem., 38, 309 (1966).

26. T. R. Lynn and C. L. Hoffman, Guide to Stationary Phases for Gas Chromatography, 8th ed., Analabs, North Haven, Conn., 1971.

27. Gas Chromatographic Data — Punched Card Index, AMD-20-1a, 1b, and 1c, American Society for Testing and Materials, Philadelphia, Pa., 1968; Gas Chromatographic Data Compilation, AMD-25A, American Society for Testing and Materials, 1967.

28. J. S. Lewis, G. T. McCloud, and W. Schirmer, Jr., J. Chromatog., 5, 541 (1961).

29. R. N. Ewell, J. M. Harrison, and L. Berg, Ind. Eng. Chem., 36, 871 (1944).

30. H. M. Mc Nair and E. J. Bonelli, Basic Gas Chromatography, 5th ed., Varian Aerograph, Walnut Creek, Calif., 1969, p. 39.

31. J. C. Giddings, J. Chem. Ed., 39, 569 (1962).

32. M. Dimbat, P. E. Porter, and F. H. Stross, Anal. Chem., 28, 290 (1956).

33. A. J. P. Martin and A. T. James, Biochem. J., 63, 138 (1956).

34. J. C. Sternberg, W. S. Gallaway, and D. T. C. Jones, Gas Chromatography, Third International Symposium, Academic, New York, 1962, pp. 9 and 231.

35. C. H. Deal, J. W. Otvos, V. N. Smith, and P. S. Zucco, Anal. Chem., 28, 1958 (1956).

36. J. E. Lovelock, J. Chromatog., 1, 35 (1958).

37. A. B. Littlewood, C. S. G. Phillips, and D. T. Price, J. Chem. Soc., 1480 (1955).

38. L. S. Ettre, Anal. Chem., 36, 31A (1964).

39. E. Kovats, Helv. Chim. Acta, 41, 1915 (1958).

40. A. Wehrli and E. Kovats, Helv. Chim. Acta, 42, 2709 (1959).

41. W. A. Dietz, J. Gas Chromatog., 5, 68 (1967).

42. D. M. Rosie and R. L. Grob, Anal. Chem., 29, 1263 (1957).

43. E. Lederer and M. Lederer, Chromatography, 2nd ed., Van Nostrand, Princeton, N.J., 1957.

44. F. W. Karasek, Res. Devel., 21, 43 (1970).

45. R. E. Leitch and J. J. Kirkland, Ind. Res., 12, 36 (1970).

46. F. Feigl, Chemistry of Specific, Selective and Sensitive Reactions, Academic, New York, 1949, p. 596.

47. E. Stahl, Thin Layer Chromatography, 2nd ed., (E. Stahl, ed., M. R. F. Ashworth, Transl.) Springer Verlag, New York, 1969, p. 86.

48. Camag Bibliography, available from Gelman Instrument Company, Ann Arbor, Michigan.

49. E. I. Stahl, Thin Layer Chromatography — A Laboratory Handbook, Academic, New York, 1965.

50. Reviews of all forms of chromatography are published biannually in even-numbered years in Anal. Chem.

51. F. Helfferich, Ion Exchange, McGraw-Hill, New York, 1962, Chapter 2.

52. C. F. Coleman, C. A. Blake, Jr., and K. B. Brown, Talanta, 9, 297 (1962).

53. Discussions of factors that affect selectivity have been presented by D. Reichenberg and by R. M. Diamond and D. C. Whitney in Ion Exchange, Vol. 1 (J. A. Marinsky, ed.), Marcel Dekker, New York, 1966, Chapters 7 and 8.

54. Price List V, Bio Rad Laboratories, Richmond, Calif., July, 1970.

PART II

AIR POLLUTION SECTION

Chapter 2

GAS CHROMATOGRAPHIC ANALYSIS IN AIR POLLUTION

Robert S. Braman

Department of Chemistry
University of South Florida
Tampa, Florida

I. INTRODUCTION

Gas chromatography is applicable to the determination of any or all materials that may be reproducibly volatilized down columns, whether or not sample matrix materials can do the same. The technique is applicable to the analysis of nearly everything in the environment given suitable sample modification procedures. This chapter gives emphasis to the analysis of air samples taken for the purpose of air pollution detection. Particulate analysis (after suitable collection procedures), permanent gas analysis, and the analysis for materials of unknown physical state — which may be vapor or particulate — are covered. Lesser coverage is given to analyses for the purpose of industrial toxicology. Locally high concentrations of industrial chemicals, such as may be found in chemical manufacturing plants, can usually be easily detected and analyzed by gas chromatography techniques. Nevertheless, the same techniques can be incapable of use remote from the source of emissions because of insufficient limits of detection.

The field of gas chromatography is important to air pollution detection research because of the capability of the gas chromatographic column to resolve complex mixtures. Detector research and development are also important to air pollution studies. Much detector research is aimed at providing a more specific response or one with reduced lower limits of detection to get at some particular analyte. This common area of applicability ties the two fields together.

Detectors are multicomponent devices for sensing materials based on the observation of some physical or chemical property that the device is capable of measuring. In order for the proper measurement to be effected, samples must frequently be modified so that the signals obtained are specific in nature or are of sufficient intensity to provide meaningful information. Many ways exist to utilize detectors at the end of a gas chromatographic column, or to utilize some type of chemical treatment system. Far more detectors are used in air pollution analysis without columns than are used with columns.

The scope of this chapter is limited to detection systems or detectors that employ gas chromatographic columns with the detectors for the specific purpose of providing separations prior to analysis. An attempt is made to indicate the current extent of development of gas chromatographic detection systems referenced, whether the method reported is simply of research interest or whether it has been put into more general use.

Finally, the reader should note that gas chromatography has not been as widely applied in the field of air pollution analysis as may have been predicted. Reasons for this include high cost, relative complexity, and in some cases technical problems with the sampling and chromatographic column. Analysis of active chemical constituents below 1 ppm has been difficult. Fortunately, some of the technical application problems are easing. The complexity and high cost of gas chromatography systems, however, are likely to slow their wide adaptation to air pollution analysis.

II. ACCURACY, PRECISION, DATA EVALUATION, AND CALIBRATION

A. Accuracy

The accuracy of a gas chromatographic air analysis is a function of factors influencing the entire detection or analysis system; the sample acquisition system, the sample modification system (gas chromatographic column), and the specificity of the detector are the main factors. A restraint is also placed on the analysis system by the analyte weight required by the detector to achieve the desired detection or quantitative analysis. The composition of the sample matrix is important because it defines the interference problem. A gas chromatography system used in air analysis must be tailored to the particular sample type to be analyzed.

Sample acquisition must be representative and reproducible. The physical equipment should not chemically alter the analyte or reduce its ambient concentration by adsorption. Some sampling devices perform a crude separation of the analyte from matrix materials by preferential selection in favor of the analyte. In preconcentration devices, the analyte is sampled in preference to the major air components. The gas chromatographic column is probably the most effective element of a system for providing selectivity of detection by separation of the analyte from matrix materials. The capability of gas chromatography columns is variable because of the great number of different column packings, column lengths, and partitioning materials available. Indeed, because of the necessity of defining column parameters, the column is the most important factor usually studied in applications to air pollution detection. The column, it must be remembered, is solely a separation element in an analysis system. It does not itself concentrate separated analytes. In fact, the concentration of analyte leaving a partitioning column is always less than the concentration that enters the column. The column provides an element of selectivity and also the capability of multiple analyses on the same sample.

The detector is almost as important as the column in providing selectivity of detection. In fact, because of the excellent selectivity available in detectors, the requirement of a gas chromatographic separation in air pollution analysis is decreasing. If analytes have some particular chemical or physical property providing high selectivity of detection, a column may not be needed at all.

Despite the best combination of sampling, column, and detector, accuracy is still at the mercy of the possibility that major sample composition changes can occur. Interferences not previously present can become a factor. The range of analyte compositions can be considerable. The dynamic range of a detection or analysis system may have to be 10^4 or more to be capable of covering all probable or even usual concentrations experienced. The accuracy of sample analysis outside the available dynamic range of a gas chromatography system is obviously nil.

B. Precision

The precision of analysis by gas chromatography is a function of the sensitivity of the detector, the actual sample sizes being analyzed, and the readout system employed. Precision of analysis can usually be determined in calibration of a particular method. Regression analysis of linear area response vs amount of analyte, for example, can be conducted. The standard deviation in slope, the individual data results, and the intercepts are quite useful in describing precision or accuracy.

C. Calibration Methods

In gas chromatographic systems the known gas concentrations are prepared by injecting known amounts of materials into the carrier gas stream by means of syringe or gas-sampling valves. Responses are noted or recorded depending on the nature of the readout system. Although the procedure appears simple and straightforward, complications arise when very low sample concentrations are required or when materials being tested exhibit chemisorption reactions. If low concentrations of reactive materials must be delivered to the gas chromatography system, care must be exercised that concentrations do not change from known prepared values.

Calibration of detectors and detection systems having moderate to low sensitivities is generally a far easier experimental exercise than calibration of high-sensitivity, low limit of detection systems. This is simply because high known concentration calibration gases may be used and these retain their calibration values better than the low concentration calibration gases. Known gas mixture compositions are available commercially in standard pressurized gas cylinders from a number of organizations — J. T. Baker, the Matheson Company, and Lif-O-Gen, Inc., are three examples. Compounds or atomic gases available from these companies in various carrier gases are given in Table 1. Concentrations range from low parts per million to high percentage values. Nonstandard gas compositions can likely be obtained on request.

D. Dynamic Calibration Gas Generator Technique

A number of techniques have been tried and used for laboratory generation of known compositions of liquid vapors for calibration. A typical laboratory gas mixing system for doing this, employed by O'Neil et al. [1], is shown in Fig. 1. Concentrations delivered are calculated simply based on vapor pressures of liquids and gas flow rates. This device was found by this writer to be slow to equilibrate, especially at low concentrations. Similar results may be expected with similar systems. Also, system

TABLE I

Some Commercially Available Calibration Gas Materials[a]

Material	Concentration range	Mfg[b]
1 Argon	50 ppm to 50%	(1)
2 Butane	50 ppm to 1%	(1)
3 Carbon dioxide	50 ppm to 30%	(1-3)
4 Carbon monoxide	50 ppm to 99%	(1-3)
5 Ethane	50 ppm to 15%	(1)
6 Ethylene	50 ppm to 50%	(1)
7 Helium	50 ppm to 50%	(1-3)
8 Hexane	50 ppm to 0.7%	(1-3)
9 Hydrogen	50 ppm to 50%	(1-3)
10 Methane	50 ppm to 99%	(1-3)
11 Neon	50 ppm to 99%	(1)
12 Nitrogen	50 ppm to 50%	(1-3)
13 Oxygen	50 ppm to 50%	(1-3)
14 Propane	50 ppm to 1%	(1)
15 Propylene	50 ppm to 1%	(1)

[a]Available in various carrier gases, i.e., air, Ar, H_2, He, N_2, and O_2.

[b](1), Lif-O-Gen; (2), J. T. Baker; (3), Matheson.

cleanout to change from one sample material to another is slow. The method appears most suitable for preparation of higher parts per million concentrations of a single material.

McKelvey and Hoelscher [2] were perhaps the first to report the use of diffusion cells for preparation of low concentrations of various volatile materials in air. Their diffusion apparatus is shown in Fig. 2. It is assumed that a saturated vapor concentration is attained in the lower flask. The diffusion rate then may be calculated from the equation

$$\gamma = 2.303 \frac{DPMA}{RTL} \log \frac{P}{P - p} \tag{1}$$

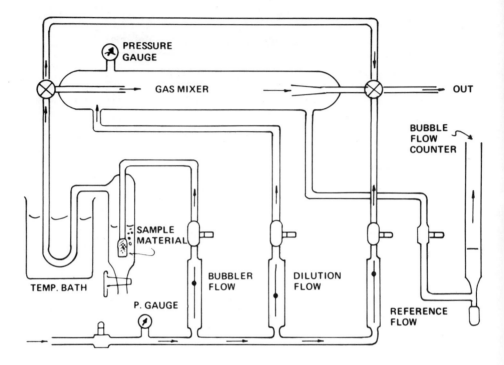

FIG. 1. Dynamics gas sample generating system.

where:

γ = diffusion rate, g/sec
M = molecular weight of diffusing liquid
P = total pressure in the chambers, atm
A = cross-sectional area of the connecting tube, cm²
L = length of the tube, cm
T = absolute temperature, K
R = gas constant, liter-atm/mole K
p = vapor pressure of diffusion liquid at T, p atm
D = diffusion coefficient

Because of the temperature dependency of p and D the cell should be in a constant temperature bath. McKelvey and Hoelscher [2] reported a 5% difference between calculated and experimental data in diffusion rate over a temperature range of 35-60°C with toluene. The experimental diffusion rate was determined by weight loss.

The diffusion rate must usually be determined by calibration using a referee method. Experimental calibration of the diffusion rate should

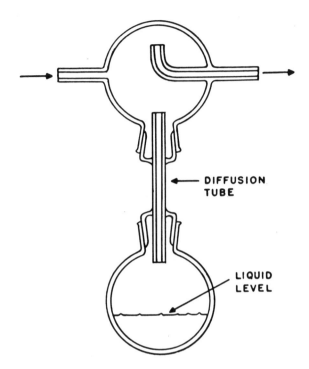

FIG. 2. Diffusion cell (reprinted from Anal. Chem. , 29, 123 (1957) by permission of the publisher).

probably be used in all cases to obtain accurate diffusion rate data. Since calibration must be rendered each time the cell is used to get the best data, the double diffusion cell may not be as convenient as some of the other techniques.

Altshuller and Cohen [3] have reported an extensive study of diffusion cells for the production of known gaseous concentrations. In initial work with separate reservoir-type cells, such as used by McKelvey and Hoelscher [2], they found that the observed diffusion rate was very substantially different from the calculated rate using the standard diffusion rate equation.

Much better results were obtained with the single-chamber diffusion cell shown in Fig. 3. Experimental data and data calculated from literature values of diffusion coefficients were in satisfactory agreement for methanol, water, toluene, n-hexane, n-heptane, n-decane, 2-methyl-1,3-butadiene, 1-hexene, 1-octene, and benzene. The single-chamber diffusion cell would appear to be more reliable than the double-chamber cell, but again calibration checks may be frequently required. The diffusion cell was recommended by Altshuller and Cohen [3] for production of concentrations in the 10-10,000 ppm range.

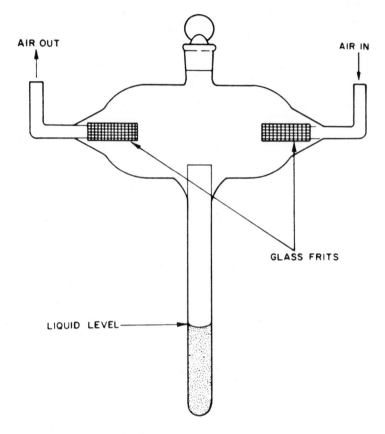

FIG. 3. Diffusion cell (reprinted from Anal Chem., 32, 802 (1960) by permission of the publisher).

Saltzman and Clemons [4] have devised a gaseous sample dilution device for use in calibrating detector response at low concentrations. The design is shown in Fig. 4. The sample to be diluted is contained in a plastic bag attached to port 1. The dilution gas (N_2) is passed through ports 3 or 4. A vacuum at port 2 keeps the sample flowing past the bottom of the aspirator. The diluted sample is available to a gas-sampling valve system at port 6 and excess diluted sample exits at port 7. Dilution ratios up to 1:3000, dilution accuracy of better than ±1%, and mixing accuracy within a few seconds of operation were obtained with this device.

The dilution system was used to reduce the sample concentrations of various air pollutants or tracer materials in a study of gaseous meterological tracers [5]. Although not specified, the level to which the sample concentrations were reduced was apparently in the fractional part per million range.

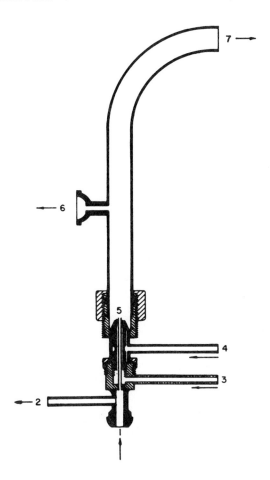

FIG. 4. Schematic drawing of quantitative dilution device for gaseous samples (reprinted from <u>Anal. Chem.</u>, <u>38</u>, 800 (1966) by permission of the publisher).

   O'Keefe and Ortman [6] have developed an elegant and very useful technique for calibration by permeation through fluorinated ethylene propylene resin (FEP Teflon, Du Pont) in tubing form. The technique was developed and used for calibration with sulfur dioxide, nitrogen dioxide, propane, butane, pentane, carbon dioxide, selected fluorcarbons, propylene, butylene, benzene, and some other light hydrocarbons. Permeation tubes were from approximately 1/16 to 3/16 in. i.d. and had wall thicknesses of 0.015–0.030 in. Permeation rates were linear with tube length and were a logarithmic function of temperature. Figure 5 gives permeation rate data for four gases

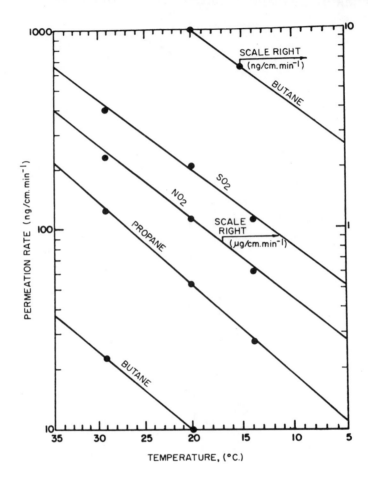

FIG 5. Permeation rate vs temperature for four gases (reprinted from
Anal. Chem. , 38, 760 (1966) by permission of the publisher).

as a function of temperature. Temperature must be carefully controlled.
An apparatus such as is shown in Fig. 6 is suggested by PolyScience
Corporation, a firm that now manufactures and sells the permeation tubes
made to order and calibrated. Calibration of the tubes may be carried out
by weighing over long periods of operation. Microcombustion and coulometric titration calibrations have also apparently been used in calibration.

Dilutions of the permeating gases permit generation of a wide range of
concentrations of trace gases, from the fractional part per million range to
the 100 ppm range as desired. The technique can be used in conjunction
with a gas-sampling valve when gas chromatography systems are to be

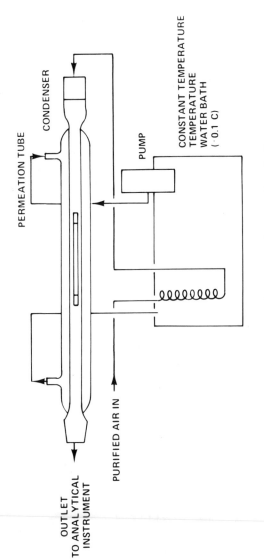

FIG. 6. Apparatus with constant temperature bath.

calibrated. It appears to be of greatest use with gases or the more volatile liquids. Permeation rates of higher boiling organic compounds may be too low to be used even at elevated temperatures. Nevertheless, a test experiment should be carried out and the permeation tube approach not summarily dismissed. Since permeation rates remain comparatively constant over the life of a permeation tube, this technique appears to be the best yet devised for dynamic calibration gas preparation.

The "exponential dilution flask" method [7] is one of the more interesting approaches to detector calibration, in that a dynamic range of concentrations is generated. This technique is based on the analysis of gas samples from a known gas composition which is being mixed and diluted at a known rate. The apparatus employed, i.e., a typical flask design, is shown in Fig. 7. An initial known gas composition is prepared in the flask before dilution by either flushing the chamber with a known gas composition or by syringe injection of a known weight of liquid or known gas volume. The magnetic stirrer and paddle wheel arrangement rapidly mixes the gases. The gas volume of the mixing chamber must be known and may be obtained by some suitable, simple technique, such as measuring the volume of water the chamber can hold. When ready for operation, the dilution gas inlet and chamber exit valves are opened. A known, constant flow of dilution gas passes into the chamber and is rapidly mixed. The concentration of gas

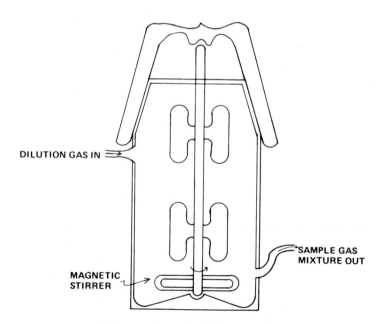

FIG. 7. Exponential dilution flask.

exiting from the chamber decreases as an exponential function of time; hence the name of the technique. The gas composition leaving the chamber obeys the equation:

$$C = C_0 \cdot e^{-tQ/V} \tag{2}$$

where C_0 is the initial gas composition, C is the composition at time t, V is the flask volume, and Q is the dilution gas flow rate (also the exiting sample gas flow rate).

The salient feature of this technique is its ability in one experiment to cover a wide range of gas composition values. For example, assuming a chamber volume of 300 ml and a dilution gas flow of 500 ml/min, four and one-half orders of magnitude of gas compositions are covered in 60 min. Thus, the technique is suitable for studying the dynamic range of detector response in a single experiment. Since precision and accuracy of gas dilutions are generally good, response curves may be obtained in a single operation.

The technique is theoretically capable of producing any low concentration level desired for calibration purposes and this may be used to obtain response data near the limits of detection. Nevertheless, this may depend on the volatility and adsorption characteristics of the compound being studied. Initial work with the exponential dilution flask technique was carried out with gases or highly volatile materials only. I have also used the technique with less volatile materials, e.g., octane, hexane, and dimethylmethyl phosphonate. A typical response curve is shown in Fig. 8. For these liquids the apparatus was placed in an oven at approximately $120^\circ C$. The usual exponential decrease in response was observed to low values but a leveling off of the log C/C_0 (response) vs time was observed after five or six orders of magnitude. This may be attributed to surface adsorption of the liquids at high concentrations and a slow desorption from the chamber walls. The detection of the desorbed concentration finally becomes significant when the gas concentration is decreased to the level produced by the desorption. Since the adsorption effect indicates a surface concentration of adsorbed material not in the vapor state, the C_0 value calculated for an injected liquid is incorrect. The amount adsorbed on surfaces is not easily determined and corrections to give a true C_0 are therefore not easy to obtain. It appears that despite the adsorption effect, linearity of response can be determined and even approximate calibration values or limits of detection if the linearity of log C vs t is exhibited over five to six orders of magnitude of C values. Further work with the technique is indicated. Construction of dilution flasks having a smaller surface adsorption effect than glass may decrease the slow desorption effect to the extent that lower concentrations may be observed.

More recently, Bruner et al. [8] have coupled permeation tubes with exponential dilution flasks in the preparation of dynamic concentrations of

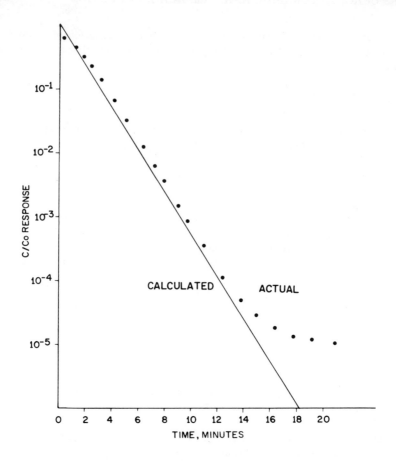

FIG. 8. Exponential dilution flask data.

sulfur dioxide, methylmercaptan, and n-butane. This is an improvement over the exponential dilution flask alone because a more accurate C_0 value may be obtained than by syringing in an amount of gaseous analyte.

E. Rotating Syringe Calibration Gas Preparation Technique

A nondiffusion type of calibration technique is carried out utilizing the apparatus shown in Fig. 9. A large syringe, preferably 100 ml (not greased), is filled with sample gas of known composition, fitted with a small gauge needle (27 gauge), and mounted vertically as shown. Vanes on the syringe inner member are air driven to rotate it. This causes a smooth injection of

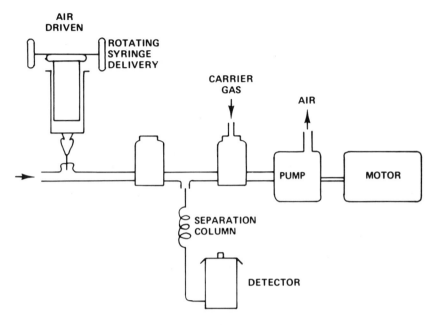

FIG. 9. Apparatus for determining electron capture detection sensitivities.

gas from the syringe into the carrier gas stream. Gas injection flow rates
are in the 10-30 ml/min range and are comparatively constant. Rotation
rate and total syringe volume delivered have little or no effect on gas delivery rate. The gas viscosity does influence flow rate through the needle.
Liquids that vaporize readily at room temperature can be injected through
the syringe tip from a microliter syringe. Dilutions of the syringe vapor
mixture may be made by injecting all but a fraction of the syringe volume
and then refilling with carrier gas or air. I have found that gas dilutions
over a range of 1000 may be made with good precision (±2-5%). A further
dilution of the syringe gas occurs in the injection port area and depends on
carrier gas flow rate. Drawbacks of the technique are the short periods of
sample delivery time and unsuitability of delivery of small concentrations
because of gas mixtures' reactive compounds, such as nitrogen dioxide,
pentaborane, chlorine, sulfur dioxide, and ammonia [9]. Fluorine oxide
and hydrogen fluoride were found to attack the syringe, as might be expected.
    This dynamic calibration gas delivery technique is best used where a
continuous monitoring of a sample stream is desired. If used in a system
employing a chromatographic column for sample modification, a gas-
sampling valve would be needed to inject sample plugs.

F. Least Squares Plotting of Calibration Data

Many good references on plotting linear data are available. The following
is from Youden's book [10]. It must be assumed at first that a linear
relationship exists between peak area and sample size. If not, one must be
found. Peak height vs sample size is frequently not linear in the range of
interest.

A straight line will fit the equation $Y = a + bX$. The experimental variables should be selected so that the real errors in X are smaller than those
in Y. Known values of calibration gases may be considered superior in
value to the response obtained. Therefore, X is in terms of concentration
and Y is area response. The least squares technique pools all variance in Y.

The slope of the response curve, b, is calculated from:

$$b = \frac{\eta \Sigma XY - \Sigma X \Sigma Y}{\eta \Sigma X^2 - (\Sigma X)^2} \tag{3}$$

where η is the number of data points. The intercept a is calculated from the
equation:

$$a = \bar{Y} - b\bar{X} \tag{4}$$

where $\bar{Y}$ and $\bar{X}$ are the average of the X and Y values. The standard deviation
of individual points in terms of Y, S_y, is calculated from the following:

$$(N - 2)Sy^2 = \bar{\Sigma}y^2 - \frac{(\bar{\Sigma}y)^2}{N} - \frac{\left(\bar{\Sigma}xy - \frac{\Sigma x \Sigma y}{N}\right)^2}{\bar{\Sigma}x^2 - \frac{(\bar{\Sigma}x)^2}{N}} \tag{5}$$

The variance for b, S_b^2, is calculated from:

$$S_b^2 = \frac{Sy^2}{\bar{\Sigma}x^2 - n\bar{X}^2} \tag{6}$$

and, finally, the standard deviation in the intercept, S_a, is calculated from:

$$Sa = Sy\sqrt{\frac{\Sigma x^2}{n\Sigma x^2 - (\Sigma x)^2}} \tag{7}$$

The value of the intercept can be tested by comparing:

$$t = a/Sa \tag{8}$$

calculated to the table values at N - 2 degrees of freedom.

It is of course convenient to use a computer program to calculate all
of these statistical data. The problem of least squares plotting is common

and many programs should be readily available.

G. Limits of Detection

Gas chromatography systems produce readout patterns in the form of Gaussian-shaped curves, as shown in Fig. 10. The shape of the signal, its height, and its duration are governed by the column and often by sampling conditions. Responses may be taken in terms of peak height or in terms of area, integrating signal over t_1 to t_2. The response of the system is usually obtained in terms of gas concentration (parts per billion or parts per million) vs peak area. It is obvious from Fig. 10 that as the sample size decreases, distinguishing between the peak and the noise becomes more and more difficult. The lower limit of detection is usually defined as that amount (or concentration) of analyte which produces a discernible signal above the noise level. The determination of a good value for the limit of detection has been treated in several publications.

Dimbat et al. [11] early recognized the need for adequate expression of limits of detection. They developed a "sensitivity" value, S, for detectors. Detector "sensitivity" was defined by the expression:

$$S = \frac{A \times C_1 \times C_2 \times C_3}{W} \tag{9}$$

where S = sensitivity, ml mv/mg
       A = peak area, cm^2

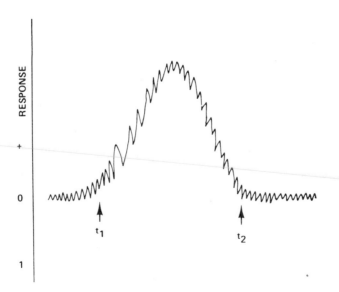

FIG. 10. Typical response, intermittent-type detection.

C_1 = recorder sensitivity, mv/cm
C_2 = chart speed, min/cm
C_3 = carrier gas flow rate at column exit, ml/min corrected to
      (atm, column temp.)
W = sample wt. introduced, mg

This expression is the response of a detection system having a concentration responsive detector. The equation may be converted by substituting mA for mV. Since some detectors are rate of flow sensitive, the expression is not suitable for all. Limits of detection require noise values in addition to S for calculation. The authors' [11] recommended noise values in millivolts is

$$LD = \frac{noise \ (mV)}{\underline{S}} \tag{10}$$

where $\underline{S}$ = the value of S as the maximum concentration of a sample W passes into the detector divided by effective detector volume. Perhaps the most illuminating work on the subject of detection and response was reported by Johnson and Stross [12]. They pointed out that the sensitivity (response) value of a detector is not a good measure of ability to detect small amounts of material because noise must be considered. These workers invented the term "limit of detection" to indicate the least detectable amount of material for a detector. Their concept of limit of detection was developed in application to the evaluation of gas chromatography. A statistical technique for evaluating detectors was developed. The major point of the method is that response value data are obtained in area units in gas chromatography. In addition, the peaks have variable, real widths. Since response is in area units, it is also necessary that noise values be so determined.

Response (sensitivity) must be determined at or near the limit of detection of the detection system. This is done primarily by determining response as a function of concentration (or mass flow rate) and extrapolating to low values. Generally, response is expected to be linear down to the limit of detection.

The determination of noise in area units requires considerable effort. The readout system is given a high gain setting until the noise of the recording tracing is quite high. Figure 11 illustrates a typical noise recording. Sets of three intervals are selected as shown. The intervals must be sufficiently wide in time to include the peak width of an eluted sample near the limit of detection in size. A noise area is determined for the three intervals from the areas of the intervals a, b, and c as in the equation:

$$noise \ area = b - \frac{(a + c)}{2} \tag{11}$$

Noise areas are determined for a large number of independent sets of three

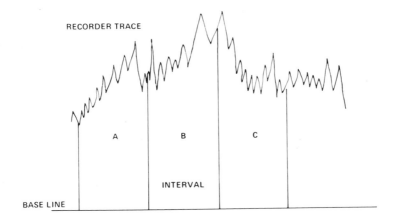

FIG. 11. Intervals and interval areas for LD Calculation.

intervals (Johnson and Stross used 25 sets) and the standard deviation of the noise areas (N_s) is calculated.

The true standard deviation of the noise areas N_s may be expressed in terms of weight by:

$$W_s = \frac{N_s}{R} \tag{12}$$

where R is response at the W_s weight of materials. For a finite number of samples the N_s value calculated must be treated statistically. A confidence interval was calculated for estimating N_S from N_s at a 95% probability level. The limit of detection, LD, was then derived as:

$$LD = \frac{2(1.96N_s \sqrt{n})}{R\sqrt{k}} \tag{13}$$

where n is the number of samples used in determining N_s and k is taken from chi square statistical tables. For n = 25 with 24 degrees of freedom, k = 13.8, and

$$LD = \frac{5.29N_s}{R} \tag{14}$$

The value of two times the estimate for the standard deviation of the noise area population in weight terms was an arbitrary selection.

The LD value perhaps may be physically described as an uncertainty in weight terms and of high probability in determination of the actual baseline under a very small gas chromatography peak. The observed uncertainty in baseline may be smaller for any particular case but the probability that LD is exceeded is only one in 20.

Johnson [13] later described a computer-based technique for the calculation of the limit of detection of a detector. The mechanization of the calculation involved would certainly be desirable and useful where a selection of different time intervals could be examined. The method could be used to frequency sort the noise and thus improve limits of detection through elimination of all noise not in the same time interval as the sample.

Perhaps the most convenient alternative to the calculations of Stross and Johnson is to use the standard deviation of individual data points calculated from the regression analysis of calibration points. The uncertainty in the individual data points in terms of concentration or sample weight should reflect the uncertainty in small signal values. The uncertainty in the intercept is also of use. In both of the above cases, nevertheless, uncertainty contributions from larger samples could be larger than those observed at low sample sizes.

Finally, many authors multiply the signal noise by some factor, such as two or three, to get a lower limit of detection. This should be noted in reported LD values, if done. I favor the use of regression analysis data.

III. SAMPLING

Gas chromatographic analysis is a point-sampling technique. Systems employing gas chromatographic systems are therefore not area monitors. Multiple-points sampling could be used in an extensive monitoring system to simulate area monitoring. In toxic vapor detection, the individual point source detector does have the advantage of providing information at the most critical point of interest, as selected by the system designer. It has the disadvantage of not giving a good value for the entire area of interest.

The direct analysis of gas samples may be carried out by direct injection of samples, using a gas-tight syringe or a gas-sampling valve. Gas-tight syringes are available commercially from a variety of sources and must be operated by hand. Precision of sample volume injection is high, but the method is actually useful only where adsorption of analytes onto syringe surfaces is not a problem and where preconcentration of above 10-20 ml of air is required.

Gas-sampling valves are automatic devices for injecting sample volumes into the carrier gas stream, which then passes the sample into a separation column. The advantages of this type of sampling are that the sampling system may be heated to reduce adsorption and that the process of sampling may be made automatic by operating the gas-sampling valve by means of a solenoid.

Regardless of the type of gas injection used, there is a limitation to the sample size that may be injected with advantage. Gas samples above 10 ml are rarely directed injected into columns. The practical upper limit depends on the column length, its diameter, its interstitial volume, and the degree of sample separation necessary. Capillary columns and short 1/8-in. o.d. columns require less than 1-ml sample volumes to prevent taking up the entire interstitial volume with matrix sample material. Sample spreading due to large sample sizes reduces the apparent number of theoretical plates in a column.

Grab sampling is used when field monitoring is not done. Evacuated sample bombs (or as an alternative, plastic bags) are taken to the sample location and then opened. They are closed and returned to the analysis facility where syringe sampling, or a displacement technique, may be used to inject the samples onto the gas chromatographic column.

If large samples are required to obtain the necessary amount of analyte for analysis, the technique of sample preconcentration must be used. Two techniques are generally used in preconcentration; freeze-out techniques or the substrate equilibrium technique. The former have been by far the most popular.

The freeze-out technique is described in Sec. IV, B, 2 on the determination of specific materials and so is not covered here. The equilibration method of Cropper and Keminsky [14] has been less extensively employed and so is covered here.

In the method of Cropper and Kaminsky, air samples are drawn through a short adsorption tube packed with the same filling as the gas chromatography column used for separation. Silicone gum rubber and polyethylene glycol on acid-washed firebrick were used. Air samples are passed through the adsorption columns for approximately 80% of the retention time for the constituent being detected. The adsorbed material is desorbed by heating and is then passed through the separation column. Individual calibration is needed.

Novak et al. [15] in a variation of this method pass sample air until the entire tube is equilibrated prior to injection. The concentration of analyte is calculated from phase equilibration data.

A. Monitoring Levels

The established National Air Quality Standards set recently by the Environmental Protection Agency [16] define the air pollutant concentration levels that a monitoring system must be capable of detecting. These are given in Table 2. The values indicated in Table 2 are simply maximum values and obviously analysis systems must be capable of detecting far lower concentrations. A good estimate for the capability of a monitoring system would be a limit of detection at 5% of the maximum levels of better.

TABLE 2

National Air Quality Standards

Pollutant	Standard
SO_2	80 μg/m^3 (0.03 ppm)[a]
	365 μg/m^3 (0.14 ppm)[b]
Particulate matter	75 μg/m^3 [c]
	260 μg/m^3 [b]
Carbon monoxide	10 mg/m^3 (9 ppm)[d]
	40 mg/m^3 (35 ppm)[e]
Photochemical oxidants	160 μg/m^3 (0.08 ppm)[e]
Hydrocarbons	160 μg/m^3 (0.24 ppm)[f]
Nitrogen oxides	100 μg/m^3 (0.05 ppm)[a]

[a] Annual arithmetic mean.

[b] Maximum 24-hr concentration not to be exceeded more than once a year.

[c] Annual geometric mean.

[d] Maximum 8-hr concentration not to be exceeded more than once a year.

[e] Maximum 1-hr concentration not to be exceeded more than once a year.

[f] Maximum 3-hr concentration (6-9 a.m.) not to be exceeded more than once a year.

IV. METHODS USED FOR VARIOUS SUBSTANCES

Many dozens of compounds — volatile liquids, solids, and gases — have maximum allowable concentration values (MAC) specified periodically by the American Conference of Governmental Industrial Hygienists. In theory, all of these can be considered to be likely objects for analysis by gas chromatography. Nevertheless, the reported work on the application of gas chromatography to air pollution analysis has dealt largely with only a few of the compounds. This has been due in part to the fact that only a few of the compounds have been considered as major air pollution threats. In addition, some of the more reactive compounds, at the low concentrations encountered, could not be easily detected by gas chromatography analysis because of adsorption reactions on columns and sampling systems. Fortunately, the problem of adsorption of reactive gases onto column materials is gradually being solved.

Contained in this section are methods for the determination of:

A. The major air components
B. Organic materials
    1. Automobile exhaust gases
    2. Low boiling hydrocarbons
    3. Polynuclear hydrocarbons
    4. Halogenated hydrocarbons
    5. Aldehydes, ketones, and alcohols
    6. Hydroperoxides and alkylpolynitrates
C. The nitrogen oxides
D. Carbonyl sulfide and phosgene
E. Sulfur compounds
F. Alkyl lead compounds
G. Tracer gas, SF_6

A. Determination of the Major Air Components

Nitrogen, oxygen, carbon dioxide, water vapor, and argon are considered the major components of air. The separation of these gases is easily accomplished by gas chromatography, but carbon dioxide is likely to be the only constituent of interest in the air pollution field.

Many authors have reported the separation and analysis of air samples for the major constituents [17-20]. The separation of all constituents on a single column, however, is difficult because of the wide range in their polarity. Oxygen and nitrogen are separable on 5A or 13X molecular sieves at ambient temperatures. Conditioned molecular sieves retain both water and carbon dioxide. Argon often appears with oxygen. Silica gel and activated carbon are used to separate out and detect carbon dioxide.

Frequently a split column and dual detector are used to determine all the major constituents.

A 5A molecular sieve column, 3 m by 2 mm, serves to separate the major constituents at $-40^{\circ}C$. When the column heated to $50-60^{\circ}C$, the nitrogen peak is observed.

Carbon dioxide is determined using an activated carbon column. A series arrangement of activated carbon, followed by a molecular sieve column, may be used if a three-way valve arrangement between the columns is used. Carbon dioxide is retained on the carbon column until the permanent gases are separated and eluted from the molecular sieve column. The effluent from the activated charcoal column may then be switched directly through the detector by using the three-way valve.

Water vapor, because of its comparatively high concentration in air, may be easily determined. Several methods have been reported. Burke et al. [21] used a Barber-Colemen Model 5340 gas chromatograph and a thermal conductivity cell. The column was 6 ft by 1/4 in. aluminum, packed with TEE-Six 110/120 mesh coated with 20% Carbowax 20M. The column was operated isothermally at 115°C with a helium carrier gas flow rate of 120 ml/min. Samples of 2.5-ml size were injected by syringe. Analysis took 5 min. The limits of detection were not reported, but samples having less than 1% H_2O were analyzed to ±1% relative. The Porapak series of polar column supports should also be quite suitable for water vapor separation prior to detection.

B. Organic Materials

1. Automobile Exhausts

The analysis of automobile exhausts from internal combustion engines has long been studied using, almost exclusively, gas chromatographic methods. The obvious reason for this is that a great many compounds are found in combustion exhausts, and separation is required if information on the presence of any one specific compound is desired.

Polynuclear aromatic and higher molecular weight hydrocarbons are usually handled differently from the lower boiling hydrocarbons. (See the pertinent subsections relating methods for them specifically.) Nitrogen oxides and nitrogenated organic compounds, also present in exhausts, are also discussed in separate sections. The analysis of the remaining organic hydrocarbons is given below. An attempt has been made to be selective, as a great number of references may be found in the literature — many nearly duplicating one another in content.

Sampling of the exhaust is usually done by cold trapping, and separation is performed on columns generally suitable for hydrocarbons, as the reader can see from the following referenced papers. The flame ionization detector (FID) has sufficient sensitivity for these analyses and has been generally used. The use of thermal conductivity detectors is not recommended because of the larger sample sizes needed to obtain reasonable signals.

Sanders and Maynard [22] have shown the complexity of motor gasoline (and exhausts) by carrying out the gas chromatographic analysis of these materials on a 200-ft capillary column with squalane liquid phase. Over 240 chromatographic peaks were observed, some 180 of which were identified. The sheer complexity of the analysis attests to the probable difficulty of obtaining good results from concentrated air samples and trying to point out their sources.

Ellis et al. [23] identified aldehydes, ketones, and alcohols in automobile exhausts by the infrared gas analysis of column effluent from the separation of oxygenate fractions. Exhaust gases were scrubbed through a 1% $NaHSO_3$

solution, which retained the oxygenates and passed the hydrocarbon fractions of the samples. Water was separated from the organic fractions by a preparative-type column. The organic components were trapped in a narrow metal tube cooled by liquid nitrogen. The process could be carried out several times to build up a cumulative sample, which could then be analyzed on an analytical gas chromatographic column. The separation column was 20 ft by 1/4 in. o.d. packed with 9% Carbowax on unsized Teflon. A thermal conductivity detector was used. Polyvinyl fluoride bags were used to collect column fractions, as effluent gas, which were then passed into 10-m IR gas cells for analysis. Acetone, acetaldehyde, methylethylketone, methanol, and ethanol were identified. Large exhaust samples had to be scrubbed; 100 liters was the average size taken. Although this method has not been used for ambient air analysis, it demonstrates the type of general approach used for the identification of trace constituents in air samples. The quantitative analytical data by this technique are probably poor because of the probable incomplete scrubbing of the air samples in water at ambient temperatures.

Williams [24] reported a method for the collection of volatile organic compounds from air for qualitative identification, using an inert substrate at -80°C, and an 8 in. by 1/4 in. o.d. stainless steel column packed with Chromosorb P partially deactivated with di-n-butyl phthalate. Deactivation was necessary to prevent retention of the polar compounds. Air samples of 2.5 liters volume were drawn through the system and injected onto the partitioning column after being warmed to 120°C. The partitioning column was 2 m by 6.5 mm o.d. glass, packed with either 33% dodecyl phthalate on acid-washed Chromosorb P or 25% tri-m-tolyl phosphate on acid-washed Chromosorb W. Both columns were used in the range 73-93°C. An all-glass, laboratory constructed electron capture detector and a flame ionization detector (FID) were used in series to permit comparison of detector signals as a qualitative identification index. Subtractive columns of 5A 30/60 mesh molecular sieves and 25% concentrated H_2SO_4 on Chromosorb G were also placed after the sampling column, prior to separation, in several experiments to determine the effect on various types of organic compounds sampled. Drying tubes of K_2CO_3 and $Mg(ClO_4)_2$ were evaluated for retention of various compounds. As expected, the electron capture detector gave good selective responses for the halogenated compounds. Alkanes were not retained by the subtractive columns or the drying agents, while the alkenes and compounds having functional groups of heteroatoms were retained to a degree by the columns and drying agents.

2. Low Boiling Hydrocarbons in Air

Gordon et al. [25] have reported a typical method for the analysis of light hydrocarbons in air. Samples were collected by pumping air into plastic bags of 50-liter capacity over 1-hr sampling periods. A neoprene diaphragm pump was used. It was found that less than 5% loss of sample was experienced on overnight standing of the samples in the plastic bags. Separation of the C_2-C_5 hydrocarbons was achieved on a 12 ft by 1/4 in. o.d. column packed with 12% dodecyl phthalate on 40/60 mesh firebrick. Samples 300 ml in size were first frozen out to preconcentrate the analytes. The freezing-out technique employed was that of Feldstein and Balestrier [26]. A dual-column arrangement was employed. One column had a mercurous perchlorate pre-column to remove olefins and aromatic compounds as an aid in identification. Using a flame ionization detector (FID) it was possible to reliably analyze samples in the 10-350 ppb range with repeatability of 2-10 ppb.

Cavanaugh et al. [27] analyzed the light hydrocarbons on a 6-ft 20% Carbowax 20M column and a 6-ft Porapak Q column, valved in tandem. Both columns were operated at 78-82°C with a helium carrier gas flow of 229 ml/min. A hydrocarbon flame ionization detector was used. A cryogenic trap packed with either Carbowax 20M or Porapak Q was used to concentrate hydrocarbons from 200-ml samples prior to separation.

The C_2-C_{10} organics were separated on the Carbowax column, but the Porapak column was used for better resolution of the C_2-C_4 alkanes and alkenes. The Carbowax column proved to be the better of the two. The light alkanes were detected as low as 0.02 ppb; n-butanol was found in much larger amounts, up to 126 ppb.

Eggersten and Nelson [28] developed a technique aimed at the analysis of air for C_2-C_5 hydrocarbons. Theirs is a good example of a technique employing a thermal conductivity detector. The apparatus arrangement is shown in Fig. 12. Bottle samples are drawn by means of a vacuum system through a trapping column cooled by liquid oxygen. Samples up to 10 liters in size were required for analysis. Samples were passed through an Ascarite drying tube and then through the trapping column. The trapping column was 12 in. by 5/16 in. o.d. packed with 6 ml of 40% dimethylsulfolane on 20/30 mesh firebrick. Trapping required from 10 to 45 min. Helium was next flushed through the cooled, trapped sample for 20-30 min to remove materials boiling below C_2 hydrocarbons. The trapped sample was then warmed to 0°C in an ice bath and passed through another Ascarite tube into the separating column, a gas chromatographic column 25 ft by 1/4 in. o.d. copper, packed with the same material as the trapping column. Calibrations were made using standard gas mixtures. Analyses of exhaust samples and city traffic air were performed. Resolution of nitrogen, oxygen, ethane, and ethylene was poor. Propane and nitrous oxide were not resolved. Good resolution of the higher hydrocarbons was obtained. Limits of detection for the hydrocarbons appear to be near 0.02 ppm in a 10-liter sample. The analysis procedure appears to take 2 hr or more.

Guerrant [29] designed, constructed, and operated a portable gas chromatograph weighing 33 lbs. The instrument consisted of the necessary battery supply and electronics package for a flame ionization-type detector. A lecture size hydrogen bottle was used. The air supply was from a 1.4-liter, 2000-psi test cylinder. Miniature valves with capillary tubing for controlling gas flow were used. Air served as the carrier gas. Several

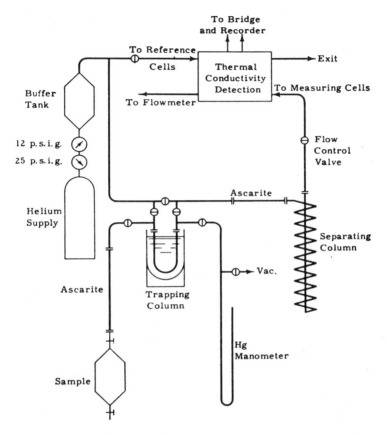

FIG. 12. Schematic diagram of gas chromatographic apparatus for determining hydrocarbons in exhaust gas and air (reprinted from <u>Anal Chem.</u>, <u>30</u>, 1040 (1958) by permission of the publisher).

different columns were selected; all were operated at ambient temperature. For the separation and analysis of CH_4, C_2H_4, and C_2H_6 in the presence of N_2, O_2, and CO_2, a 6 ft by 1/8 in. o.d. column packed with 100/200 mesh activated silica gel was used. For determination of organics C_3-C_5 the same size column packed with 10% hexadecane on firebrick was used. For the higher boiling organic compounds a capillary column, 300 ft by 0.02 in., was used coated with hexadecane or squalane.

Sampling was done using a 1.6-ml sample loop on a gas-sampling valve. Good detection of parts per million levels of various hydrocarbons was obtained. Calibration was by peak height compared to standard mixtures of hydrocarbons in air. The detector had to be maintained every 6 hr of battery operation, 8 hr of air use, or 50 hr of hydrogen gas use. Benzene and

toluene specifically, at the parts per billion level, have been detected in
1-liter air samples by a combination of cryogenic trapping and gas chromatography [30]. Benzene and toluene were detected after separation from
some 26 other eluted components on a 6 ft × 1/8 in. stainless steel column
packed with 8% SF96 on 60/80 mesh Chromosorb W.

3. Polynuclear Aromatic Hydrocarbons

Because of the carcinogenic activity of certain polynuclear aromatic hydrocarbons considerable work has been done on their analysis. Much of the
work has dealt with specific sources, such as tobacco smoke or soot.
Applications of the following methods to air analysis require modification of
sampling techniques to preconcentrate amounts sufficient for analysis.
Limits of detection are near 1 ng. The analysis problem is not a simple one.
Over 100 polycyclic aromatic hydrocarbons have been separated and identified by combined gas chromatography-mass spectrometry in one recent
elegant, comprehensive study [31].

Bowman and Beroza [32] have combined gas chromatography with
spectrofluorimetry, using the latter as a detector. Solute from a gas
chromatographic column is picked up by a slowly moving stream of ethanol
and passed through an Aminco-Bowman spectrophotofluorimeter. Conventional chromatograms are recorded. The detector monitors the intensity of
fluorescent emission as a function of time. This provides a degree of
selectivity. It is also possible to scan the fluorescent emission spectrum of
detected materials separated on the column. Polynuclear hydrocarbons,
such as anthracene, fluorene, chrysene, and benzo[a]pyrene, could be
detected with limits of detection near 1 ng in favorable cases. Some methyl-
enedioxyphenyl compounds, such as safrole, were also detected with similar
limits of detection. The above compounds were separated on a 1.25 m by
6 mm o.d. glass column packed with 5% w/w QF-1 on 80/100 mesh Gas
Chrom Q. The compound 1-naphthol, a hydrolysis product of Carbaryl, was
chromatographed on 5% DC 200 using the same support. The limit of detection was 2 ng.

Although this method has been used for a research study only, it shows
promise for particulate analysis after proper sampling. Sample sizes must
be in the fractional milligram range.

Individual hydrocarbons in soot samples have been determined by programmed temperature gas chromatography, using dual flame ionization
detectors (FID) [33]. Peaks were identified by comparing the ultraviolet
spectra and retention times to those of known compounds.

Soot is extracted with chloroform for 8 hr in a soxhlet apparatus under
nitrogen atmosphere. After room temperature evaporation of the chloroform, the extract is separated on a silica gel column with benzene solvent.
The eluate, after evaporation of the benzene solvent, is taken up in CS_2 and

separated by gas chromatography. Two columns were used; one, 12 ft by
1/4 in. o.d. and the other 50 ft by 1/8 in. o.d. Both were packed with 10%
SE 52 in 60/80 mesh Chromosorb W. Some 15 components were identified,
including fluorene, anthracene, pyrene, benzo[α]pyrene, benzo[e]pyrene,
and chrysene.

Acetylene-oxygen flames (premixed) have been analyzed for content of
polycyclic aromatic hydrocarbons [34]. Soot collected on stainless steel
wire grids was extracted with chloroform and analyzed on a 12 ft by 1/4 in.
o.d. column packed with 10% SE 52 silicone gum rubber on 60/80 mesh
Chromosorb W at 190-300°C (programmed temperature GC). Flame ionization and thermal conductivity detectors were used. Ultraviolet spectra were
obtained on trapped fractions. Anthracene, phenanthracene, methyl phenan-
threnes, fluoranthene, pyrene, methyl pyrenes, and 2,13-benzfluoranthene
were detected. Diphenylene oxide and cyclopentacenaphthylene(s) have also
been detected in flame soots by a gas chromatographic method [35].

Davis [36] obtained a lower limit of detection of 1 ng for benzo[a]pyrene
using an electron capture detector. His work was exclusively with the
analysis of cigarette smoke. A 9 ft by 1/8 in. o.d. stainless steel column
packed with 3% SE 30 on 60/80 mesh Chromosorb W conditioned at 280°C
for 3 hr was used with carrier gas at 70 ml/min. Separations were temperature programmed from 100°C to 280°C. Samples were separated into
fractions by liquid chromatography prior to GC of the benzo[a]pyrene
fraction. This would appear to be a good combined chromatography technique
for air analysis if properly designed. Lane et al. [37] have developed a
single column suitable for the separation of some 20 polynuclear hydrocarbons. OV-7, 1% on 80/100 mesh Chromosorb W, was found to be optimum
packing for the 5.4 m x 6 mm o.d. glass column. The column was also
suitable for analysis of alkanes in the C_{11}-C_{33} range.

Hauser and Pattison [38] have developed a method for the analysis of air
particulate matter for the C_{15}-C_{36} n-alkane range. Analyses lead to the
conclusion that gasoline and diesel fuel combustion can produce this range of
aliphatic hydrocarbons in the particulate, although they are not present in the
fuels. Analyses were made of benzene-soluble fractions of soxhlet-extracted
particulate. A gas chromatograph with a dual hydrogen flame ionization
detector was used. A matched pair of 20 ft x 1/8 in. diameter stainless
steel columns packed with 3% SE 30 on 100/120 mesh Chromosorb W was
used. The column was programmed at a rate of 27.5°C/min from 150°C to
315°C. The 20 ft column length was optimum. Analysis time was 20 min.

4. Halogenated Hydrocarbons

Halogenated hydrocarbons from CH_3Cl to $CCl_2=CCl_2$ in composition (including several Freons) and having boiling points ranging from -24°C to 121°C
were preconcentrated onto Porapak Q and S columns [39]. A Microtex
MT-220 gas chromatograph fitted with a Dohrmann Microcoulometer

detector was used. The columns were 6 ft by 1/4 in. stainless steel packed
with Porapak Q or S. The operating procedure consisted of passing 100 or
500 ml of sample air in the sample loop of a gas sample injection valve into
the column at 30°-50° C. Helium carrier gas was used at approximately 100
ml/min. After sample injection, the column was raised in temperature to
100° C and held at that temperature for 5 min. Depending on the composition
of the sample, the column was then programmed at 10° C/min from 100° to
120° C, at 2° C/min from 120° to 136° C, and 10° C/min from 136° to 166° C.
To drive off tetrachloroethylene the column was heated to 210° C. Standard
gas mixtures were obtained from commercial sources and diluted to prepare
calibration standards. Liquid compound standards were prepared by injecting known amounts into evacuated stainless steel bottles pressurizing to 300
psi with helium, and by then using the mixture in the same way as the gases.
Components were easily determined in the 10 ppb range.

   This on-column concentration technique apparently has the single disadvantage of requiring cool-down time before reuse of a column. Bis(chloromethyl) ether, a possible carcinogenic impurity in the chloromethylating
reagent chloromethyl methyl ether, is a hazard in production facilities.
This compound has been detected and determined at the part per billion level
using a gas chromatography-mass spectrometer instrument [40].

   Dichloroacetylene (DCA) and a number of other chlorinated hydrocarbons
have been detected by gas chromatography coupled with a microcoulometric
detector [41]. By controlling the temperature of the pyrolysis furnace prior
to the microcoulometric cell interferences not separated from DCA are
avoided. The chromatographic column used was 6 ft X 0.25 in. stainless
steel packed with polyethylene glycol 400, 15% by weight on 80/100 mesh
Chromosorb P.

5. Aldehydes, Ketones, and Alcohols

Ellis et al. [42] identified aldehydes, ketones, and alcohols in automobile
exhausts by the infrared gas analysis of column effluent from the separation
of oxygenate fractions. Exhaust gases were scrubbed through a 1% NaHSO$_3$
solution that retained the oxygenates and passed the hydrocarbon fractions of
the samples. Water was separated from the organic fractions by a preparative column. The organic components were trapped in a narrow metal
tube cooled by liquid nitrogen. The process could be carried out several
times to build up a cumulative sample, which could then be analyzed on an
analytical gas chromatographic column. The separation column was 20 ft by
1/4 in. o.d. packed with 9% Carbowax on unsized Teflon. A thermal conductivity detector was used. Polyvinyl fluoride bags were used to collect
column fractions as effluent gas which were then passed into 10-m IR gas
cells for analysis. Acetone, acetaldehyde, methylethylketone, methanol,
and ethanol were identified. Large exhaust samples had to be scrubbed; 100
liters was the average size taken. Although this method has not been used

for ambient air analysis, it demonstrates the type of general approach used for the identification of trace constituents in air samples. The quantitative analytical data by this technique is probably poor because of the probable incomplete scrubbing of the air samples in water at ambient temperatures.

6. Hydroperoxides and Alkylpolynitrates

A flowing liquid colorimeter detector (see Fig. 13) [43] was used for the detection of hydroperoxides and nitrogen dioxide in laboratory samples after treatment on a gas chromatography column. Samples were prepared solutions of the hydroperoxides. Ferrous thiocyanate was used as the colorimetric reagent. The flow cell was fitted into the light path of a Bausch and Lomb Spectronic 20 colorimeter. An electrical output to a 1 mV recorder was made from the output of the colorimeter amplifier. Limit of detection was 9 μg for the methylhydroperoxide. Several stationary phases were used, the best of which was Polyethylene Glycol 400 on Chromosorb G, treated with H_2O_2. More than 90% of the NO_2 was absorbed by all columns studied.

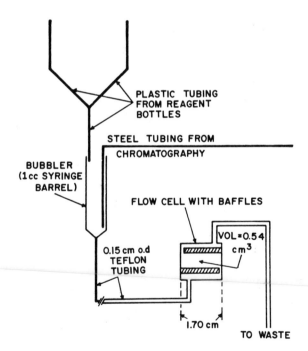

FIG. 13. Diagram of flow cell for colorimetric detector (reprinted from Anal. Chem., 41, 1777 (1969) by permission of the publisher).

Although not used directly for air analysis, the method is a good specific approach because of the nature of the detector. Work on the columns appears to be needed.

Three of the more industrially important alkylpolynitrates — ethylene glycol dinitrate, 1,2-propylene glycol dinitrate, and glycerol trinitrate — have been determined in air [44]. A gas chromatograph fitted with an electron capture-type detector was used. The column was of glass, 25 cm by 2 mm i.d., packed with 10% Igepal Co-880 on 80/120 mesh siliconized Chromosorb P. The column was operated at 120°-160° C with a nitrogen carrier gas flow rate of 133 ml/min. Sampling by syringe was unsuccessful, and air had to be passed through 10 ml of ethyl alcohol in each of two micro-impingers in series. Some 10 liters of air were passed through impingers; 5 μl of this solution was injected onto the column for analysis, which took 6 min. Limits of detection on a weight basis were determined from calibration data and found to be 5×10^{-11} g for propylene glycol dinitrate, 2×10^{-10} g for nitrotoluene and ethylene glycol dinitrate, and 2×10^{-9} g for glycerol trinitrate. Calibration was done by preparing standard solutions of the nitro compounds in ethyl alcohol and then analyzing them. Using the scrubbing procedure described and the 5-μl fraction, the air concentration limits of detection are in the 0.02-0.08 mg/m^3 range. Obviously, improvements by factors of 10-100 should easily be possible.

C. The Nitrogen Oxides, N_2O, NO, and NO_2

All three of the nitrogen oxides are normal constituents of air at low concentrations, below the parts per million level. Nitrous oxide is by far the least toxic of the three and consequently of less toxicological interest. Reactions of nitrous oxide with hydrocarbons in air are slow or nonexistent. Nitric oxide at low concentrations reacts only very slowly with atmospheric oxygen to produce nitrogen dioxide. Nitrogen dioxide and nitric oxide are both important chemicals in the production of photochemical smog and thus are of great importance in air pollution research and monitoring.

Several gas chromatographic methods have been reported for the nitrogen oxides. They are not necessarily as good as the current chemiluminescent methods.

Perhaps the best method is that of Morrison and Corcoran [45]. A plane-parallel electron capture detector was constructed and studied for the detection of NO_2. Operated in the pulsed voltage mode, limits of detection appear to be near 0.2 ppm using 0.5 cm^3 samples of gas mixtures. The operating range of the detector was shown to be at least up to 75 ppm. A 20 ft long, 1/8 in. o.d. stainless steel column packed with 10% w/w SF-96 on 40/80 mesh Fluoropak 80 was used to separate NO_2 from air at 22° C. It is apparent that improved limits of detection are required to get to the parts per billion level necessary for monitoring. This method is more sensitive than the following ones described, where a thermal conductivity detector is used.

Nitric oxide may be determined in the presence of NO_2. Dietz [46] developed a technique for treating molecular sieve 5A to prevent the tailing of nitric oxide peaks in gas analyses. The column, 6 ft long by 1/4 in. o.d. stainless steel, was heated to 300° C under vacuum for 20 hr to remove water, filled with He gas, and then treated by slowly passing NO gas at 300° C for 1 hr. After cooling, with the NO still flowing, He was then passed through to remove the NO. Oxygen was then passed through to convert bound NO to NO_2. Nitrogen dioxide, which was not eluted from the column, did not interfere in the NO analysis. It would appear that the use of a more sensitive detector with the treated column, and perhaps the use of some preconcentration column, would permit analysis of part per million levels of NO in air. At least the column treatment reported looks promising. A more sensitive detector should permit detection of concentrations below parts per million.

Soil atmosphere was analyzed for O_2, N_2, CO_2, NO, N_2O, and NO_2 by Van Cleemput's method [47]. A three-column system was used with a thermal conductivity detector. The first column was 1 ft by 1/8 in. o.d. stainless steel, packed with 0.5% Carbowax 1500 on 60/80 mesh silanized glass beads. The second was 18 ft by 1/16 in. o.d. stainless steel, packed with Porapak Q 80/100 mesh, activated before use by heating to 230° C for 1-2 hr. The third was 3 ft by 1/4 in. o.d. stainless steel, packed with 0.2-0.5 mm molecular sieve 5A, activated at 220° C for 24 hr. The first column was cooled to liquid air temperature. This retained NO_2, NO, CO_2, and N_2O. The remaining gases, O_2 and N_2, go to the second and third columns. O_2 and N_2 are separated by the molecular sieve column. The remaining gases are separated on the Porapak Q column after heating the glass column. Good separation of all gases was obtained. A thermal conductivity detector was used giving parts per million limits of detection.

A three-column arrangement has been used [48] for the separation of H_2, O_2, N_2, NO, CO, N_2O, CO_2, C_2H_6, C_2H_4, C_2H_2, and NO_2 (see Fig. 14). The first column was 1 ft by 1/8 in. stainless steel, packed with 0.5% Carbowax 1500 on 60/80 mesh silanized glass beads. Column 2 was 20 ft by 1/4 in. stainless steel, packed with 40% DMSO on 60/80 mesh Gas Chrom RZ. Column 3 was 8 ft by 1/4 in. stainless steel, packed with 30/60 mesh 13X molecular sieves.

Injected samples were passed through column 1 at -76° C; NO_2 and H_2O are removed by this column. Column 2 separates N_2O and CO_2. The third column separates H_2, O_2, N_2, NO, and CO. Finally, column 1 is heated to desorb and detect NO_2. The hydrocarbons are eluted from column 2. Limits of detection reported were in the parts per million range with the thermal conductivity detector used.

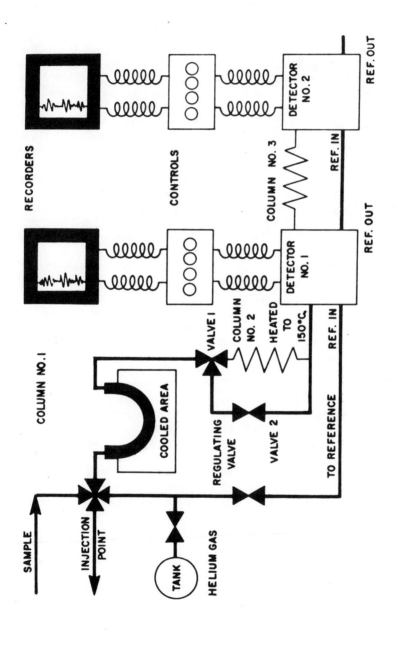

FIG. 14. System schematic. Column 1, ½% Carbowax on glass beads; Column 2, 40% DMSO on Gas Chrom RZ; Column 3, 13X molecular sieve. (Reprinted from Anal. Chem., 37, 1152 (1965) by permission of the publisher.)

D. Gas Chromatographic Determination of Carbonyl Sulfide and Phosgene

Carbonyl sulfide (COS) has been determined down to the 25-ppm level [49] in natural gas samples. A 12 ft by 1/4 in. o.d. copper tubing column was packed with Kromat FB support coated with 30% w/w N,N-di-n-butyl-acetamide. The column was operated at 28°C with a helium carrier gas flow of 50 ml/min. A thermal conductivity detector was used, which accounts for the rather poor lower limit of detection. Gas samples of 5 ml were injected into the column; a gas-sampling valve was also used, on occasion. Propylene was the only natural gas component providing any interference. Results of analyses compared well to results by mass spectrometry.

Phosgene in air was determined in the 1 ppb to 2 ppm range [50]. An Aerograph model A-350-B gas chromatograph equipped with an electron capture detector was used. The column was 2 m by 4.7 mm i.d. aluminum packed with 30% w/w Flexol plasticizer 10-10 (didecyl phthalate) on 100/200 mesh GC-22 Super Support (Coast Engineering Company). The column was operated at 50°C, with a N_2 carrier gas flow rate of 50 cm^3/min. Samples of 0.5 ml were used. Calibration samples were prepared using a triple dilution system. All sampling equipment had to be kept free from moisture because of the reactivity of phosgene. Samples were prepared down to 4 ppb in size. The limit of detection appeared to be near 1 ppb. A linear response of peak area vs phosgene concentration was obtained up to 2 ppm. Hydrochloric acid gas, which elutes near phosgene, was not an interference up to at least 1% HCl in air. The method has been used on prepared samples only.

E. Gas Chromatographic Determination of Sulfur Compounds

The sulfur compounds, and sulfur dioxide in particular, have long been recognized as major air pollutants. Some of the earliest analytical procedures aimed at air analysis were used to determine sulfur dioxide or sulfides in air. The maximum permissible concentrations recommended for all of the sulfur compounds are in the parts per million range. Monitoring, nevertheless, requires detection and analysis in the range below parts per million as this is the usual ambient concentration experienced.

The application of gas chromatographic methods to the analysis of air for SO_2, H_2S, and the mercaptans was not immediately successful because of the reactivity of these compounds and their adsorption onto column packings, onto column walls, and in the sampling system. The work of Koppe and Adams [51] is particularly noteworthy. These authors found that even with the best columns, Porapak Q and 10% Triton 305 on Chromosorb W, unacceptable losses of H_2S and SO_2 were encountered at concentrations below 1 ppm. Preconcentration by factors of 10-50 were required from the ambient level to avoid losses.

More recently, Cook and Ross [52] have accomplished the separation of CO_2, H_2S, and water in wet air samples by using a dual column, 6 ft of Porapak Q treated with $\frac{1}{4}\%$ of Triton 305 followed by 6 ft by 1/8 in. column of 50% Carbonase 1500 on Teflon 6.

Inert column materials, which also perform the separation, are key to the successful analysis of SO_2, H_2S, CH_3SH, and CH_3SCH_3. A special graphatized carbon black column material has been developed [53] and applied [54] to air analyses.

Stevens et al. [55] have solved the column materials problem and developed an excellent method for the analysis of H_2S, SO_2, and CH_3SH down to approximately 2 ppb in air without preconcentration. Their analysis system and calibration system is shown in Fig. 15. Samples were pulled by vacuum into a six-port rotary Chromatronix automatic gas-sampling valve equipped with a Teflon 10-ml sampling loop. Samples were then passed through a 36 ft by 1/8 in. o.d. (0.085 in. i.d.) fluorinated ethylene propylene (FEB, Teflon, DuPont) tubing packed with 40/60 mesh treated with Haloport-F. The packed columns were treated by passing through them 50 ml of an acetone solution containing 12 g of polyphenyl ether (five-ringed polymer, F and M Scientific Co.) and 500 mg of reagent grade orthophosphoric acid at a rate of 20 ml/hr under N_2 pressure. Nitrogen was passed through the column until it appeared dry, usually requiring 7-8 hr. The treated column was then conditioned by heating to 140° C for 6 hr while passing a N_2 carrier gas at 50 ml/min through the column. The column is operated at a carrier gas flow rate of 100 ml/min. With this column, a separation of the sulfur compounds at several concentrations below 1 ppm is experienced, as shown in Fig. 16. Analyses of ambient air in Cincinnati, Ohio, were performed. The SO_2 component appeared to be the only sulfur compound present. The SO_2 analysis followed almost exactly the results of a total sulfur analysis by the flame photometric detector (without gas chromatography).

The detector employed was the flame photometric detector developed by Brody and Chaney [56]. This device monitors the sulfur emission bands obtained in a hydrogen-oxygen flame when SO_2 or any other sulfur compound passes through the flame. It is highly selective for the detection of sulfur compounds versus ordinary organic hydrocarbons even without gas chromatographic separation. The hydrogen flow rate to the detector was 80 ml/min and the oxygen flow rate was 16 ml/min.

Calibration of the instrument was carried out using the permeation tube technique [6] in Section II, D, p. 87 (see also the discussion of system calibration). Since the detector senses the S_2 band system, plots of the log of the peak area vs log of the concentration are linear with a slope of two. All the sulfur compounds except dimethyl sulfide exhibited this. A careful study of the detector response to sulfur compounds has been made [57]. Limits of detection for the analysis system were found to be 2 ppb for H_2S and SO_2, 3 ppb for CH_3SH, and 10 ppb for CH_3SCH_3 when a 10-ml air sample is taken.

A modification of the technique was made to avoid the possibility of sulfur compounds having three or more carbon atoms being detected. A ten-port

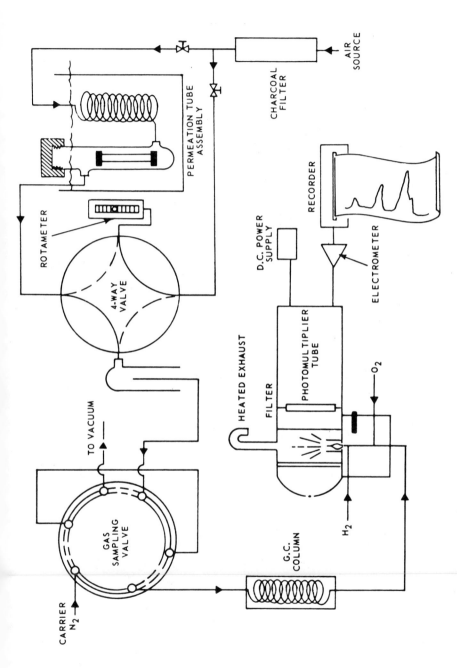

FIG. 15. Automated gas chromatographic–FPD atmospheric sulfur gas analyzer and calibration apparatus (reprinted from Anal. Chem., 43, 827 (1971) by permission of the publisher).

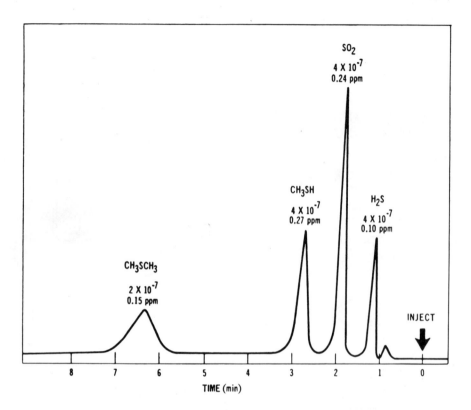

FIG. 16. Chromatogram of a mixture of SO_2, H_2S, CH_3SH and CH_3SCH_3 at concentrations below 1 ppm (reprinted from <u>Anal. Chem.</u>, <u>43</u>, 827 (1971) by permission of the publisher).

gas-sampling valve was used and a stripping column added. After the H_2S, SO_2, CH_3SH, and CH_3SCH_3 were passed through the stripping column onto the separation column, the stripping column was back flushed to remove any materials still in the stripping column. The higher molecular weight sulfur compounds would still be on the stripping column.

   This method is certainly the best available at this time, especially where the possibility of detecting the various sulfur compounds is desired. It is being used currently.

F.   Alkyl-lead Compounds

Lead is present in air both as solid PbO and as uncombusted alkyllead compounds. While total lead analysis is done by a variety of chemical methods, gas chromatography appears to be the best suited to identification of the alkyl compounds. The maximum recommended concentrations for the alkyl-lead compounds are near 1 mg/m^3 and so fractional microgram samples will likely be the usual sample size for analysis.

The method of Cantuti and Cartoni [58] is probably typical. Tetraethyllead (TEL) in air was determined by gas chromatographic separation using an electron capture detector and sample preconcentration. Samples were passed through a short, cooled sampling column and desorbed onto a gas chromatographic column. A glass column, 1 m by 0.3 mm i.d. packed with Chromosorb P, 80/100 mesh, was operated at 80°C with nitrogen carrier gas. The method was used for the analysis of air near a tetraethyllead plant. Samples analyzed were in the 0.05-0.31 ppm TEL range. Limits of detection were near 0.1 ppm. Soulages [59] and Bonelli and Hartman [60] have also reported methods using different detectors.

It would appear for this particular application that the emission-type detector for lead should be more sensitive and selective.

G. Tracer Gas, SF$_6$

Sulfur hexafluoride has been widely used for tracing the movements of air masses [61-63]. This compound was selected because of its nontoxic, highly stable character. It does not occur in the ambient air in significant concentrations. In addition, the detection of this compound by electron capture detectors can be accomplished at the 1 part per 10^{14} level [64]. Although SF$_6$ is not necessarily an air pollutant, it is included here because of its use in the air pollution analysis field. Two typical analysis methods are given below.

Turk et al. [64] sampled into 1-liter stainless steel tanks. These were then pressurized to 45 psig with purified air prior to injection of 5-ml samples by a gas-sampling valve into the gas chromatograph. A 1 m by 1/8 in. silica gel packed column followed by a 1 m by 1/8 in. activated charcoal column was required to separate SF$_6$ from air components. The columns were operated at 120°C in a Perkin Elmer model 810 gas chromatograph fitted with an electron capture detector. Prepurified nitrogen carrier gas was used at a flow rate of 60 ml/min. The direct injection of air samples without preconcentration was found suitable for concentrations down to 1 ppb. For lower concentrations it was necessary to freeze out SF$_6$ in the sample loop of the gas-sampling valve. Concentrations down to 0.01 ppb were apparently determinable using the preconcentration technique. The use of this method for a stack discharge study was reported.

Clemons et al. [65] used an activated charcoal trap to collect SF$_6$ and the analysis system shown in Fig. 17. Air samples in Saran bags were pulled through a gas-sampling valve and a trapping column of 1/8 in. stainless steel packed with 1 g of activated charcoal. The trapping column was actually connected to the gas-sampling valve in the position of what would usually be the sample loop. After the desired sample volume was pulled through the trap at

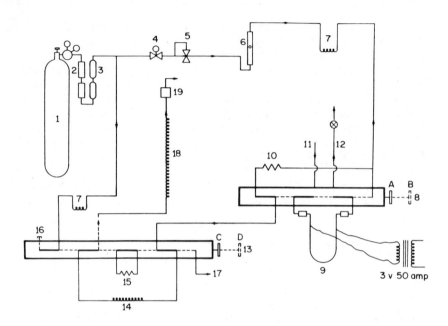

FIG. 17. Flow schematic of inlet and trapping system. (1) Carrier gas,
(2) canisters of BTS catalyst, (3) canisters of molecular sieve 5A,
(4, 5) flow controls, (6) flowmeter, (7) charcoal traps for purification of
carrier, (8) seven-port valve, (9) sample trap with heater, (10) matched
flow resistance, (11) sample entry, (12) exit to vacuum source through
needle valve control, (13) nine-port valve, (14) cutter column, (15) flow
resistance matched to cutter column, (16) capped port, (17) exit for back-
flush, (18) analytical column, (19) detector. (Reprinted from Env. Sci. and
Tech., 2, 551 (1969) by permission of the publisher.)

ambient temperature, a heavy current of electricity was passed through the
metal column to heat it and desorb the SF_6. The desorbed SF_6 was then
passed through another gas-sampling valve and a cutter column. The function of the cutter column in the sample loop of the second sampling valve
was to remove part of the O_2 in the desorbed SF_6 sample prior to passing
it through the separation column and the electron capture detector. The
cutter column was 3 ft by 1/8 in. o.d. stainless steel packed with Baymal-
14. The separation column was 12 ft by 1/8 in. o.d. stainless steel packed
with 40/60 mesh Baymal [66]. The column and detector were at ambient
temperature. The carrier gas was 5% hydrogen in argon. The flow rate was
75 ml/min. Carrier gas purification was required. Backflushing was used
to clear the cutter and analysis columns. Sample gas volumes of up to
several hundred milliliters were analyzed. Limits of detection were as low

as 1 part per 10^{14}. The method is capable of detecting the ambient concentration of SF_6 resulting from leaks in manufacturing processes. SF_6 could be detected over 100 miles from release points.

The separation of SF_6 from oxygen is the major problem in air analysis. The methods using molecular sieve columns of Simmonds et al. [67] and Dietz and Cote [68] are the most recent developments.

Dietz and Cote have improved the separation of SF_6 from oxygen by using a nitric oxide-treated molecular sieve column. Both laboratory-based and portable instruments have been developed, with a lower limit of detection for SF_6 of 4×10^{-13} cm^3/cm^3.

REFERENCES

1. C. T. O'Neil, A. Dynako, and R. Scholz, private communication, 1972.

2. J. M. McKelvey and H. E. Hoelscher, Anal. Chem., 29, 123 (1957).

3. A. P. Altshuller and I. R. Cohen, Anal. Chem., 32, 802 (1960).

4. B. E. Saltzman and C. A. Clemons, Anal. Chem., 38, 800 (1966).

5. B. E. Saltzman, A. E. Coleman, and C. A. Clemons, Anal. Chem., 38, 753 (1966).

6. A. E. O'Keefe and G. C. Ortman, Anal. Chem., 38, 760 (1966).

7. J. E. Lovelock, Anal. Chem., 33, 162 (1961).

8. F. Bruner, C. Canulli, and M. Possanzini, Anal. Chem., 45, 1790 (1973).

9. R. S. Braman and E. S. Gordon, IEEE Trans., Instrumental and Measurement, IM-14, 11-19 (1965).

10. W. J. Youden, Statistical Methods for Chemists, Wiley, New York, 1951.

11. M. Dimbat, P. E. Porter, and F. H. Stross, Anal. Chem., 28, 290 (1956).

12. H. W. Johnson, Jr., and F. H. Stross, Anal. Chem., 31, 1206 (1959).

13. H. W. Johnson, Jr., Anal. Chem., 37, 1581 (1965).

14. F. R. Cropper and S. Kaminsky, Anal. Chem., 35, 735 (1963).

15. J. Novak, V. V. Vasak, and J. Janak, Anal. Chem., 37, 660 (1965).

16. S. S. Miller, Env. Sci. Tech., 5, 503 (1971).

17. K. Abel, Anal. Chem., 36, 953 (1964).

18. G. S. Vizard and A. Wynne, Chem. Ind., 1959, 196.

19. G. Kyryacos and C. E. Boord, Anal. Chem., 29, 787 (1957).

20. E. W. Lard and R. C. Horn, Anal. Chem., 32, 878 (1960).

21. D. E. Burke, G. C. Williams, and C. A. Plank, Anal. Chem., 39, 544 (1967).

22. W. N. Sanders and J. B. Maynard, Anal. Chem., 40, 527 (1968).

23. C. F. Ellis, R. F. Kendall, and B. H. Eccleston, Anal. Chem., 37, 511 (1965).

24. I. H. Williams, Anal. Chem., 37, 1723 (1965).

25. R. J. Gordon, H. Mayrsohn, and R. M. Ingels, Env. Sci. Tech., 2, 1117 (1968).

26. M. Feldstein and S. Balestrier, J. Air Poll. Contr. Assoc., 15, 117 (1965).

27. L. A. Cavanagh, C. F. Schadt, and E. Robinson, Env. Sci. Tech., 3, 251 (1969).

28. F. R. Eggersten and F. M. Nelsen, Anal. Chem., 30, 1040 (1958).

29. G. O. Guerrant, Anal. Chem., 37, 516 (1965).

30. S. Pilar and W. F. Graydon, Env. Sci. Tech., 7, 628 (1973).

31. R. C. Lao, R. S. Thomas, H. Oja, and L. Dubois, Anal. Chem., 45, 908 (1973).

32. M. C. Bowman and M. Beroza, Anal. Chem., 40, 535 (1968).

33. B. B. Chakraborty and R. Long, Env. Sci. Tech., 1, 828 (1967).

34. R. Long and E. E. Tompkins, Nature (London), 213, 1011 (1967).

35. B. D. Crittenden and R. Long, Env. Sci. Tech., 7, 742 (1973).

36. H. J. Davis, Anal. Chem., 40, 1583 (1968).

37. D. A. Lane, H. K. Moe, and M. Katz, Anal. Chem., 45, 1776 (1973).

38. T. R. Hauser and J. N. Pattison, Env. Sci. Tech., 6, 549 (1972).

39. F. W. Williams and M. E. Umstead, Anal. Chem., 40, 2232 (1968).

40. L. A. Shadoff, G. J. Kallos, and J. S. Woods, Anal. Chem., 45, 2341 (1973).

41. F. W. Williams, Anal. Chem., 44, 1317 (1972).

42. C. F. Ellis, R. F. Kendall, and B. H. Eccleston, Anal. Chem., 37, 511 (1965).

43. T. E. Healy and P. Urone, Anal. Chem., 41, 1777 (1969).

44. E. Camera and D. Pravisani, Anal. Chem., 39, 1645 (1967).

45. M. E. Morrison and W. H. Corcoran, Anal. Chem., 39, 255 (1967).

46. R. N. Dietz, Anal. Chem., 40, 1576 (1968).

47. O. Van Cleemput, J. Chromatog., 45, 315 (1969).

48. J. M. Trowell, Anal. Chem., 37, 1152 (1965).

49. J. A. Schols, Anal. Chem., 33, 359 (1961).

50. L. J. Priestly, Jr., F. E. Critchfield, N. H. Ketcham, and J. D. Cavender, Anal. Chem., 37, 70 (1965).

51. R. K. Koppe and D. F. Adams, Env. Sci. Tech., 1, 479 (1965).

52. W. G. Cook and R. A. Ross, Anal. Chem., 44, 641 (1972).

53. A. DiCorcia, P. Ciccioli, and F. Bruner, J. Chromatog., 62, 128 (1971).

54. F. Bruner, A. Liberti, M. Possanzini, and I. Allegrini, Anal. Chem., 44, 2070 (1972).

55. R. H. Stevens, J. D. Mulik, A. E. O'Keeffe, and K. J. Krost, Anal. Chem., 43, 827 (1971).

56. S. Brody and E. Chaney, J. Gas Chromatog., 4, 42 (1966).

57. D. G. Greer and T. J. Bydalek, Env. Sci. Tech., 7, 153 (1973).

58. V. Cantuti and G. P. Cartoni, J. Chromatog., 32, 641 (1968).

59. N. L. Soulages, Anal. Chem., 39, 1340 (1967).

60. E. J. Bonelli and H. Hartman, Anal. Chem., 35, 1980 (1963).

61. C. A. Clemons and A. P. Altshuller, Anal. Chem., 38, 133 (1966).

62. G. F. Collins, F. E. Bartlett, A. Turk, S. M. Edwards, and H. L. Mark, J. Air Poll. Contr. Assoc., 15, 54 (1965).

63. B. E. Saltzman, A. I. Coleman, and C. A. Clemons, Anal. Chem., 38, 753 (1966).

64. A. Turk, S. M. Edwards, H. L. Mark, and G. F. Collins, Env. Sci., Tech., 2, 44 (1968).

65. C. A. Clemons, A. I. Coleman, and B. E. Saltzman, Env. Sci. Tech., 2, 551 (1968).

66. J. J. Kirkland, Anal. Chem., 35, 1295 (1963).

67. P. G. Simmonds, G. R. Shoemake, J. E. Lovelock, and H. C. Lord, Anal. Chem., 44, 860 (1972).

68. R. N. Dietz and E. A. Cote, Env. Sci. Tech., 7, 338 (1973).

Chapter 3

LIQUID CHROMATOGRAPHY ANALYSIS IN AIR POLLUTION

Kingsley Kay

Department of Environmental Medicine
Mt. Sinai School of Medicine
New York, N.Y.

I. INTRODUCTION

In any consideration of the uses of liquid chromatography for air pollution
analysis, it has become important to explain that to date all work has been
carried out according to the original procedure developed by Tzvyet [1] in
1911 — the slow process of linear elution adsorption chromatography. The
new high-speed process, with operating conditions similar to those in gas
chromatography (GC), has just recently been tested experimentally on auto
exhaust [2] and mixtures of polycyclic aromatic hydrocarbons [2-4]. It
appears to be the method of the future for low volatility compounds and is
considered in detail later in the chapter.

Liquid chromatography (LC) has been widely used in air pollution study
for the detection and determination of polycyclic aromatic hydrocarbons
(PAH). This important contribution of LC originated in 1934 from investigations by Winterstein and Schön [5-7] into the complex of cancer-producing
constituents of coal tar. These workers applied the chromatogen adsorption
analysis technique developed by Tzvyet [1] for chlorophyll studies. A
number of PAH's was successfully separated on alumina. Despite lengthy
elution times and other shortcomings, to be discussed in this chapter,
liquid chromatography became the method of choice for some 30 years in
separating PAH from coal tar, atmospheric particulates, automobile exhaust,
and other source materials of interest to cancer research workers. In the
course of time, the problem of separating other associated carcinogenic
chemicals, such as aza heterocyclics, certain primary aromatic amines,
and heterocyclic imines, was also successfully handled by LC. Owing to
the high volatility or gaseous state of other common air pollutants, LC did
not find application in their detection. This position may change now that
high-speed LC has been developed.

Chimney sweep cancer, resulting from occupational exposure to soot,
was first described by Sir Percival Pott [8] in England in 1775. It was not
until 1922 that experimental soot cancer was produced on the skin of mice
by application of an ether extract of soot [9]. The excessive occurrence of
skin cancers among coal-tar and pitch workers was reported in 1892 [10, 11]
and experimentally reproduced in 1915 [12]. Eventually an association
between ultraviolet-fluorescing fractions of tar and their skin carcinogenicity was established [13]. This led in 1933 [14] to the separation of

characteristically fluorescing and carcinogenic 3,4-benzpyrene (benzo-[a]pyrene). The conventional abbreviation for this compound is BaP. The 1,2 isomer[1] was also recovered but was found noncarcinogenic. It is noteworthy that these separations were effected by formation of picrates. With a chrysene fraction, creation of an acidic adduct with maleic anhydride led to the isolation of 1,2-benzanthracene [14], not actively carcinogenic but the parent substance for many carcinogens [15-17].

Clearly, the possibilities for isolating the expected variety of polycyclic hydrocarbons from the complex tar mixtures by chemical means were limited, and it might be suggested that the separation of the many related PAH in tar could not have been accomplished without introduction of liquid chromatography. This development [5, 6] occurred the year after Cook and associates isolated the benzpyrene isomers as picrates [14]. It stimulated research on PAH in products of combustion and the assessment of their carcinogenicity.

II. CHEMICAL CONSTITUTION AND CARCINOGENIC ACTIVITY

The subject has been reviewed by a number of authorities in the field [18-21]. The biological basis for the major studies on carcinogenicity as related to chemical structure has generally been the action on mouse skin. Many factors may influence development of skin neoplasias. As a result, the various systems of grading are not fully comparable [18]. Nevertheless, it has been possible to develop a conception of structure in relation to carcinogenicity. It is important to bear in mind that a total evaluation of carcinogenicity demands assessment by other routes of administration and in a variety of species.

---

[1] These compounds were referred to in Ref. 14 as 1,2-benzopyrene and 4,5-benzopyrene, respectively, in accordance with the Patterson system for the numbering of the pyrene molecule. Subsequently, the older Richter system was adopted and has been widely used up to the present. In 1957, a lettering system was designed by the International Union of Pure and Applied Chemistry, under which system 3,4-benzopyrene (Richter) became benzo[1]pyrene and 1,2-benzopyrene (Richter) became benzo[e]pyrene (BeP). In this chapter, the Richter and IUPAC system are employed as these have been most frequently used in the literature on which the chapter is based.

A. Unsubstituted Polycyclic Aromatic Hydrocarbons

In skin tests of one monocyclic (benzene), one dicyclic (naphthalene), and two tricyclics (anthracene and phenanthrene), results have proved negative. It is, of course, known that inhalation of benzene can cause leukemia in man [22]. Four fused benzene rings can arrange to yield six compounds. Among these, weak activity for 3,4-benzophenanthrene [16, 23] and 1,2-benzanthracene [16] has been claimed. Five rings may be fused to produce 15 entities. The entity 1,2,5,6-dibenzoanthracene was, in fact, the first synthetic PAH identified as carcinogenic [13, 15, 24]. The activities of the 14 remaining pentacyclics were measured by the same investigating group [16]. The most potent carcinogen was 3,4-benzopyrene (BaP). Active but much less potent were 1,2,5,6-dibenzoanthracene, 1,2,7,8-dibenzoanthracene, 1,2,5,6-dibenzophenanthrene, and 1,2,3,4-dibenzo-phenanthrene; and according to Daudel and Daudel [19], so also is picene. Badger [18] reports that the hexacyclics 1,2,3,4-dibenzopyrene and 3,4,8,9-dibenzopyrene have activity. Subsequently shown to be active [25-28] were 3,4,9,10- and 1,2,4,5-dibenzopyrenes. The heptacyclic coronene has been found in air pollution samples and is reported to be inactive [29].

B. Substituted Polycyclic Aromatic Hydrocarbons

Methyl substituents frequently endow carcinogenic activity, depending on the position taken in the hydrocarbon. The subject was exhaustively reviewed by Badger [18]. Higher alkyl substitutions become progressively less effective as the chain lengthens. Carcinogenic methyl derivatives of active PAH compounds have been found [30, 31] as they have for inactive parents [32].

Fluorene is inactive, whereas 1,2,5,6-dibenzofluorene is active. Chol-anthrene is another case of double substitution with a methylene group into benzoanthracene to produce a highly active carcinogen from an inactive parent substance [33]. Among the heterocyclics, the aza compounds with nitrogen substitution generally lack activity but on methylation may yield high-activity derivatives. This is particularly true for the methyl-substituted 3,4-benzacridines [34-36].

III. GENESIS OF THE APPLICATION OF LIQUID CHROMATOGRAPHY IN AIR POLLUTION STUDY

In view of the substantial number of carcinogens found among the PAH's in the 1930-1940 period, attention was directed to their possible role as etiological agents in lung cancer.

The first experimental evidence of a possible link between PAH and the lung was derived from a 1934 study of dust from tarred roads. The dust was claimed by Campbell [37] to produce, besides cancer of the skin, an increased incidence of primary tumors in the lungs of mice. Experiments by other investigators with chimney soot [38, 39], city street dust, and dust from an air-purifying unit [40] confirmed the occurrence of an excess of pulmonary tumors in dust-exposed mice, although the excesses were small. The identification of the carcinogens was not made by these investigators but, in 1946, Hieger [41] exposed nonfluorescent benzene to London air and found BaP contamination. Concurrently it was suggested [42] that coal smoke might account for an observed excess of lung cancer deaths in urban populations over rural. As a result, in 1949 Goulden and Tipler [43] undertook to identify the potent carcinogen BaP in domestic soot. They used column chromatography with alumina for separation and fluorescence spectroscopy for identification. This work constituted the first use of liquid chromatography in an air pollution study.

For almost 20 years following the introduction of chromatographic separation of PAH on alumina by Winterstein and Schön [5-7], the use of the technique was largely for identification of PAH entities in tar-bearing combustion products or in tissues of test animals. BaP became the most frequently sought PAH because it was both carcinogenic and a strong fluorescer.

When interest arose in Great Britain about air pollution as an etiological agent in the genesis of human lung cancer, efforts were begun to develop methods for the quantitative assessment of the PAH content of air. The object of quantitative assessment was to explore geographical, climatological, and technological influences on air concentrations of BaP and to test for correlation with the incidence of lung cancer in population groups.

The first quantitative method used in air survey was developed by the English investigators, Wedgwood and Cooper, in 1951 [44]. This consisted of elution from alumina with cyclohexane and cyclohexane-benzene. The detection method was absorption spectroscopy. As health authorities in other countries undertook air surveys, variations were introduced when inadequacies of the method came to light. The result has been that much published data on PAH levels cannot be cross-compared except possibly in a qualitative way. To improve this situation an intersociety committee in the United States has recently issued standard liquid chromatographic methods [45-47].

According to current medical thought, lung cancer occurs some 20 years or more after onset of carcinogenic exposure. However, it should be noted that in spite of a massive and lengthy analytical effort to catalog and estimate levels of PAH, no correlation has yet been found between the amount of PAH in the air and the incidence of lung cancer [48, 49]. This may be the result of inadequate analytical methodology as related to the selection of the PAH estimated failure to reflect the sum total of carcinogenic potency of the air pollution sampled or because the role of cigarette smoking in the genesis of pulmonary neoplasms has imposed itself in the epidemiological picture [50]. It has also been reported [51] that atmospheric smoke and cigarette smoke exert a complimentary action in cancer production in mice. Furthermore, it has been known since 1945 that noncarcinogenic PAH can inhibit the action of the carcinogenic entities [29, 52, 53]. Finally, it is now known that there is a microsomal enzyme system in the body that provides for metabolism of exogenous chemical stressors at portals of entry to the body. The subject was reviewed in 1972 by Gelboin [54]. For instance, there exists in lung tissue an aryl hydrocarbon-hydrolyzing enzyme activity that can be induced not only by BaP but by other substances entering the body by other routes [55-57]. For these reasons many investigators have estimated airborne PAH entities, both carcinogenic and noncarcinogenic, in anticipation that such information ultimately may become meaningful in terms of the role of air pollution in lung cancer.

IV. LIQUID CHROMATOGRAPHY IN AIR POLLUTION STUDY

There were exhaustive reviews of the PAH analytical literature in 1964 [58] and 1970 [59]. A comprehensive list of PAH entities reported in the literature was assembled and published in 1967, along with a compilation of BaP levels measured in urban atmospheres throughout the world from 1949 to 1963 [60]. A list of PAH compounds found in gasoline exhaust by Hoffman and Wynder [61-63] is shown in Table 1 with levels of skin carcinogenicity. Reference [63] gives the structural formulas for the compounds. Table 2 lists those commonly assayed in air by LC. The extensive range of application of LC to PAH analysis of air pollution and other sources is presented in Table 3.

Four basic steps in LC analysis of air samples for PAH content have evolved: (1) Extraction of PAH from collected particulate matter or other source material; (2) column chromatography of extracts to effect linear adsorption and separation of PAH entities; (3) linear-based elution of adsorbed entities by solvents; and (4) identification and quantitative estimation of material in eluent fractions by ultraviolet absorption and emission or excitation fluorescence.

Tables 4 and 5 show the combinations of procedures that have been used since 1951 [44] to identify and quantitate BaP and associated PAH by LC. Modifications which have been introduced as sources of error have come to light. Improvements in the procedure have been effected but the four basic steps have remained intrinsic to the methodology. In particular it has not been possible to shorten the elution stage (until the recent developments of high-speed LC). Methods eliminating steps [139, 178, 179] have been presented but have not been widely adopted in the field.

TABLE 1

Polycyclic Aromatic Hydrocarbons in Gasoline Exhaust[a]

PAH	Skin carcinogenicity[b]
Benzo[a]pyrene	+++
Dibenzo[a,h]anthracene	+++
Benzo[j]fluoranthene	++
Benzo[b]fluoranthene	++
Benzo[a]anthracene	+
Benzo[e]pyrene	+
Chrysene	+
Alkylbenzo[a]pyrene	?
Alkylbenzo[a]anthracene	?
Alkylchrysene	?
Indeno-(1,2,3-cd)-fluoranthene	?
Benzo[ghi]perylene	-
Indeno-(1,2,3-cd)-pyrene	+
Anthanthrene	-
Benzo[k]fluoranthene	-
Benzo[m,n,o]fluoranthene	-
11H-benzo[b]fluorene	-
Perylene	-
Triphenylene	-
Pyrene	-
Alkylpyrene	?
Fluoranthene	-
Alkylfluoranthenes	?
Coronene	-

[a]According to Hoffman and Wynder [61-63]; by liquid and paper chromatography.

[b]Relative activity on mouse epidermis [62] +++, active; ++, moderate; +, weak; -, inactive.

TABLE 2

Commonly Assayed Polycyclic Aromatic Hydrocarbons in Air[a,b]

PAH	Melting point[c] (°C)	Skin carcinogenicity[d]
Anthracene	217	-
Phenanthrene	100	-
Fluoranthene	110	-
Pyrene	150	-
Benz[a]anthracene	160	+
Chrysene	245	+
Benzo[a]pyrene	179	+++
Benzo[e]pyrene	179	+
Perylene	274	-
Benzo[k]fluoranthene	215	-
Benzo[g,h,i]perylene	273	-
Anthanthrene	257	-
Coronene	435	-

[a]Roughly arranged according to order of elution from alumina using cyclohexane-ether or pentane-ether.

[b]References to presence in air are given by Sawicki et al. [64].

[c]As reported in Refs. 64 and 65.

[d]Relative activity on mouse epidermis according to Hoffman and Wynder [62]; +++, active; ++, moderate; +, weak; -, inactive.

A. Extraction

PAH's are soluble in such solvents as cyclohexane, benzene, chloroform, acetone, and methanol but the rapidity of solution depends on associated material, notably carbon black. Since PAH's are strongly adsorbed by carbon black [148, 149], many hours of solvent extraction are required on samples of airborne particulates and other source material containing carbon black.

Another consideration with PAH-containing mixtures has been to dissolve out the minimum amount of non-PAH material, so as to avoid deleterious effects on column adsorptivity and complications at the later detection and

TABLE 3

Some Applications of Liquid Chromatography to Analysis for
Polycyclic Aromatic Hydrocarbons in Air Pollution and Related Fields

	Period	References (in chronological order)
Urban air	1951-1971	44, 66-88, 64, 89-107, 45-47, 108
Engine exhaust	1954-1971	109, 110, 75, 111, 77, 78, 112, 63, 84, 86, 113, 114, 62, 61, 64, 94, 115-119, 108, 120
Cigarette smoke	1955-1972	121-124, 75, 125-133, 63, 134-139
Coal-tar pitch and fume	1945-1962	140-145
Carbon black	1951-1965	146-151, 65
Fuel	1943-1971	152-157, 75, 158-163, 64, 164-170
Miscellaneous	1951-1971	Asbestos, 171; incinerators, 167; iron and steelwork, 172; kraft mill, 173; meteorites, 174; rubber, 146, 29; waxes, 175, 176

quantitation stage in the form of high background. This has indicated that
solvents of the lowest possible polarity be employed consistent with PAH
dissolving efficiency. As seen in Table 4, cyclohexane and benzene soon
became the extractants of choice for air samples after some early use of
chloroform and acetone. Cyclohexane has been specified for extraction
in two recommended methods adopted by the United States Intersociety
Committee on Recommended Methods [46, 47]. Benzene has been specified
in another [45]. The length of time needed to fully extract samples depends
on sample size and solvent but predominately on associated materials,
particularly carbon black. For some years it has been standard practice
to take air samples on glass fiber sheets through which about 2500 m^3 of
air have been drawn over 24 hr, a procedure first described by Tabor et
al. [180]. The collection will run toward 0.25 g, depending on atmospheric
loading. Whole sheets or disks cut therefrom are subjected to soxhlet
extraction, in thimbles, using 300 ml of solvent.

The first users of cyclohexane for air samples [77, 78, 86] extracted
glass fiber sheets for 3 hr but provided no data on the extent of completeness
of extraction. Lindsey and Stanbury [87] claimed 2 hr was sufficient for
complete removal of soluble matter but Cleary and Sullivan [97] claimed 12

TABLE 4

Extractants for Polycyclic Aromatic Hydrocarbons
from Air Samples

Extractant	Reference and year
Cyclohexane	[77]/1957, [78]/1958, [85]/1960, [86]/1961, [87]/1962, [93]/1964, [96]/1965, [97]/1965, [100]/ 1965, [101]/1966, [102]/1967, [103]/1967, [104]/1967, [105]/ 1967, [46]/1970, [47]/1970
Benzene	[71]/1954, [75]/1956, [81]/1960, [82]/1960, [83]/1960, [84]/1960, [88]/1962, [64]/1962, [99]/1965, [105]/1967, [107]/1968, [45]/ 1970
Acetone	[66]/1952, [68]/1953, [69]/1954, [73]/1955, [79]/1958, [80]/1959, [89]/1962, [90]/1963, [94]/1964, [97]/1965
Chloroform	[44]/1951, [67]/1953, [72]/1955, [74]/1956, [95]/1964, [98]/1965
Benzene-petroleum ether (1:1)	[66]/1952
Benzene-methanol (4:1)	[108]/1971
Methylene chloride	[177]/1967

hr to be necessary, as judged by the disappearance of fluorescence of
successive extracts exposed to ultraviolet radiation. Del Vecchio et al.
[119] specify 20-30 hr. The Intersociety Committee has adopted 24 hr
with cyclohexane for one method [46] and 6-8 hr for another [47]. Stanley
et al. [181] enriched air samples and recovered 76% BaP after 6 hr of
cyclohexane refluxing.

With regard to benzene, Sawicki [58] reported fluorescent material
coming out of samples after 4 hr. Stanley et al. [181] recovered 95%
from BaP-enriched samples in 6 hr. The Intersociety Committee has
adopted 5-6 hr [45] for benzene extraction.

Only in 1967 was a comprehensive study of factors in extraction made
[181], using quantitative measurement methods for BaP, benzo[c]acridine
(BcAcR) and 7H-benzo[d,e]anthracen-7-one (BO) in enriched air samples.

TABLE 5

Liquid Chromatography Systems Used to Analyze for Polycyclic
Aromatic Hydrocarbons in Air over the Period 1951-1971

Column	Eluent	Detector	References (chronologically)
Alumina	Cyclohexane	Absorption	76, 77, 78, 86, 87, 91, 96, 100
Silica gel[a]	Cyclohexane	Absorption	93
Alumina[a]	Cyclohexane	Absorption and fluorescence	85
Alumina[a]	Cyclohexane-benzene	Absorption	44, 67, 68, 69, 72, 73, 74, 79, 80
Silica gel Alumina	Benzene Cyclohexane-benzene	Absorption	84
Alumina	Cyclohexane-ether	Absorption	88[a], 89[b], 90[b], 94[b], 97[a,b], 101
Alumina	Cyclohexane-ether	Absorption and fluorescence	103
Silica gel Alumina	Benzene Cyclohexane-ether	Absorption and fluorescence	102, 46
Alumina-Silica gel[a]	Cyclohexane-ether	Fluorescence	107
Alumina	Pentane-ether	Absorption	95[c], 98[c], 45, 81, 83, 64
Alumina	Pentane-ether	Absorption and fluorescence	82
Alumina	Pentane-ether	Fluorescence	99[c]
Alumina	Petroleum ether-benzene	Fluorescence	66
Alumina	Petroleum ether benzene	Absorption and fluorescence	75
Alumina	Petroleum ether-ether	Absorption	71

TABLE 5. — Continued

Column	Eluent	Detector	References (chronologically)
Activated alumina	Toluene	Fluorescence	104, 47
Activated alumina	Benzene	Fluorescence	105
Alumina	Hexane–benzene	Absorption	63[a], 134[a], 108

[a]Prior to paper chromatography.

[b]Long column.

[c]Prior to thin-layer.

It was clearly shown that extraction efficiency studies are necessary to establish the conditions for use of each solvent. Time, temperature, and method were notable factors determining extraction efficiency for a particular solvent. Cyclohexane and benzene, after 6 hr refluxing, recovered 36% and 63% BcAcR, respectively, and 46% and 42% BO. Chloroform, acetone, and benzene–diethylamine recovered 100% BcAcR in 6 hr. In the light of these observations, speculation is aroused as to whether significant differences in extractability occur as between BaP and other PAH compounds of current interest. If this is so, either recommended extraction times [45-47] may have to be reevaluated to insure that the least extractable entities are efficiently recovered, or correction factors must be assigned.

In view of the long extraction periods necessary to dissolve out PAH from air samples, a recent report on a rapid ultrasonic method is of interest [182]. By this method of extraction, time can be cut to 30 min.

Losses during the extraction process can occur due to decomposition of the PAH fraction if overheating of soxhlet flasks occurs [177]. Reduction in volume or drying of extracts in preparation for chromatography should be conducted under vacuum at or around room temperature. A temperature of 60°C for drying benzene extracts to accelerate the step has been recommended [45]. The use of lower boiling solvents, such as methylene chloride (b. p. 40°C), has also been recommended [177]. Direct transfer of the extract to the chromatographic column is ideal [47]. Since the melting point of BaP is 179°C, and since others of the class have lower melting points, it is obvious that vapor pressures above room temperature are high enough to introduce the risk of losses [91, 183-185]. Collected samples should be extracted without delay [185] but are stable in cyclohexane (and probably in other solvents). Air, light, and smog [70]; ozone [186]; and ultraviolet radiation [187-189] have been shown to be destructive to various PAH entities.

Augmentation of PAH levels of samples can occur due to the widespread dissemination of these chemicals throughout the environment and in reagents. This was recognized early in chromatographic work with PAH compounds [41, 66, 189]. Therefore, microgram detection required that measures be taken to eliminate the possibility of contaminating analytical material. In 1954 Cooper [69] prepared extraction solvents by redistillation until distillate and residue were nonfluorescing under ultraviolet radiation. Then all apparatus was washed with the purified solvent and the washings were checked with ultraviolet radiation. Filter paper, used for collection of particulates from air at that time, was also examined. Soxhlet thimbles were pre-extracted. Comparable procedures were used by later investigators and have been incorporated into the existing standard methods [45-47]. Redistillation of solvents or passage through activated carbon columns is now standard procedure to obtain "clean" extraction material.

It is cautioned [47] that technical cyclohexane produced from benzene is unsuitable as it may contain benzene to the extent of 10%. Muel and Lacroix [190] established a series of sanitary precautions against contamination by laboratory dust for use when extremely low PAH concentrations are being assayed. To offset the effect of unknown losses and gains on accuracy, Hoffmann and Wynder [63] introduced the use of [^{14}C]-labeled BaP as an internal standard in an isotope dilution technique for analysis of cigarette smoke condensate. Recently, this method has been used in auto exhaust analysis by injecting the [^{14}C]-labeled BaP into the exhaust gas [191]. Subsequently, Davis et al. [192] used the noncarcinogenic PAH, perylene, at concentrations around 25 times those found in the materials under investigation, so that the correction factor derived by the procedure would be largely unaffected by the perylene content of the smoke sample.

B. Liquid-Liquid Partition

In 1958, Tabor et al. [180] undertook to characterize urban air pollution in various locations according to organochemical composition, in the belief that the composition might be related to physiological and epidemiological findings. Earlier methods used in other fields were modified [193, 194]. A benzene extract of the air sample is prepared. Partition between ether and water yields water solubles. Ether-hydrochloric acid and ether-sodium hydroxide partition leads to separation of weak-acid, strong-acid, basic, and neutral groups. The neutral group is chromatographed on silica gel with isooctane to elute the aliphatic fraction, with benzene to elute aromatics, and with chloroform-methanol to elute oxygenated compounds, according to the method of Rosen and Middleton [195]. The procedure obviously provides for an aromatic fraction (containing PAH compounds) of high purity and a low background factor in ultraviolet assay. The large sample required and the extended procedure have mitigated against the use of the method in routine air testing. However, the Intersociety Committee has adopted a method for PAH determination [46] that employs silica gel chromatography and elution with isooctane to eliminate aliphatic compounds, then benzene elution to separate the aromatic material.

Liquid-liquid partition found application widely in cigarette smoke research, where the BaP content in the early days was being reported in the range of 1-20 µg per 100 cigarettes smoked [80, 121, 122, 124]. This quantity was too low for the practicalities of the methods of smoke making and the interfering impurities that led to a high background mask in ultra-violet absorption estimation employed in those early days of the field. At that time, 1955, Cooper and Lindsey [121] acid washed and base washed cyclohexane solutions of cigarette smoke condensate to secure a neutral fraction. The procedure was successful and was used by other investigators of the same period [125, 129, 131].

Another early method, extraction of PAH content of petroleum ether-ethyl ether eluates with cold concentrated sulfuric acid [140], was applied to air pollution samples by Kotin et al. [71].

Hoffman and Wynder [134] partitioned the tobacco smoke condensate solution three times with cyclohexane and methanol-water (4:1). Then, cyclohexane extract was partitioned with nitromethane five times to yield 99% of the PAH content. Separation of hydrophilic compounds, separation of paraffins, and enrichment of PAH content in a nitromethane concentrate was accomplished. The method was later used on gasoline engine exhaust with the original subsequent steps of chromatography on alumina, elution with hexane and benzene, and finally paper chromatography [62]. Davis et al. [192] partitioned with lower boiling solvents, hexane and acetonitrile, to facilitate vacuum removal with less likelihood of loss of active constituents.

Liquid-liquid partition has proved effective in enriching extracts in class separation of the PAH group but has not proved selective in terms of individual compounds. Golumbic [196] tried countercurrent distribution with the solvent systems cyclohexane, 80% ethanol-cyclohexane, 98% acetic acid-benzene, 80% acetic acid-isooctane, 87% ethanol-n-heptane, and aniline but results were minimal. Subsequently, Mold et al. [197] attempted countercurrent distribution of benzo[g,h,i]perylene, BeP, BaP, chrysene, and benzo[a]anthracene between cyclohexane and 0.83% tetramethyluric acid (TMA) in methanol-water (9:1 by volume). This procedure was based on an early observation by Brock et al. [198] that improved solubilization of PAH compounds resulted from complexing with purines. Weil-Malberbe [199] later found that TMA gave superior solubilization among several purines tested, including caffeine. Mold et al. [197] did not succeed in attaining a fully discreet separation.

C. Health Hazards of Extractions

The high toxicity of benzene toward the hematopoietic system and its
reported carcinogenicity in man [22] may be mentioned at this juncture.
Use of the compound should really be avoided since its high volatility makes
control of the contamination of the laboratory atmosphere difficult. A
method substituting less toxic toluene has been evolved [104] and adopted
as a standard [47]. Additionally it may be cautioned that all organic solvents are capable of producing narcotic or anesthetic effects on overexpo-
sure. Thus, all solvents should be contained and used under conditions of
good ventilation. Likewise, skin contamination in handling PAH compounds
should be prevented.

D. Column Adsorption Chromatography

Reference to Table 5 establishes that alumina (Al_2O_3) has been the material
of choice for linear adsorption of PAH compounds extracted from air
samples during the past 20 years. It was employed when liquid chromatography was first introduced into PAH analysis in 1934 [5-7], at which
time relationships between chemical structure, physical properties, and
adsorbability on alumina were elucidated. Fifteen years later, alumina
was used in the first quantitative procedure for determining PAH content
of air pollutants [44, 153].

In 1944, silica gel was used, along with alumina, in quantitative analysis
of the tissues of experimental animals for BaP content [200]. Prior chromatography on silica gel served to separate the BaP by adsorption from
some of the ingredients of the highly complex tissue extract that passed the
silica gel column. Silica gel was later used to adsorb the aromatic content
of the neutral fractions of refinery wastes [195] and cigarette tar [123].
Tabor et al. [180] chromatographed the neutral fraction of the organic
material in air pollution samples on silica gel, eluting the aliphatics with
isooctane and thereafter the aromatics with benzene. A quantitative method
for determining PAH compounds in the organic fraction of diesel exhaust
particulates from railway tunnels and urban air [84] employed the same
procedure as Tabor et al. [180] but called for further chromatography, on
alumina, as originally described by Weil-Malherbe [200]. The latter
procedure [84] has lately been adopted by the United States Intersociety
Committee to provide a method whereby 12 PAH compounds can be quantified
in air pollution samples [46].

It is the purpose of the column chromatography of extracts from air
pollution samples to place the PAH compounds on an adsorbent, linearly,
according to chemical structure and to remove them linearly by elution,

as Winterstein and Schön [5] showed to be possible in 1934. Furthermore, since microquantities are involved, the PAH adsorption and elution must be as complete as possible. The separation and recovery process must be as short as possible and the order of adsorption and elution must be as discrete as possible. Fundamental exploration of factors influencing the attainment of these ideals has been relatively limited. Methods tend to be empirically evolved and derivative from early work in the field.

In 1943, Weil-Malherbe [201] studied the factors involved in quantitative chromatographic determination of BaP in nonsaponifiable extracts of animal tissues in order to attain the best possible separation and recovery of the BaP. Activated alumina and activated silica gel were examined in relation to recovery of a quantity of 300 μg BaP. Some of the early observations from this research influenced the design of later liquid chromatography methods for pollution study. Among these observations were the following:

1. The ratio of length to diameter of columns may be varied between 5 and 20 (for the same quantity of adsorbent) without affecting elution. Very short columns (below ratio of five) required excessively large amounts of solvent for full recovery of the BaP. Very long columns (over ratio of 20) filtered very slowly but required less solvent than columns having ratios of between 5 and 20.
2. Particle size 20/30 mesh yielded "tailed" elution curves and required large amounts of solvent, as did very short columns, whereas 100/160 mesh particles gave elution similar to columns having length to diameter ratios between 5 and 20.
3. For column ratios of 10 to 15, the amount of solvent required for recovery of the 300 μg of BaP increased in relation to the amount of adsorbent, whether alumina or silica gel was used. For instance, 1 g of alumina required 80 ml solvent; 2 g required 200 ml; and 5 g required 550 ml. The corresponding threshold volumes (filtrate passing before detectable BaP) were 20, 50, and 170 ml.

The first methods for quantitative determination of PAH compounds in air pollution samples [44, 71, 74, 75, 153] used columns having length to diameter ratios in the 5 to 20 range. In 1962 Cleary [89] assessed the problem of separation of the complex of PAH compounds in air pollution samples and found columns of ratio 52 superior. However it is reported [58] that two to three times as many fractions must be collected, increasing the analysis time. In the 1970 standard methods [45-47] lengths of adsorbent are not given but tubes of ratio 25-30 are specified. Particle size for alumina or silica gel in the range 100-200 mesh has been most commonly employed since the inception of the field as related to air pollution assessment.

E. Activity of Adsorbents

In 1940, Trappe [202] noted that combination of highly active adsorbents and strongly polar solvents created conditions for possible chemical change, whereas weak adsorbents (partially deactivated silica gel) in combination with solvents of low polarity were inactive chemically. The following year Brockmann and Schodder [203] described improved separation of some PAH compounds using partially deactivated alumina. This was first attributed to occupation of the most active adsorption centers by water [204], resulting in more uniform adsorption [205], but more recently Snyder [206] has concluded that it results from removal of strong sites as sites on a distribution tail to give a closer approach to isothermal linearity.

Early methods for quantitation of PAH compounds in air pollution involved the use of activated alumina with cyclohexane [153] or petroleum ether-ethyl ether [71]. In 1956, Lindsey et al. [76] introduced standardization of alumina in chromatographic separation of polycyclic aromatic hydrocarbons by equilibration over 70% sulfuric acid, with cyclohexane as eluent. Subsequent methodology provided for the use of deactivated alumina prepared either by equilibration over sulfuric acid or by deactivation through the addition of known amounts of water to active material. This technique was first described by Muller [207] and incorporated into the method of Sawicki et al. [81], where alumina of 13.7% water content was used with pentane and ether. The latter method was adopted in a 1970 standard [45].

Weil-Malherbe [200] drew attention to differences in adsorbability of different batches of activated material. Cahnmann [208] washed silica gel successively with acetone, methanol, and water before gel activation. Sawicki [81] washed alumina with ether. The importance of establishing the chromatographic characteristics of each batch of deactivated alumina has been stressed [45], particularly with respect to variations in the order of elution of PAH compounds. For instance, the R_f value, as defined by Consden et al. [209], can be used to follow batch variations using BaP [208].

The quantitative effect of deactivation was assessed in detail by Cahnmann [208] using silica gel with cyclohexane and cyclohexane-benzene to recover BaP from a solution containing 1.25 μg/ml. The chemical was estimated by recording spectrophotometry at 296 nm. The findings were presented in such a way as to be comparable with the Weil-Malherbe study on activated silica gel [201]. Increasing water content of the gel increased rate of movement of the chromatographic zone. It was found that the BaP zone was narrower and better defined with gel containing 3% water than with fully activated or strongly deactivated gel (18-24% water). Similarly, less eluent was required at 3% water content. "Tailing" began to appear with gel over 12% water content. This effect can also be corrected by adding polar solvent in limited quantity to nonpolar. It was established that the rate of movement of the BaP zone was similar among the following combinations: silica gel of 14-15% water content with cyclohexane; silica gel of 6% water content with cyclohexane and benzene (8%); fully activated gel and cyclohexane containing 16% benzene. It is clear from these findings how adsorption elution chromatography depends on state of activation of adsorbents and polarity of solvents used for elution. Finally, Cahnmann drew attention to the fact that partial deactivation of adsorbents makes it possible to employ nonpolar solvents, for instance pentane and cyclohexane, which do not absorb ultraviolet radiation and thus permit direct spectrophotometry with no need to evaporate polar solvents and redissolve fractions into a nonabsorbing medium. Ether, used in small quantities to improve eluting power, is also nonultraviolet absorbing.

In the studies of Weil-Malherbe [201] and Cahnmann [208], using simple models with adsorbent and BaP, negligible amounts of the chemical were retained by the adsorbent. When complex extracts of air pollution samples are handled, column retention is substantial, around 25% [177]. As a result, columns must be calibrated in this respect for each batch of adsorbent. At the same time the solvent composition (for instance, cyclohexane-ether) required for elution of each hydrocarbon is established. The 1970 standard method [45] details a procedure for column calibration with individual hydrocarbons, synthetic mixtures, and actual air pollution extracts that include normally occurring interfering substances. The alternative to determining column retention factors is to use a $[^{14}C]$-labeled internal standard [63] or perylene [192].

The adsorptives that can be handled by the adsorbent charge in columns while retaining linear capacity bear a relationship to the size of the charge. This has been clearly established experimentally using BaP [201]. More recently Snyder [206] calculated, on the basis of observations on pentane elution of 1,2,3,5-tetramethylbenzene from alumina, that the maximum allowable column loading to retain linear capacity would be $10^{-4}M$ for elution from alumina. Hence, the specifications of measurement methods for PAH compounds have had to be developed so as to prevent the likelihood of PAH overloading.

Another overloading factor relates to non-PAH constituents of air pollution extracts having first order of adorption on adsorbents. These include aliphatics, dicyclics, and possibly other material as yet unidentified. The prechromatography of extracts on silica gel [84] has been provided for in a recent standard method [46] and has as its purpose the elimination of nonaromatic constituents to the greatest extent possible. The threshold volume (volume before appearance of PAH entities) showed aliphatics and dicyclics in another standard method [45], in which prechromatography on silica gel was not done. Clearly, column overloading with aliphatics, dicyclics, and other non-PAH material could also occur in circumstances where the alumina charge was inadequate. This would take place at the expense of PAH adsorption from extracts being chromatographed. Hence, determination of the appropriate adsorbent charge for the type of samples being handled is necessary.

F. Chemical Structure and Adsorbability

Fundamental studies on the structure of PAH compounds in relation to adsorbability have been made by Winterstein and Schön [5] and by Klemm et al. [210, 211] to elucidate the factors underlying the order of adsorption and elution (the same) observed for the PAH. These investigators, among others [69, 81], established that order is reproducible. It was later established as a general rule — the constant eluent noncrossing rule — that for a given eluent, order was the same as adsorption activity varied [212]. Differences in order of elution have been observed between cyclohexane and benzene [80] and pentane with ether [81].

It was noted by Winterstein and Schön [5] and confirmed by Klemm et al. [211] that adsorbability increases with the number of double bonds (so-called "unsaturation rule"). The latter investigators also established the validity or rationale of the following generalizations:

1. Coplanarity rule. It was shown in an earlier study [210] from the same laboratory that conjugated isomeric biaryls, arylalkenes, and conjugated iso-π-electronic (not isomeric) aromatic hydro- carbons, adsorbability of the most nearly coplanar was highest. Adsorbability of trans isomers was higher than cis. This rule was expanded by the newer examples studied [211].

2. The symmetry rule (adsorbability proportional to symmetry number) considers condensed molecules as two dimensional. This rule was not found dominant in adsorbabilities of six possible benzenoid tetra- cyclics.

3. It had been reported [5] that adsorbability for a linearly condensed arene (acene) is greater than for an angularly condensed arene (phene) or a cata-condensed compound with the same number of rings. This was called the "acene rule."

4. The adsorbability findings on six possible benzenoid tetracyclic compounds, napthacene, chrysene, benzo[a]anthracene, triphenylene, benzo[c]phenanthrene, and pyrene, led to the speculation that the order of rule precedence was unsaturated, coplanarity, and symmetry. The acene rule was viewed as a corollary.

5. It can be speculated, on the basis of the findings, that among isomeric unsaturated hydrocarbons differing only in degree of conjugation, the extent of conjugation determines the most adsorbable.

6. Alkyl or alkylene substitution increases adsorbability so long as steric hindrance to flatwise adsorption of the arene moiety to a surface does not occur.

7. Steady state processes were occurring on their alumina columns as evidenced by the complete conformity of the PAH compounds used to the law of inequalities (if adsorbability of A>B and B>C, then A>C).

8. The chromatographic process observed in their experiments featured a reversible steady state occurring between monomolecularly adsorbed compounds on the adsorbent surface, in π-type (outer or charge transfer) complexing, and molecules of the compounds in the eluent.

The significance of these advances in knowledge of the fundamentals of PAH adsorbability on alumina will undoubtedly influence future analytical developments, and possibly the conceptualization of the basis of the carcinogenic activity of this class of chemicals.

G. Miscellaneous Adsorbents and On-column Complexing

Some column adsorbents other than silica gel and alumina have been used for PAH separations. Lijinsky [213] used magnesium oxide and Celite for separating the PAH content of waxes. The mixture had been found to permit easier filtration [214]. A powerful adsorbent for tetracyclic and pentacyclic aromatic hydrocarbons was needed since large volumes of benzene were used to wash the wax through the columns. Magnesium oxide retained these compounds even on elution with the polar solvents acetone and alcohol. Therefore, the magnesia was dissolved by dilute acid and the PAH adsorptive eluted easily from the residual Celite.

Tye and Bell [215] found pyrene and fluoranthene well separated by liquid-liquid chromatography of s-trinitrobenzene complexes developed on Columnpak (Fisher Scientific Co.) ground to 100/200 mesh. Acetylated cellulose has proved effective in separating complex PAH mixtures [216] and phenanthrene-phenylphenanthrene mixtures [217]. Sawicki [218] used the π-complexing agent 2,4,5,7,-tetranitrofluorenone in an amount of 15 mg/g of alumina to obtain BaP minus its usual chromatographic associate BkF. Buu-Hoi and Jacquignon [219] exploited the affinity of anhydrides and imides of tetrahalogenophthalic acids for aromatic and heterocyclic polycyclic compounds by column chromatography on these anhydrides and imides.

Porapak T (a porous polymer based on ethylvinylbenzene) has proved effective in separating an extensive range of PAH compounds and their partially hydrogenated derivatives when used with n-hexane [220].

H. Eluents for PAH Compound Separation

Sawicki [58] reviewed the extent of use of various eluent solvents in LC of
PAH compounds occurring in extracts of air pollution samples. A further
review is presented in Table 5 and indicates that cyclohexane, cyclohexane
and benzene, cyclohexane and ether, n-pentane, n-pentane and ether have
been used for eluting PAH compounds from the favored adsorbent alumina.
The rationale underlying the superior effectiveness of these eluents relates
to the low polarity of the primary eluents and to the power of the secondary
constituents, benzene or ether, to develop the chromatographic separation
by successive increases in their proportionality. This aspect has been
mentioned in the discussion of deactivated adsorbents in Sec. IV, E. To
restate the position, chromatographic separation depends, apart from the
deactivation level of the adsorbent, on the polarity of the eluting solvent.
Various combinations of adsorbent and eluotropic power can be equally
effective.

Trappe [202] demonstrated in 1940 that an eluotropic (or developer) series
could be set up for separating biological fats from natural mixtures. In
1948, Smit [221] effectively separated a complex mixture of pure hydrocarbons (hexadecane, dodecylbenzene, α-propylnaphthalene, and 1, 2-
diphenylpropane), using silica gel and an eluotropic series consisting of
pentane, carbon tetrachloride, and chloroform. In another set of experiments
Smit [222] used the method on lubricating oil distillation fractions and
effected separations in order of elution as follows: with pentane paraffins
and polycyclic naphthenes; with carbon tetrachloride, mainly dicyclics;
with chloroform, polycyclic aromatics; with ether, to elute a remaining 10%
of the oil, sulfur compounds and other non-hydrocarbons. In 1954, Cooper
and Lindsey [67] used an eluotropic series consisting of cyclohexane and
cyclohexane-benzene in increasing proportions of benzene. The same year
Kotin et al. [71] evolved a series with petroleum ether and ethyl ether.

There are practical aspects to consider in selecting eluents. The analytical time factor is involved. Inept selection of eluents can lead to "tailing"
and increase greatly the volume of the eluent that must be passed to effect
high efficiency of desorption [208]. Also, the volume of solvent preceding
appearance of active material in column effluents, so-called "threshold
volume" [201], will be greatly augmented or diminished according to the
desorbing power of eluents [125]. The sharpness of the chromatographic
zones, basic to separation of mixtures, is influenced by eluent composition
[208]. The wrong composition can lead to eluent fractions that are inter-
constituent contaminated, which may result in detection and quantitation
problems. Cahnmann [208] has noted that most polar solvents absorb
ultraviolet radiation in the range used for detecting PAH compounds (220-
420 nm). In the early days such solvents, of which benzene is an example,
were used with activated adsorbents. These solvents had to be evaporated

and the fraction residues redissolved in low polarity solvent not absorbing in the critical range. This increased analysis time and introduced the possibility of losses of the compounds being estimated. With deactivated adsorbents, elution can be effected with low polarity material, such as cyclohexane and pentane, in combination with nonultraviolet absorbent ethyl ether as secondary eluent. Thus, qualitative absorption spectroscopy can be applied directly to eluent fractions. Sawicki et al. [58, 223] have drawn attention to the superiority of pentane, pointing to its low blank in fluorimetric analysis, and to advantages in relation to absorption spectroscopy.

In the LC procedures for separating PAH compounds from air pollution extracts, it is necessary to collect large numbers of eluent fractions if the compounds listed in Table 2 or as embraced in the recent standard methods [45, 46] are to be well enough separated for quantitation. For instance, the standard method [45] anticipates 36 tube collections (20 ml each) using for elution pentane and pentane-ether mixtures increasing to 25% ether. The extent of separation is shown in Table 6. It is evident that discrete separation of the pure compounds is not effected. However, quantitation of individual PAH compounds by ultraviolet absorption spectrometry is possible and is discussed later in Sec. IV, J.

In another standard method [46], using cyclohexane and ether on alumina, sharper separations are sought by prechromatography on silica gel with isooctane for removal of aliphatics, dicyclics, and other loading entities [84]. Around 70 fractions of 8 ml each are taken. Solute elution positions are claimed to be accurate to several milliliters. The same general grouping of PAH compounds and the same order of elution is attained as in the method given in Ref. 45.

In spite of the trends to deactivated alumina combined with low polarity of eluents, Dubois et al. [104], with activated alumina and polar toluene, succeeded in separating BaP and BkF as a pair in several eluent fractions from typical air pollution extracts. This permitted quantitation by spectrofluorescence through the elimination of other strong fluorescers. This method was adopted by the Intersociety Committee [47].

The standard methods [45-47] are the culmination of many years investigation throughout the world. The degree of separation attainable (see Table 6) is not discrete, compound by compound. However, the use of ultraviolet absorption and fluorimetric spectroscopy for identification and quantitation has largely obviated the need for chromatography that separates pure PAH compounds from mixtures. Nevertheless, obstacles have been encountered in spectrometry due to similarities in the spectra of certain members of the PAH class. Ingenious methods have been worked out over the years to overcome these obstacles and are discussed in Sec. IV, I. Concurrent with the spectrometric research other investigators have attempted to effect discrete separation, some by use of paper, thin-layer, and gas chromatography (not the concern of this chapter) and others by chemical complexing to improve the sharpness of chromatographic separation. The attempts with countercurrent distribution have been mentioned [196, 197].

TABLE 6

The Eluent Fractions and PNA Content for Determination of
Eleven PAH Compounds in Air Pollution Extracts[a]

Eluent (% ether in pentane)	Eluent tube numbers[b]	PAH content
0 and 3	1-8	Aliphatic and small cyclics
6	9	Pyrene
6	10	Pyrene, fluoranthene
6	11-13	Nil
9	14-15	Chrysene, benzo[a]anthracene
9	16	Chrysene
12	17	Nil
12	18	BaP, BeP, benzo[k]fluoranthene
12	19-21	BaP, BeP, benzo[k]fluoranthene, perylene
12	22	BaP, benzo[k]fluoranthene
15	23	Nil
15	24-28	Benzo[g,h,i]perylene, anthanthrene
15	29	Benzo[g,h,i]perylene
⎡18⎤	30	Nil
⎣25⎦	31-35	Coronene
	36	Nil

[a]According to the United States Intersociety Committee standard method
[45]; adsorbent, alumina, 13.7% water content; detection by ultraviolet
absorption spectral measurement on tube contents taken to dryness and
redissolved in 3 ml of pentane.

[b]Collection volume 20 ml per tube.

Tye and Bell [215] evolved a liquid-liquid chromatographic technique based
on formation of complexes with s-trinitrobenzene (TNB) on columns charged
with Columnpak and 0.25M TNB in polyethylene glycol (Carbowax 400) as
stationary phase, and a mobile phase of isooctane saturated with the TNB in

Carbowax 400 solution, so that the stationary phase was not depleted. The rationale was the knowledge that the complexes were more polar with a corresponding increase in retention volume and with differences in distribution coefficients substantially higher than the uncomplexed material. By this method, pyrene and fluoranthene, not separated by conventional column chromatography (Table 6), came off the TNB column in two discrete fractions.

In 1967, Rothwell and Whitehead [224] separated purine complexes of PAH compounds by applying a potential of 1200-1500 V to silica gel columns, using an eluent having the composition 2 g purine in 90 ml water, 10 ml ethanol, and 2 ml ammonia. This method was based on the conception advanced by Liquori et al. [225] that a dipole-induced dipole interaction occurs between the polar purine (caffeine) and the polarizable PAH.

In 1955 a method [156] for separating BaP from other PAH material in high boiling products of petroleum operations was designed to follow conventional chromatography on alumina by catalytic iodination on activated alumina, producing 6-iodobenzo[a]pyrene. The iodo derivative was quantitated by ultraviolet absorption spectrometry, without some of the interferences besetting the conventional methods of the period.

Spotswood [216] followed alumina column chromatography using n-hexane and benzene, with chromatography on long (70 × 2.5 cm) columns of partially acetylated cellulose powder (24% and 32% acetyl content). The same solvents as in acetylated paper chromatography [226] were employed, viz., ethanol-benzene-water (17:4:1 by volume), ethanol-toluene-water (17:4:1 by volume), or methanol-toluene-water (10:1:1 by volume). It was possible to separate complex PAH mixtures mostly into single PAH entities. Not more than two components occurred in eluent fractions. The latter component pairs were not reported. However complete resolution of BaP from perylene, BeP, chrysene, and BkF is claimed.

R_F values were determined and were found to vary between the three solvent systems. The order of elution was the same but differed from that occurring in the conventional prechromatography. It was observed that the ethanol constituent provided a solvent of greater eluting power than that containing methanol. The methanol constituent improved resolution of some compounds of high R_F value on columns of the lower (24%) acetyl content. For instance, the following range of R_F values was found with the ethanol-toluene-water system and a column of 28% acetylated cellulose power:BaP, 0.093; anthanthrene, 0.163; chrysene, 0.206; fluorene, 0.401; coronene, 0.410; penanthrene, 0.501; and acenapthylene, 0.576. Water, as a constituent, increased rate of passage of compounds through the column and maintained sharply defined bands. "Tailing" occurred without water.

The Spotswood method was successfully employed by Beckwith and Thompson [217] to resolve and quantitate mixtures of phenanthrene, 1-phenylphenanthrene, and 9-phenylphenanthrene.

It will be noted that for all the previously mentioned separations, substantial amounts of eluents are passed through columns in the course of an analysis. Eventually the eluent has to be evaporated in preparation for quantitation by ultraviolet absorption. The occurrence of impurities in solvents used for elution is therefore important, since their concentration is inherent in the procedure. PAH compounds pervade the environment and must be expected to occur in laboratory reagents and to contaminate at all open analytical stages [41, 69, 190]. Trace contamination cannot be ignored in analysis at the microgram level of quantitation, which is involved in work with air pollution extracts. The occurrence of other impurities in eluents must also be considered. As a result of these problems it has long been the practice in LC of PAH-containing material to subject the eluent solvents to purification by redistillation or passage through activated carbon columns as previously described for extraction solvents.

I. Identification and Quantitation of PAH Compounds in Chromatographic Eluent Fractions

This section of the chapter deals with the ultraviolet absorption fluorescence and phosphorescence spectra in PAH detection and quantitation. Spectrometry has been the only quantitative measurement method that has proved practical with the simplified mixtures of PAH compounds that adsorption-elution chromatography has been able to produce from the complex of substances in extracts of air pollution samples. Attempts to exploit classical properties, such as melting points and refractive indices, have been defeated by the inability of liquid chromatography to isolate pure substances of the PAH class. R_F determination has been largely inadequate in LC except perhaps for the separations attained by Spotswood [216] with columns of acetylated cellulose powder and the solvent complexes he used in an attempt to exploit the superior separating power of acetylated paper [226]. The LC application of this technique evolved in 1960 has not been taken up in the field of air pollution analysis. Spectrometry has lasted as the method of choice since the first PAH air pollution assays in 1951 [44].

J. Ultraviolet Absorption Spectrometry

The characteristic spectra for PAH compounds were identified well before the first efforts to assess the types and amounts of those chemicals in air pollution samples [80, 223, 227-232]. The spectra embraced the wavelength range of roughly 220-420 nm. However, below 300 nm, background absorption interferes with recognition by overwhelming the characteristic peaks [153]. Also, peaks over 300 nm are more characteristic and absorption is stronger, providing for peaks that stand out more clearly against background.

Table 7 shows ultraviolet absorption peaks for the PAH compounds listed in Table 2. As pointed out in Sec. IV,H on eluents, strongly polar solvents absorb within the PAH range. Therefore, absorbance characteristics came to be determined in cyclohexane [69, 86, 153] or in even less polar

TABLE 7

The Ultraviolet Absorption Spectra of PAH Compounds in Cyclohexane and Pentane Solutions

PAH compound	Peaks for identification[a] (as wavelength, λ, in nm)		Recommended wavelength for quantitation[d]
	PAH in Cyclohexane[b]	PAH in Pentane[c]	
Anthracene	253, 325, 340, 357, 376	<221, 246, 252, 290S, 308, 322, 334S, 338, 351S, 355, 368, 374	374
Phenanthrene	252, 275, 282, 293	<221, 244, 251, 258S, 264S, 267S, 275, 281, 290S, 293, 310, 314S, 317, 324, 331, 339, 347	293
Fluoranthene	237, 254, 259, 263, 272, 277, 282, 288	230S, 235, 243S, 248S, 252, 258, 262, 264S, 271, 276, 281, 287, 309, 323, 341, 351S, 358	287
Pyrene	273, 306, 320, 335	231, 237S, 240, 252, 261, 272, 293, 305, 318, 330S, 334, 351, 357, 361, 365S, 371	287
Benzo[a]anthracene	250, 255, 260, 269, 278, 289, 300	<222, 227, 242S, 250S, 255, 267, 276, 287, 297S, 299, 314, 325, 340, 358, 364, 374, 382	287

Compound			
Chrysene	269, 282, 294, 306, 320,	241, 249S, 257, 267, 282, 294, 306, 320, 333S, 343, 350, 352, 360	320
Benzo[a]pyrene	266, 273, 285, 297, 331, 347, 364, 385, 404	~226, 254, 265, 271, 283, 296, 313, 330, 345, 362, 377, 379, 382, 391, 401	382
Benzo[e]pyrene	257, 267, 278, 290, 317, 332	236, 256, 266, 277, 289, 302, 305, 316, 320S, 327S, 331, 348, 356, 358, 365	331
Perylene	253, 339, 355, 388, 409, 437	~226, 238S, 244, 252, 262, 291, 365, 385, 406, 410S, 429, 434	434
Benzo[k]fluoranthene	Not reported	237, 247, 268, 283, 296, 308, 321S, 336S, 341, 352, 358, 370, 377, 383S, 392, 401	—
Benzo[g,h,i]perylene	249, 255, 261, 276, 289, 300, 313, 392, 345, 363, 384	~222, 230S, 245S, 253, 269, 276, 287, 299, 312, 324, 329, 344, 361, 377S, 382, 390, 394, 405	382
Anthanthrene	233, 259, 294, 307, 380, 401, 406, 422, 430	234, 242S, 246S, 255, 258, 271, 283S, 294, 307, 321S, 365S, 374S, 378, 382, 393S, 399, 405, 414, 421, 429, 434S	420

TABLE 7. — Continued

PAH compound	Peaks for identification [a] (as wavelength, λ, in nm)		Recommended wavelength for quantitation [d]
	PAH in Cyclohexane [b]	PAH in Pentane [c]	
Coronene	302, 317, 324, 334, 339, 345	~251, 267, 274S, 278, 285S, 289, 292, 297, 301, 308S, 312S, 316, 323, 328, 332, 338, 344	338

[a] Strong absorption wavelengths are underlined.

[b] From Ref. 80; instrument, Beckman Dk2 ratio recording spectrophotometer.

[c] From Ref. 223; S = shoulder; instrument, Cary recording spectrophotometer, model 11.

[d] From Ref. 45; Solvent, pentane; BkF quantitation can be approached by using $\lambda = 401$. Bap also absorbs at this wavelength. Its contributions must be subtracted.

pentane [81, 82, 223]. Sawicki [58, 223] showed that absorption spectra in pentane were sharper and revealed more fine structure than in cyclohexane (Table 7). For instance, a triplet at 377, 379, and 382 nm was discernible.

When Tables 6 and 7 are compared, it can be seen that the chromatographic separation of the various PAH compounds in air pollution extracts simplifies the use of the ultraviolet absorption spectra, which clearly overlap for some of the compounds. For instance, the close similarity of the bands for fluoranthene and benz[a]anthracene does not present an identification problem, because these two compounds are well separated by chromatography. Quantitation of both can be effected using the strong absorption that each displays at 287 nm. In contrast, BaP and BkF chromatograph together (Table 6) and were found in 1960 to interfere at 382 nm [83, 105], the current peak of choice for BaP estimation [45]. Their interference at 401 nm, the peak of choice for BkF estimation, was found the same year [83]. These discoveries have engendered concern about the reliability of BaP and BkF estimations in the literature. Ramifications of this situation yield to adjunct fluorimetric analysis under certain circumstances [82, 104, 223, 233]. This is discussed in Section IV, L on fluorescence spectrometry.

Attention has also been drawn to the occurrence of the so-called "bathochromic" effect detectable in the spectrograms of some air pollution extracts [102]. Methyl derivatives of the PAH compounds pyrene, chrysene, and fluoranthene have absorption spectra closely similar to the parents but shifted higher, although so slightly as to constitute a possible source of error in estimating the parents. Although most recording spectrometers are accurate to a few Ångstrom units, investigators are cautioned that the shifts (84 Å maximum in the case of 4-methylpyrene) may be overlooked [102].

K. Quantitation of PAH Entities by Ultraviolet Absorption Spectrophotometry

Apart from the above-described difficulties arising in differentiation of some PAH entities by ultraviolet absorption spectrophotometry, Table 7 shows that it has been possible to select specific wavelengths for quantitation. This is because of the relative effectiveness of the chromatographic separation process.

Quantitation requires that there be a standard against which to compare the absorbability of eluent fractions in respect to peaks identifiable with the particular hydrocarbon for which quantitation is sought. The procedure to effect quantitation involves a calibration with each pure PAH compound, whereby absorbance of a solution of known concentration is measured at a wavelength at which the compound has been found to absorb strongly, and if possible uniquely, in relation to other PAH entities. Since the concentration in air pollution extracts is at the microgram level, the calibration of the spectrophotometer is made with solutions containing micrograms per cuvet filling (around 3 ml).

To calibrate, the peak height at the selected wavelength caused by the absorbance of the calibrating solution must be determined so that the index, micrograms substance per unit of peak height, can be calculated. Frequently contiguous peaks overlap so that the peak for measurement does not have a horizontal base. Furthermore, with test samples it is found that the absorption peaks lie on top of a substantial background absorbance, contributed to by non-PAH material that the column adsorbent failed to retain during elution. As pointed out in Sec. IV,D, prechromatography with silica gel reduces background (compare standard methods [45, 46]) which appears to consist of aliphatics, dicyclics, oxygenated material, and aromatics of petroleum origin [234]. In light of the foregoing, Cooper [69] evolved the so-called "baseline" technique. This provides for drawing the peak baseline between the wavelength at either side of the peak base. The peak height is then measured perpendicularly from the center of the baseline (usually sloping). Baselines ran to 20 nm with the original hand-operated equipment used by Cooper [69] but were reduced to 8 nm when recording equipment of higher resolution was introduced [78]. This had the effect of reducing spectral interference by other material. In the range of absorbances employed, the Beer-Lambert law applies and linearity of background absorption is assumed for the baseline wavelength intercept.

The quantitation procedure for the pertinent PAH compound can be summarized as follows:

1. Solution of the PAH from all eluent fractions containing the substance into the cuvet volume of cyclohexane or pentane.
2. Absorbance calibration (micrograms pure material per unit of peak height) multiplied by peak height of the sample being analyzed. This gives the micrograms of hydrocarbon found.
3. Correction for incomplete column elution. This is required [45, 177].
4. Correction for incompletion extraction. Some aspects have yet to be elucidated [181].
5. Calculation of the concentration in the extract, in the original solid before extraction, and in the volume of air filtered.

It seems evident that there must be a substantial accumulation of possible errors in this quantitation procedure unless, of course, internal standards are used [63, 192]. The reliability of this additional step in the analysis is complicated [63] and it has not been established that procedural losses of all PAH entities are the same.

Sensitivity of ultraviolet spectrophotometry (limit of detection) as stated previously is at the microgram level: for BaP, 2.1 μg per 3 ml (a cuvette volume), least sensitively estimated; for BghiP, 0.03 μg per 3 ml, most sensitively estimated [235]. The extent to which sensitivity of the complete treatment depends on the concentration in the original sample has been estimated [181]: pure BaP, 2 μg; for benzene-soluble extracts containing 500 μg/g, 20 μg; for particulate matter containing 50 μg/g, 240 μg; and for air containing 5 μg BaP per 1000 m^3, 2000 μg.

Accuracy of the spectrophotometers is high compared to that of extraction and chromatography. In the final analysis, it must be taken into account that the significance of the estimations rests on the extent to which chemical levels correlate with biological effects, and this is far from being understood.

Because the ultraviolet absorption method depends on comparison of the absorbances of unknowns and pure substances, the purity of the PAH material for calibration purposes is crucial. Some can be obtained commercially but require recrystallization. For others, synthesis is necessary. Brown and Kelliher [236] discussed the role of reference compounds in the study of PAH content of automobile exhausts. They acquired 37 examples and found 32 of above 95% purity. It was noted in 1970 [46] that BkF was one example not commercially available. Methods of synthesis [237, 238] were mentioned. Obviously, the purity of the standards can greatly influence the accuracy of estimations.

L. Fluorescence Spectrometry

Since PAH compounds fluoresce under ultraviolet radiation, the phenomenon was used for identification and quantitation before ultraviolet absorption spectrometry was introduced. Hieger [13] observed in 1930 that for a number of PAH compounds, there were characteristic banded fluorescent spectra. Shortly thereafter, Cook et al. [14] identified BaP in gas works pitch by fluorimetry. Weil-Malherbe [200] estimated BaP in mouse tissue fluorimetrically with a photocell spectrometer after column chromatography of tissue extracts.

The first use of the fluorescence spectra on air pollution material was made in 1946 by Hieger [41], who exposed nonfluorescing benzene to London air and identified BaP in the solvent. Subsequently, Goulden and Tipler [43] separated BaP from domestic soot by LC and used the fluorescence spectrum for quantitation. BaP in British air pollution samples was estimated by Waller [66] in 1952 using spectral photography of the fluorescence. Gurinov et al. [239] in the U.S.S.R. and, later, Kuratsune in Japan [75] used the same method for air pollution samples. Apparently instrumental inadequacies and concentration effects (quenching) at the period diverted air pollution investigators to ultraviolet absorption.

Fluorescence spectra for many PAH entities were provided by the work of Berenblum and Schoental [240] and Schoental and Scott [241]. These workers used mercury vapor lamp radiation with appropriate glass filters to secure exciting bands. Fluorescence bands were photographed. When commercial automatic recording spectrofluorimeters became available a decade later, Van Duuren expanded knowledge of fluorescence of the PAH class through work on cigarette tar [128, 131, 135]. His fluorescence findings were summarized in 1960 [242]. Concurrently, Sawicki et al. [223] examined aspects of fluorescence relating to its use for air pollution analysis, in particular respect to its usefulness where ultraviolet absorption was then

TABLE 8

Ultraviolet Excitation and Fluorescence Maxima for Some Polycyclic Aromatic Hydrocarbons in n-Pentane[a]

PAH	Ultraviolet absorption	Excitation	Fluorescence	Fluorescence intensity (K_Q)[b]
Anthracene	252, 338, 355, (374)[c]	249, 334, 350[d], 370	376, 398, 420, 447	1.64
Phenanthrene	251, 275, 293, (347)	252, 274, 282	348, 362, 382	0.52
Fluoranthene	235, 287, 334, 351, (358)	238, 283, 340, 354	445, 464 (equal intensity)	0.68
Pyrene	240, 272, 318, 334, (371)	239, 269, 314, 330	382, 392	0.53
Benzo[a]anthracene	287, 325, 340, 358, (382)	284, 327, 338, 352	382, 406, 430	0.87
Chrysene	267, 282, 306, 320, (360)	264, 282, 303, 315	362, 381, 401, 427	0.81
Benzo[a]pyrene	296, 345, 362, 382, (401)	295, 345, 361, 380	403, 427, 454	4.1
Benzo[e]pyrene	289, 316, 331, (365)	290, 315, 329	389, 398, 409	0.48
Perylene	252, 406, 429, (434)	252, 408, 430	438, 464, 497	20

Compound	Absorption maxima[c]	Fluorescence maxima[d]	Intensity[b]
Benzo[k]fluoranthene	247, 296, 352, 377, (401)	245, 302, 354, 372, 397 400, 428, 457	13.4
Benzo[g,h,i]perylene	287, 324, 344, 361, 377S[e] (405)	292, 324, 340, 354, 374 393, 404, 419, 440	0.67
Anthanthrene	258, 307, 382, 399, 421, (429, 434S)	257, 304, 380, 400, 420 430, 457, 487	18.5
Coronene	292, 301, 338, (344)	293, 302, 337, 378 440, 450, 460, 480, 490, 510	0.63

[a] Selected from data reported by Sawicki et al. [223]. Comparable data may be found in Refs. 241 and 242.

[b] Fluorescence intensity of pertinent band relative to a quinine standard.

[c] Longest wavelength absorption maximum.

[d] Underlined values represent highest intensity.

[e] S = shoulder.

proving inadequate (notably for determination of BaP in the presence of
BkF). With the automatic recording equipment, fluorescence spectra may
be identified for PAH material in far lower concentration than is possible
with absorption spectrometry [223, 242].

Two possibilities for identification and quantitation arise out of the fluorescing property of PAH compounds: excitation spectra and emission
(fluorescence) spectra. These are shown in Table 8 for the PAH compounds
of current interest in the air pollution field. The excitation and fluorescence
bands are characteristic (when the same solvent is used) but intensities of
fluorescence vary according to excitation wavelengths. Hence there is the
possibility of selecting combinations designed to overcome interferences.
Sawicki et al. [223] have pointed out that compounds should be analyzed at
excitation wavelengths, for which mixture associates are only weakly
activated.

Schoental and Scott [241] found, for many members of the PAH class,
an approximate mirror image relation between the shortest wavelength
fluorescence spectrum and the longest wavelength ultraviolet absorption
maxima which had been noted for other chemicals. This relationship appears
in data reported by Van Duuren [242] and Sawicki et al. [223] (see Table
8). The principle was found to apply also to the heterocyclic aza compounds,
dibenzo[a,h]acridine and dibenzo[a,j]acridine [242].

Van Duuren [242] established experimentally, with dibenzo[a,i]pyrene,
that the characteristic absorption spectra of the PAH class are the same,
within a few nanometers, as the characteristic excitation (activation) spectra.
This investigator noted that in cases where background absorption obscures
characteristic absorption maxima, the excitation bands may assist identification. The two spectra of BAP showed the same agreement [223]. This
point is also clear from Table 8.

Schoental and Scott [241] showed that small shifts to longer fluorescence
wavelengths occurred on methylation of benzo[a]anthracene, as had been
found in the bathochromic shifts in ultraviolet absorption spectra of methylated pyrenes to which attention has been previously drawn [102]. Effects
of methylation on the fluorescence spectra of other PAH entities have been
recorded [242]. The effects of replacement of isoelectronic nitrogen atoms
in benzene rings (producing aza compounds) was examined [241, 242] and
the fluorescence spectrum was found to be only altered slightly. However,
intensity of fluorescence was augmented [241].

1.   Factors in Intensity and Band Changes

Weil-Malherbe [200] calibrated the photocell used in his 1944 measurements
of BaP fluorescence against a standard solution of quinine sulfate (micrograms per milliliter) in 1N-sulfuric acid. Quinine in sulfuric acid is still
employed for the purpose [223, 242]. K_Q values shown in Tables 8 and 9

TABLE 9

Comparison of the Excitation and Fluorescence Spectra of PAH and Aza Heterocyclics in Relation to Solvents

PAH[a]	In pentane		In sulfuric acid	
	Excitation[b]	Fluorescence[b]	Excitation[b]	Fluorescence[b]
Benzo[a]pyrene	295, 345, 361, 380[c]	403, 427, 454	283, 385, 470, 493, 525	548(K_Q = 3.5)[d]
Anthanthrene	257, 304, 380, 400, 420	430, 457, 487	323, 400, 432, 469, 512, 563	582(K_Q = 0.006)

Aza heterocyclics[e]	In cyclohexane		In concentrated sulfuric acid	
	Excitation	Fluorescence	Excitation	Fluorescence
Dibenzo[a,h]acridine	294, 319, 332, 353, 374, 384, 393	395, 403, 418, 445, 470	310, 400, 430, 442, 457	460, 485, 500
Dibenzo[a,j]acridine	300, 321, 337, 355, 375, 393	395, 400, 416, 440, 465	300, 425, 445	452, 478

[a]Reported by Sawicki et al. [223].

[b]Wavelength, (nm).

[c]Highest intensity bands are underlined. Sulfuric acid concentration not specified.

[d]Fluorescence intensity of pertinent band relative to a quinine standard.

[e]Reported by Van Duuren [242].

represent the fluorescence intensity of the pertinent PAH band and relative to the quinine standard. The procedure, used at the microgram level of concentration, is consistent with Beer's law.

Solvents in which the PAH compounds are contained for fluorospectrometry influence intensity and location of bands. For instance, fluorescence efficiency has been reported to increase from cyclohexane to ethyl alcohol to water to 10% sulfuric acid and to concentrated sulfuric acid [242]. At the same time most bands of highest intensity move to higher and higher wavelengths (see Table 9). These effects of change of solvent can be exploited, as seen below. Weil-Malherbe [200] eliminated oxygen from solvents and claimed almost complete independence from solvent effects.

Emission spectra of aromatic hydrocarbons were determined in crystalline paraffins by Bowen and Brocklehurst [243, 244] at -180°C. Personov [245, 246] and Fedoseeva and Khesina [247] have also used low temperatures, which have been found to sharpen fluorescence bands.

A factor of major importance in fluorescence intensity of PAH compounds is "quenching," caused by dissipation of exciting radiation through intramolecular collisions of PAH compounds, when concentration exceeds a level permitting full excitation efficiency. The result is a reduction in concentration — fluorescence response. Sawicki et al. [223] demonstrated the phenomenon for BaP in dimethyl formamide (molarity range 10^{-8} to 10^{-2}). Intensity was reduced to zero around 10^{-3}M but remained maximal from 10^{-8} to 10^{-6}M showing that fluorescence intensity measurements are only "quenching-free" at very low concentrations of this PAH compound. Obviously in calibration of spectrofluorimeters, either the no-quenching concentration range must be delineated or correction factors must be applied.

Fluorescent impurities represent another factor in intensity estimation that influences relative band intensities [242]. Solvents must therefore be purified to remove fluorescing contaminants. The interference of mixture associates of the PAH class is a problem in identification and quantitation as in ultraviolet absorption spectrometry.

2. Quantitation by Fluorescence Spectrometry

Weil-Malherbe [200] determined the accuracy of his fluorimetric method for determining BaP in mouse tissue by an instrumental control calibration with quinine sulfate, and a calibration with BaP in oxygen-free benzene over a linear range of concentration vs fluorescence intensity. The accuracy of the method was determined by adding amounts of 3-300 μg BaP to mouse tissue and determining the amount present at the end of the analytical process. It was found that recovery of 85-88% was made in 36 tests. A correction factor of 1.145 was therefore applied to analysis of unknowns, this procedure introducing an error of 6.3% in 5% of the cases. The limit of sensitivity was the reagent blank of 0.1-0.15 μg BaP. Subsequent instrumental improvements have made it possible to detect some PAH compounds in nanogram per milliliter quantities [223] in pure solutions.

The analysis of synthetic mixtures and air pollution extracts introduces problems of interferences in the fluorimetry of mixture associates, and these were eventually found to be of comparable magnitude and complexity to those encountered in ultraviolet absorption spectrometry [177]. Nevertheless, as Table 5 shows, many investigators have tried the spectroflourimetric technique in analysis of air pollutants for PAH content. However, Sawicki et al. [82] have demonstrated that BkF interferes in the spectrofluorimetric determination of BaP as it does in ultraviolet spectrometry. Solutions to this crucial issue have been advanced. For instance, Sawicki et al. [82, 223] have proposed to estimate BaP in sulfuric acid, to take advantage of the finding that in sulfuric acid BaP fluorescence is not interfered with at 545 nm. Only anthanthrene fluoresces nearby (582 nm) but at 1/600 of the BaP intensity. No simple solution to the related problem of high fluorescence background of mixtures [177] has been advanced, nor has a simple way of determining "no-quenching" test conditions been evolved. These two considerations are fundamental to accurate methodology.

A scheme was conceived by Dubois and Monkman [233] to estimate BaP and BkF in the presence of one another. It was purported that at its maximum excitation wavelength, 308 nm, BkF flouresces 25 times as strongly as BaP and therefore the BaP contribution could be ignored. BaP at its excitation maximum, 385 nm, was found to fluoresce at the same intensity as BkF. Therefore, the apparent concentration from excitation at 385 nm need only be corrected by the extent of BkF found at 308 nm. Attempts to confirm the reliability of the scheme with air pollution samples led to values ten times too high [177] due to background fluorescence.

Elution of air pollution extracts from activated alumina with toluene has been advanced [104] as giving superior results to the deactivated alumina-cyclohexane system on air pollution extracts when the comparative fluorescence method [233] is used to estimate BaP and BkF. A tentative standard method for BaP-BkF determination in atmospheric particulate matter has been adopted [47]. This method is described as able to (a) detect 0-0.25 μg of BaP or BkF per milliliter of solution, (b) estimate 0.25 μg BaP per milliliter with an accuracy of better than 0.002 μg, and (c) estimate 0.25 μg BkF to better than 0.001 μg; however, if BaP concentration is more than twice BkF, results will be in error by 10%. The error arises from the BaP fluorescence contribution at excitation wavelength 308 nm, which is used for BkF estimation [233].

Use has been made of very low temperatures to sharpen fluorescence spectra in quantitation of PAH compounds by investigators in the U.S.S.R. from 1953 [239] to 1972 [120] and by Muel and Lacroix [190] in France. The possibility of BaP-BkF interference does not appear to have been resolved in these low-temperature techniques.

The phosphorescence spectra represent another low-temperature phenomenon. The spectra for BaP were reported in 1958 by Muel and Hubert-Habart [248]. However, quantitation using phosphorescence spectra has not been exploited in LC of PAH compounds. Sawicki [58], in reviewing the literature on the subject, has noted that a number of weakly fluorescing PAH compounds are in fact intense phosphorescers.

V. AZA HETEROCYCLICS

Some aza heterocyclic hydrocarbons (AHH) and methyl derivatives have been incriminated as carcinogenic [34-36, 249]. The presence of such compounds in the basic fraction of cigarette smoke condensates was reported by Van Duuren et al. [135] in 1960. Sawicki et al. [116] demonstrated their presence in the basic fraction of air pollution samples and automobile exhaust. The extent to which AHH compounds may participate as environmental carcinogens has not yet been elucidated. There has been little interest in their levels in air pollution to date. The class received only passing reference in an exhaustive review on the chemical composition of tobacco smoke by Stedman [137] in 1968. Nevertheless, much work has been done on their estimation, particularly by the Sawicki group, and this has been reviewed to 1970 [59].

Essentially, the analytical approach to the estimation of the AHH class has been the same as for PAH compounds: chromatographic separation on one medium or another, followed by absorption and fluorescence spectrometry. Longest wavelength ultraviolet absorption bands and isolated high-intensity bands for a lengthy range of AHH compounds were assembled from the literature range of AHH compounds were assembled from the literature through 1958 by Karr [250]. Sawicki et al. [251] have provided ultraviolet absorption, fluorescence excitation, and emission spectra for an extensive range of compounds. Apparently the AHH compounds present a more complex separation problem than the PAH compounds, at least in coal-tar pitch [252]. In this circumstance it may be supposed that analysis of the AHH group will more likely yield to high-pressure column chromatography than to the conventional LC techniques so exhaustively exploited for PAH analysis over the past 20 years. This in fact, has just eventualized [253]. The method is discussed in Sec. VII, on high-speed chromatography.

A. Aromatic Amines and Heterocyclic Imines

The same chromatographic system as used for aza heterocyclic analysis [252] has been applied to the separation of a number of carcinogenic aromatic amines and heterocyclic imines [254]. Aza heterocyclics were eluted by pentane-ether and aromatic amines by pentane-acetone. For the separation of heterocyclic imines, extraction of basic compounds should be made to avoid the interference of aza and amino compounds. Column decomposition of some aromatic amines was detected. Identification of eluent constituents required thin-layer chromatography.

B. Polynuclear Ring Carbonyls

Sawicki et al. [254] succeeded in separating a number of polynuclear (PN) ring carbonyls in a prepared mixture using alumina (13% water) and pentane-ether eluents. They were identified by ultraviolet absorption spectroscopy. It had been shown [255] that carbonyl compounds were a major component of various fractions of the organic fraction of air pollution samples. Furthermore, a large variety were found, including the PN ring carbonyls. It was proposed that there should be prior extraction of basic compounds from samples. This would be followed by alumina column chromatography of the nonbasic fraction and then by a scheme of thin-layer chromatography.

VI. LIQUID CHROMATOGRAPHY DETERMINATION OF PHENOLS

Some phenolic compounds have been found to accelerate tumor formation and to be ciliostatic [62].

In 1951, Zahner and Swann [256] separated phenol from cresylic acid using partition chromatography with water on silica gel as the stationary phase and cyclohexane as the mobile phase. Subsequently Pearson [257] used the same method with very long columns (500 and 900 $\times$ 18 mm) to separate xylenols, cresols, and phenols, with identification by ultraviolet absorption spectroscopy. These methods were employed for macroscale analysis.

In 1956, Commins and Lindsey [258, 259] separated microgram quantities of such phenols as occur in smoke by their conversion to the methyl ethers, followed by column chromatography. Identification of eluent constituents was made by ultraviolet absorption spectroscopy. Quantitative determinations on wood smoke [260] and cigarette smoke [261, 262] were made.

The conversion to methyl ethers was effected with dimethyl sulfate in alkaline solution. The chromatographic separation was made in columns (30 $\times$ 1 cm) using alumina with 1-5% water content and cyclohexane as eluent.

VII. MODERN HIGH-SPEED LIQUID CHROMATOGRAPHY

In 1963 [263], the first reported attempt was made to adapt to a liquid chromatographic scheme for PAH analysis those features of gas chromatography (GC) that had given the technique a preeminence in particular respect to readily volatile substances and gases. The distinctive features of gas chromatography are the use of long pressurized columns, a long-lived adsorbent, one carrier throughout the procedure, and the highly reproducible retention volumes that the system provides. Karr et al. [263] used a 25-ft column of 0.25 in. i.d. copper tubing providing a length-diameter ratio of 1200. The adsorbent was alumina (80/100 mesh) with 4% water content. Cyclohexane was used as a carrier under 50 psig pressure. Synthetic

mixtures of acenaphthene and acenaphthylamine and pure 1,2-dimethylnaph-
thalene were injected into the column in amounts of 5-50 mg. Retention
volumes were measured as drops of cyclohexane (35 drops per milliliter
cyclohexane). It was found that the retention volumes were highly reproducible and could be read to ±50 drops. The packing lasted for large numbers
of runs. Concurrently, Giddings [264] noted that the system of Karr et al.
[263] was analogous to GC but that the operating conditions were not, since
plate heights, the measure of column efficiency, were not analogous to GC.
Giddings used his universal reduced plate height equation to evolve parameters
for LC that were analogous to optimum operation in GC.

Since 1963, there have been many developments in liquid chromatography
speed, resolution, and detection that have largely eventualized from attempts
to exploit the principles of gas chromatography. These developments have
been reviewed recently [265-267]. A book on modern practice of liquid
chromatography has been edited by Kirkland [4]. Currently available
equipment and costs have been summarized in Chemical and Engineering
News [268].

A. Main Features of Modern Liquid Chromatography

1. High Pressures and Small Adsorbent Particles

Giddings [264] proposed high pressures and small adsorbent particles
(micron diameter). Hamilton [269] had used small particles advantageously
in ion-exchange chromatography. In 1969, Kirkland [270] made separation
of urea herbicides within minutes using 30-μ particles, pressures to 3000
psi, and flow rates of 1.14 cm^3/min.

2. Column Diameter and Particle Size

Huber and Hulsman [271] established that resolution and speed are functions
of column diameter and particle size of the adsorbent filling. Kirkland [270]
used columns of 2.1 mm.

3. Solvents

The solvents (carriers) are pumped in steadily during the course of a run.
Small particles and small diameter columns account for the high pressures
that are necessary to move solvent through, even at the slow velocities that
have been found necessary for high resolution (fractions of a millimeter to
several millimeters per second). Degassing of solvents is desirable because
of the high pressures, to avoid bubble formation at the detection stage.

Solvents are moved by pressurization with compressed gases or pumps.
With pumps, pulsing must be avoided. For liquid-solid systems, water
content of adsorbents may have to be maintained by adding water to solvents.

4. Injection of Samples

Samples must be injected against high pressures. The injection port design
is important.

5. Column Packings

Deformation of packing adsorbent material at high pressures stimulated
development of alternatives to the conventional LC adsorbents such as
alumina. Dense materials were introduced by Kirkland [270] with a view
to overcoming the pressure effects. Kirkland [272] also showed how to
optimize adsorbent power of packings by varying not only particle dimensions
but also surface porosity.

6. Detectors

In modern high-speed LC, the solvent (carrier or eluent) flows continuously
during runs. Hence the system is adaptable to continuous automatic recording
detection of solutes. It has been found that dead volume in the detector, as
well as the volume of the connection between column and detector (Ref. 4,
pp. 30-31) must be kept as small as possible to minimize band broadening.
Now available are detectors for ultraviolet absorption, refractive index,
and heat of adsorption, and ionization detectors where the eluent is transported through a solvent evaporator in preparation for detection. These
detectors offer limits at the microgram and nanogram level. A polarographic
detector with a dropping mercury electrode for analysis of organic phosphate
pesticides was described in 1970 [273]. It was claimed that concentrations
of 10^{-8} M could be determined in a few minutes with standard deviation of less
than 2%. Contributions to the literature in regard to criteria for design of
high-speed systems continue to appear [274-277].

B. Application to PAH Separation

The principles derived from GC have proved applicable to most forms of LC;
liquid-solid, liquid-liquid (LL), gel, and ion exchange. In respect to air
pollution analysis for the PAH class, thermal instability precluded use of
GC. The high-speed LC may provide a solution to the lengthy conventional

LC procedures. Experience with the new procedure has been confined to mixtures of PAH compounds put together to test degree of resolution and speed. However, it may be observed that none of the test mixtures included BaP and BkF, which together present an interference in conventional LC, a problem already discussed in this chapter (Sec. IV, L, 2).

Karr et al. [278] extended their original work [263] related to analysis of aromatic hydrocarbons in pitch oils. The GC analog was modified by increasing pressure to 125 psig.

In 1968, Jentoft and Gouw [279] applied a high-resolution LL chromatography system [280] to the study of the isolation and identification of mixtures of model PAH compounds. Later, the same investigators claimed increased speed of elution and better resolution using a variety of super-critical fluids as the moving phase [3]. There was no evidence that these systems offered a solution to the PAH resolution problems of conventional LC, but speeds of separation were in minutes.

It has been established that in LLC, polar solutes require the use of polar stationary phases and relatively nonpolar carriers, whereas nonpolar solutes may be separated most effectively by using nonpolar liquid stationary phases with polar carrier (Ref. 4, pp. 186). Entirely discrete separation of seven PAH compounds (nonpolar) was effected using a hydrocarbon polymer stationary phase (of unstated composition) with an aqueous methanolic carrier. This is shown in Fig. 1. Subsequently, Schmit et al. [2] employed a more stable and effective packing (octadecyl silicone bonded to Zipak) to separate the same compounds.

The first modern LC systems that gave high resolution and high speed were technically complex, because many more factors than in conventional LC had to be covered. Bombaugh et al. [281] studied column packing and elution volume, noting that detector sensitivity decreased as peak width in the column increased. As a result of their investigations a design for a simplified high-performance system was evolved. This included a high-pressure pulse damper (pump pressures to 1000 psi), on-column injection, and low dead volume in a refractometer detector. Separations were made of a mixture (in decalin) containing benzene, naphthalene, azulene, o-quaterphenyl, and m-quaterphenyl. With deactivated alumina (6% water) of particle size range 70-150 μm in columns 100 cm $\times$ 2.3 mm i.d., and solvent n-hexane flowing at 0.9 ml/min, separation was complete in 25 min. With particles 38-53 μm and flow rate 2.9 ml/min, separation was completed in 4 min, confirming earlier observations on the advantage of very small particle size in column resolution.

Adsorbents consisting of particles with solid cores and thin porous coatings had been suggested for GC some years before [282]. In 1969, Kirkland [272] prepared spherical siliceous particles with a porous surface controlled in respect to thickness and pore size, so-called "controlled surface porosity." This preparation was called Zipax. It was adapted to high-speed reversed-phase liquid chromatography by Schmit et al. [2], who

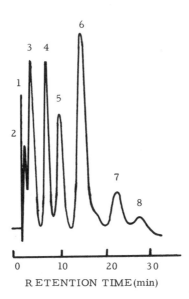

RETENTION TIME(min)

Figure 1. Separation of fused-ring aromatics by reverse-phase liquid chromatography. Column length, 1m; 2.1 mm i.d., hydrocarbon polymer on <37 μm Zipax; carrier, 60% water/40% methanol (v/v); temperature, 40°C; column input pressure, 1200 psi; UV detector. Peak identity: (1) Solvent; (2) Benzene; (3) Naphthalene; (4) Anthracene; (5) Pyrene; (6) Chrysene; (7) Benzo[e]pyrene; (8) Benzo[a]pyrene. From Ref. 4, reproduced by permission of Wiley Interscience.

permanently bonded octadecyl silicone to the Zipak to produce a durable stationary phase. The instrument used, a DuPont 820 liquid chromatograph, has been described by Felton [283]. A test mixture consisting of benzene and eight polycyclic aromatics was effectively resolved in around 30 min. BkF was not a constituent. Gradient elution was used with a mobile phase running from 20% methanol in water to 100% methanol, increasing at the rate of 2% methanol per minute. Detection was by ultraviolet absorption at 254 nm.

Pressure-assisted reverse-phase liquid chromatography in separation of polynuclear hydrocarbons from engine oils has been described by Vaughan et al. [284]. Using Corasil/C_{18} and methanol-water solvents, sharp separations of a variety of PAH compounds were attained. Corasil/C_{18} displayed no deterioration due to irreversible adsorption of polar materials, suggesting that this material might be suitable for the separation of PAH compounds from samples of solids collected from urban air.

The PAH mixtures employed to date in development and assessment of modern LC methods have shown that fast separations can be accomplished. It remains to test mixtures more typical of those occurring in air pollutants and to effect column resolution of PAH entities that available detection methods have not been able to differentiate due to the closeness of their properties.

Some progress in this direction has been reported by Klimisch [285] who has separated benzo[a]pyrene from other substances of the so-called "benzopyrene" fraction. High-pressure liquid chromatography with cellulose acetate columns and ethanol/dichloromethane solvent (2:1) was used.

As previously noted, Vivilecchia et al. [253] have separated complex mixtures of aza heterocyclics by high-pressure liquid chromatography using Zipax adsorbent, the surface of which was silver impregnated. It was claimed that acridine could be detected at the 1 ng level per injection. The compounds are separated by donor-acceptor complexing between the adsorbed silver ion and the heterocyclic nitrogen atom. Aliphatic and aromatic hydrocarbons were not retained. Amines were not eluted by 1% acetonitrile in n-hexane, which was the solvent used in the method. See Addendum for further information, p. 675.)

REFERENCES

1. M. Tzvyet, Bot. Inst. Polytech. (Warsaw), 44, 1124 (1911).

2. J. A. Schmit, R. A. Henry, R. C. Williams and J. F. Dieckman, J. Chromatog. Sci., 9, 645 (1971).

3. R. E. Jentoft and T. H. Gouw, J. Chromatog. Sci., 8, 138 (1970).

4. J. J. Kirkland, ed., Modern Practice of Liquid Chromatography, Wiley-Interscience, New York, 1971.

5. A. Winterstein and K. Schön, Z. Physiol. Chem., 230, 146 (1934).

6. A. Winterstein and K. Schön, Naturwissenshaften, 22, 237 (1934).

7. A. Winterstein and K. Schön, Z. Physiol. Chem., 230, 158 (1934).

8. P. Pott, Chirurgical Observations, Hawes, Clarke and Collings, London, 1775, p. 63.

9. R. D. Passey, Brit. Med. J., 2, 1112 (1892).

10. H. T. Butlin, Brit. Med. J., 2, 1 (1892).

11. H. T. Butlin, Brit. Med. J., 2, 66 (1892).

12. K. Yamagiwa and K. Ischikava, Mitteil. Med. Facultat, Kaiser, (Univ. Tokyo), 15, 296 (1915).

13. I. Hieger, Biochem. J., 24, 505 (1930).

14. J. W. Cook, C. L. Hewett, and I. Hieger, J. Chem. Soc., Part 1, 395, (1933).

15. J. W. Cook, I. Hieger, E. L. Kennaway, and W. V. Mayneord, Proc. Roy. Soc. (London), B111, 455 (1932).

16. G. Barry, J. W. Cook, G. A. D. Haslewood, C. L. Hewett, I. Hieger, and E. L. Kennaway, Proc. Roy. Soc. (London), B117, 318 (1935).

17. J. W. Cook, Proc. Roy. Soc. (London), B111, 485 (1932).

18. G. M. Badger, Brit. J. Cancer, 2, 309 (1948).

19. P. Daudel and R. Daudel, Chemical Carcinogenesis and Molecular Biology, Interscience, New York, 1966, p. 158.

20. R. Schoental, in Polycyclic Hydrocarbons, Vol. 1, (E. Clar, ed.), Academic, London and New York, 1964, pp. 133-160.

21. L. F. Fieser, Amer. J. Cancer, 34, 37 (1938).

22. E. C. Vigliani and G. Saita, New Engl. J. Med., 271, 872 (1964).

23. G. M. Badger, J. W. Cook, C. L. Hewett, E. L. Kennaway, N. M. Kennaway, and R. H. Martin, Proc. Roy. Soc. (London), B131, 170 (1942).

24. E. L. Kennaway, Biochem. J., 23, 497 (1930).

25. A. Lacassagne, F. Zajdela, N. P. Buu-Hoi, and H. Chalvet, Compt. Rend., 244, 273 (1957).

26. A. Lacassagne, N. P. Buu-Hoi, and F. Zajdela, Compt. Rend., 246, 1477 (1958).

27. A. Lacassagne, N. P. Buu-Hoi, F. Zajdela, D. Lavit, and J. J. Lamy, Compt. Rend., 256, 2728 (1963).

28. E. L. Wynder and D. Hoffmann, Nature (London), 192, 1092 (1961).

29. H. L. Falk, P. Kotin, and A. Miller, Int. J. Air Poll., 2, 201 (1960).

30. M. J. Shear, Amer. J. Cancer, 33, 499 (1938).

31. M. J. Shear, Amer. J. Cancer, 36, 211 (1939).

32. A. Lacassagne, N. P. Buu-Hoi and F. Zajdela, Compt. Rend., 245, 991 (1957).

33. W. E. Bachmann, J. W. Cook, A. Dansi, C. G. M. de Worms, G. A. D. Haslewood, C. L. Hewett, and A. M. Robinson, Proc. Roy. Soc. (London), B123, 343 (1937).

34. A. Lacassagne, N. P. Buu-Hoi, J. Lecocq, and G. Rudali, Bull. Assoc. Franc. Etude Cancer, 33, 48 (1946).

35. A. Laccassagne, G. Rudali, N. P. Buu-Hoi and J. Lecocq, Compt. Rend. , 139, 955 (1945).

36. A. Lacassagne, N. P. Buu-Hoi, J. Lecocq, and G. Rudali, Bull. Assoc. Franc. Etude Cancer, 34, 22 (1947).

37. J. A. Campbell, Brit. J. Exp. Pathol. , 15, 287 (1934).

38. M. G. Sellig and E. L. Benignus, Amer. J. Cancer, 28, 96 (1936).

39. M. G. Sellig and E. L. Benignus, Amer. J. Cancer, 33, 549 (1938).

40. S. McDonald and D. L. Woodhouse, J. Pathol. Bact. , 54, 1 (1942).

41. I. Hieger, Cancer Res. , 6, 657 (1946).

42. E. L. Kennaway and N. M. Kennaway, Brit. J. Cancer, 1, 260 (1947).

43. F. Goulden and M. M. Tipler, Brit. J. Cancer, 3, 157 (1949).

44. P. Wedgwood and R. L. Cooper, Chem. Ind. , 1066 (1951).

45. E. Sawicki, R. C. Corey, A. E. Dooley, J. B. Giselard, J. L. Monkman, R. E. Neligan, and L. A. Ripperton, Health Lab. Sci. , 7(1), Pt. 2, 31 (1970).

46. E. Sawicki, R. C. Corey, A. E. Dooley, J. B. Giselard, J. L. Monkman, R. E. Neligan, and L. A. Ripperton, Health Lab. Sci. , 7(1), Pt. 2, 45, (1970).

47. E. Sawicki, R. C. Corey, A. E. Dooley, J. B. Giselard, J. L. Monkman, R. E. Neligan, and L. A. Ripperton, Health Lab. Sci. , 7(1), Pt. 2, 60 (1970).

48. W. C. Hueper, P. Kotin, E. C. Tabor, W. W. Payne, H. Falk, and E. Sawicki, Arch. Pathol. , 74, 89 (1962).

49. P. Kotin and H. L. Falk, Proceedings National Conference on Air Pollution, U. S. Dept. Health, Education and Welfare, Public Health Service, Washington, D.C. , 1963, p. 140.

50. E. L. Wynder and E. C. Hammond, Cancer, 15, 79 (1962).

51. G. R. Clemo and E. Miller, Brit. J. Cancer, 14, 651 (1960).

52. A. Lacassagne, N. P. Buu-Hoi, and G. Rudali, Brit. J. Exp. Pathol. , 26, 5 (1945).

53. H. L. Falk, P. Kotin, and S. Thompson, Arch. Env. Health, 9, 169 (1964).

54. H. V. Gelboin, Rev. Can. Biol. , 31, 39 (1972).

55. L. W. Wattenberg, J. L. Leong and A. R. Galbraith, Proc. Soc. Exp. Biol. Med. , 127, 467 (1968).

56. L. W. Wattenberg and J. L. Leong, Cancer Res., 30, 1922 (1970).

57. E. Schlede, R. Kuntzman, and A. H. Conney, Cancer Res., 30, 2898 (1970).

58. E. Sawicki, Chemist-Analyst, 53, 24(1964); 53, 56 (1964); 53, 88 (1964).

59. E. Sawicki, Crit. Rev. Anal. Chem., 1, 275 (1970).

60. E. Sawicki, Arch. Env. Health, 14, 46 (1967).

61. D. Hoffmann and E. L. Wynder, Cancer, 15, 93 (1962).

62. D. Hoffmann and E. L. Wynder, Nat. Cancer Inst. Monog., 19, 91 (1962).

63. D. Hoffmann and E. L. Wynder, Cancer, 13, 1062 (1960).

64. E. Sawicki, T. R. Hauser, W. C. Elbert, F. T. Fox, and J. E. Meeker, Amer. Ind. Hyg. Assoc. J., 23, 137 (1962).

65. D. J. Von Lehmden, R. P. Hangebrauck and J. E. Meeker, J. Air Poll. Contr. Assoc., 15, 306 (1965).

66. R. E. Waller, Brit. J. Cancer, 6, 8 (1952).

67. R. L. Cooper and A. J. Lindsey, Chem. Ind., 1177 (1953).

68. R. L. Cooper, Chem. Ind., 1364 (1953).

69. R. L. Cooper, Analyst, 79, 573 (1954).

70. H. L. Falk, I. Markul, and P. Kotin, Arch. Ind. Health, 13, 13 (1956).

71. P. Kotin, H. L. Falk, P. Mader, and M. Thomas, Amer. Med. Assoc. Arch. Ind. Hyg., 9, 153 (1954).

72. J. A. S. Gilbert and A. J. Lindsey, Chem. Ind., 1439 (1955).

73. P. Stocks and J. M. Campbell, Brit. Med. J., 2, 923 (1955).

74. V. C. Shore and M. Katz, Anal. Chem., 28, 1399 (1956).

75. M. Kuratsune, J. Nat. Cancer Inst., 16, 1485 (1956).

76. A. J. Lindsey, E. Pash, and J. R. Stanbury, Anal. Chim. Acta, 15, 291 (1956).

77. B. T. Commins, R. E. Waller, and P. J. Lawther, Brit. J. Ind. Med., 14, 232 (1957).

78. B. T. Commins, Analyst, 83, 386 (1958).

79. B. T. Commins, Int. J. Air Poll., 1, 14 (1958).

80. A. J. Lindsey, Anal. Chim. Acta, 20, 175 (1959).

81. E. Sawicki, W. Elbert, T. W. Stanley, T. R. Hauser, and F. T. Fox, Anal. Chem., 32, 810 (1960).

82. E. Sawicki, W. Elbert, T. W. Stanley, T. R. Hauser, and F. T. Fox, Int. J. Air Poll., 2, 273 (1960).

83. E. Sawicki, W. C. Elbert, T. R. Hauser, F. T. Fox, and T. W. Stanley, Amer. Ind. Hyg. Assoc. J., 21, 443 (1960).

84. G. E. Moore and M. Katz, Int. J. Air Poll., 2, 221 (1960).

85. L. Dubois, A. Corkery, and J. L. Monkman, Int. J. Air Poll., 2, 236 (1960).

86. P. Stocks, B. T. Commins, and K. V. Aubrey, Int. J. Air Water Poll., 4, 141 (1961).

87. A. J. Lindsey and J. R. Stanbury, Int. J. Air Water Poll., 6, 387 (1962).

88. A. Frank and G. Gerhardson, Nord. Hyg. T., 43(1), 42 (1962).

89. G. J. Cleary, J. Chromatog., 9, 204 (1962).

90. G. J. Cleary, Int. J. Air Water Poll., 7, 753 (1963).

91. D. Rondia, Arch. Belg. Med. Soc. Hyg. Med. Travail Med. Legale, 21, 73 (1963).

92. D. Rondia, Arch. Belg. Med. Soc. Hyg. Med. Travail Med. Legale, 21, 578 (1963).

93. H. O. Hettche, Int. J. Air Water Poll., 8, 185 (1964).

94. J. L. Sullivan and G. J. Cleary, Brit. J. Ind. Med., 21, 117 (1964).

95. E. Sawicki, T. R. Stanley, J. D. Pfaff and W. Elbert, Chemist-Analyst, 53, 6 (1964).

96. C. W. Louw, Amer. Ind. Hyg. Assoc. J., 26, 520 (1965).

97. G. J. Cleary and J. L. Sullivan, Med. J. Australia, 52, 758 (1965).

98. E. Sawicki, J. E. Meeker, and M. J. Morgan, Int. J. Air Water Poll., 9, 291 (1965).

99. E. Sawicki, S. P. McPherson, T. W. Stanley, J. Meeker, and W. C. Elbert, Int. J. Air Water Poll., 9, 515 (1965).

100. P. Valori and A. Grella, Nuovi Ann. Ig. Microbiol., 17(4), 264 (1966).

101. G. E. Moore, M. Katz, and W. B. Drowley, J. Air Poll. Contr. Assoc., 16, 492 (1966).

102. G. E. Moore, R. S. Thomas, and J. L. Monkman, J. Chromato., 26, 456 (1967).

103. A. Zdrojewski, L. Dubois, G. E. Moore, R. S. Thomas, and J. L. Monkman, J. Chromatog., 28, 317 (1967).

104. L. Dubois, A. Zdrojewski, C. Baker, and J. L. Monkman, J. Air Poll. Contr. Assoc., 17, 818 (1967).

105. L. Dubois, A. Zdrojewski, and J. L. Monkman, Mikrochim. Acta, 903 (1967).

106. L. Dubois, A. Zdrojewski, and J. L. Monkman, Mikrochim. Acta, 834 (1967).

107. G. Lindstedt, Atm. Env., 2, 1 (1968).

108. J. M. Colucci and C. R. Begeman, Env. Sci. Tech., 5, 145 (1971).

109. P. Kotin, H. L. Falk, and M. Thomas, Arch. Ind. Hyg., 9, 164 (1954).

110. P. Kotin, H. L. Falk, and M. Thomas, Arch. Ind. Health, 11, 113 (1955).

111. M. J. Lyons and H. Johnston, Brit. J. Cancer, 11, 60 (1957).

112. M. J. Lyons, Brit. J. Cancer, 13, 126 (1959).

113. A. Vercruysse, N. Nordvik, and A. Heyndrickx, Arch. Belg. Med. Soc., 19, 157 (1961).

114. E. L. Wynder and D. Hoffmann, Cancer, 15, 103 (1962).

115. L. R. Reckner, W. E. Scott, and W. F. Biller, Proc. Amer. Petrol. Inst., 45, 133 (1965).

116. E. Sawicki, J. E. Meeker, and M. J. Morgan, Arch. Env. Health, 11, 773 (1965).

117. R. P. Hangebrauck, R. P. Lauch, and J. E. Meeker, Amer. Ind. Hyg. Assoc. J., 27, 47 (1966).

118. G. A. Smirnov, Vap. Onkol., 16(5), 83 (1970).

119. V. Del Vecchio, P. Valori, C. Melchiorri, and A. Grella, Pure Appl. Chem., 24(4), 739 (1970).

120. L. M. Shabad and G. A. Smirnov, Atm. Env., 6, 153 (1972).

121. R. L. Cooper and A. J. Lindsey, Brit. J. Cancer, 9, 304 (1955).

122. R. Latarjet, J. L. Cusin, M. Hubert-Habart, B. Muel, and R. Royer, Bull. Assoc. Franc. Etude Cancer, 43, 180 (1956).

123. J. Bonnet and S. Neukomm, Helv. Chim. Acta, 39, 1724 (1956).

124. S. Z. Cardon, E. T. Alvord, H. J. Rand, and R. Hitchcock, Brit. J. Cancer, 10, 48 (1956).

125. M. J. Lyons and H. Johnston, Brit. J. Cancer, 11, 554 (1957).

126. E. L. Wynder and G. Wright, Cancer, 10, 255 (1957).

127. B. Muel, M. Hubert-Habart, and N. P. Buu-Hoi, J. Chim. Phys. , 54, 483 (1957).

128. B. L. Van Duuren, J. Nat. Cancer Inst. , 21, 623 (1958).

129. H. R. Bentley and J. G. Burgan, Analyst, 82, 442 (1958).

130. E. Kennaway ánd A. J. Lindsey, Brit. Med. Bull. , 14, 124 (1958).

131. B. L. Van Duuren, J. Nat. Cancer Inst. , 21, 1 (1958).

132. J. Ahlmann, Arch. Pathol. Microbiol, Scand. , 43, 379 (1958).

133. A. J. Lindsey, Brit. J. Cancer, 13, 195 (1959).

134. D. Hoffman and E. L. Wynder, Anal. Chem. , 32, 295 (1960).

135. B. L. Van Duuren, J. A. Bilbao, and C. A. Joseph, J. Nat. Cancer Inst. , 25, 53 (1960).

136. M. Hubert-Habart, R. Latarjet, D. Lavalette, B. Muel, R. Loic, and R. Royer, Bull. Cancer, 53, 53 (1966).

137. R. L. Stedman, Chem. Rev. , 68, 153 (1968).

138. J. H. Bell, S. Ireland, and A. W. Spears, Anal. Chem. , 41, 310 (1969).

139. D. Hoffman and G. Rathkamp, Anal. Chem. , 44, 899 (1972).

140. I. Berenblum, Nature (London), 156, 601 (1945).

141. H. Wieland and W. Muller, Ann. , 564, 199 (1949).

142. N. P. Buu-Hoi, Nature (London), 182, 1158 (1958).

143. J. Bonnet, Nat. Cancer Inst. Monograph, 9, 221 (1962).

144. E. Sawicki, F. T. Fox, W. C. Elbert, T. R. Hauser, and S. Meeker, Amer. Ind. Hyg. Assoc. J. , 23, 482 (1962).

145. H. Wyszynska and J. Zwolinski, Roczniki Panstwowego Zakladu Hig. , 13(5), 503 (1962).

146. H. L. Falk, P. E. Steiner, S. Goldfein, A. Breslow, and R. Hykes, Cancer Res. , 11, 318 (1951).

147. H. L. Falk, P. E. Steiner, and S. Goldfein, Cancer Res. , 11, 247 (1951).

148. H. L. Falk and P. E. Steiner, Cancer Res. , 12, 30 (1952).

149. H. L. Falk and P. E. Steiner, Cancer Res. , 12, 40 (1952).

150. P. E. Steiner, Cancer Res., 14, 103 (1954).

151. A. J. Lindsey, M. A. Phillips, and D. S. Wilkinson, Chem. Ind., 1365 (1958).

152. I. Berenblum and R. Schoental, Brit. J. Exp. Pathol., 24, 232 (1943).

153. P. Wedgwood and R. L. Cooper, Analyst, 78, 170 (1953).

154. E. M. Charlet, K. P. Lanneau, and F. B. Johnson, Anal. Chem., 26, 861 (1954).

155. H. E. Lumpkin and B. N. Johnson, Anal. Chem., 26, 1719 (1954).

156. R. Tye, M. J. Graf, and A. W. Horton, Anal. Chem., 27, 248 (1955).

157. P. Wedgwood and R. L. Cooper, Analyst, 80, 652 (1955).

158. B. D. Tebbens, J. F. Thomas, E. N. Sanborn, M. Mukai, Amer. Ind. Hyg. Assoc. Quart., 18, 165 (1957).

159. G. M. Badger and T. M. Spotswood, J. Chem. Soc., 1635 (1959).

160. G. M. Badger and T. M. Spotswood, J. Chem. Soc., 4420 (1960).

161. T. M. Spotswood, J. Chem. Soc., 4427 (1960).

162. G. M. Badger and T. M. Spotswood, J. Chem. Soc., 4431 (1960).

163. L. R. Snyder, Anal. Chem., 33, 1535 (1961).

164. G. M. Badger and J. Novotny, Australian J. Chem., 16, 623 (1963).

165. S. T. Cuffe, Arch. Env. Health, 6, 422 (1963).

166. S. K. Ray and R. Long, Combust. Flame, 8(2), 139 (1964).

167. R. P. Hangebrauck, D. J. Von Lehmden and J. E. Meeker, J. Air Poll. Contr. Assoc., 14(7), 267 (1964).

168. P. J. Lawther, B. T. Commins, and R. E. Waller, Brit. J. Ind. Med., 22, 13 (1965).

169. P. P. Dikun, N. D. Krasnitskaya, E. G. Levitina, L. S. Sarochinskaya, G. N. Severinets, and N. L. Staskevich, Gig. Sanit., 35(9), 12 (1970).

170. P. P. Dikun, N. D. Krasnitskaya, G. P. Komina, E. G. Levitina, L. S. Sarochinskaya, G. N. Severinets, and N. L. Staskevich, Vop. Onkol, 17(6), 77 (1971).

171. J. S. Harington, Nature (London), 193, 43 (1962).

172. H. Tanimura, Arch. Env. Health, 17, 172 (1968).

173. E. R. Hendrickson, C. G. Walker, and V. D. Chapnerkar, Amer. Ind. Hyg. Assoc. J., 24, 121 (1963).

174. B. T. Commins and J. S. Harington, Nature (London), 212, 273 (1966).

175. W. Lijinsky, C. R. Raha, and J. Keeling, Anal. Chem., 33, 810 (1961).

176. H. O. Hettche, Staub, 31(2), 34 (1971).

177. E. Sawicki, T. W. Stanley, W. C. Elbert, J. Meeker, and S. McPherson, Atm. Env., 1, 131 (1967).

178. E. Balgairies and C. Claeys, Rev. Méd. Minière, 10(34/5), 20 (1957).

179. J. L. Monkman, G. E. Moore, and K. Katz, Amer. Ind. Hyg. Assoc. J., 23, 487 (1962).

180. E. C. Tabor, T. R. Hauser, J. P. Lodge, and R. H. Burtschell, Arch. Ind. Health, 17, 58 (1958).

181. T. W. Stanley, J. E. Meeker and M. J. Morgan, Env. Sci. Tech., 1, 927 (1967).

182. G. Chatot, M. Castegnaro, J. L. Roche, R. Fontanges, and P. Obaton, Anal. Chim. Acta, 53(2), 259 (1971).

183. B. T. Commins and P. J. Lawther, Brit. J. Cancer, 12, 351 (1958).

184. D. Rondia, Int. J. Air Water Poll., 9(3), 113 (1965).

185. B. T. Commins, Nat. Cancer Inst. Monograph, 9, 225 (1962).

186. Z. Morlin and K. M. Saringer, Nature (London), 191, 907 (1961).

187. J. Borneff and R. Knerr, Arch. Hyg. Bakt., 143, 405 (1959).

188. M. Kuratsune and T. Hirohata, Nat. Cancer Inst. Monograph, 9, 117 (1962).

189. B. D. Tebbens, M. Mukai, and T. F. Thomas, Amer. Ind. Hyg. Assoc. J., 32, 365 (1971).

190. B. Muel and G. Lacroix, Bull. Sci. Chim. France, 2139 (1960).

191. M. E. Griffing, A. R. Maler, and D. G. Cobb, Amer. Chem. Soc. Div. Petrol. Chem. Preprints, 14(3), B162 (1969).

192. H. J. Davis, L. A. Lee, and T. R. Davidson, Anal. Chem., 38, 1752 (1966).

193. R. L. Shriner, R. C. Fuson, and D. Y. Curtin, The Systematic Identification of Organic Compounds, 4th ed., Wiley, New York, 1956, p. 347.

194. H. Braus, F. M. Middleton, and C. C. Ruchhoft, Anal. Chem., 24, 1872 (1952).

195. A. A. Rosen and F. M. Middleton, Anal. Chem., 27, 790 (1955).

196. C. Golumbic, Anal. Chem., 22, 579 (1950).

197. J. D. Mold, T. B. Walker, and L. G. Veasey, Anal. Chem., 35, 2071 (1963).

198. N. Brock, H. Druckrey, and H. Hamperl, Arch. Exp. Pathol. Pharmakol., 189, 709 (1938).

199. H. Weil-Malherbe, Biochem. J., 40, 351 (1946).

200. H. Weil-Malherbe, Biochem. J., 38, 135 (1944).

201. H. Weil-Malherbe, J. Chem. Soc., 303 (1943).

202. W. Trappe, Biochem. Z., 305, 150 (1940).

203. H. Brockmann and H. Schodder, Berichte, 74B, 73 (1941).

204. H. Brockmann, Angew. Chem., A59, 199 (1947).

205. G. Hesse, I. Daniel, and G. Wohlleben, Angew. Chem., 64, 103 (1952).

206. L. R. Snyder, J. Chromatog., 5, 430 (1961).

207. P. B. Muller, Helv. Chim. Acta, 26, 1945 (1943).

208. H. J. Cahnmann, Anal. Chem., 29, 1307 (1957).

209. R. Consden, A. H. Gordon, and A. J. P. Martin, Biochem. J., 38, 224 (1944).

210. L. H. Klemm, D. Reed, and C. D. Lind, J. Org. Chem., 22, 739 (1957).

211. L. H. Klemm, D. Reed, L. A. Miller, and B. T. Ho, J. Org. Chem., 24, 1468 (1959).

212. L. R. Snyder, J. Chromatog., 6, 22 (1961).

213. W. Lijinsky, Anal. Chem., 32, 684 (1960).

214. W. Lijinsky, U. Saffiotti, and P. Shubik, J. Nat. Cancer Inst., 18, 687 (1957).

215. R. Tye and Z. Bell, Anal. Chem., 36, 1612 (1964).

216. T. M. Spotswood, J. Chromatog., 3, 101 (1960).

217. A. L. J. Beckwith and M. J. Thompson, J. Chem. Soc., 73 (1961).

218. E. Sawicki, Nat. Cancer Inst. Monograph, 9, 201 (1962).

219. N. P. Buu-Hoi and P. A. Jacquignon, Experientia, 13, 375 (1957).

220. V. Martinu and J. Janak, J. Chromatog., 65, 477 (1972).

221. W. M. Smit, Anal. Chim. Acta, 2, 671 (1948).

222. W. M. Smit, Disc. Faraday Soc., 7, 248 (1949).

223. E. Sawicki, T. R. Hauser, and T. W. Stanley, Int. J. Air Poll., 2, 253 (1960).

224. K. Rothwell and J. K. Whitehead, Nature (London), 213, 797 (1967).

225. A. M. Liquori, B. DeLerma, F. Ascoli, C. Botre, and M. Trasciatti, J. Mol. Biol., 5, 521 (1962).

226. T. M. Spotswood, J. Chromatog., 2, 90 (1959).

227. R. N. Jones, Chem. Rev., 32, 1 (1943).

228. C. Karr, Jr., Appl. Spectroscopy, 13, 15 (1959).

229. R. N. Jones, J. Amer. Chem. Soc., 67, 2127 (1945).

230. C. Sandorfy and R. N. Jones, Can. J. Chem., 34, 888 (1956).

231. R. N. Jones, Chem. Rev., 41, 353 (1947).

232. R. N. Jones and E. Spinner, Spectrochim. Acta, 16, 1060 (1960).

233. L. Dubois and J. L. Monkman, Int. J. Air Water Poll., 9, 131 (1965).

234. L. Dubois, A. Zdrojewski, P. Jennawar, and J. L. Monkman, Atm. Env., 4, 199 (1970).

235. A. Zdrojewski, L. Dubois, G. E. Moore, R. Thomas, and J. L. Monkman, Amer. Chem. Soc. Div. Water Waste Chem. Preprints, 6, 40 (1966).

236. R. A. Brown and J. M. Kelliher, Proc. Amer. Petrol. Inst. Div. Refining, 51, 349 (1971).

237. H. Moureu, P. Chovin, and G. Rivoal, Compt. Rend., 223, 951 (1946).

238. N. P. Buu-Hoi, D. Lavit, and J. J. Lamy, J. Chem. Soc., 1845 (1959).

239. B. P. Gurinov, V. A. Zore, A. A. Iljina, and L. M. Shabad, Gig. Sanit., 2, 10 (1953).

240. I. Berenblum and R. Schoental, J. Chem. Soc., 1017 (1946).

241. R. Schoental and E. J. Y. Scott, J. Chem. Soc., 1683 (1949).

242. B. L. Van Duuren, Anal. Chem., 32, 1436 (1960).

243. E. J. Bowen and B. Brocklehurst, J. Chem. Soc., 3875 (1954).

244. E. J. Bowen and B. Brocklehurst, J. Chem. Soc., 4320 (1955).

245. R. I. Personov, Zh. Anal. Khim, 17, 506 (1962).

246. R. I. Personov, Izvest. Akad. Nauk. S.S.S.R. Ser. Fiz. , 24, 624 (1960).

247. G. E. Fedoseeva and A. Ya. Khesina, Zh. Prikl. Spectrosk. , 9(2), 282 (1968).

248. B. Muel and M. Hubert-Hobart, J. Chim. Phys. , 55, 377 (1958).

249. Survey of compounds which have been tested for carcinogenic activity, U.S. Department of Health, Education and Welfare, Public Health Service Publ. 149 (1951) and suppl. 1 (1957), supp. 2 (1969), Washington, D.C.

250. C. Karr, Jr. , Appl. Spectrosc. , 14, 146 (1960).

251. E. Sawicki, T. W. Stanley, J. D. Pfaff, and W. C. Elbert, Anal. Chim. Acta, 31, 359 (1964).

252. E. Sawicki, J. E. Meeker, and M. J. Morgan, J. Chromatog. , 17, 252 (1965).

253. R. Vivilecchia, M. Thiebaud, and R. W. Frei, J. Chromatog. Sci. , 10, 411 (1972).

254. E. Sawicki, H. Johnson, and K. Kosinski, Microchem. J. , 10, 72 (1966).

255. E. Sawicki and T. R. Hauser, Anal. Chem. , 31, 523 (1959).

256. R. J. Zahner and W. B. Swann, Anal. Chem. , 23, 1093 (1951).

257. R. M. Pearson, Analyst, 80, 656 (1955).

258. B. T. Commins and A. J. Lindsey, Anal. Chim. Acta, 15, 446 (1956).

259. B. T. Commins and A. J. Lindsey, Anal. Chim. Acta, 15, 551 (1956).

260. B. T. Commins and A. J. Lindsey, Anal. Chim. Acta, 15, 554 (1956).

261. B. T. Commins and A. J. Lindsey, Brit. J. Cancer, 10, 504 (1956).

262. B. T. Commins and A. J. Lindsey, Anal. Chim. Acta, 15, 557 (1956).

263. C. Karr, Jr. , E. E. Childers, and W. C. Warner, Anal. Chem. , 35, 1290 (1963).

264. J. C. Giddings, Anal. Chem. , 35, 2215 (1963).

265. S. G. Perry and R. J. Maggs, Lab. Practice, 19, 802 (1970).

266. S. G. Perry, Chem. in Britain, 7, 366 (1971).

267. J. N. Done, G. J. Kennedy, and J. H. Knox, Nature (London), 237, 77 (1972).

268. Liquid chromatography becomes important quantitative analysis technique. Chem. Eng. News, p. 22, March 27, 1972.

269. P. B. Hamilton, Anal. Chem., 30, 914 (1958).

270. J. J. Kirkland, J. Chromatog. Sci., 7, 7 (1969).

271. J. K. F. Huber and J. A. R. Hulsman, Anal. Chim. Acta, 38, 305 (1967).

272. J. J. Kirkland, Anal. Chem., 41, 218 (1969).

273. J. G. Koen, J. F. K. Huber, H. Poppe, and G. der Boef, J. Chromatog. Sci., 8, 192 (1970).

274. J. F. K. Huber, J. Chromatog. Sci., 9, 72 (1971).

275. R. P. W. Scott, J. Chromatog. Sci., 9, 449 (1971).

276. L. R. Snyder, J. Chromatog. Sci., 10, 200 (1972).

277. L. R. Snyder, J. Chromatog. Sci., 10, 369 (1972).

278. C. Karr, Jr., E. E. Childers, W. C. Warner, and P. E. Estep, Anal. Chem., 36, 2105 (1964).

279. R. E. Jentoft and T. H. Gouw, Anal. Chem., 40, 1787 (1968).

280. R. E. Jentoft and T. H. Gouw, Anal. Chem., 40, 923 (1968).

281. K. J. Bombaugh, R. F. Levangie, R. N. King, and L. Abrahams, J. Chromatog. Sci., 8, 657 (1970).

282. M. J. E. Golay, Gas Chromatography 1960, Edinburgh Symposium, (R. P. W. Scott, ed.), Butterworths, London, 1960, p. 139.

283. H. R. Felton, J. Chromatog. Sci., 7, 13 (1969).

284. C. G. Vaughan, B. B. Wheals, and M. J. Whitehouse, J. Chromatog., 78, 203 (1973).

285. H. J. Klimisch, Anal. Chem., 45, 1960 (1973).

Chapter 4

PAPER CHROMATOGRAPHIC ANALYSIS IN AIR POLLUTION

Gordon Chesters and D. A. Graetz*

Water Resources Center
Department of Soil Science and
Water Chemistry Program
University of Wisconsin-Madison
Madison, Wisconsin

I. INTRODUCTION

Paper chromatographic techniques have not been applied extensively to air
pollutants, and the majority of the literature deals with the separation and
identification of polynuclear aromatic hydrocarbons. Specific sampling
techniques and the sensitivity of particular methods are discussed in the

---

*Present Address: Soil Science Department
                   Institute of Food and Agricultural Sciences
                   University of Florida
                   Gainesville, Florida

appropriate section of the chapter. For general information on methods of
taking air samples, the reader is referred to Hendrickson [1] and Ruch [2].

II. METHODS AND USES OF PAPER CHROMATOGRAPHIC ANALYSIS IN AIR POLLUTION INVESTIGATIONS

A. Polynuclear Aromatic Hydrocarbons

Organic contaminants are among the major components of atmospheric
pollution [3]. The polynuclear aromatic hydrocarbons have been of greatest
concern because several display carcinogenic properties. The presence of
these compounds in the atmosphere has been linked generally with incomplete
combustion of solid and liquid fuels [4, 5]. Methods of analysis of gaseous
[6] and particulate [7] organic air pollutants have been reviewed. An
extensive review of separation and identification techniques for polynuclear
aromatic hydrocarbons has been prepared by Sawicki [8, 9]. Among these
methods, paper chromatography has found considerable application.

Paper chromatographic separation of polynuclear aromatic hydrocarbons
using aqueous solutions of CH_3COOH and Whatman papers has been developed
by Tebbens and co-workers [3-5]. Samples were collected during burning
of fuels under conditions resulting in incomplete combustion. Filtration of
the particulate component of the combustion effluent followed by extraction
of this material by a soxhlet technique using 300 ml of diethyl ether (24 hr)
produced a reddish-brown solution. The intensity of color provides a rough
indication of aromatic hydrocarbon concentration.

Ascending and descending one-dimensional methods, two-dimensional
methods, and circular methods of paper chromatography were investigated.
Except for development time, i.e., 16 hr for descending and 24 hr for
ascending development, little difference was found between ascending or
descending paper chromatographic methods. Because the chromatograms
can be developed for a longer time with the one-dimensional methods, such
methods allow better resolution than do circular methods. However, the
circular method is adequate for most practical applications and the relatively
short development time (4 hr) is a distinct advantage [5]. Complete separation by one-dimensional methods requires a series of chromatograms,
whereas two-dimensional methods allow separation on a single sheet.

Whatman filter papers Nos. 40, 41, 42, and 44 have been used satisfactorily. Each of the papers can be obtained in a double acid-washed condition
that prevents metal contaminants from interfering with fluorometric detection
[5]. Whatman No. 44 paper is recommended frequently because of its lighter
weight and finer texture, which is an advantage in obtaining smooth absorption
curves directly from the paper.

Acetic acid was used for development of the chromatograms. However,
any inert liquid may be used that possesses a low dielectric constant, is
relatively nonvolatile at ambient temperature, contains no fluorescent

material, and is readily removed during drying of the developed chromatogram [5]. It is necessary to use a series of acid concentrations on separate
chromatograms to obtain complete resolution of the compounds. Generally,
40, 60, and 80% aqueous CH_3COOH solutions are used. For two-dimensional
chromatography 80% CH_3COOH is used in one direction, followed by a 40%
solution in the second direction. Dilute CH_3COOH does not allow movement
of high molecular weight components but provides good resolution of low
molecular weight components. When high CH_3COOH concentrations are
used, the low molecular weight components move essentially with the solvent
front while the high molecular weight components are adequately resolved.

The circular method of paper chromatography used by Tebbens et al. [5]
was developed by Rutter [10, 11]. A wick was cut in a 15-cm diameter
filter paper disk and the ether extract (3-5 μl) was placed on the paper at
the point where the wick bent away from the disk. The paper and developer
were enclosed in a petri dish. Development time was approxmately 4 hr.

Developed chromatograms were dried and irradiated with ultraviolet
light; polynuclear aromatic hydrocarbons were detected by their natural
fluorescence. Ultraviolet absorption spectra of components separated by
dilute CH_3COOH solutions were determined directly on the paper by placing
the isolated spot in a Beckman Model DU spectrophotometer fitted with a
photomultiplier attachment. About 1-10 μg of material per spot is necessary
to obtain the absorption curve. High molecular weight compounds separated
by 80% CH_3COOH cannot be determined by this method because quantities
are generally small and background interferences large.

Of the 16 components initially separated, seven were tentatively identified
and two of the components exhibited carcinogenic activity.

As an aid to the identification of the remaining components the original
ether extract was separated into three solubility classes: (1) those soluble
in dilute NaOH and $NaHCO_3$, which likely consisted of carboxylic acid
derivatives of polynuclear aromatic compounds;   (2) those soluble in dilute
NaOH but not $NaHCO_3$ and believed to be phenolic derivatives; and (3) those
insoluble in NaOH and $NaHCO_3$, consisting largely of neutral hydrocarbons.
The solutions were converted to their original form (i.e., acid salts were
reconverted to acid salts) and extracted in ether. The three solutions were
concentrated to approximately equal concentrations as indicated by the
intensity of the red-brown colors. Because the phenolic and carboxylic
derivatives were present initially in much lower amounts than the neutral
hydrocarbons, they had to be concentrated to a much greater extent.

N, N-Dimethylformamide-impregnated paper has been employed for the
paper chromatography analysis of 3,4-benzopyrene and several other polycyclic aromatic hydrocarbons [12-15]. The papers were developed with
hexane saturated with N, N-dimethylformamide as originally described by
Zaffaroni et al. [16, 17] for the chromatography of oxygenated steroids.
Whatman No. 1 filter paper was saturated with redistilled N, N-dimethylformamide and dried for 1 hr. The sample, 1 μg in benzene, was applied to
the paper approximately 5 cm from the lower end. The developing chamber

was lined with filter paper and maintained in a saturated condition with the developing solvent. The paper, allowed to equilibrate overnight, was developed by the ascending technique. To prevent photodecomposition of the sample, chromatography was conducted in the dark [12]. The procedure was modified to allow for relative humidity control, which provided improved chromatographic resolution and improved reproducibility of R_f values [13]. Prior to saturation with N,N-dimethylformamide, the paper was dried for 1 hr at 120°C and stored in a desiccator over $CaCl_2$. Calcium chloride also was placed in the developing chamber to control humidity.

Mitchell [18] employed aqueous and nonaqueous immobile/mobile solvent systems for the separation of eight carcinogenic polynuclear aromatic and related compounds by paper chromatographic methods. The nonaqueous solvent system consisted of 175 ml of N,N-dimethylformamide diluted to 500 ml with diethyl ether (35%, v/v) as the immobile phase and 2,2,4-trimethylpentane as the mobile phase. The aqueous system employed soybean oil 5% (v/v) in diethyl ether as the immobile phase and 35% isopropanol or 35% propanol as the mobile phase. Generally, the isopropanol and propanol mobile solvents were used on separate chromatograms because some components were resolved better by one solvent and some by the other. Approximately 1-μg sample sizes were used; larger samples tended to produce overlapping spots and impurities in the samples caused significant interference. The compounds were located and identified on the chromatogram by their fluorescence in ultraviolet light at 366.0 and 253.7 nm. A chromogenic agent (0.25% tetrachlorophthalic anhydride in acetone) was sprayed on the chromatograms to aid in identification [19].

The nonaqueous system was the preferred method for several reasons: shorter development times were required ($1\frac{1}{4}$ hr compared with 8 hr); improved separation of impurities was achieved; sensitivity was improved; both solvent systems were volatile; the compounds could be eluted from the paper with smaller amounts of impurities; and the chromatograms could be blueprinted. Blueprinting is accomplished by placing the chromatogram, with the front side of the paper facing upward, on the sensitized surface of blueprint paper and exposing it to germicidal light (253.7 nm) for 30 min [20].

Determination of 3,4-benzopyrene in air was accomplished by the combination of Al_2O_3 column chromatography and paper chromatography [21]. The 3,4-benzopyrene was extracted by a soxhlet technique with benzene, eluted from the Al_2O_3 column with cyclohexane, and further separated by paper chromatography [22]. The column effluent was evaporated to 0.5-1.0 ml in vacuo at 40°C and applied to N,N-dimethylformamide-acetone (2:1) impregnated paper of the FN2 type produced by the K. A. Bodenstein Plant, Kupferhammer, Germany. Descending chromatography using dimethylformamide-decalin (1:1) was used for separation. The chromatogram was allowed to develop for 8 hr and the benzopyrene spot was located by ultraviolet irradiation, cut from the paper, and extracted with cyclohexane. Quantitative determination was accomplished by measuring the ultraviolet absorption spectrum in the range 375-400 nm. Using this method, 0.5-1 μg/ml can be determined with a high level of reproducibility.

Acetylated paper chromatographic methods also have been used in the analysis of polynuclear aromatic hydrocarbons. Spotswood [23, 24] has described an acetylation technique that produces papers with highly reproducible properties and of excellent mechanical strength. It was found that other methods of acetylating paper for chromatography [25-28] produced insufficiently acetylated paper to provide adequate resolution of polycyclic compounds. The recommended acetylating mixture consists of acetic anhydride (redistilled, 700 ml), thiophene-free benzene (1500 ml), and H_2SO_4 (2-3 g, 92%). Whatman No. 1 paper (16 strips of dimensions 6 × 24 in.) was immersed in the mixture for 24 hr at 18°C with frequent stirring and separation of the papers. The paper was drained and immersed in ethanol for 24 hr. After an additional ethanol wash, followed by a 6-hr rinse in running water and a 2-hr rinse in distilled water, the paper was air dried. Best reproducibility was obtained by using a newly prepared batch of acetylating mixture for each treatment, although three batched of paper could be acetylated with satisfactory results if 0.5 g of H_2SO_4 was added prior to each treatment.

The resolving power of the paper was varied by changing the amounts of H_2SO_4 added, which caused a change in the acetyl content of the paper. The use of 1.6, 2.0, 2.5, and 3.4 g of 92% H_2SO_4 provided acetyl contents of 21.2, 23.7, 25.1, and 28.0%, respectively. In general, R_f values decrease with increasing acetyl content.

Several solvent systems provided high resolution of the sample components, e.g., ethanol-toluene-water (17:4:1); ethanol-benzene-water (12:6:1); methanol-toluene-water (12:6:1 or 10:1:1); methanol-benzene-water (12:6:1); and methanol-diethyl ether-water (4:4:1). The solvent systems containing methanol were most useful with paper of low acetyl content and those containing ethanol with paper of high acetyl content. However, when ultraviolet absorption spectra are to be determined, methanol-ether-water (4:4:1) is the most appropriate solvent because benzene or toluene interferes with the absorption measurements.

Fluorescent compounds were located on the papers by viewing under ultraviolet light. Nonfluorescent compounds were located by spraying the paper with 2% tetrachlorophthalic anhydride in an acetone-chlorobenzene (10:1) mixture, followed by viewing under ultraviolet light [23]. With this technique, all nonfluorescent compounds examined except naphthalene gave yellow or orange spots.

This technique has been used by several investigators. Dubois et al. [29] used the acetylated paper produced by Schleicher and Schüll, Dassel, Kreis Einbeck, Germany. Among the papers available are No. 2043a, which is completely acetylated, and Nos. 2043b, 2045, and 2095, which are partially acetylated. The papers can be obtained with three different types of finish, mat, glazed, and machine. Most satisfactory separations were achieved on paper No. 2043b with the machine finish. Solvent systems used in these investigations included methanol-ether-water (4:4:1) and methanol-chloroform (3:1), as described by Kracht [30], and ethanol-toluene-water (17:4:1) and ethanol-benzene-water (12:6:1), as described by Spotswood [23]. Dubois et al. [29] recommended that as many criteria as possible be used in identifying the compounds. Among the criteria suggested were R_f values, color produced under ultraviolet light, and ultraviolet absorption and fluorescence spectra.

Hoffmann and Wynder [31-33] have used a similar method to that of Spotswood [23] to determine the contents of polynuclear aromatic hydrocarbons in tobacco smoke [31] and in automobile exhausts and polluted air [32, 33]. Tobacco smoke condensate was extracted with benzene and the extract fractionated on silica gel columns [31].

For the analysis of automobile exhaust condensates [32, 33], dry benzene extracts were partitioned between 100 ml methanol-water (4:1) and an equal volume of cyclohexane [34]. Polynuclear aromatic hydrocarbons partitioned into the cyclohexane phase. The methanol-water was extracted three times with 100 ml of cyclohexane and the combined cyclohexane layers were reduced to a 100 ml volume and extracted five times with 100 ml of nitromethane to remove most of the paraffins. Dried residues of the nitromethane extract were further separated on an Al_2O_3 (activity I, near neutral) column. Fractions displaying fluorescence were separated into individual polynuclear aromatic hydrocarbon components by chromatography on acetylated papers [24]. Papers acetylated to different extents were tried, and best results were obtained with paper containing 24-26% acetyl using methanol-diethyl ether-water (4:4:1) as the mobile solvent phase. To insure reproducibility of the order of compound separation, N, N-dimethylformamide (2%) was added to the solvent. Apparently, variations in acetyl content of the paper can cause changes in the order of development. Because variations in R_f values of as much as 15% were noted, the movement of the hydrocarbons was reported relative to the movement of benzo[a]pyrene. For effective purification of the compounds it was necessary to repeat the chromatographic procedure two to five times. After each chromatogram was developed, the fluorescent bands were cut from the paper and extracted with benzene-ethanol (4:1) mixture under nitrogen by a soxhlet technique. Final purification was accomplished by chromatography on silica gel columns.

Determination of 3,4-benzopyrene in benzene extracts of dust samples [35] has been accomplished using acetylated paper [23] and descending development. The chromatogram was allowed to develop for 20-24 hr. For samples containing large amounts of oil or tar it was necessary to repeat the chromatographic procedure in order to obtain a pure compound. Fluorimetric measurement in H_2SO_4 was used to determine 3,4-benzopyrene [36].

Cleary [37] developed a column chromatographic technique for the separation of polycyclic hydrocarbons in airborne particulates using long alumina columns. For many samples this method alone gave sufficient resolution; however, in some fractions it was necessary to use paper chromatographic methods to obtain satisfactory resolution. Acetylated paper type Schleicher and Schüll 2043b (machine finished) with a solvent system consisting of acetic acid-diethyl ether-water (12:5:20) or n-propanol-diethyl ether-water (20:9:4) provided the most satisfactory resolution.

Using paper chromatographic techniques similar to those described by Tarbell et al. [12] and Conway and Tarbell [13], it was possible to identify 18 polynuclear aromatic hydrocarbons in cigarette smoke condensate and to obtain the compounds in a high state of purity [38, 39]. Among the compounds positively identified were pyrene, 4-methylpyrene, fluoranthene, 8-methylfluoranthene, benzo[a]pyrene, benzo[e]pyrene, benzo[ghi]perylene, chrysene, perylene, benzo[mno]fluoranthene, benzo[j]fluoranthene, 1,2-benzofluorene, dibenz[ah]anthracene, a methylbenzo[a]pyrene, benzo[c]phenanthrene, an alkylfluoranthene, and an alkylchrysene. Four heterocyclic components, namely 1,8,9-perinaphthoxanthene, 7H-dibenzo-[cg]carbazole, dibenz[aj]acridine, and dibenz[ah]acridine, were isolated from cigarette tar [39, 40].

Extensive use has been made of paper chromatographic separation and identification of polynuclear aromatic hydrocarbons and, with the use of acetylated papers, excellent resolution of closely related compounds can be attained.

B.   Carbonyl Compounds

As early as 1949, Cavallini et al. [41, 42] described a paper chromatographic method for the separation and identification of the 2,4-dinitrophenylhydrazones of aldehydes and ketones. Rice et al. [43] adapted the capillary-ascent test tube method of Rockland and Dunn [44] for separating 2,4-dinitrophenylhydrazones. The 2,4-dinitrophenylhydrazones were applied to the paper in chloroform solution and it was found that no drying was necessary before or after development of the chromatogram. After the solvent boundary has been reached (1 cm from the top of the paper), the paper strip is removed from the tube and sprayed with a 10% aqueous solution of KOH. While still wet the strips are placed on a white background and the colored spots formed from the reaction of the hydrazones with KOH are outlined. By this method the 2,4-dinitrophenylhydrazones can be separated in microgram quantities. The authors [43] state that a difference of 0.05 between the R_f values of two compounds is required before proper resolution of the spots can be achieved. Typical solvents used for separation of the 2,4-dinitrophenylhydrazones of aldehydes and ketones are [43]:   5% diethyl ether in petroleum ether (boiling range $65°-110°C$); 50 ml of 30% aqueous acetone containing 1 ml of petroleum ether (boiling range $20°-40°C$); and 30% tetrahydrofuran in petroleum ether (boiling range $65°-110°C$). Separations achieved for carbonyl compounds on Whatman No. 1 filter paper using diethyl ether in petroleum ether as solvent include the 2,4-dinitrophenylhydrazones of salicaldehyde ($R_f = 0.30$), cinnamaldehyde (0.48), propionaldehyde (0.57), furfural (0.57), benzaldehyde (0.68), formaldehyde (0.75), acetone (0.85), n-butyraldehyde (0.85), methylethylketone (0.90), and acetaldehyde (0.79). Resolution of some of these compounds was improved by using Schleicher and Schüll No. 598 paper impregnated with silicic acid [45].

The method of separating 2,4-dinitrophenylhydrazones described by Meigh [46, 47] reputedly gave improved separations of volatile aldehydes and ketones dissolved in petroleum ether at levels of 10^{-7}M. Whatman No. 1 paper was used in these systems with a solvent system consisting of methanol-heptane. The method of descending irrigation was used and the solvent front was allowed to move 20 cm, which took approximately 1 hr. The solvent system used by Meigh [46, 47] was modified by Wallgren and Nordlund [48] by the addition of 10% of glacial CH_3COOH to methanol-99% heptane in a ratio of 1:4 by volume after the papers had equilibrated in the two-phase solvent system. This system was used to identify carbonyl compounds as their 2,4-dinitrophenylhydrazones in automobile exhaust [49]. The CH_3COOH was added through a 0.25-in. diameter hole in the top of a center glass covering the chromatographic jar so that the papers could be developed by the ascending method [50].

Automobile exhaust samples were obtained by Barber and Lodge [49] from an automobile on a chassis dynamometer that was operating in such a way as to closely simulate normal driving conditions. Samples were collected in smog-type bubblers at a rate of 500-1000 ml/min for a total sample volume of 60-90 liters. The bubblers each contained 15 ml of a 15% solution of $NaHSO_3$ at the temperature of ice water. If the samples had to be stored prior to analysis they were maintained at $5°C$. Free carbonyl compounds were released from the bisulfite solutions by the addition of Na_2CO_3 in a closed flask fitted with a side arm that was dipped into ether. Vapors released as a result of the neutralization were trapped in the ether and the aqueous solution was extracted for 3-4 hr with ether. The ethereal solution was evaporated to a volume of approximately 100 ml and the 2,4-dinitrophenylhydrazones were synthesized by two different methods, namely, by adding 25-30 ml of saturated 2,4-dinitrophenylhydrazine in 2N HCl [51] or by adding 2,4-dinitrophenylhydrazine in concentrated H_2SO_4 and ethanol [52]. The reaction was completely by stirring the solution for 4-6 hr and allowing it to stand overnight in the refrigerator. The aqueous solution was distilled to about one-third of its original volume and the distillate was collected in ice water [53]. The distillate was reacted with 2,4-dinitrophenylhydrazine as described above. Following formation of the phenylhydrazones the ethereal and aqueous solutions were extracted separately with 75-100 ml benzene for a total of eight to ten extractions. The extracts were washed with 2N HCl or 4N H_2SO_4 — depending on the method of preparation of the 2,4-dinitrophenyl-hydrazones — to remove excess 2,4-dinitrophenylhydrazine and concentrated to a known volume.

Paper chromatographic analysis was conducted using the Meigh [46, 47] system modified by Wallgren and Nordlund [48] described above. For identification of the 2,4-dinitrophenylhydrazone derivatives of some carbonyls the DeJonge system [54] was used. The solvent system consisted of cyclohexane-80% methanol (4:1) and was useful for separating the carbonyl derivatives from formaldehyde to butyraldehyde but was not capable of separating higher carbonyl derivatives. It was found [49] that the carrying capacity of the paper could be increased by impregnating it with 10% propylene glycol in 20% acetone as described by Horner and Kirmse [55].

Buyske et al. [56] have developed a method of general applicability to the determination of lower molecular weight aldehydes and ketones in impure mixtures. The method was applied to cigarette smoke samples obtained by the use of an automatic smoker. The smoke was rapidly and completely condensed by passage through four cold traps placed in series and maintained in a liquid air bath. After removal of the traps from the liquid air, the frozen smoke was dissolved in 200 ml of methanol containing 2 g of recrystallized 2,4-dinitrophenylhydrazine and 0.5 ml of 1N HCl. The resulting dark colored solutions were heated under reflux for 2 hr. One milliliter of the solution was streaked across a sheet of 20 × 40 cm Whatman No. 1 filter paper of chromatographic grade. Paper preparation involved immersion for 30 sec in N,N-dimethylformamide, followed by drying for 45 min in a forced draft hood. By use of a small orifice pipet it was possible to keep the streak width to <1 cm if a hair dryer was held over the paper to volatilize methanol at such a rate that 1 ml could be applied in 10 min. The papers were developed by descending chromatography in a glass cylinder (30 × 46 cm) lined with filter paper that was kept saturated with n-hexane by immersion to a 2 cm depth. The papers were allowed to equilibrate overnight at 15°-18°C in the saturated hexane atmosphere, at which time the chromatograms were developed in hexane previously saturated with N,N-dimethylformamide and cooled to 15°-18°C. After 4 hr, the streak was resolved clearly into eight yellow bands of width 1-2 cm. These bands were clearly visible in artificial light and the boundaries showed improved definition in ultraviolet light. The bands were cut from the paper and eluted with methanol in a soxhlet extractor. Following extraction and concentration, the bands were individually rechromatographed and band 5 was found to divide into two streaks. The bands were again cut from the paper, re-extracted, and tested for homogeneity by chromatographing in three solvent systems — n-hexane on N,N-dimethylformamide-treated paper, methanol-isooctane (1:1) on untreated paper, and water-methanol (95:5) on silane-treated paper. All unknowns were found to be homogeneous in all three systems. Qualitative identification of the phenylhydrazones of the unknown carbonyls was made by comparison with known pure compounds. Quantitative determinations were made by absorption spectrometry.

By quantitative conversion of carbonyls to their 2,4-dinitrophenylhydra-zone derivatives, satisfactory separation of carbonyl compounds in polluted air, cigarette smoke, and car exhaust samples can be achieved.

C. Alcohols, Phenols, and Amines

A method for separating volatile aliphatic alcohols as their 3,5-dinitrobenzo-ate derivatives has been described for quantities in excess of 10^{-7}M [57]. The alcohols, dissolved in benzene or petroleum ether, were chromatographed on Whatman No. 1 paper soaked in a methanolic solution of Rhodamine 6 GBN.500 (20 mg/liter) as supplied by I.C.I., Ltd. The papers were allowed to drain and were dried at $100°C$. The benzoates were revealed as dark spots on a yellow fluorescent background. Spots containing $10^{-8}M$ were just visible, while those containing $10^{-6}M$ tended to overload the paper. A descending irrigation technique was used and arrangements were made to maintain a saturated solvent atmosphere in the chamber. The benzoates in benzene solution were applied to the paper and the paper was placed inside the developing chamber, which contained three troughs side by side. The paper was hung in the center trough overnight and the outer troughs were filled, one with methanol saturated heptane and the other with heptane saturated methanol. After the paper was equilibrated overnight, the center trough was filled with the heptane phase through a hole in the chamber lid and the chromatograms were developed. Development time for the solvent front to move 25 cm was approximately 90 min. Separations were achieved for methanol ($R_f = 0.24$), ethanol (0.39), 1-propanol (0.46), 2-propanol (0.51), 1-butanol (0.57), 2-butanol (0.61), 2-methyl-1-propanol (0.55), 1-pentanol (0.66), 3-methyl-1-butanol (0.65), and 1-hexanol (0.72).

The method of determining aliphatic alcohols in air samples as their 3,5-dinitrobenzoate derivatives was utilized by Pinigina [58]. The air samples were collected on activated carbon by an aspiration technique. The activated carbon was extracted with benzene at $55°-60°C$ for 1 hr. To a 1-ml aliquot of the benzene solution, 1.2 ml of 10% pyridine in benzene and 2 ml of the acid chloride of 3,5-dinitrobenzoic acid were added and the solution was diluted to 10 ml with benzene. After 30 min the excess acid chloride was removed by extraction with 50% NaOH. The benzene solution was evaporated to dryness and the residue was dissolved in 0.15 ml of benzene. Chromatographic separation of the C_1-C_6 alcohols was achieved on one 0.05-ml portion and the C_7-C_{10} compounds on a second 0.05-ml portion. For the higher molecular weight alcohols, the paper (type not stated) was dried at $100°-120°C$ for 20-30 min, and the solvent system consisted of a stationary phase of 10% petrolatum in hexane and a moving phase of formamide. The lower molecular weight alcohols were separated by descending chromatography using a stationary phase of hexane or cyclohexane and a moving phase of 50% N,N-dimethylformamide in either ethanol or acetone. The spots were developed by spraying with a solution consisting of 0.7 g of $SnCl_2$ dissolved in 5 ml of H_2O and 15 ml of concentrated HCl and diluting to 100 ml followed by a second solution containing 1 g of p-dimethylaminobenzaldehyde and 5 ml of concentrated HCl in 95 ml of ethanol. For the saturated aliphatic alcohols in the range C_1-C_{10}, the R_f values were C_1, 0.11; C_2, 0.20; C_3, 0.30; C_4, 0.38; C_5, 0.45; C_6, 0.51; C_7, 0.54; C_8, 0.40; C_9, 0.22; C_{10}, 0.13. With use of the two different chromatographic systems it appears that excellent resolution of aliphatic alcohols is achieved.

A method has been described for the separation of simple phenols in automobile exhaust by paper chromatography [59]. Raw automobile exhaust samples (3-5 m^3) were collected from an automobile mounted on a chassis dynamometer and operated to simulate normal driving conditions. The samples were collected in four impingers connected in series. Three impingers contained 250 ml of NaOH ranging in concentrations from 0.1N to 1.0N at ice-water temperature, while the fourth was used as a cold trap. The phenols were released from the NaOH by adding HCl until the solution was acidic. The solution in each impinger was extracted twice with 100 ml of chloroform; the fourth impinger (the cold trap) was rinsed with 50 ml of chloroform. Portions of 50-70 ml of the chloroform extract from each of the three impingers were combined and reacted with 25 ml of 0.0105M of diazotized o-nitroaniline. After thorough shaking, the mixture was allowed to stand for 2 min and was then made alkaline with 25 ml of 25% by weight Na_2CO_3 as described by Crump [60]. The solutions were acidified and the azo dyes were extracted with diethyl ether until no further color was evident in the aqueous solution. The ethereal chloroform solution dried over $MgSO_4$ was evaporated to dryness by a stream of air. The azo dye extract was dissolved in chloroform and made up to a specific volume. Paper chromato-graphic separations of phenols in which the 4 position is unsubstituted have been made on these solutions by the method of Crump [60] using benzene-cyclohexane-dipropylene glycol (30:70:3) as solvent and papers impregnated with 20% formamide. For phenols possessing substituents in the 4 position, papers impregnated with acetone-N,N-dimethylformamide (3:1) were used [61] and a solvent system consisting of N,N-dimethylformamide-hexane (1:4) [62]. By these methods excellent separations of a wide variety of phenols were obtained.

Aliphatic amines in air have been identified gas chromatographically as their benzamides [62]. Samples of airborne particulates (3-8 g) were moistened with NaOH and extracted with 3 × 250 ml portions of anhydrous ether (11-12 hr). After the extract was decanted through a filter it was acidified with HCl gas. The volume of extract was reduced to 50 ml in an air stream and a slight excess of 10% NaOH and 0.10-0.15 ml of radioactive benzoyl chloride were added; the reaction mixture was then shaken for 20 min. The benzamides thus formed were extracted three times with a total volume of 40 ml of benzene; the extract was reduced in volume to 5 ml and chromatographed. Satisfactory separations were achieved by the systems described by DeJonge [54], i.e., methanol-cyclohexane (1:4), or propanol-cyclohexane (1:4). The spots were located in ultraviolet light and the lower limit of detectability was 2.5 μg of the benzamides. Radioautographs of the chromatograms were prepared by laying the chromatogram on X-ray film for 8-10 days, at which time the radioautographs were developed.

Excellent resolution was achieved for the benzamides formed from a variety of aliphatic amines, and the radioisotope method provided a high degree of sensitivity.

Paper chromatographic techniques have not been used extensively in air pollution investigations with the exception of analyses for polynuclear aromatic hydrocarbons. For qualitative analysis and for purification of samples the technique has proved useful but, because of the low carrying capacity of paper, the method is restricted as a quantitative technique. Furthermore, the lengthy development time needed for paper chromatograms will perhaps continue to make other chromatographic methods of analysis more attractive.

REFERENCES

1. E. R. Hendrickson, in Air Pollution, Vol. II (A. C. Stern, ed.), Academic, New York, 1968, pp. 3-52.

2. W. E. Ruch, Quantitative Analysis of Gaseous Pollutants, Ann Arbor-Humphrey, Ann Arbor, Mich., 1970, 241 pp.

3. J. F. Thomas, B. D. Tebbens, M. Mukai, and E. N. Sandborn, Anal. Chem., 29, 1835 (1957).

4. B. D. Tebbens, J. F. Thomas, and M. Mukai, Amer. Med. Assoc. Arch. Ind. Health, 14, 413 (1956).

5. B. D. Tebbens, J. F. Thomas, and M. Mukai, Amer. Med. Assoc. Arch. Ind. Health, 13, 567 (1956).

6. A. P. Altshuller, in Air Pollution, Vol. II (A. C. Stern, ed.), Academic, New York, 1968, pp. 115-145.

7. D. Hoffmann and E. L. Wynder, in Air Pollution, Vol. II (A. C. Stern, ed.), Academic, New York, 1968, pp. 187-247.

8. E. Sawicki, Chemist-Analyst, 53, 24 (1964).

9. E. Sawicki, Chemist-Analyst, 53, 56 (1964).

10. L. Rutter, Nature (London), 161, 435 (1948).

11. L. Rutter, Analyst, 75, 37 (1950).

12. D. S. Tarbell, E. G. Brooker, A. Vanterpool, W. Conway, C. J. Claus, and T. J. Hall, J. Amer. Chem. Soc., 77, 767 (1955).

13. W. Conway and D. S. Tarbell, J. Amer. Chem. Soc., 78, 2228 (1956).

14. J. Ahlmann, Acta Pathol. Microbiol. Scand., 43, 379 (1958).

15. J. Gasparič, Mikrochim. Acta, 1958, 68 (1958).

16. A. Zaffaroni, R. B. Burton, and E. H. Keutmann, Science, 111, 6 (1950).

17. A. Zaffaroni, R. B. Burton, and E. H. Keutmann, J. Biol. Chem., 188, 763 (1951).

18. L. C. Mitchell, J. Assoc. Offic. Agr. Chemists, 42, 161 (1959).

19. N. P. Buu and P. Jacquignon, Experimentia, 13, 375 (1957).

20. H. T. Gordon, Science, 128, 414 (1958).

21. K. Wettig, Hyg. Sanit., 29.1, 66 (1964).

22. G. Grimmer, Beitr. Tabakforsch., 3, 107 (1961).

23. T. M. Spotswood, J. Chromatog., 2, 90 (1959).

24. T. M. Spotswood, J. Chromatog., 3, 101 (1959).

25. J. V. Koster and K. Slavik, Coll. Trav. Chim. Tchecoslov., 15, 17 (1950).

26. J. E. Scott and L. Golberg, Chem. Ind. (London), 1954, 48 (1954).

27. H. S. Burton, Chem. Ind. (London), 1953, 1229 (1953).

28. F. Micheel and H. Schweppe, Mikrochim. Acta, 1954, 53 (1954).

29. L. Dubois, A. Corkery, and J. L. Monkman, Int. J. Air Poll., 2, 236 (1960).

30. W. Kracht, Inaugural Dissertation, Johann Wolfgang Goethe Universität, Frankfurth/Main, Germany.

31. D. Hoffmann and E. L. Wynder, Cancer, 13, 1062 (1960).

32. D. Hoffmann and E. L. Wynder, Nat. Cancer Inst. Monograph, 9, 91 (1962).

33. D. Hoffmann and E. L. Wynder, Cancer, 15, 93 (1962).

34. D. Hoffmann and E. L. Wynder, Anal. Chem., 32, 295 (1960).

35. G. Linstedt, Atm. Env., 2, 1 (1968).

36. E. Sawicki, W. Elbert, T. W. Stanley, T. R. Hauser, and F. T. Fox, Int. J. Air Poll., 2, 273 (1960).

37. G. J. Cleary, J. Chromatog., 9, 20 (1962).

38. B. L. van Duuren, J. Nat. Cancer Inst., 21, 1 (1958).

39. B. L. van Duuren, J. Nat. Cancer Inst., 21, 623 (1958).

40. B. L. van Duuren, J. A. Bilbao, and C. A. Joseph, J. Nat. Cancer Inst., 25, 53 (1960).

41. D. Cavallini, N. Frontali, and G. Toschi, Nature (London), 163, 568 (1949).

42. D. Cavallini, N. Frontali, and G. Toschi, Nature (London), 164, 792 (1949).

43. R. G. Rice, G. J. Keller, and J. G. Kirchner, Anal. Chem., 23, 194 (1951).

44. L. B. Rockland and M. S. Dunn, Science, 109, 539 (1949).

45. J. G. Kirchner and G. J. Keller, J. Amer. Chem. Soc., 72, 1867 (1950).

46. D. F. Meigh, Nature (London), 169, 706 (1952).

47. D. F. Meigh, Nature (London), 170, 579 (1952).

48. H. Wallgren and E. Nordlund, Acta Chem. Scand., 10, 1671 (1956).

49. E. D. Barber and J. P. Lodge, Jr., Anal. Chem., 35, 348 (1963).

50. A. P. DeJonge and A. Verhage, Rec. Trav. Chim., 76, 221 (1957).

51. O. L. Brady, Analyst, 51, 77 (1926).

52. C. F. H. Allen, J. Amer. Chem. Soc., 53, 2955 (1930).

53. F. E. Huelin, Australian J. Sci. Res. Ser. B, 5, 328 (1952).

54. A. P. DeJonge, Rec. Trav. Chim., 74, 760 (1955).

55. L. Horner and W. Kirmse, Ann. Chem., 597, 48 (1955).

56. D. A. Buyske, L. H. Owen, P. Wilder, Jr., and M. E. Hobbs, Anal. Chem., 28, 910 (1956).

57. D. F. Meigh, Nature (London), 169, 707 (1952).

58. I. A. Pinigina, Gig. Sanit., 30, 65 (1965); in Chem. Abstr., 64, 7265e (1966).

59. E. D. Barber, E. Sawicki, and S. P. McPherson, Anal. Chem., 36, 2442 (1964).

60. G. B. Crump, J. Chromatog., 10, 21 (1963).

61. J. Borecky, Mikrochim. Acta, 5, 824 (1962).

62. J. P. Lodge, Jr., and E. D. Barber, Anal. Chim. Acta, 24, 235 (1961).

Chapter 5

THIN-LAYER CHROMATOGRAPHIC ANALYSIS IN AIR POLLUTION

Daniel F. Bender and Walter C. Elbert*

National Fields Investigations Center — Cincinnati
U.S. Environmental Protection Agency
Cincinnati, Ohio

*This chapter was prepared by the authors in their private capacity and does not reflect the opinion or policy of the U.S. Government.

I. INTRODUCTION

This chapter contains information concerning the thin-layer chromatographic systems that have been applied to air pollutants. It is a field that is
approximately 10 years old. In general, the chapter is limited to the
separation of organic constituents from the particulate material collected
from either the air of the community environment in general, in close
proximity to an air pollution source, in a close research chamber (often
automobile exhaust), or within a closed working or living area (as a sub-
marine for example). Inorganic constituents in the air have usually been
determined by methods other than thin-layer chromatography.

II. ACCURACY, PRECISION, AND DATA EVALUATION

For the most part air pollution thin-layer chromatography has been concerned with the qualitative identification of constituents with some semiquantitative estimations. Because of the extremely complicated nature of
the particulate fraction it is rare that an individual separated area consists
of a pure compound, even with cleanup procedures and multiple development
techniques. This has been overcome by employing a supplemental means of
identification, usually fluorescence spectrophotometry and occasionally
absorption spectrophotometry, after the separated area has been extracted
with a solvent. The sensitivity and ability of fluorescence spectrophotometry to be manipulated so as to eliminate interfering spectra have made
it an extremely useful method of characterizing and estimating organic
constituents in air particulate matter.

III. SAMPLING

The particulate material to be analyzed generally consists of tar that has
been extracted via soxhlet extraction with benzene from the collector. In
some cases it has been "cleaned up" by fractionating on a chromatographic
column after extraction. The collector for most experiments has been a
glass fiber filter pad through which air was pulled by a vacuum cleaner
motor. A roof over the horizontally placed pad prevented settleable dust
from being collected. Other methods of particulate collection, such as
dustfall jars, electrostatic precipitation, and the like, are also available.
Further details can be obtained from the literature [1-7]. An average
urban sample is obtained by pulling approximately 2000 m^3 of air through a
glass fiber filter over a 24-hr period. It usually contains about 236 mg of
particulates [8]. This in turn gives around 20 mg of benzene solubles, containing approximately 10 μg of benzo[a]pyrene [4, 6]. From 0.004 to 0.13%
of known compounds, mostly polynuclear aromatics, have been found in the
total airborne particulates of urban communities [5].

IV. SEPARATION OF INTERFERENCES

Cleanup procedures using column chromatography have often been used. These consist of a class separation: neutrals (aromatic hydrocarbons), acids, bases, oxygenates, and the like. Reference to these techniques is made in Sec. VIII in connection to instances where they have been applied. It was mentioned in Sec. II that fluorescence spectrophotometry can be manipulated in some cases to eliminate or reduce interfering spectra; thus, in a sense, it removes interferences in the identification step. Quite often it is possible to use the benzene extract directly without prior cleanup.

V. CONCENTRATION TECHNIQUES

Class separation (Sec. IV) is a method of concentrating the sought after constituent. In a sense multiple separation provides a means of concentration as well. Multiple separation either means two-dimensional thin-layer chromatography with two different solvent systems or it means extracting a given separated area and rechromatographing it.

Another often used technique is to spot the plate with a series of sample spots close together so that a band rather than a small circle is obtained, thus providing more of the constituent when extracted.

VI. DETECTION OF SEPARATED AREAS

Most of the air particulate analyses have been for substances hazardous to health. These are generally aromatic in character. The unsaturation involved is the theoretical basis for these compounds to fluoresce when exposed to ultraviolet radiation. This has two distinct advantages: it allows for the use of the extremely sensitive methods of fluorescence spectrophotometry for identifying and estimating extracted materials and it provides a convenient relatively nondestructive means of locating compounds on the thin-layer plate. In addition, the fluorescence characteristics can be viewed with various quenching reagents and used as a characterization method [9-11].

VII. TRACE ANALYSIS

Essentially all air pollution thin-layer chromatography can be regarded as trace analysis. It has been this way because the samples contain only trace quantities of any individual compound and it would thus be of no air pollution interest to do any macro work.

VIII. METHODS USED FOR VARIOUS SUBSTANCES

This section is divided into separations involving (a) aromatic hydrocarbons, (b) aromatic hydrocarbons containing carbonyl oxygen, (c) aromatic compounds containing aza nitrogen, (d) aromatic compounds containing amino and imino nitrogen, and (e) naturally occurring pollutants. Separations demonstrating community or source-related air pollution samples appear throughout where applicable.

The original work should be examined for details before research is commenced. Plate thickness is usually not mentioned although it generally should be 250 μm. Details of plate and solvent system preparation are not within the scope of this chapter and may be found in Part I of the monograph.

What this section attempts is to familiarize the reader with a system (substrate and solvent) by which he or she can accomplish the separation of

TABLE 1

Eluting Power of Solvents on Florisil Thin Layers[a]

Solvent	R_f values of standards[b]					
	A	B	C	D	E	F
Furan	0.96	0.00	0.94	0.55	0.39	0.21
Acetone	0.94	0.92	0.92	0.92	0.92	0.92
Acetonitrile	0.94	0.92	0.94	0.92	0.93	0.92
Ethyl acetate	0.93	0.87	0.91	0.86	0.91	0.88
Nitromethane	0.93	0.87	0.93	0.93	0.90	0.91
Chloroform	0.88	0.04	0.69	0.15	0.16	0.20
Toluene	0.80	0.00	0.60	0.09	0.00	0.00
Methanol	0.74	0.57	0.81	0.76	0.61	0.65
Acetic acid	0.48	0.00	0.67	0.00	0.71	0.68
Triethylamine	0.44	0.27	0.47	0.13	0.61	0.27
1-hexane	0.37	0.00	0.11	0.00	0.00	0.00
Carbon disulfide	0.30	0.00	0.08	0.00	0.00	0.00
Carbon tetrachloride	0.09	0.00	0.05	0.00	0.00	0.00
Hexachlorobutadiene	0.08	0.00	0.00	0.00	0.00	0.00
Hexane	0.03	0.00	0.00	0.00	0.00	0.00
Cyclohexane	0.02	0.00	0.00	0.00	0.00	0.00
Pentane	0.02	0.00	0.00	0.00	0.00	0.00
Trifluoroacetic acid	0.00	0.00	0.00	0.00	0.00	0.00

[a]From Ref. 12 with permission from the J. T. Baker Chemical Co.

[b]Standards: A, benzo[a]pyrene; B, pyrenoline; C, 4H-benzo[def]-carbazole; D, 1-aminopyrene; E, anthraquinone; F, benzanthrone.

interest. Some details of what each system can do and examples of its application are given along with references to these studies and applications.

Some systems are mentioned throughout that provide more of a class separation than the separation of individual components. In addition to those which appear throughout the rest of this section, there is a study in which representative members of classes of compounds were separated on florisil using a large number of solvents (Table 1). This study points out possibilities for wide applications of florisil after further research [12]. The proportions of substrate are expressed in w/w and proportions of solvents are expressed in v/v.

A. Separations Involving Aromatic Hydrocarbons

1. Aluminum Oxide with Pentane-Ether (19:1)

Twenty aromatic hydrocarbons were separated using this system (Table 2). The compounds exhibited R_B values from 0.10 to 1.25 but more than half of them had R_B values from 0.89 to 1.14 (R_B is the ratio of the distance traveled by the compound to the distance traveled by benzo[a]pyrene). Although this is considered a relatively poor system for the separation of polynuclear aromatic compounds from one another, it has been found to give superior separation of fluorescent areas with actual organic airborne particulate samples. Benzo[a]pyrene and benzo[ghi]perylene are separated. Benzo[a]pyrene, benzo[e]pyrene, and benzo[k]fluoranthene are more difficult to separate. Benzo[a]pyrene and perylene are separated [13].

Figure 1 shows a section of the bands that are obtained when a number of closely applied spots of a composite benzene-soluble fraction is developed in this system. By increasing the percentage of ether each band is brought closer to the solvent front so that further unknown groups may be eluted from the origin [13]. Figures 2 and 3 show the fluorescent areas obtained from the separation of various samples of interest in research involving air pollution chemistry, polynuclear aromatic hydrocarbon chemistry, and potential carcinogen studies. In Fig. 3 the spots applied at the origin are the subfractions of the benzene-soluble fraction of airborne particulates [2].

2. Cellulose with N,N-Dimethylformamide-water (1:1)

When 20 polynuclear aromatic hydrocarbons were separated with this system it was found to give a wide range of R_B values (Table 2). The following pairs of aromatic hydrocarbons are difficult to separate by aluminum oxide column chromatography but have been separated by this system: fluoranthene and pyrene, chrysene and benzo[a]anthracene, benzo[a]pyrene and benzo[e]pyrene, and benzo[a]pyrene and perylene. In addition, benzo[a]pyrene and

TABLE 2

Thin-Layer Chromatographic Separation of Aromatic Hydrocarbons[a]

Compound	Cellulosedimethyl-formamide-water (1:1)			Acetylated cellulose methanol-toluene-water (17:4:4)			Aluminum oxide pentane-ether (19:1)		
	R_B[b]	Fluor. color[c]		R_B	Fluor. color		R_B	Fluor. color	
		Wet	Dry		Wet	Dry		Wet	Dry
Phenanthrene	1.99	lB	lB	3.74	lB		1.13	lB	lB
Anthracene	1.99	lB	lB	3.33	dB		1.14	B	B
Fluoranthene	1.89	B	B	2.92 d	B		1.09	B	lB
Chrysene	1.75	B	Pk				1.10	B	Pk
Pyrene	1.72	B	B	3.16 d	B	lB	1.25	B	B
Triphenylene	1.49	dB	dB				1.07	lB	lB
Benz[a]anthracene	1.47	B	B	2.70	B	B	1.03	B	Pk
11H-Benzo[b]fluorene	1.33	B		3.54	dB		1.08	lB	lB
Benzo[e]pyrene	1.16	B	B	2.94	B	B	1.04	GB	GB
Perylene	1.14	BG	BG	2.86	GB	GB	0.91	B	lY
Benzo[k]fluoranthene	1.03	B	B	2.40	B	B	0.98	B	B
Benzo[a]pyrene	1.00	B	B	1.00	B	B	1.00	B	Pk
Anthanthrene	0.70	YG	YG	2.17	BG	lBG	0.71	B	lY
Benzo[ghi]perylene	0.69	BG	BG	3.04	B	BG	0.89	B	YG
Dibenz[a,h]anthracene	0.66	dB	dB	2.92	B	B	0.74	B	B
Naphtho[1,2,3,4-def]chrysene	0.48	B	dV	1.85	BG		0.78	B	lY
Benzo[rst]pentaphene	0.45	B	B	2.41	B	B	0.68	B	Pk

Coronene	0.37	BG	BG	2.87	BG	lBG	0.46	G	G
Benzo[a]coronene	0.15	dB	PkB	2.48	BG	BG	0.10	BG	1YO
Dibenzo[h,rst]pentaphene	0.14	BG	BG	2.35	BG	lBG	0.12	BG	lYO

[a] From Ref. 13 with permission from the publisher.

[b] 0.02 to 1 μg of hydrocarbon per spot.

[c] l, light; d, dark; B, blue; Pk, pink; G, green; Y, yellow; O, orange; and V, violet. Spots read under long-wavelength ultraviolet light.

[d] Could not find at 2 μg of hydrocarbon per spot.

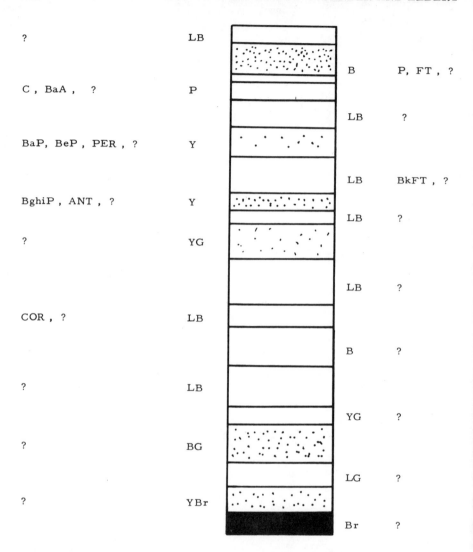

FIG. 1. Thin-layer chromatogram of bands separated on aluminum oxide of a composite benzene-soluble airborne particulate fraction. Symbols in use for Figs. 1-3 are L, light; B, blue; P, purple; Y, yellow; G, green; Br, brown; R, red; P, pyrene; FT, fluoranthene; C, chrysene; BaA, benz[a]anthracene; BaP, benzo[a]pyrene; BeP, benzo[e]pyrene; PER, perylene; BkFT, benzo[k]fluoranthene; BghiP, Benzo[ghi]perylene; ANT, anthanthrene. (From Ref. 13 with permission from the publisher.)

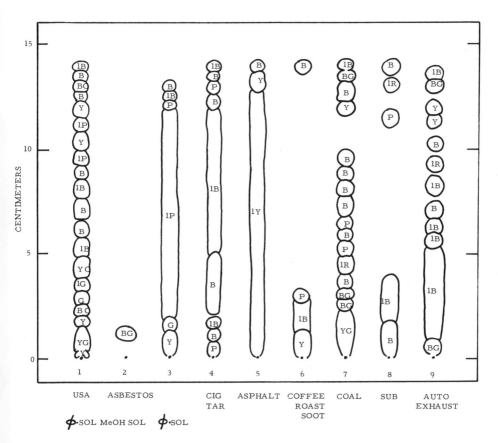

FIG. 2. Thin-layer chromatograms of various fractions. See legend of Fig. 1 for definition of symbols. (From Ref. 13 with permission from the publisher.)

benzo[ghi]perylene are separated. Benzo[a]pyrene and benzo[k]fluoranthene are not separated. R_B values range from 1.99 to 0.14 [13].

3. Cellulose Acetate with Ethanol-Toluene-Water (17:4:4)

Twenty polynuclear aromatic hydrocarbons were studied in this system (Table 2). The R_B values ranged from 3.74 to 1.0. It comes closest to complete separation of the components of the benzopyrene fraction, which are benzo[a]pyrene, benzo[k]fluoranthene, benzo[e]pyrene, and perylene. (The last two developed together [13].) Two difficult components of the benzopyrene fraction to separate, benzo[a]pyrene and benzo[k]fluoranthene,

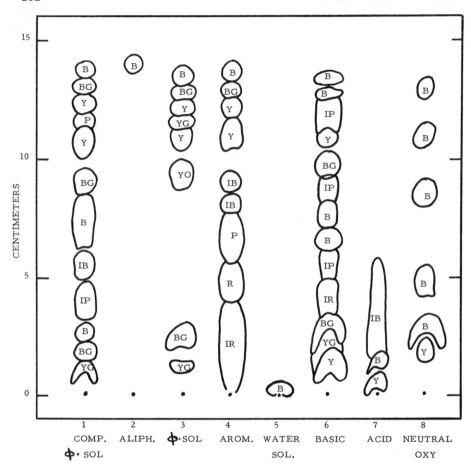

FIG. 3. Thin-layer chromatograms of various fractions and subfractions. See legend of Fig. 1 for definitions of symbols. (From Ref. 13 with permission from the publisher.)

can be separated from one another by this system [14]. This system was also used in a study of direct fluorometric scanning techniques [15].

4. Aluminum Oxide-Cellulose (2:1) with Ethanol-Toluene-Water (17:4:4)

Used for the separation of components in an urban atmospheric sample extracted with methylene chloride in semidarkness to prevent a photochemically initiated interfering reaction. It is primarily intended for use in the quantitative determination of benzo[a]pyrene [16].

5. Aluminum Oxide-Cellulose Acetate (2:1) with Pentane (First Dimension) and Ethanol-Toluene-Water (17:4:4) (Second Dimension)

Also used in the study of urban atmospheric samples and involves different extraction methods than ordinarily used to obtain fractions [16]. It was also used in a study involving the benzene-soluble fraction in which it achieves separation of benzo[a]pyrene from benzo[k]fluoranthene [17]. Also see Sect. VIII, D, 10.

6. Aluminum Oxide with Pentane-Nitrobenzene (9:1)

This system was used primarily for the separation of aza heterocyclics. Polynuclear aromatic hydrocarbons can be distinguished from aza hetero-cyclics using quenching techniques with this system [18, 19]. It is discussed in Sec. VIII, C.

7. Aluminum Oxide with Pentane-2-Nitropropane-Triethylamine (9:1:0.01)

The comments in Sec. VIII, A, 6 apply to this system also [18]. This is the system discussed in Sec. VIII, C, 22 on aza heterocyclics.

8. Aluminum Oxide-Cellulose (2:1) with Pentane (First Dimension) and N,N-Dimethylformamide-Water (35:65) (Second Dimension)

This is further discussed under aza heterocyclics, although pyrene has been detected in a composite benzene-soluble fraction from urban particulate by this method [17].

9. Aluminum Oxide-Silica Gel (9:1) with Pentane-Ether (19:1)

A simple method to determine benzo[a]pyrene, benz[c]acridine, and 7H-benz[de]anthracene-7-one from crude benzene extracts [20].

10. Cellulose with Ethanol-Toluene-Water (17:4:4)

Used in a study of direct fluorometric scanning of thin-layer chromatograms to determine the concentration of various polynuclear aromatic hydrocarbons and aza heterocyclics [15].

11. Aluminum Oxide with Pentane-2-Nitropropane (19:1)

The system may be used to study quenching effects and adds to the ready differentiation of benzo[a]pyrene from benzo[k]fluoranthene [9,19].

12. Aluminum Oxide-Cellulose Acetate (2:1) with Cyclohexane

See Sec. VIII,A,15.

13. Aluminum Oxide-Cellulose Acetate (2:1) with N,N-Dimethylformamide-Water (3:1) Saturated with Ether

See Sec. VIII,A,15.

14. Aluminum Oxide-Cellulose Acetate (2:1) with Cyclohexane (First Dimension) and N,N-Dimethylformamide-Water (3:1) Saturated with Ether (Second Dimension)

See Sec. VIII,A,15.

15. Aluminum Oxide-Cellulose Acetate (2:1) with Cyclohexane Saturated with N,N-Dimethylformamide and Then Further Diluted with Cyclohexane (First Dimension) and N,N-Dimethylformamide-Water (3:1) Saturated with Ether (Second Dimension)

Dibenzo[a,e]pyrene was identified and estimated spectrophotofluorometrically in urban air particulate using these systems. The method discussed in Sec. VIII,A,12 was run with the benzene-soluble fraction and a standard. The area of corresponding R_f value was extracted and run on the system in Sec. VIII,A,13. Figure 4 shows a typical separation using the system in Sec. VIII,A,14, which can be compared to Fig. 5, which gives the separation described in this section. Saturation with N,N-dimethylformamide in the first dimension produces a significant difference [21] in the separation.

B. Separations Involving Aromatic Compounds Containing Carbonyl Oxygen

Infrared spectroscopy indicates that carbonyl compounds are present in the benzene-soluble fraction and in the aromatic, neutral oxygenated, acid, water-soluble, and basic subfractions of the benzene-soluble fractions [11]. Aluminum oxide systems do not give good reproducibility but are valuable in separating and identifying ring-carbonyl compounds when standards are run simultaneously.

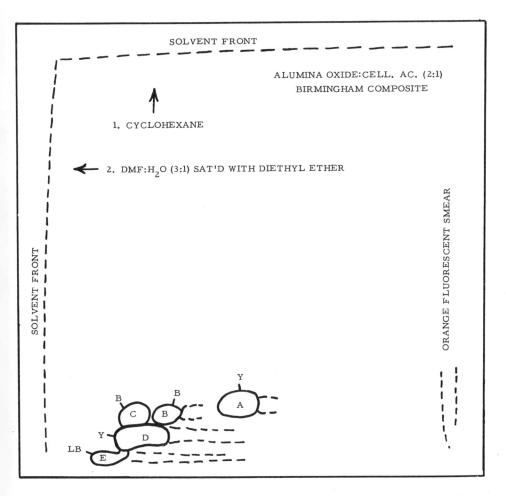

FIG. 4. Two-dimensional, mixed-substrate, thin-layer chromatogram of urban airborne particulate sample. Y, yellow fluorescence; B, blue fluorescence; LB, light blue fluorescence. Fluorescence spectra of each lettered fluorescent area were run. B was identified as dibenzo[a,e]pyrene. (From Ref. 21 with permission from the publisher.)

1. Aluminum Oxide with Toluene

A large series of ring-carbonyl compounds may be separated as shown in Fig. 6, especially compounds 3 and 4 [23]. Because of the complicated structures and nomenclature, the figures and tables will be heavily relied on to describe the compounds. Figure 7 shows the results of the separation of the separation of a neutral oxy fraction of airborne particulates using this system [24].

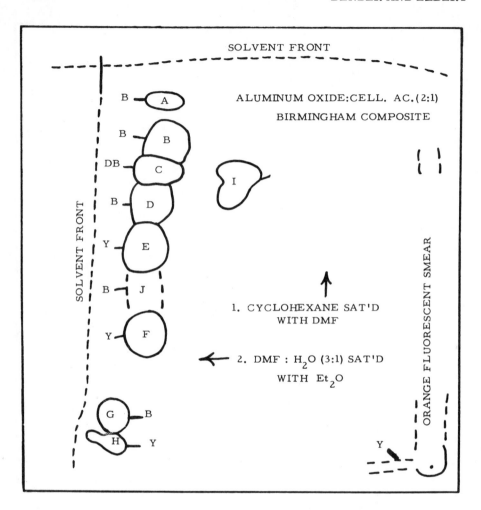

FIG. 5. Two-dimensional, mixed-substrate, thin-layer chromatogram of urban airborne particulate sample. Y, yellow fluorescence; B, blue fluorescence; DB, dull blue fluorescence. Fluorescence spectra of each lettered area were run. (From Ref. 21 with permission from the publisher.)

2. Aluminum Oxide with Toluene–Ether (19:1)

Of the systems shown in Fig. 6 this one produces the best overall separation [23].

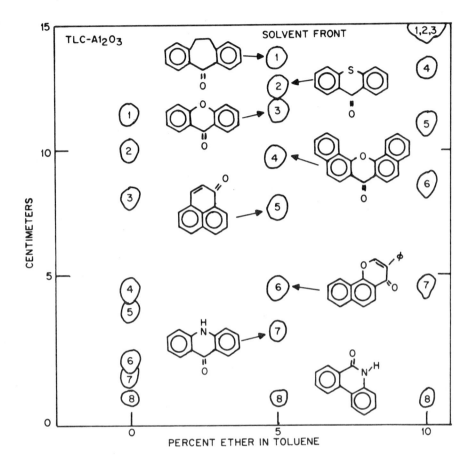

FIG. 6. The effect of an increasing percentage of ether in toluene on the alumina thin-layer-chromatographic separation of ring-carbonyl compounds. (From Ref. 23 with permission from the publisher.)

3. Aluminum Oxide with Toluene-Ether (9:1)

Figure 6 also demonstrates the type of separation obtained in this system; superior separation of compounds 4-8 is obtained [23].

4. Aluminum Oxide with Pentane-Ether (19:1)

Aromatic hydrocarbons, aza heterocyclics, ring carbonyls, and aromatic amines may be separated using this system. Figure 8 shows the results.

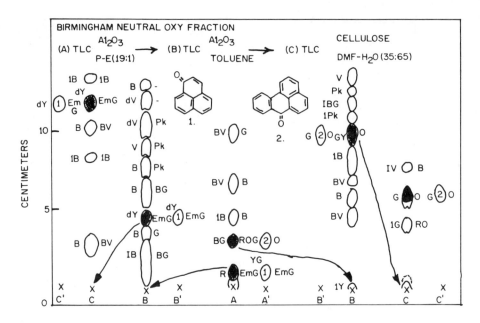

FIG. 7. One-dimensional thin-layer chromatography of a neutral oxy fraction
of the airborne particulates collected from the Birmingham downtown atmosphere. (A) Thin-layer chromatography on alumina with pentane-ether (19:1)
followed by (B) thin-layer chromatography of appropriate spots on alumina
with toluene, followed by (C) thin-layer chromatography of appropriate spots
on cellulose with dimethylformamide-water (35:65). A′, B′, and C′ are the
standards, phenalen-1-one and 7 H-benz[de]anthracen-7-one, separated as
in A, B, and C, respectively. Fluorescence colors: B, blue; d, dull; Em,
emerald; G, green; l, light; O, orange; Pk, pink; R, red; V, violet; and Y,
yellow. Letters on left of spot represent fluorescence colors of wet spot
before treatment; letters on right, after treatment with trifluoroacetic acid
fumes. (From Ref. 24 with permission from the publisher.)

Carbonyl-type compounds are the main interference in analyzing for ring
carbonyls when present cleanup procedures are employed [23]. Figure 7
shows the separation of a neutral oxy fraction of airborne particulate using
this system [24].

5. Cellulose with N,N-Dimethylformamide-Water (65:35)

Again, many classes of compounds are separable in this system with carbazoles being the main interference in ring-carbonyl identification (Fig. 9) [23].
Figure 7 shows the results of the separation of a neutral oxy fraction of
airborne particulate using this system [24].

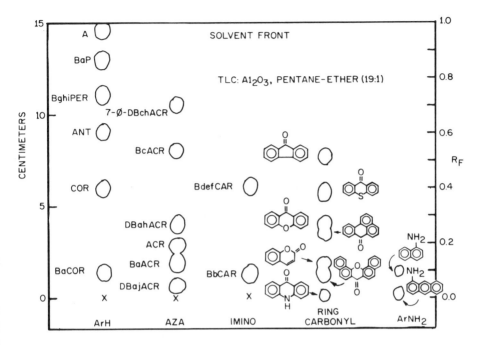

FIG. 8. A comparison of the alumina thin-layer chromatographic separation of various families of polynuclear compounds with pentane-ether as developing solvent (all separations with benzo[a]pyrene, $R_f = 0.85$, as internal standard). (From Ref. 23 with permission from the publisher.)

6. Aluminum Oxide with Toluene-Methanol (19:1)

The system produces a class separation as follows: ring carbonyls (R_f from 0.87 to 1.0), aromatic hydrocarbons (R_f from 0.96 to 1.0), aza heterocyclic compounds (R_f from 0.93 to 1.0), 6(5H)-phenanthridone (R_f approximately 0.60), and 9-acridone (R_f approximately 0.52) [23].

7. Aluminum Oxide with Cyclohexane-Ethylacetate [1:1] (First Dimension) and Toluene (Second Dimension)

In this system the ring carbonyls usually appear on the diagonal (Fig. 10) with aromatics above the diagonal and with amines, phenols, and acids below the diagonal. Using this information it was possible to identify some ring carbonyls from a composite organic airborne particulate fraction as shown in Fig. 11 [24]. Standards must be run because of the previously mentioned poor reproducibility of the R_f values. This system was used in a study

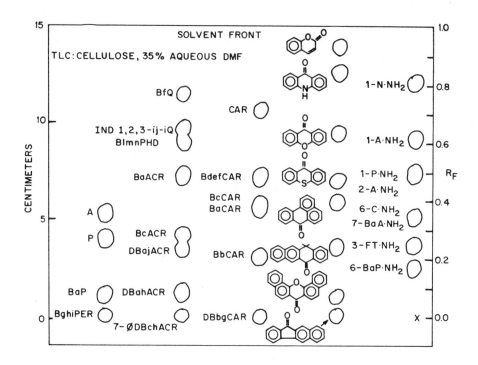

FIG. 9. A comparison of the cellulose thin-layer chromatographic separation of various families of polynuclear compounds with 35% aqueous dimethylformamide as developing solvent. (From Ref. 23 with permission from the publisher.)

comparing methods for the assay of 7H-benz[de]anthracene-7-one in airborne particulates and air pollution source effluents [25].

8. Aluminum Oxide with Methylene Chloride

The study of a number of methods for the assay of 7H-benz[de]anthracene-7-one in airborne particulate and in air pollution source effluents [25] is made possible by this system.

9. Aluminum Oxide:Cellulose Acetate (2:1) with Pentane (First Dimension) and Ethanol:Toluene:Water (17:4:4)

This system was used to separate and identify phenalen-1-one and 7H-benz-[de]anthracene-7-one in a nonbasic subfraction of the benzene-soluble fraction of airborne particulate (Fig. 12) [17].

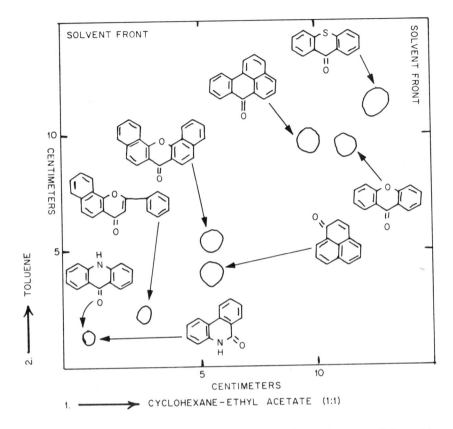

FIG. 10. Two-dimensional alumina thin-layer chromatogram of 9-acridone, 6[5H]phenanthridone, 2-phenyl-4H-naphtho(1,2-b)pyran-4-one, phenalen-1-one, 7H-dibenzo[c,h]xanthen-7-one, 7H-benz[de]anthracen-7-one, xanthen-9-one, and thiaxanthen-9-one. (From Ref. 24 with permission from the publisher.)

10. Silica Gel with Benzene

The systems in Sections VIII,B,10 through VIII,B,15 were used to separate azine derivatives of aromatic carbonyls [26]. The systems in Sec. VIII,B,10 and VIII,B,11 were also used in the separation of 4-nitrophenylhydrazone derivatives of aromatic carbonyls [27].

11. Silica Gel with Ethyl Acetate-Hexane (1:1)

See Sec. VIII,B,10.

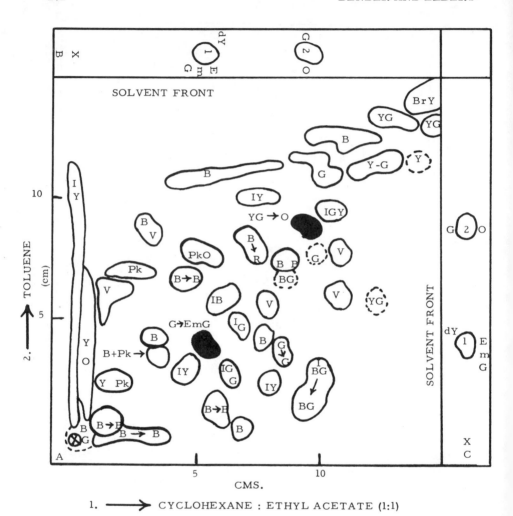

FIG. 11. Two-dimensional thin-layer chromatogram of (A) a composite organic airborne particulate fraction obtained from the polluted atmospheres of 100 large American cities; (B) Phenalen-1-one and 7H-benz[de] anthracen-7-one separated one dimensionally by solvent system 1; and (C) the same two standards separated by solvent system 2. Fluorescence colors defined in Fig. 7. Letters alone or preceding an arrow are the fluorescence colors of a freshly separated chromatogram. Letters to which arrows point and letters in a spot circumscribed by a dashed line represent the fluorescence colors after treatment with trifluoroacetic acid fumes. (From Ref. 24 with permission from the publisher.)

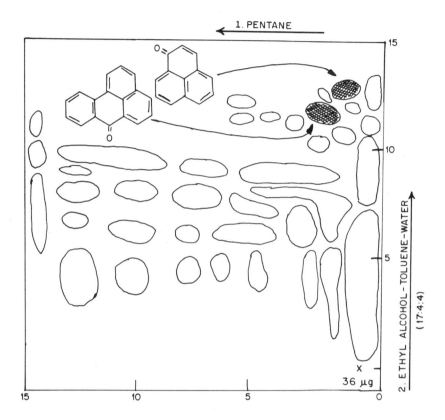

FIG. 12. Two-dimensional thin-layer chromatographic separation on alumina-cellulose acetate (2:1) of a nonbasic subfraction of the benzene-soluble fraction of the airborne particulates collected from the open burning of auto tires, floor mats, and seat covers. Separation of phenalen-1-one and 7H-benz[de]anthracen-7-one shown. (From Ref. 17 with permission from the publisher.)

12. Silica Gel with Benzene:Paraffin-Acetic Acid-Water (55:20:45:30)

See Sec. VIII,B,10.

13. Silica Gel with Tetrahydrofuran-Dibutyl Ether (4:85)

See Sec. VIII,B,10.

14. Silica Gel with Dichloromethane

See Sec. VIII, B, 10.

15. Silica Gel with Decalin-Dichloromethane-Methyl Alcohol (5 : 4 : 1)

See Sec. VIII, B, 10.

16. Silica Gel with Dichloromethane

This system, as well as those in Sec. VIII, B, 10, Sec. VIII, B, 11, and Sec. VIII, B, 17, was used in the separation of 4-nitrophenylhydrazone derivatives of aromatic carbonyls and applied to automobile exhaust effluent samples [27].

17. Silica Gel with Benzene-Methanol (19 : 1)

See Sec. VIII, B, 16.

18. Glass-Fiber Paper Impregnated with Silica Gel with Pentane-Trifluoroacetic Acid (50 : 1)

The term "instant thin-layer chromatography" is often used here because of its increased speed of development over conventional thin-layer chromatography without destruction of resolution. The system was used for the rapid assay of phenalen-1-one (Fig. 13) [28].

19. Glass-Fiber Paper Impregnated with Silica Gel with Pentane-Methylene Chloride (3 : 1)

Used to assay for phenalen-1-one and 7H-benz[de]anthracene-7-one in urban airborne particulate as shown in Fig. 14 [28].

20. Aluminum Oxide-Silica Gel (1 : 1) with Pentane-Ether (19 : 1)

The rapid analysis of the benzene-soluble fraction of samples from a large number of American cities for benzo[a]pyrene, benz[c]acridines, and 7H-benz[de]anthracene-7-one [20] is possible utilizing this sytem.

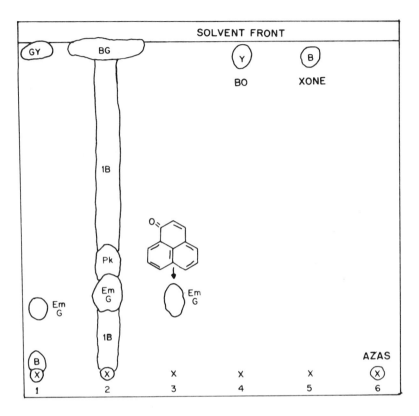

FIG. 13. Thin-layer chromatogram on silica gel glass fiber paper with pentane-trifluoroacetic acid (50 : 1, v/v) as the developer. The fluorescence colors of the compounds shown appeared after the chromatogram had been sprayed with trifluoroacetic acid fumes. Color abberviations are as follows: B, blue; Em G, emerald green; GY, green yellow; O, orange; BG, blue green; Pk, pink; Y, yellow; l, light. Amounts of the compounds separated are as follows: (1) 1 mg of coal-tar pitch, (2) 1 mg of a benzene-soluble fraction of urban airborne particulates, (3) 1 µg of phenalen-1-one; (4) 1 µg of 7H-benz[de]anthracen-7-one, (5) 1 µg of 9-xanthenone, and (6) a mixture containing 1 µg each of benzo[f]quinoline, benzo[h]quinoline, benz[a]acridine, and benz[c]acridine. (From Ref. 28 with permission from the publisher.)

21. Aluminum Oxide with Ethyl Acetate-Cyclohexane (1 : 4) (First Dimension) and Pentane-Toluene (1 : 1) (Second Dimension)

This system was used to identify 7H-benz[di]anthracene-7-one in the non-basic fraction of coal-tar pitch [17] (see also Sec. VIII, D, 11).

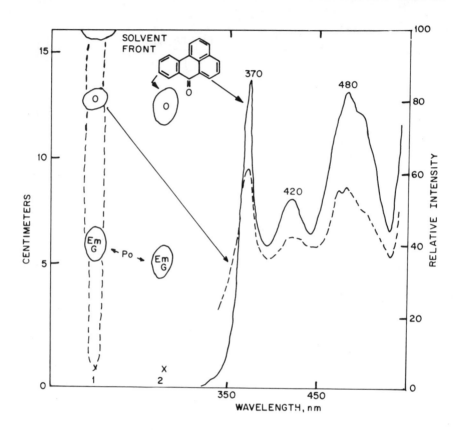

FIG. 14. On the left side appears the separation on silica gel glass-fiber paper of (1) 1 mg of a benzene-soluble fraction of urban airborne particulate from Greenville, S.C., and (2) a mixture containing 0.8 μg of phenalen-1-one and 0.8 μg 7H-benz[de]anthracen-7-one, with pentane-methylene chloride (3:1, v/v) as the developer. The fluorescence colors shown appeared after TFA fuming. For color abbreviations refer to Fig. 13. On the right side are shown the excitation spectra at emission λ 560 and meter multiplier (MM) 0.01 of (——) the extract from the BO standard spot in 0.5 ml concentrated sulfuric acid and (– – –) the extract from the unknown spot opposite the BO standard spot in 0.5 ml concentrated sulfuric acid. (From Ref. 28 with permission from the publisher.)

C. Separations Involving Aromatic Compounds
Containing Aza Nitrogen

In this section aza arenes and compounds that contain both nitrogen and oxygen are covered. Aza heterocyclic hydrocarbons and other basic compounds, such as imines and amines, can be separated by extraction of the benzene-soluble

TABLE 3

Thin-Layer Chromatographic Separation of Aza Heterocyclic
Hydrocarbons from Polynuclear Aromatic Hydrocarbons[a]

Adsorbent	Solvent	R_f				
		P	BaP	Cor	ACR	Pyre-noline
Cellulose	Formic acid–water (3:7)	0.02	0.00	0.00	0.8	0.3
Cellulose	Formic acid–water (1:1)	0.1	0.00	0.00	0.9	0.6
Cellulose	Acetic acid–water (2:3)	0.2	0.04	0.00	0.9	0.6
Carboxymethyl-cellulose	Acetic acid–water (3:7)	0.2	0.04	0.00	0.9	0.4
Cellulose acetate	Ethanol–toluene–water (17:4:4)	0.5	0.2		0.8	0.6
Alumina	Hexane–benzene (1:1)	0.9	0.9		0.4	0.7
Adsorbil	Hexane–benzene (1:1)	0.97	0.97		0.04	0.3
Silica gel	Hexane–benzene (1:1)	0.99	0.96		0.09	0.3
Florisil	Hexane–benzene (1:1)	0.9	0.9		0.00	0.02
Florisil	Pentane–benzene (19:1)	1.0	1.0	0.96	0.00	0.00
Florisil	Pentane–chloroform (19:1)	0.8	0.5	0.00	0.00	0.00
Florisil	Pentane–furan (19:1)	0.7	0.4	0.3	0.00	0.00
Florisil	Hexane	0.7	0.4		0.00	0.03
Florisil	Carbon tetrachloride	0.9	0.8	0.3	0.00	0.05
Florisil	Benzene–methanol (99:1)	1.0	1.0	1.0	0.2	0.7

[a]From Ref. 29 with permission from the publisher.

fraction with dilute aqueous acid [29]. A review of thin-layer chromatographic procedures for the analysis of aza heterocyclic compounds has been
published [30].

1 Through 15

Table 3 shows 15 systems that could be utilized to obtain an aza heterocyclic
fraction. In particular, with cellulose and formic acid–water (1:1) the polynuclear aromatic hydrocarbons remain at the origin, while the aza heterocyclic hydrocarbons have R_f values from 0.6 to 0.9. The florisil with
pentane–benzene (19:1) system gives polynuclear aromatic hydrocarbons at
the solvent front, aza heterocyclic hydrocarbons at the origin, and aromatic
amines and imines in between [29].

TABLE 4

Thin-Layer Chromatographic Separation of
Aza Heterocyclic Compounds[a]

| | R_f values | | | Color change on Al_2O_3 (on addition of TFA)[b] |
| | Cellulose | | | |
Compound	DMF-H_2O (35:65)	AcOH-H_2O (3:7)	Alumina P-E (19%1)	
7-Phenyldibenz[c,h]acridine	0.00	0.00	0.70	B → G
8,12-Dimethylbenz[a]acridine	0.20	0.62	0.60	B → B
8,10-Dimethylbenz[c]acridine	0.13	0.40	0.54	B → B
7-Methylbenz[c]acridine	0.20	0.61	0.53	B → BG
Benz[c]acridine	0.26	0.55	0.53	B → BG
Benzo[h]quinoline	0.71	0.77	0.51	lB → B
7,10-Dimethylbenz[c]acridine	0.10	0.53	0.46	B → BG
7,9-Dimethylbenz[c]acridine	0.10	0.46	0.45	B → BG
Dibenz[a,h]acridine	0.06		0.26	Y → B
Pyrenoline	0.20	0.26	0.22	B → O
Indeno[1,2,3-ij]isoquinoline	0.64	0.71	0.19	Y → O
Acridine	0.74	0.83	0.18	B → G
3-Methylbenzo[f]quinoline	0.75	0.79	0.18	B → B
Phenanthridine	0.72	0.78	0.16	B → B
Benzo[f]quinoline	0.76	0.79	0.14	B → B
Acenaphtho[1,2-b]pyridine	0.60	0.75	0.13	B → B
Benzo[l,m,n]phenanthridine	0.57	0.71	0.12	BG → B
Benz[a]acridine	0.46	0.53	0.12	B → B
12-Methylbenz[a]acridine	0.41	0.61	0.11	B → B
9,12-Dimethylbenz[a]acridine	0.28	0.49	0.11	B → B
14-Phenyldibenz[a,j]acridine	0.00	0.12	0.08	BG → BG
Dibenz[a,j]acridine	0.22		0.04	Y → B

[a]From Ref. 29 with permission from the publisher.

[b]Symbols mean same as in Table 2.

16. Cellulose with N,N-Dimethylformamide-Water (35:65)

Table 4 compares the R_f value of 23 aza heterocyclic hydrocarbons. Anthracene would be at R_f 0.35, with larger polynuclear aromatic hydrocarbons being lower. The rationale behind the selection of this system was the old

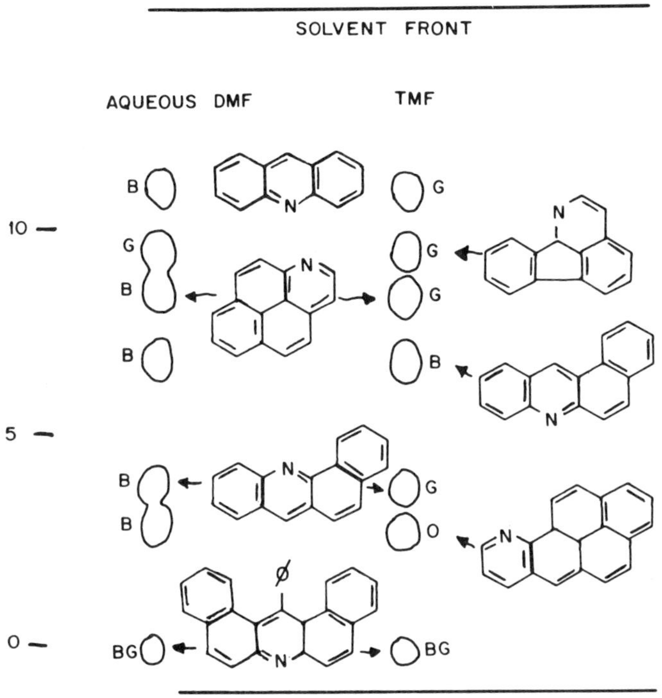

FIG. 15. TLC separation of some aza heterocyclic compounds (cellulose; DMF-H$_2$O, 35:65). (From Ref. 29 with permission from the publisher.)

rule of thumb "like dissolves like," with the nitrogen-containing solvent pulling the nitrogen-containing polynuclear hydrocarbon from the origin and leaving the polynuclear aromatic hydrocarbons behind. An interference of imines, carbazole, and benzocarbazoles can result since their R$_f$ values range from 0.7 to 0.2. Benz[c]acridine is readily separated from benz[a]-acridine as shown in Fig. 15 [29].

Figure 16 shows the separation of the basic fraction of incinerator effluents and standards using this system [31]. Figures 17a and 17b show separations of a large number of basic fractions from domestic coal burning furnaces with identification included [31]. Direct spectrophotofluorometric scanning techniques have been studied with this separations system [15]. Figures 18a and 18b show the separation of various fractions (by tube numbers) from the aluminum oxide column chromatographic separation of the basic fraction of a composite airborne particulate fraction [32]. This gives a very complete profile of a typical air pollution sample and shows just how complicated the problem of identifying constituents of polluted air can be. An example of the application of this method to coal-tar pitch has been published in which five of six unknown aza heterocyclics were identified [33].

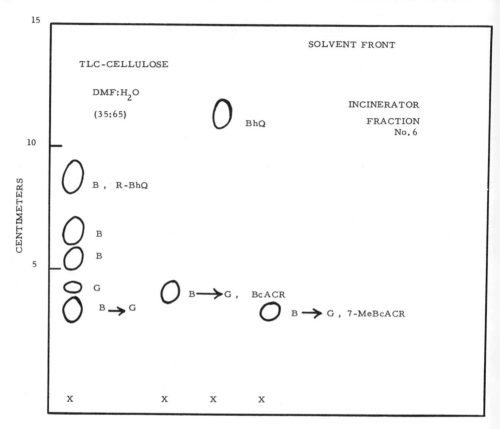

FIG. 16. Thin-layer chromatogram of a column chromatographic fraction obtained from the basic fraction of the particulates collected from incinerator effluents, benzo[h]quinoline, benz[c]acridine, and 7-methylbenz[c]acridine. Fluorescent colors of spots: B, blue; G, green. (From Ref. 31 with permission from the publisher.)

17. Cellulose with Acetic Acid-Water (3 : 7)

This system is also shown in Table 4. Dibenzacridines and pyrenoline are readily separated from other aza heterocyclic polynuclear compounds with the carbazole and benzocarbazoles at R_f values from 0.24 to 0.05 as interferences to the larger aza heterocyclic compounds. Interference would come from carbonyl compounds, but these can be eliminated during the identification stage by obtaining spectra in solvents of varying pH. Polynuclear aromatic hydrocarbons remain very near the origin with anthracene at R_f 0.10 and benzo[a]pyrene at R_f 0.00 [29].

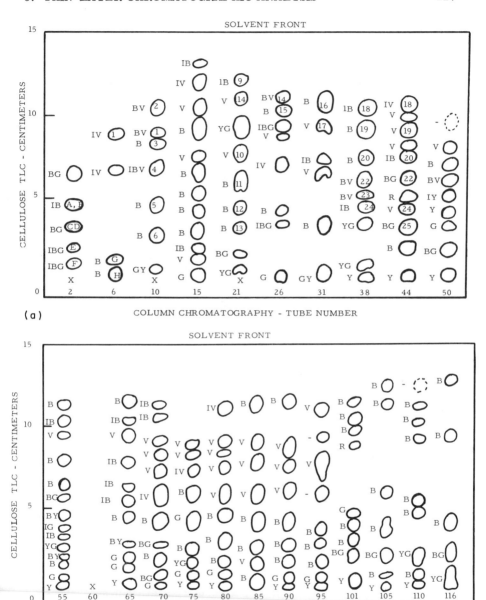

FIG. 17. Thin-layer chromatograms (cellulose; dimethylformamide-water, 35:65) of the alumina column chromatographic fractions of the basic fraction of particulates collected from the chimney effluent of a domestic coal burning furnace. A, anthracene; B, phenanthrene; C, pyrene; D, fluoranthene; E,

FIG. 17 (Continued)

an alkylpyrene; F, another alkylpyrene; G, benzo[a]pyrene; and H, benzo-
[k]fluoranthene. The fluorescent colors of spots, B, blue; G, green; I,
light; V, violet; Y, yellow. The aza heterocyclic compounds, numbers
1-25 are: 1, R_2 benzo[h]quinoline; 2, benzo[h]quinoline; 3, $R_b$benzo[h]
quinoline; 4, R_3benzo[h]quinoline; 5, benz[c]acridine; 6, $R_a$benz[c]acri-
dine; 7, $R_b$benz[c]acridine; 8, dibenz[ah]acridine; 9, 5H-indeno[1,2-b]-
pyridine(?); 10, 11H-indeno[1,2-b]quinoline; 11, indenoquinoline I(?);
12, indenoquinoline II(?); 13, indenoquinoline III(?); 14, phenanthridine;
15, R phenanthridine; 16, acridine; 17, indeno[1,2,3-ij]isoquinoline; 18,
benzo[f]quinoline; 19, $R_a$benzo[f]quinoline; 20, $R_b$benzo[f]quinoline; 21,
benzo[lmn]phenanthridine; 22, benz[a]acridine; 23, $R_a$benz[a]acridine;
24, $R_b$benz[a]acridine; 25, dibenz[aj]acridine. (From Ref. 31 with permission from the publisher.)

   Figure 19 shows the separation of a basic fraction of coal-tar pitch
polluted air and of the acridine subfraction of that basic fraction [31].

18. Aluminum Oxide with Pentane-Ether (19:1)

See Table 4. Compounds with nonsterically hindered aza nitrogen are more
strongly attracted to the adsorbent than compounds with sterically hindered
aza nitrogen and therefore have low R_f values. Thus, the achievable separation does not depend on molecular weight or aromatic basicity (as measured
by extent of conjugation). This can be extremely useful for producing unusual
separations. The six-ring 7-phenyldibenz[c,h]acridine has the highest R_f
value at 0.70 where the six-ring polynuclear aromatic hydrocarbon, anthanthrene, also has an R_f value of approximately 0.70 (see Table 2). This effect
is obvious in the separations of such isomers as benzo[f]quinoline and
benzo[h]quinoline, benz[a]acridine and benz[c]acridine, dibenz[a,h]acridine
and dibenz[a,j]acridine, as well as 7-phenyldibenz[c,h]acridine and 14-
phenyldibenz[a,j]acridine.
   The R_f values of polynuclear aromatic hydrocarbons of two to five rings
are greater than 0.7. The R_f values of six- to eight-ring polynuclear
aromatic hydrocarbons range from 0.6 to 0.8. Interferences could result
from seven-ring or more hydrocarbons as well as compounds that are more
polar, such as carbonyl-containing polynuclear aromatic hydrocarbons. This
system was used to assay for benz[c]acridine and benzo[h]quinoline in urban
atmospheres and source effluents [34].

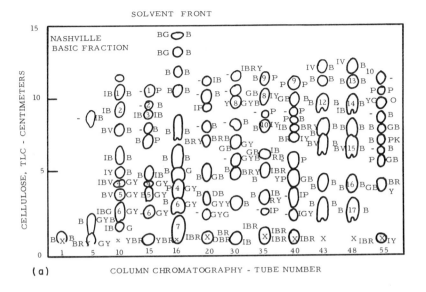

(a)

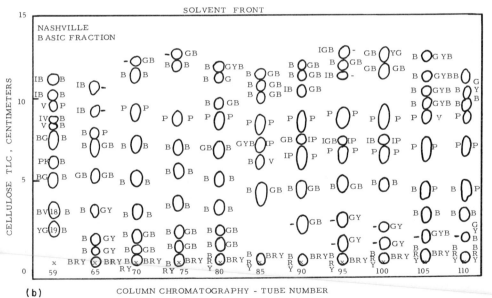

(b)

FIG. 18. Thin-layer chromatograms (cellulose; dimethylformamide-water, 35 : 65) of the alumina column chromatographic fractions of the basic fraction of a 1-year composite airborne particulate sample from downtown Nashville. The letter(s) at the left of a spot represents the fluorescence color of the spot wet with solvent; the letter(s) at the right, the fluorescence color of

FIG. 18 (Continued)

the spot treated with trifluoroacetic acid fumes. B, blue; Br, brown; d, dull; G, green; Gy, gray; l, light; O, orange; P, purple; Pk, pink; V, violet; Y, yellow; and — no fluorescence. The aza heterocyclic compounds numbered from 1 to 19 are: 1, benzo[h]quinoline; 2, R_a benzo[h]quinoline; 3, R_b benzo[h]quinoline; 4, benz[c]acridine; 5, R_a benz[c]acridine; 6, R_b benz[c]acridine; 7, dibenz[a,h]acridine; 8, indeno[1,2,3-ij]isoquinoline; 9, phenanthridine; 10, 11H-indeno[1,2-b]quinoline; 11, acridine; 12, R_a benzo[f]quinoline; 13, benzo[f]quinoline; 14, R_b benzo[f]quinoline; 15, benz[a]acridine; 16, R_a benz[a]acridine; 17, R_b benz[a]acridine; 18, dibenz[a,j]acridine; 19, R_a dibenz[a,j]acridine. (From Ref. 31 with permission from the publisher.)

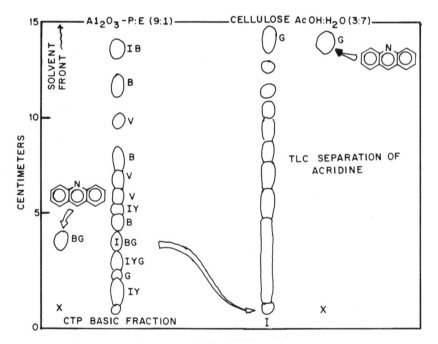

FIG. 19. Thin-layer chromatogram on alumina of the basic fraction of a coal-tar pitch polluted air sample followed by separation of the acridine fraction, I, on a cellulose thin-layer chromatogram. For explanation of letters see legend to Fig. 13. (From Ref. 31 with permission from the publisher.)

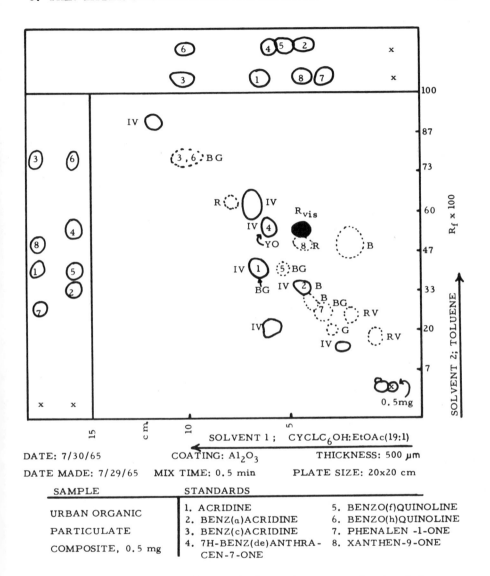

FIG. 20. Two-dimensional thin-layer chromatogram on alumina-2,4-dinitrophenoxystarch (2:1) of the composite airborne particulate sample. Definitions as in Fig. 22; in addition R, red; V, violet. (From Ref. 35 with permission from the publisher.)

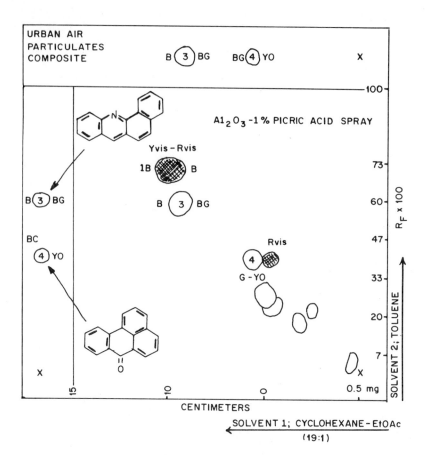

FIG. 21. Two-dimensional alumina thin-layer chromatogram of 0.5 mg of composite airborne particulate sample separated as in Fig. 22. Following development the plate is sprayed with 1% aqueous picric acid. Benz[c]acridine and 7H-benz[d,e]anthracen-7-one standards run one dimensionally. Definitions as in Fig. 22. (From Ref. 35 with permission from the publisher.)

19. Aluminum Oxide 2,4-Dinitrophenoxy Starch (2:1) with Cyclohexane-Ethyl Acetate (19:1) (First Dimension) and Toluene (Second Dimension)

Quenching techniques in thin-layer chromatography may be studied by this system. Figure 20 shows the results that have been obtained from a composite airborne particulate sample [35].

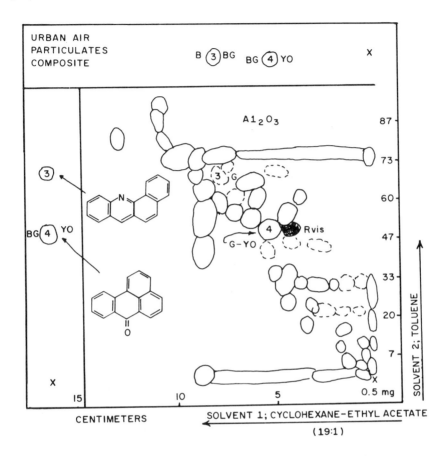

FIGURE 22. Two-dimensional alumina thin-layer chromatograms of 0.5 mg of the benzene extract of a composite of airborne particulate samples collected from about 100 American cities. Standards run one dimensionally: (3) benz[c]acridine and (4) 7H-benz[d,e]anthracen-7-one. B, blue; G, green, l, light; O, orange; Y, yellow. Letters at left of spot represent fluorescence color on the wet chromatograms; letters at right, fluorescence colors after fuming with trifluoroacetic acid. Spots encircled by dashed line appeared only after fuming. Rvis, spot with red visible color. (From Ref. 35 with permission from the publisher.)

20. Aluminum Oxide with Cyclohexane-Ethyl Acetate (19:1) (First Dimension) and Toluene (Second Dimension)

Figures 21 and 22 show the separation of a composite airborne particulate sample and standards. The utility of the quenching reagent spray is pointed out in the legend [35].

21. Aluminum Oxide with Pentane:Nitrobenzene (9:1)

This system is used in a study of fluorescence-quenching techniques. Its value can best be assessed from the results shown in Table 5 and Fig. 23 [18].

22. Aluminum Oxide with Pentane-2-Nitropropane-Triethylamine (9:1:0.01)

This system may be used in the study referred to in Sec. VIII, C, 21. Table 5 and Fig. 24 show typical results [18].

TABLE 5

R_f Values and Fluorescence Colors of Some Aza
Heterocyclic Compounds on Alumina[a]

Compound	Pentane-nitrobenzene (9:1, v/v)			Pentane-2-nitropropane-triethylamine (9:1:0.01)			
	R_f	Color[b]		R_f	Color[b]		
		Wet plate	After TFA		Wet plate	Dry plate	After TFA
7-Phenyldibenz[c,h]acridine	0.94	—[c]	YG	0.93	BG	BG	YG
Benz[c]acridine	0.92	—	G	0.92	—	B	G
Pyrenoline	0.45	—	RO	0.82	B	B	RO
Acridine	0.37	lG	G	0.66	—	B	G
Benzo[h]quinoline	0.35	—	B	0.90	—	B	B
Benzo[f]quinoline	0.30	—	B	0.53	—	B	B
14-Phenyldibenz[a,j]acridine	0.28	G	G	0.67	B	B	B
Indeno[1,2,3-i,j]isoquinoline	0.26	1B	dRO	0.65	YG	YG	R
Benz[a]acridine	0.23	1G	BG	0.61	B	B	BG
Acenaphthol[1,2-b]pyridine	—	—	—	0.59	B	B	BG
Benzo[l,m,n]phenanthridine	0.15	—	G	0.48	B	B	G

[a] From Ref. 18 with permission from the publisher.

[b] B, blue; G, green; O, orange; Pk, pink; R, red; Y, yellow; d, dull; l, light.

[c] —, quenched.

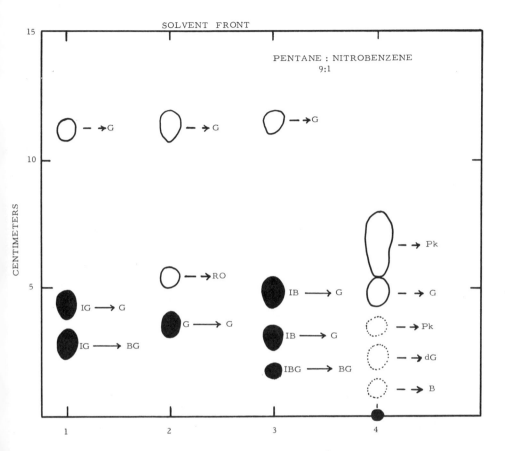

FIG. 23. Thin-layer chromatogram on alumina with pentane-nitrobenzene (9:1, v/v) as developer. All compounds were present in 2-μg amounts. Reading from top to bottom: (1) Benz[c]acridine, acridine, and benz[a]-acridine. (2) 7-Phenyldibenz[c,h]acridine, pyrenoline, and 14-phenyl-dibenz[a,j]acridine. (3) A basic fraction obtained from an airborne particulate sample in which the main source of pollution was coal-tar pitch fumes. (4) A benzene-soluble fraction of urban airborne particulates. The notation −→G signifies that the compound was not fluorescent on the wet plate but on treatment with trifluoroacetic acid fumes it fluoresced with a green color. The following compounds were also separated on this plate and were nonfluorescent on the wet plate or after treatment with trifluoroacetic acid fumes: anthracene, phenanthrene, pyrene, benzo[a]pyrene, benzo[e]-pyrene, fluoranthene, benzo[k]fluoranthene, and acenaphtho[1,2-b]pyridine. Color abbreviations are: B, blue; G, green; O, orange; Pk, pink; R, red; Y, yellow; d, dull; l, light; −, quenched. (From Ref. 18 with permission from the publisher.)

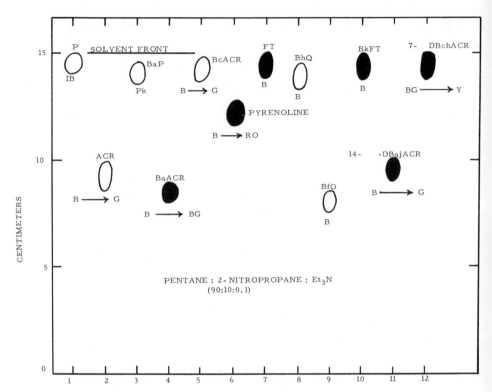

FIG. 24. Thin-layer chromatogram on alumina with pentane-2-nitropropane-triethylamine (9:1:0.01, v/v) as developer. (1) Pyrene, (2) acridine, (3) benzo[a]pyrene, (4) benz[a]acridine, (5) benz[c]acridine, (6) pyrenoline, (7) fluoranthene, (8) benzo[h]quinoline, (9) benzo[f]quinoline, (10) benzo[k]-fluoranthene, (11) 14-phenyldibenz[a,j]acridine, (12) 7-phenyldibenz[c,h]-acridine. Darkened spots are fluorescent on the wet or dry plate. Light spots are nonfluorescent on the wet plate and become fluorescent after approximately 1 hr of standing. Spots on dry plate show no change in fluorescent color with trifluoroacetic acid fumes except as shown after an arrow. For color abbreviations see Fig. 23. (From Ref. 18 with permission from the publisher.)

23. Aluminum Oxide with Pentane-Ether (9:1)

This system also was used in the study referred to in Sec. VIII,C,21 with the results shown in Fig. 25 [18]. Refer also to the system in Sec. VIII,C,18 for information on expected results.

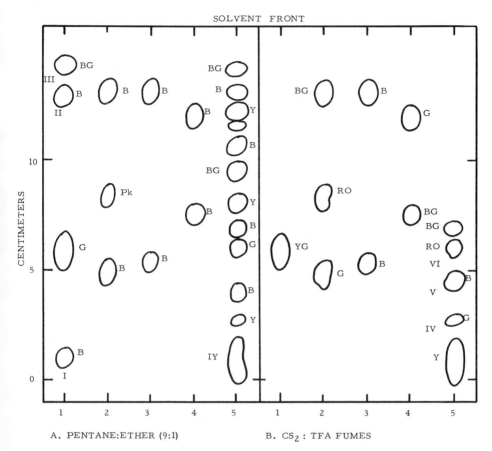

FIG. 25. Thin-layer chromatogram on alumina with pentane-ether (9 : 1, v/v) as developer. All compounds were present in approximately 2 μg amounts. Reading from top to bottom: (1) Pyrene, benzo[a]pyrene, benzanthrone, and 1-aminopyrene. (2) Fluoranthene, pyrenoline, and benz[a]acridine. (3) 7-Phenyldibenz[c,h]acridine and benzo[f]quinoline. (4) Benz[c]acridine and acridine. (5) Benzene-soluble fraction of urban airborne particulates. (A) Read under ultraviolet light after development; (B) then spray with fumes of carbon disulfide saturated with trifluoroacetic acid. For color abbreviations see Fig. 23. (From Ref. 18 with permission from the publisher.)

24. Aluminum Oxide with Pentane-2-Nitropropane (3 : 1)

This system was used in the previously mentioned study, producing results comparable to those found in the system discussed in Sec. VIII,C,18.

25. Aluminum Oxide-Silica Gel (1 : 1) with Pentane-Ether (19 : 1)

See Sec. VIII, A, 9.

26. Silica Gel-Impregnated Glass-Fiber Paper-Toluene-Methanol-Water (93 : 5 : 2)

This "instant TLC" system was used in the fluorimetric estimation of acridine in airborne particulate [36].

27. Aluminum Oxide-Cellulose (2 : 1) with Pentane-Ether (19 : 1) (First Dimension) and N, N-Dimethylformamide-Water (35 : 65) (Second Dimension)

This system was used to assay for benz[c]acridine and benzo[h]quinoline in urban air and source effluents [17].

28. Aluminum Oxide-Cellulose (2 : 1) with Toluene-Methanol (9 : 1)

See comments in Sec. VIII, C, 32 below.

29. Aluminum Oxide-Cellulose (2 : 1) with N, N-Dimethylformamide-Water (35 : 65)

See comments in Sec. VIII, C, 32 below.

30. Aluminum Oxide-Cellulose (2 : 1) with Toluene-Ether (7 : 3)

See comments in Sec. VIII, C, 32 below.

31. Aluminum Oxide with Ethyl Acetate-Cyclohexane (1 : 1)

See comments in Sec. VIII, C, 32 below.

32. Aluminum Oxide-Cellulose (2 : 1) with Toluene-Ether (7 : 3) (First Dimension) and N, N-Dimethylformamide-Water (35 : 65) (Second Dimension)

The above five systems were used in an assay for 9-acridanone in urban atmosphere by thin-layer chromatography-fluorimetric procedures [37]. Figure 26 shows results for the two-dimensional system. Standard 9-acridanone was run and the areas of the samples with corresponding R_f values were extracted for fluorescence analysis.

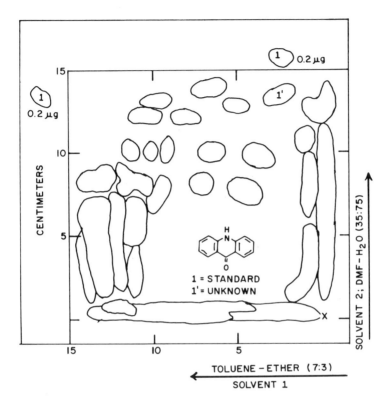

FIG. 26. Two-dimensional thin-layer chromatogram on alumina-cellulose (2:1), toluene-ether (7:3) in first direction and dimethylformamide-water (35:65) in second direction. 9-Acridanone (0.2 μg)run in each direction: 2 mg of benzene-soluble fraction of airborne particulates separated. (From Ref. 37 with permission from the publisher.)

D. Separations Involving Aromatic Compounds Containing Amino and Imino Nitrogen

1. Aluminum Oxide with Pentane-Ether (19:1)

This system, which has applications in separations involving many classes of compounds (see Sec. VIII, C, 18), is a useful starting system because it separates complex mixtures into classes. Carbazoles have low R_f values, 0-0.074. However, aromatic hydrocarbons larger than coronene, aza heterocyclics, and phenolics also have similar R_f values and can be present as interferences [38]. Table 6, column A, shows the R_f values.

TABLE 6

Thin-Layer Chromatographic Separation of Carbazole
and Polynuclear Carbazoles [a]

Compound	Approximate R_f values [b,c]					
	A [d]	B [e]	C	D	E	F
Carbazole	≤0.07	0.50	<0.02	0.8	0.4 [f]	0.3
11H-Benzo[a]carbazole	≤0.07	0.46	<0.02	0.8	<0.14	<0.07
5H-Benzo[b]carbazole	≤0.07	0.39	<0.02	0.0 [f]	<0.14	<0.07
7H-Benzo[c]carbazole	≤0.07	0.37	<0.02	0.8	<0.14	<0.07
4H-Benzo[d,e,f]carbazole	≤0.07	0.49	0.6 [f]	0.8	0.24 [f]	0.3
7H-Diebnzo[c,g]carbazole	≤0.07	0.30 [f]	<0.02	0.8	<0.14	<0.07

[a] From Ref. 38 with permission from the publisher.

[b] R_f values change with source and lot number of adsorbent employed, particularly in case of aluminum oxide.

[c] Systems described in experimental section.

[d] This system readily separates these carbazole-type compounds from aromatic hydrocarbons with fewer than seven fused rings.

[e] Average of six runs.

[f] Readily separated.

This system was used as a first separation, followed by the system discussed in Sec. VIII,D,2, to identify carbazole and polynuclear carbazoles in urban air and air polluted by coal-tar pitch fumes [39] and in conjunction with various characterization tests for carbazole in urban air [40].

2. Aluminum Oxide with Pentane-Chloroform (3:2)

Table 6, column B, shows the R_f values. Polynuclear aromatic hydrocarbons appear at the solvent front. The six- and seven-ring polynuclear aromatic hydrocarbons traveled 1.7 to 1.9 times farther than carbazoles. Large aza heterocyclic hydrocarbons traveled as far as polynuclear aromatic hydrocarbons and indole traveled 1.3 times as far as carbazole. Phenolic compounds remained near the origin, 0.2 times as far as carbazoles. The value of this system is in the separation of carbazoles as a group from polynuclear aromatic hydrocarbons, aza heterocyclic compounds, and phenolics [38].

The previously difficult to separate pairs, 5H-benzo[b]carbazole and chrysene, can be readily separated by this system [38].

This system was used as a second system, after collection from the system described in Sec. VIII,D,1, and followed by fluorescence spectrophotometry, to identify carbazole and polynuclear carbazoles in urban air and in air polluted by coal-tar pitch fumes [39].

3. Aluminum Oxide with Commercial Grade Ammonium Hydroxide

Figure 27 and Table 6, column C, show that this system is especially useful for the separation of carbazoles of the 4H-benzo[def]carbazole type from

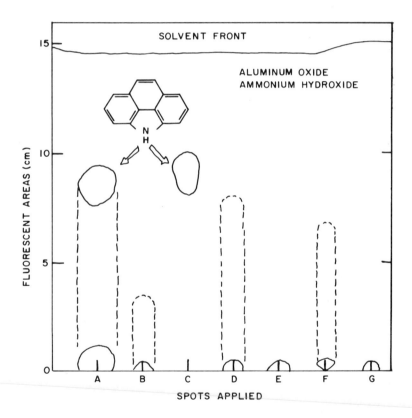

FIG. 27. Thin-layer chromatographic separation of 4H-benzo[def]-carbazole from other carbazoles. Aluminum oxide with ammonium hydroxide. Spots applied contained: A, mixture of carbazoles; B, 7H-dibenzo[c,g]carbazole; C, 4H-benzo[def]carbazole; D, 7H-benzo[c]-carbazole; E, 5H-benzo[b]carbazole; F, 11H-benzo[a]carbazole; G, carbazole. (From Ref. 38 with permission from the publisher.)

other types of polynuclear carbazoles. Some "smearing" occurs due to partial solubility, but the smears do not extend as high as the 4H-benzo-[def]carbazole area. Diluting the developing agent with water lowers the R_f value of 4H-benzo[def]carbazole and may therefore be useful in separations involving isomers [38].

4. Aluminum Oxide with Ethanol-Commercial Grade Ammonium Hydroxide (Various Proportions)

Table 6, column D, demonstrates that 5H-benzo[b]carbazole can be separated from the other carbazoles readily with this system. Figure 28

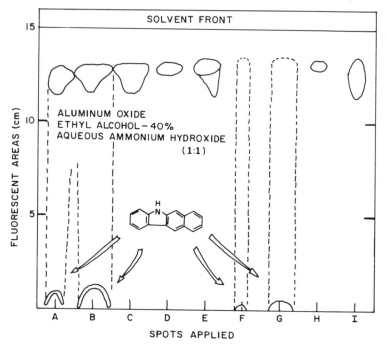

FIG. 28. Thin-layer chromatographic separation of 5H-benzo[b]carbazole from other carbazoles. Aluminum oxide with ethanol-40% aqueous ammonium hydroxide (1:1). Spots applied contained: A,B, mixture of carbazoles; C, 7H-benzo[cg]carbazole; D, 4H-benzo[def]carbazole; E, 7H-benzo[c]-carbazole; F, 0.96 μg of 5H-benzo[b]carbazole; G, 4.8 μg of 5H-benzo[b]-carbazole; H, 11H-benzo[a]carbazole; I, carbazole. (----) Smears containing approximately 0.01 μg of 5H-benzo[b]carbazole per cm² according to fluorescent spot test data. 5H-Benzo[b]carbazole in mixtures: A, 0.98 μg; B, 4.8 μg. (From Ref. 38 with permission from the publisher.)

demonstrates this with mixtures and the pure compounds. Approximately 0.01 $\mu g/cm^2$ of 5H–benzo[b]carbazole is carried up as a smear. The percentage of material carried up was constant, despite different original amounts spotted at the origin, indicating that a solubility phenomenon was involved. Usually 95% of the 5H–benzo[b]carbazole can be expected to remain at the origin [38].

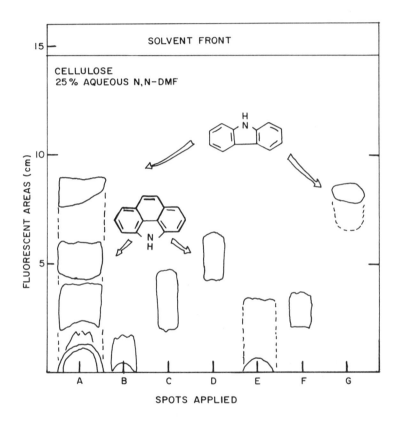

FIG. 29. Thin-layer chromatographic separation of carbazole and 4H–benzo[def]carbazole from other carbazoles. Cellulose with 25% aqueous N,N-dimethylformamide. Spots applied contained: A, mixtures of carbazoles; B, 7H–benzo[cg]carbazole; C, 7H–benzo[c]carbazole; D, 4H–benzo[def]carbazole; E, 5H–benzo[b]carbazole; F, 11H–benzo-[a]carbazole; G, carbazole. (----) Weakly fluorescent smear. (From Ref. 38 with permission from the publisher.)

5. Cellulose with N,N-Dimethylformamide-Water (1:3)

Figure 29 and Table 6, column E, show that this system can be used to
separate carbazole and 4H-benzo[def]carbazole from other types of
polynuclear carbazoles [38].

6. Florisil with Pentane-Ether (3:1)

In this system (Fig. 30 and Table 6, column F) carbazole and 4H-benzo-
[def]carbazole have approximately equal R_f values, while the other
polynuclear carbazoles have lower values. Although carbazole is not

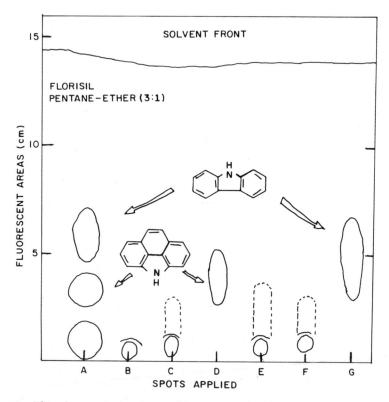

FIG. 30. Thin-layer chromatographic separation of carbazole and 4H-
benzo[def]carbazole from other carbazoles. Florisil with pentane-ether
(3:1). Spots applied contained: A, mixture of carbazoles; B, 7H-benzo-
[c,g]carbazole; C, 7H-benzo[c]carbazole; D, 4H-benzo[def]carbazole;
E, 5H-benzo[b]carbazole; F, 11H-benzo[a]carbazole; G, carbazole.
(----) Weakly fluorescent smears. Reprinted from Anal. Chem., 36, 1011
(1974) with permission of the publisher.

separated from 4H-benzo[def]carbazole, the separation distance between these and the other polynuclear carbazoles is greater than that described in Sec. VIII, D, 5, thereby ascribing it a certain usefulness even though the system in Sec. VIII, D, 5 exists [38].

7. Cellulose with N, N, Dimethylformamide-Water (35 : 65)

The R_f values of over 60 aromatic amines and imines have been reported in this system [41]. Many useful separations are possible among isomers, with the resulting R_f values ranging from 0.17 to 0.81. Many useful separations in the carbazoles series are also shown [41]. Electronegatively substituted aromatic amines have high R_f values and are therefore separated from other amines and heterocyclic imines by this system [41]. Some interference from polynuclear aromatic hydrocarbons and aza heterocyclics can be expected. However, aluminum oxide column chromatography was used to produce complete removal of these interferences [41].

8. Aluminum Oxide with Toluene-Methanol (19 : 1)

The electronegatively substituted aromatic amines, which can be separated as a class by the system described in Sec. VIII, D, 7 can be separated from one another by this system [41]. Reproducibility is a problem with this system. The use of standards is usually necessary for any system involving aluminum oxide as a substrate.

9. Aluminum Oxide with Pentane Saturated with Aniline

This has been used to separate amines and imines. The aniline quenches the fluorescence of polynuclear aromatic hydrocarbons and aza heterocyclic compounds so that the amines and imines can be more readily detected and characterized [41].

10. Aluminum Oxide-Cellulose Acetate (2 : 1) with Pentane (First Dimension) and Ethanol-Toluene-Water (17 : 4 : 4) (Second Dimension)

This system, which is mentioned in Sec. VIII, A, 5, was also used to identify carbazole in the benzene-soluble fraction of urban airborne particulate [17].

11. Aluminum Oxide with Ethyl Acetate-Cyclohexane (1 : 4) (First Dimension and Pentane-Toluene (1 : 1) (Second Dimension)

This system, mentioned in Sec. VIII, B, 21, was used to identify carbazole and 4H-benzo[def]carbazole in coal-tar pitch [17].

E. Separations Involving Naturally Occurring Pollutants

Only recently has interest been stirred in the identification of specific compounds of biological origin as pollutants of the atmosphere [42-44]. Very

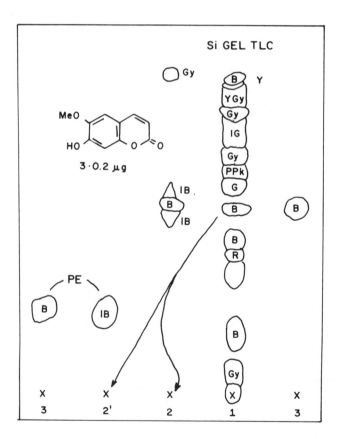

FIG. 31. Thin-layer chromatographic separation, 1, followed by a second thin-layer chromatographic separation of the spot opposite scopoletin, 2. Thin-layer chromatographic separation of pure scopoletin, 3 (on right). Paper electrophoretic separation of pure scopoletin, 3 (on left), and TLC spot from 1, 2. (From Ref. 42 with permission from the publisher.)

little thin-layer chromatography has been done in this area because the high polarities of the compounds make them easily separable by electrophoresis. In one case silica gel was used with a developing solvent of toluene-ethyl acetate-formic acid (5:4:1). Figure 31 shows the characterization of scopoletin in an urban airborne particulate sample using this system [42].

REFERENCES

1. American Conference of Governmental Industrial Hygienists, Air Sampling Instruments for Atmospheric Contaminants, 3rd ed., 1014 Broadway, Cincinnati, Ohio, 1967 (RA576.A62).

2. E. C. Tabor, T. R. Hauser, J. P. Lodge, and R. H. Burtschell, Arch. Ind. Health, 17, 58 (1958).

3. E. Sawicki, W. C. Elbert, T. W. Stanley, T. R. Hauser, and F. T. Fox, Int. J. Air Poll., 2, 273 (1960).

4. E. Sawicki, W. C. Elbert, T. W. Stanley, T. R. Hauser, and F. T. Fox, Anal. Chem., 32, 810 (1960).

5. E. Sawicki, T. R. Hauser, W. C. Elbert, F. T. Fox, and J. E. Meeker, Amer. Ind. Hyg. Assoc. J., 23, 137 (1962).

6. E. Sawicki, W. C. Elbert, T. R. Hauser, F. T. Fox, and T. W. Stanley, Amer. Ind. Hyg. Assoc. J., 21, 443 (1960).

7. E. Sawicki, Chem. Anal., 53, 24-26, 28-30, 56-62, 88-91 (1964).

8. PHS Publication No. 978, Air Pollution Measurements of the National Air Sampling Network, Analysis of Suspended Particulates — 1957-1961, U.S. Government Printing Office, Washington, D.C., 1962.

9. E. Sawicki, T. W. Stanley, and W. C. Elbert, Talanta, 11, 1433 (1964).

10. E. Sawicki, T. W. Stanley, and H. Johnson, Mikrochim. Acta, 1, 178 (1965).

11. E. Sawicki and J. D. Pfaff, Microchim. Acta, 1-2, 322 (1966).

12. W. C. Elbert and T. W. Stanley, Chem. Anal., 54, 68 (1965).

13. E. Sawicki, T. W. Stanley, W. C. Elbert, and J. D. Pfaff, Anal. Chem., 36, 497 (1964).

14. E. Sawicki, T. W. Stanley, J. D. Pfaff, and W. C. Elbert, Chem. Anal., 53, 6 (1964).

15. E. Sawicki, T. W. Stanley, and W. C. Elbert, J. Chromatog., 20, 348 (1965).

16. E. Sawicki, T. W. Stanley, W. C. Elbert, J. E. Meeker, and S. McPherson, Atm. Env., 1, 131 (1967).

17.   E. Sawicki, T. W. Stanley, S. McPherson, and M. Morgan, Talanta, 13, 619 (1966).

18.   E. Sawicki, W. C. Elbert, and T. W. Stanley, J. Chromatog., 17, 120 (1965).

19.   E. Sawicki, Pure Appl. Chem., 10, 101 (1965).

20.   T. W. Stanley, M. J. Morgan, and E. M. Grisby, Env. Sci. Tech., 2, 699 (1968).

21.   D. F. Bender, Env. Sci. Tech., 2, 204 (1968).

22.   E. Sawicki and T. R. Hauser, Anal. Chem., 31, 523 (1959).

23.   E. Sawicki, T. W. Stanley, W. C. Elbert, and M. J. Morgan, Talanta, 12, 605 (1965).

24.   E. Sawicki, T. W. Stanley, and W. C. Elbert, Mikrochim. Acta, 5-6, 1110 (1965).

25.   E. Sawicki, H. Johnson and M. J. Morgan, Mikrochim. Acta, 2, 20 (1967).

26.   E. D. Barber, J. Chromatog., 27, 398 (1967).

27.   E. D. Barber and E. Sawicki, Anal. Chem., 40, 984 (1968).

28.   C. R. Engel and E. Sawicki, J. Chromatog., 37, 508 (1968).

29.   E. Sawicki, T. W. Stanley, J. D. Pfaff, and W. C. Elbert, Anal. Chim. Acta, 31, 359 (1964).

30.   E. Sawicki, T. W. Stanley, and W. C. Elbert, Occ. Health Rev., 16, 8 (1964).

31.   E. Sawicki, T. W. Stanley, and W. C. Elbert, J. Chromatog., 18, 512 (1965).

32.   E. Sawicki, S. P. McPherson, T. W. Stanley, J. E. Meeker, and W. C. Elbert, Int. J. Air Water Poll., 9, 515 (1965).

33.   E. Sawicki, T. W. Stanley, and H. Johnson, Microchem. J., 8, 257 (1964).

34.   E. Sawicki, T. W. Stanley, and W. C. Elbert, J. Chromatog., 26, 72 (1967).

35.   E. Sawicki and H. Johnson, J. Chromatog., 23, 142 (1966).

36.   E. Sawicki and C. R. Engel, Mikrochim. Acta, 1, 91 (1969).

37.   E. Sawicki, T. W. Stanley, and W. C. Elbert, Talanta, 14, 431 (1967).

38.   D. F. Bender, E. Sawicki, and R. M. Wilson, Anal. Chem., 36, 1011 (1964).

39.   D. F. Bender, E. Sawicki, and R. M. Wilson, Int. J. Air Water Poll.,
      8, 633 (1964).

40.   E. Sawicki and H. Johnson, Mikrochim. Acta, 2-4, 435 (1964).

41.   E. Sawicki, H. Johnson, and K. Kosinski, Mikrochem. J., 10, 72
      (1966).

42.   E. Sawicki and C. Golden, Mikrochem. J., 14, 437 (1969).

43.   E. Sawicki, C. R. Engel, W. C. Elbert, and K. Gerlach, Talanta, 15,
      803 (1968).

44.   E. Sawicki, C. R. Sawicki, C. C. Golden, and T. Kober, Mikrochem.
      J., 15, 25 (1970).

PART III

SOIL CHEMISTRY SECTION

Chapter 6

GAS CHROMATOGRAPHIC ANALYSIS IN SOIL CHEMISTRY

Robert L. Grob

Chemistry Department
Villanova University
Villanova, Pennsylvania

I. INTRODUCTION

"Soil" may be considered a broad term that describes the large variety of substances found on the surface of the earth. The soil is a loose textured material containing minerals, organic matter (5%), water, and air. The organic portion results from the decay of plant biomass. It has been estimated that 1 lb of rich farm soil may harbor about 200 million fungi, 25 million algae, 15 million protozoa, and billions of bacteria, with a certain number of worms, insects, and mites also being present [1].

Several of the soil microorganisms are capable of removing carbon monoxide by oxidation, utilizing their enzyme catalysts to speed up this very slow reaction. Certain microorganisms are also capable of breaking down the persistent insecticides in the soil. It has been shown that these insecticides break down more rapidly in soils that have been previously treated, possibly because of the large amounts of the specific microorganisms present.

Hundreds of organic pesticides are now on the market, being sold in thousands of various formulations. Insecticides are the largest seller of this group, followed by herbicides. Contrasted to inorganic pesticides, organic pesticides may be decomposed by the action of biological or physiochemical processes. The significance of these pesticides in the environment can only be judged from the viewpoint of toxicity, their rate of detoxification, and decomposition.

Pesticides may be detected throughout our environment. They have been introduced by many means, e.g., agricultural control, forest pest control, and human health control. They are transported to untreated areas of the environment by wind, water, transportation of food, and the distribution of farm feeds. Accumulation levels of pesticide residues in the soil may be predicted by the use of kinetics as well as by the large tabulation of analytical data presently available.

Large quantities of phosphorus are found in agricultural fertilizers. This type of phosphorus becomes chemically bound to the soil and only a small amount leaches out into our water systems. However, if agricultural phosphorus-containing fertilizers are applied incorrectly or, especially, to frozen farmland they find their way to our water systems in large amounts.

The artificial recycling and fertilization of farmlands are essential to remove vital nutrients from the land in the form of grains, vegetables, and fruits. This land must also be irrigated in dry areas to increase the acreage production of farmland. Waste water is a good source of the necessary water needed for land irrigation. If this water contains only organic material and common inorganics there is little problem because the soil is capable of degrading most organic materials and also acts as a good cation exchanger; e.g., it is capable of removing the NH_4^+ from water. Problems arise when the waste water contains persistent toxic organic compounds and dangerous heavy metals.

The soil acts as a reservoir for pesticides until they are taken up by invertebrates, escape into the atmosphere, run off into the waterways, or are degraded. When taken up into invertebrate tissue the residues may kill or have sublethal effects. Persistence is greatly influenced by soil type; e.g., heavy clays retain insecticides much longer than do lighter, sandier soils.

Since most data of insecticide residues refer specifically to agricultural soils, more information is also needed concerning the residues present in woodlands, grasslands, and the large acreage of natural areas.

DDT is the most common residue found in agricultural soils (mean value of several parts per million), followed by dieldrin, lindane, chlordane, heptachlor, and aldrin [2]. DDT presents the greatest potential hazard because of its affinity for fatty tissue and its persistence. The chlorinated hydrocarbon insecticides are not as great a nuisance if they remain on the surface; once incorporated into the soil they are readily adsorbed and persist. Their persistence has caused concern as to whether they are phytotoxic to plants. Morrison et al. [3], Foster [4], and Dennis and Edwards [5] showed that, generally, insecticide residues did not decrease crop yields except when present in large amounts; stimulated growth was evidenced in some cases. It has been suggested [6] that one way of breaking down DDT would be by creating anaerobic conditions in the soil.

Chemical structure and its resultant intrinsic stability seem to be the most important factor for the controlling of pesticide residue persistence in the soil. Volatility decreases the residue's residence time in the soil and the effectiveness tends to be inversely proportional to its solubility [7]. The presence of Fe^{2+}, Al^{3+}, Mg^{2+}, and moisture in acid soils affects the adsorption and persistence of insecticides in the soil.

Herbicides, in contrast, break down more rapidly; however, their persistence in the soil is influenced by the same factors as insecticides. Since herbicides are more water soluble than insecticides they can be more easily leached into the subsoil. They do not affect many soil animals or wild animals and are less toxic to fish except where high concentrations are used to control aquatic weeds.

Organic fungicides are biodegradable and only have a few days to a few weeks persistence in the soil. Inorganic fungicides persist for much longer times, especially if they contain a heavy metal.

Gas-liquid chromatography (GLC) provides a very accurate, sensitive, and rapid technique for determining pesticides extracted from soils, sediments, and plant tissue. The technique of GLC also provides rapid routine analyses for degradation studies of organophosphate insecticides [8-12]. Some of the difficulties encountered in identification of organochlorine compounds are not nearly so troublesome for organophosphate analysis due to the greater specificity of detectors. The GLC technique is capable of the separation and the detection of these insecticides; however, extensive cleanup of the extracts is often required prior to analysis. Correct identification is favored in GLC by the use of specific detectors and two or more columns.

II. ACCURACY AND PRECISION

It is difficult to make an all inclusive statement regarding sensitivity limits for all types of analysis in soil samples. Sensitivities as low as 0.05 ppm of the methyl ester of picloram in soil have been reported using a ^{226}Ra

TABLE 1

Characteristics of Some Gas Chromatographic Detectors

Detector type	Sensitivity (g/sec)	Minimum detectable amount (g)
Flame ionization (FID)	9×10^{-13} for alkanes	2×10^{-11} for alkanes
Electron capture (ECD)	2×10^{-14} for CCl_4[a]	10^{-13} for lindane[a]
	5×10^{-14} for CCl_4[b]	4×10^{-12} for lindane[b]
Alkali flame (for P)	5×10^{-14}	2×10^{-12} for parathion
Microcoulometric (MCD)	6×10^{-14}	5×10^{-12} for bromine
Thermal conductivity (TCD)	6×10^{-10}	10^{-5} for CH_4

[a]For ^3H source.

[b]For ^{63}Ni source.

electron capture detector (ECD) [13]. Dacthal has been determined in soils and plants using ECD with essentially quantitative recovery at levels of <1 ppm [14]. Developments in detector systems and the rapid interchangeability of both columns and detectors have made GC both a selective and a sensitive technique for pesticide residue analysis. Chromatographic and sensitivity data have also been reported for several herbicides (esters of 2,4-D, 2,4-DB, and 2,4,5-T) [15]. The cleanup procedures necessary often dictate the accuracy of subsequent analyses. Several types of detector systems are available in GC and their use, in combination with other considerations, will govern the sensitivity to be expected. General statements can be made regarding accuracy and precision but I consider a discussion of some detectors used and the sample types analyzed to be more informative. The sensitivities (grams per second) and minimum detectable quantities (in grams) for some representative detectors are given in Table 1.

Several types of detectors are available for use in conjunction with GLC. The electron capture detector (ECD) and microcoulometric detector (MCD) possess the best sensitivity and stability in the analysis of organochlorine insecticides [16, 17]. The ECD is 10^6 to 10^7 times more responsive to organochlorine insecticides than to nonchlorinated insecticides [18, 19]. It is possible to detect 10^{-10} to 10^{-12} g of an organochlorine insecticide in the presence of other components at much higher concentrations [15, 20].

ECD quantitations may be obtained by a comparison of peak heights [21, 22] or peak areas [17, 23, 24] obtained from standard samples to those obtained from samples of unknown concentration. Peak height measurements are comparable to disk integration or triangulation measurement in the prediction of quantitative amounts of aldrin, heptachlor epoxide, and dieldrin [25]. Sensitivity of the ECD generally increases with increasing number of chlorine atoms [18]. The level of contamination, rather than the polarity of the organochlorine insecticide, may limit detectability where extremely high sensitivity is required. Efficient sample cleanup prior to the GLC analysis is essential in these cases [26].

The operation of the MCD involves combustion of the organochlorine insecticide to CO_2 and HCl followed by precipitation of the chloride ion as silver chloride in a cell containing silver acetate. An electrical imbalance is detected by the MCD and an amount of electrical energy is regenerated equal to the silver equivalent that precipitated. This regenerated current defines the chromatographic peak. The MCD is more specific than the ECD because only chloride and bromide ions precipitate the silver ion in the extremely dilute solution employed. This specificity of the MCD eliminates the need for rigorous sample cleanup and minimizes the effect of column "bleed" [21, 27, 28]. Although the MCD is specific, it is less sensitive than the ECD by a factor of 10-100 [21, 28]. The MCD has been modified to include detection of sulfur and phosphorus [29]. It is sensitive to approximately 0.1-1.0 μg for organophosphate pesticides [20], whereas the ECD has a low response to organophosphates. However, malathion and diazinon have been detected at the 5-10 ng levels [30] and parathion at the 1-2 ng level [31]. Since all organophosphate insecticides contain methoxy and ethoxy groups the corresponding alkyl iodates can be liberated by the Zeisel reaction [32] and determined by EC [33]. This requires additional time for sample preparation and thereby defeats the principal advantage of gas chromatography.

The thermionic detector, which is highly specific to phosphorus, is a standard hydrogen flame detector modified by placing a platinum helix coated with sodium sulfate in the flame. The detector was reportedly 600-1000 times more responsive to phosphorus-containing compounds than to non-phosphorus-containing hydrocarbons. Its response to interfering halides can be repressed and the sensitivity increased by replacing the sodium sulfate with KCl or CsBr [34, 35]. The response can also be increased by using He as the carrier gas [36]. The characteristics and operating parameters of this detector have been reviewed by Beckman and Gauer [37].

A flame photometric detector (FPD) specific to sulfur and phosphorus has been developed by Brody and Chaney [38] and independently for metal chelates by Juvet and Durbin [39]. This detector is extremely selective and sensitive; 0.1 ng of a phosphorus-containing compound can be determined when a proper filter is employed. Phosphorus is detected at 526 nm and sulfur at 394 nm. The FPD is an easy detector to use and does not require periodic reactivation.

III. COLUMN AND INSTRUMENT CONSIDERATIONS

Resolution of organochlorine insecticides contained in soil and water is
dependent on column type, operational conditions of the gas chromatograph,
and the chemical and thermal stabilities of the compounds being determined.
The degree of resolution of a particular column is expressed in terms of
theoretical plates (see Chap. 1). Approximately 2-m columns are most
frequently used, which allows detection of organochlorine insecticides with
reasonable retention times [28]. Without column overloading, the minimum
column diameter capable of accepting a 10-μl sample is 4 mm [40]. The
percentage amount of stationary phase is determined by the particular
analysis and the need to entirely cover the support. One should keep in
mind that as stationary-phase thickness increases there is a concomitant
increase in particle size and a reduction in the number of theoretical plates.
Five to twenty percent, by weight, of the column packing is used for nearly all
stationary phases. However, if too little stationary phase is used and the
support is inadequately covered, insecticidal degradation may result. Among
the most commonly used column supports are Anakrom ABS [20, 23, 41],
Chromosorbs P and W [42], Celite and Kieselguhr [22, 43], and silanized
supports [17]. Support materials are not completely inert and do contain
some reactive sites that may lead to insecticide degradation and the appearance of extraneous peaks [21]. These reactive sites are largely removed
by the coating with the stationary phase, but in the case of small amounts
of insecticides or reactive compounds an additional support treatment is
needed to prevent the degradation, e.g., the silane treatment for many
supports.

The parameters that most often cause nonresolution of organochlorine
insecticides because of instability are (1) the use of metal columns, (2)
incomplete coverage of the support by the stationary phase, (3) excessively
high column temperatures, and (4) low carrier gas flow rates or highly
contaminated insecticides. Degradation on metal columns is attributed to
a catalytic degradation at metal surfaces [17, 40, 44-46]. It has been shown
that recoveries of organochlorine insecticides on copper, stainless steel,
aluminum, and quartz are extremely low [44]. Such an effect may also
occur at the metal injection blocks surface; thus, all-glass, one-piece
injection ports and columns are desired. Decomposition of p,p'-DDT,
p,p'-DDE, and p,p'-DDD have been ascribed to insufficient sample cleanup
prior to analysis [23]. At temperatures in excess of 230°C, decomposition
of endrin [21, 47] and p,p'-DDT [46] also has been noted. Low gas chromatographic recoveries of heptachlor, o,p'-DDT, and p,p'-DDT have been
reported when low carrier gas flow rates were used [23, 26]. Since endrin
and p,p'-DDT are subject to degradation on poorly prepared columns, these
insecticides are often used to test columns for proper "conditioning."

Degradation of sample is readily recognized by the appearance of extraneous peaks on the chromatograms [23, 41], tailing peaks [20], or an
increased sensitivity following the repeated injection of samples onto a

column [20, 23]. A procedure recommended for minimizing degradation losses of organochlorine insecticides is to pass relatively large amounts of insecticides or other inactivators through the column prior to analysis.

Gas chromatography can be made more versatile by combining it with other analytical instruments. Gas chromatographs have been combined with infrared and mass spectrometers, pyrolyzers, and ultraviolet, atomic absorption, and nuclear magnetic resonance spectrometers [48-50]. The oxygenates found in automobile exhausts have been determined with a combined IR-GC technique. This technique shows great promise for the identification of pesticide metabolites [51].

Mass spectrometry is the most sensitive of the analytical methods used in combination with gas chromatography (GC-MS). The fact that compounds eluted by GC have sufficient vapor pressure to be easily analyzed by mass spectrometry makes the two instruments very compatible [52]. This GC-MS combinations has been used successfully in the analysis of steroids, hormones, lipids, isoprenoids, urinary acids, terpenes, and drugs and their metabolites [53-61]. Complex mixtures of hydrocarbons also have been analyzed using open-tube chromatography in conjunction with mass spectrometry [62]. Furuya et al. [63] used GC-MS for the analysis of furanocoumarins from root tissue.

Pyrolytic techniques have mostly been applied to polymeric substances and their degradation products [64-67]. Barlow et al. [64] developed a pyrolysis-gas chromatographic technique that allowed the study of degradation kinetics. In this way it was possible to obtain quantitative results providing experimental variables were precisely controlled. A modification such as this should find some applications in the analysis of pesticide residues. Jones and Reynolds [68] stated that four basic criteria must be followed in developing a quantitative technique: (1) rapid and complete pyrolysis, (2) very rapid or "flash" heating to stimulate normal liquid-sample injection, (3) minimum gaseous diffusion of the pyrolysis products, and (4) prevention of condensation on the column walls. Pyrolysis-gas chromatography shows promise for the analysis of micro amounts of nonvolatile compounds and macromolecules [69].

Organic analysis of martian soil by automated pyrolysis-gas chromatography-mass spectrometry has been suggested for an early Mars lander [70]. The technique is capable of distinguishing material of present biological origin from fossil or metabolite organic matter.

Kolb et al. [71] developed a technique for the separation and analysis of volatile metal compounds utilizing atomic absorption-gas chromatography.

The measurement of radioactivity in the effluent of a gas-liquid chromatograph has made many new kinds of metabolic studies possible. The radioactivity may be measured by flow-through ionization chambers, scintillation counters, and proportional counters [72-76]. Combustion products from organic materials have been analyzed for CO_2 and H_2O with the water being reduced to molecular hydrogen for ^3H analysis [77]. Neutron activation analysis also has been used as a gas chromatographic detector [78-80]. Short-lived isotopes may be analyzed by this technique by quantitatively resolving the pulse-height scintillation spectra.

The combination of GC with other instrumental techniques is expanding and becoming more sophisticated. Efforts in this general direction will result in a greater variety of GC applications. Westlake and Gunther [81] have reviewed the subject of the various detectors and their application to pesticide analysis.

IV. METHODS, TECHNIQUES, AND SAMPLE TYPES

A. Insecticides

Dual chromatographic columns, dual detectors, and/or formation of derivatives that can be readily determined by GLC have been suggested for the improved identification of insecticides [17, 26, 82-85]. Aldrin has been used as the internal standard to correct for variations in chromatographic conditions during the analysis [17, 23, 42, 82]. A "multicolumn" system using five different columns also has been described [40]. Retention times based on two-column systems (one polar and the other nonpolar) are usually satisfactory because order of elution is dependent largely on vapor pressure for the nonpolar column and on polarity for the polar column [23, 26, 28, 42, 84, 86-88]. The use of derivatives is best exemplified by the dehydrohalogenation of p, p'-DDD, o, p'-DDT, p, p'-DDT, and methoxychlor by a potassium hydroxide-ethanol reagent to form derivatives that are easily detectable by ECD gas chromatography [83, 85]. Methods of derivatization are neither time consuming nor difficult and may be readily incorporated into routine insecticide determinations.

A great deal of interest has developed regarding the effect of polychlorinated biphenyls (PCB) in the environment. One problem is the way they may affect the determination and identification of organochlorine insecticides. Polychlorinated biphenyls have high chemical stability, resist degradation, and are readily detected by an ECD system [89-94]. Their composition is difficult to describe because a range of compounds varying in molecular weight and halogen content are synthesized simultaneously.

Burke [20] obtained acceptable separation of several organophosphate insecticides with 10% DC-200 on 80/90 mesh Anakrom ABS using a column temperature of $200°$-$210°C$ and a gas flow rate of 120 cm^3/min. Other columns have also been effective in the separation of organophosphates; e.g., 15% DC-200 on 30/40 mesh Chromosorb P [95], 10% QF-1 on HMDS-treated Chromosorb W [95], 5% SE-30 on 60/80 mesh DMCS Chromosorb G [96], 10% silicone rubber on 80/100 mesh Chromosorb W [96], 5% OV-101, 5% OV-17, 5% OV-225, and 5% OV-210, respectively, on 80/100 mesh Gas Chrom Q [97] and 10% DC-200 on 60/80 mesh Gas Chrom Q [10, 11, 98-100]. The quality of some columns may be improved by conditioning for as long as 2-4 weeks [20]. Ogata and Beckman [101] evaluated the effect of column support material on the efficiency of organophosphate separations. The data indicated that significant variability exists between different batches of the same support.

Heptachlor and its epoxides have been extracted from soils [102] and the extract has been purified on activated Florisil and subsequently analyzed by GC. Schnorbus and Phillips [103] described an efficient extraction system for pesticides. Not only is it adaptable to soil samples but also to fruits, vegetables, and animal products. A novel approach to insecticide removal in soils was tried by Lichtenstein et al. [104]. They added 1000-4000 ppm of carbon to various types of soils for the purpose of adsorbing the insecticides. The end result was not removal but inactivation of the insecticide. This technique can be applied in other circumstances but its cost makes it prohibitive at the present time. Table 2 summarizes other means of treating insecticides in soil.

B. Herbicides and Fungicides

The principles employed for the determination of herbicides, fumigants, and fungicides are very similar to those governing organochlorine insecticide determinations. Picloram residues can be determined in soil samples by methylating the extracted free acid (4-amino-3,5,6-trichloropicolinic acid) with BF_3-CH_3OH [13]. Merkle and Davis [139, 140] described a gas chromatographic method for determining picloram and 2,4,5-T in plants. The pros and cons of the GLC technique were discussed and compared to the radiotracer technique. Amiben was determined chromatographically after being methylated by refluxing with CH_3OH in an acid catalyzed system. The recovery did not exceed 85% over a concentration range of 0-40 ppm [141]. Low recovery values reflect an incomplete methylation rather than GLC error. A method has been described for diallate (S-2,3-dichloroallyl-N,N-diisopropylthiol carbamate), a herbicide used extensively for the control of wild oats in cereal and other crops on the Canadian prairies [142]; its recovery from clay and loam soils ranged from 86 to 97% at the 0.5-5.0 ppm level.

Triallate (S-2,3,3-trichloroallyl-N,N-diisopropylthiol carbamate) can be determined in soils and plant tissue samples by EC gas chromatography with a sensitivity of 0.05 ppm [143]. Recovery values ranged from 73 to 104% for the direct determination of eight substituted urea herbicides, benzomarc, chlorbromuron, fluometron, linuron, metobromuron, metoxymarc, neburon, and diuron, in soils by EC gas chromatography [144].

The determination of triazines in soils has been described using low loading of ethylene glycol adipate polyester on glass beads and an argon ionization detector [145]. Trietazine, simazine, atrazine, simetone, atratone, prometone, desmetryne, simetryne, and prometryne were resolved by this method in the parts per million range.

TABLE 2

Analysis of Insecticides in Soil

Sample	Analysis	Column	Detector	Reference
Oregon soil	Insecticide, 1-hydroxy-2, 3-epoxy-chlordene	3% DC-200 on 100/120 mesh Gas Chrom Q, 3% SE-30 on 60/70 mesh Chromoport XXX, 11% QF-1 + OV-17 on 80/100 mesh Gas Chrom Q, and 9% QF-1 on 100/120 mesh Gas Chrom Q	—	105
Soils and water	DDD and DDT	SE-30 (5%) on 100/120 mesh Chromosorb W-DCMS	ECD	106
Soil, grass, and milk	Endrin residues	DC-11 silicone on Chromo-sorb W	—	107
Soil	Dasanit (Fensulfo-thion) and three metabolites	XE-60 on Gas Chrom Q	Flame photometric	108
Air-dried soils	Endrin	7.5% QF-1 + 5% DC-200 on 80/100 mesh Gas Chrom Q	ECD	109
Lake sediments	Parathion	10% DC-200 on 60/80 mesh Gas Chrom Z	Potassium therm-ionic	110
Crops and tissue	Dichlorvos	Phenyldiethanolamine succinate (10%) on 100/120 mesh Gas Chrom Q	Flame photometric, thermionic, and ECD	111

Sample	Compound	Column	Detector	Ref.
Soil	Chlordane residues	3% OV-17, 2% DEGS, 10% DC-200	Flame photometric	112
Soil	Parathion	10% DC-200 on Anakrom ABS	–	113
Black clay on three watersheds	DDT, Toxaphene	0.25 mm Polisorb 1 0.25-0.5 mm molecular sieve 5A	Dual tritium + Katharometer	114
Soils, soil fractions, and biological materials	DDT	2% DC-11 on Gas Chrom Q	ECD tritium	115
Field sprayed agricultural crops, soils, and olive oils	Diazinon	3% SE-30 on Chromosorb W, 5% XE-60 on Diatoport, 5% QF-1 on 60/80 mesh Chromosorb W	Microcoulometric ECD	116
Soil	Diazinon	5% DC-200 on 80/100 mesh Chromosorb W-HMDS	ECD	117
Soil	Parathion	10% DC-200 on Anakrom ABS	Thermionic flame	118
Soil	DDT	3% SE-30 on 80/90 mesh Chromport	ECD	119
Corn leaves	Dieldrin and hepta-chlor	10% DC-200 and 15% QF-1 on 100/200 mesh Gas Chrom Q	ECD	120
Soils	Aldrin, dieldrin, and DDT residues	DC-11 and QF-1(3:2) 4% on 60/80 mesh Chromosorb W	ECD	121
Vegetables, crops and soils	Terraclor	2% SE-30 on 90/100 mesh Anakrom ABS	ECD	122

TABLE 2. — Continued

Sample	Analysis	Column	Detector	Reference
Soil and rutabaga	Heptachlor, heptachlor epoxide, and gamma chlordane residues	5% SE-30 on 80/100 mesh Chromosorb W	ECD	123
Soil	Nemagon fumigant, Nemafume, DBCP	100/120 mesh Porapak Q	ECD	124
Tomatoes and subsequent crops	Aldrin and dieldrin residues	5% QF-1 and 4% SE-30 on 60/80 mesh Chromosorb W-AW	ECD	125
Light sandy soil	Dieldrin, lindane, DDT, and parathion	5% DOW-11 on 70/80 mesh Chromosorb W, 10% FS-1265 on 80/100 mesh Chromosorb W, 10% DOW-200 on Gas Chrom Z, 5% Apiezon L on 70/80 mesh Aeropak 30	ECD or phosphorus detector	126
Silt, soil, and water	Granular dieldrin	2% SE-30 on Anakrom ABS	ECD	127
Crops, soil, and olive oil	Diazinon	3% SE-30 on Chromosorb W, 5% XE-60 on Diatoport, 5% QF-1 on 60/80 mesh Chromosorb W	Sodium therm-ionic, microcoulo-metric, ECD	128
Soils and crops	Chlorgenuimphos insecticide residues	2% Cyclohexane dimethanol succinate, 3% phenyldiethanolamine succinate	ECD	129

Soil	Lindane residues	5% SE-30 on 60/80 mesh Chromosorb W, 5% Epon 1001 on 60/80 mesh Gas Chrom Z, 10% QF-1 on 60/80 mesh Chromosorb W, 5% QF-1 on 60/80 mesh Chromosorb W	ECD	130
Soil	Insecticides	3% DC-200 on 100/120 mesh Gas Chrom Q	ECD	131
Wheat plants and grain	Insecticide residues	4% SE-30 on 80/100 mesh Chromosorb W	ECD	132
Soil	Organochlorine insecticides and their metabolites	1–10% mixture 4% SE-30 and 6% QF-1, 5% Dow 11, 10% DC-200, 15% QF-1, 60/80 mesh Chromosorb W-AW, 80/100 mesh Chromosorb, W-AW-DMCS	ECD	133
Agricultural soil legume crops	Organochlorine insecticides	4% SE-30 on 80/100 mesh Chromosorb W, 2% QF-1 on 80/100 mesh Chromosorb W	FID	134
Soil, cucumbers, alfalfa	Insecticide residues	5% DOW 11 on 60/80 mesh Chromosorb W	ECD	135
Crops	Malathion	10% DC-200 on 80/100 mesh Gas Chrom Q or 2% DEGS on 80/100 mesh Gas Chrom Q, 20% DC-200 on 80/100 mesh Gas Chrom Q	KCl thermoionic	136

TABLE 2. — Continued

Sample	Analysis	Column	Detector	Reference
Soils	Organochlorine insecticides	1m 10% diethylene glycol on 60/80 mesh Gas Chrom Q 2m 10% DC-200 on 60/80 mesh Gas Chrom Q	ECD	137
Nonsterile, heat sterilized soil	Degradation of organophosphorus insecticides	QF-1 on 60/80 mesh Gas Chrom Q	Phosphorus detector	138, 329

Gas chromatographic methods have been described for the detection and determination of dichloropropene-dichloropropane (D-D mixture) [146] and ethylene dibromide in soils [147]. Recoveries of the fumigants D-D mixture, ethylene dibromide, 1,2-dibromo-3-chloropropane, chloropicrin, propargyl bromide, and methyl bromide ranged from a low of 44% for methyl bromide to 93% for 1,2-dibromo-3-chloropropane [148]. The low recovery values may result from incomplete extraction or volatilization from soil, rather than loss during the GLC procedure. Katz and Strusz [149] devised a procedure for the separation of several substituted urea herbicides, their anilines, and other known or unknown metabolites. The separations was developed on a XE-60 liquid phase with temperature programming from 75° to 230°C. The separations were useful in following the transformations of soil compounds under various environmental conditions. Keanrey et al. [106] utilized high-energy UV radiation as a source to induce photodecomposition of herbicides in soils. Since complete removal of herbicides from soils is not always feasible, the photodecomposition may help in reducing the existing residues to a level where the significance of plant uptake becomes minimal. Beynon and Wright [150] studied the persistance, penetration, and breakdown of the herbicides chlorothiamid (2,6-dichlorothiobenzamide) and dichlobenil (2,6-dichlorobenzonitrile) in fields of clay, medium loam, sandy loam, and peat. Unincorporated granules and wettable powders were applied. The conversion of the chlorthiamid to dichlobenil in the soil was very fast. The original materials and residues were determined using a 5% SE-30 liquid phase on 100/120 mesh Chromosorb W-DMCS and an electron capture detector.

Bartha and Pramer [151] showed that the herbicide 3',4'-dichloropropionanilide (DCPA) decomposes in the soil to carbon dioxide, 3,4-dichloroaniline (DCA), and 3,3',4,4'-tetrachloroazobenzene (TCAB). Soil microorganisms were involved in the conversion of DCPA to DCA and the conversion of DCA to TCAB. The herbicides were separated and identified using a 5% SE-30 loading on Chromosorb W and a flame ionization detector. Bartha [152] also determined aniline herbicides in soil samples by extracting with acetone, cleaning up by a TLC procedure, and accomplishing the final separation with a 5% VCW 98 loading on Chromosorb W. Detection was made with a FID. Residues of s-triazine herbicide in crop materials are extracted with 500 cm³ chloroform and filtered and the residue is taken up in benzene [153]. Aluminum oxide (activity V) is used for the cleanup. Triazine to 0.25 μg can be quantitatively detected using this procedure and performing the final measurement by microcoulometric gas chromatography. Table 3 gives a survey of the various other techniques and determinations that apply to herbicide analyses in soil. Table 4 lists the various procedures available for the analysis of fungicides in soil.

TABLE 3

Analysis of Herbicides in Soils

Sample	Analysis	Column	Detector	Reference
Crops, fish, soil, and water	Dichlobenil and its metabolite, 2,6-di-chlorobenzoic acid	DC-200 on Gas Chrom A	—	154
Soil	Picloram			155
Soil	Benthiocarb	SE-30 on Chromosorb W	Flame photometric	156
Soil	Dyrene (anilazine) residues	OV-17 + QF-1 on Chromosorb W	ECD	157
Soil and alfalfa	Uracil herbicides (bromacil and ter-bacil)	10% Ucon Polar on 80/100 mesh Gas Chrom Q		158
Natural soil environment	Isooctyl ester of 2,4-D	5% DOW 11 + 5% Carbowax 20M on 60/80 mesh Gas Chrom Q	ECD	159
Crops and soils	Methidathion (Supracide), its oxo analogs, and its heterocyclic moiety	3% OV-17 on 80/100 mesh Chromosorb G. 2% NPGS on 80/100 mesh Chromosorb G	Sodium therionic phosphorus; Coulson electrolytic conductance	160
Soil	Fenac residues	10% DC-200 on 100/210 mesh Gas Chrom Q, 1.3% Veramid 900 on 60/80 mesh Diatoport S	ECD	161
New York surface and subsurface soils	Dichlobenil	10% DC-200 on 80/100 mesh Gas Chrom Q	ECD	162

Soil	Picloram	3% FFAP on 60/80 mesh Gas Chrom Q		163
Soil, plant, and animal tissue	Chloronet (Tillam) and its metabolites	10% DC-560 + 0.2% Epon resin 1001 on 80/100 mesh Chromosorb W-AW DMCS	Microcoulometric	164
Soil and plant tissue	Bromacil	5% Neopentyl glycol adipate on 60/80 mesh Chromosorb P, HMDS	ECD	165
Alfalfa and soil	Terbacil and bromacil	10% Ucon Polar on 80/100 mesh Gas Chrom Q		166
Soils	Fenac residues	1.3% Versamid 900 on 60/80 mesh ECD Diatoport S		167
Diuron and linuron-treated soils	3,3',4,4'-Tetra-chloroazobenzene	5% XE-60 + 0.2% Epon resin 1001, on 60/80 mesh Gas Chrom Q		168
Soil	Vernam	3% Apiezon L on 70/80 mesh Chromosorb G	Sulfur specific flame photometric	169
Crops and soil	Dichlobenil and chlorthiamid	10% Silicone Oil + 0.5% Epikote 1001 on 100/120 mesh Celite, 2% neopentyl glycol succinate on 100/120 mesh Celite, 2% PDES on 100/120 mesh Celite		170
Soil	Isooctyl ester of 2,4-D	5% DOW 11 + 5% Carbowax 20M on 60/80 mesh Gas Chrom Q, 2% QF-1 on Chromosorb W	ECD	171 330-339

TABLE 4

Analysis of Fungicides in Soil

Sample	Analysis	Column	Detector	Reference
Crops and soil	Chemagro 2635	QF-1 on Chromosorb W	ECD	172
Fruits and vegetables	Organomercurial Fungicide residues	2% ethylene glycol succinate on Chromosorb G		173
Soil	chloronitrobenzene fungicides	5% SE-30 on Chromosorb W-AW, DMCS, 4% XE-60 on Chromosorb W, 4% DOW-11 on 60/80 mesh Chromosorb G-AW, 5% XE-60 on Chromosorb W	ECD ECD TCD TCD	174
Paints and soils	2,6-Dichloro-4-nitroaniline (DCNA)	5% Carbowax 20M on Anakrom ABS	TCD	175
Soil, water, and fish	Pentachlorophenol residues	4% SF-96 + 8% QF-1 on 100/120 mesh silanized Gas Chrom P	ECD	176

C. Amino Acids, Carbohydrates, Fatty Acids, Steroids, and Vitamins

The GC determination of amino acids has always proved difficult due to the nonvolatility displayed by these compounds. One way around this direct analysis is the preparation of volatile derivatives. Zomely [177] has separated 22 naturally occurring amino acids by gas chromatographic techniques by converting them to their n-butyl-N-trifluoroacetyl derivatives and separating them on a column of 1.0% neopentylglycol succinate polyester coated on 60/80 mesh Gas Chrom Q using a FID. The column temperature was programmed from 75° to 220°C. The addition of ammonia to the nitrogen carrier gas has increased the sensitivity and resolution of the ethyl or n-butyl ester hydrochlorides of 15 amino acids [178]. This separation was achieved on a column of 22% polyethylene glycol adipate on 50/100 mesh Chromosorb W. Elution of amino acid esters is in the order methyl, ethyl, propyl, butyl, and aryl [179]. Twenty-two N-trifluoroacetyl amino acid esters have been separated on a 2% neopentylglycol succinate column utilizing Chromosorb W as the solid support [180].

Resolution of racemic mixtures of amino acids has been obtained by gas chromatography of the various ester derivatives [181-183]. Capillary columns in which Carbowax 20M and diethylene glycol succinate were the liquid phases were found to be the most satisfactory. Resolution of the majority of racemic amino acid diastereomers also has been accomplished [181].

The Coulson electrolytic conductometric detector, which is sensitive to nitrogen, should provide the increased sensitivity and selectivity presently lacking in the analysis of organic nitrogen compounds. The sample is oxidized to form electrolytes, which are dissolved in deionized water in the detector cell. The change in conductivity between the two platinum electrodes is measured by a DC (direct current) bridge and recorded. Sensitivity to organic nitrogen compounds is reported to be several orders of magnitude greater than those determined with the hydrogen flame ionization detector.

Gas chromatographic methods have been used extensively for the analysis of carbohydrates in biological materials. The problems of low volatility of the carbohydrates have largely been overcome by preparation of the trimethylsilyl ether derivatives. Investigations of gas chromatographic methods for carbohydrate analysis have been reviewed by several authors [184-188].

Polyhydroxylated compounds and even hexose tetrasaccharide derivatives become sufficiently volatile for GC because these derivatives are thermally stable and can withstand temperatures as high as 300°C [189]. Improved separations of these O-trimethylsilyl ether derivatives also have appeared in the literature [190-192]. Optimum conditions for the derivative reaction of sugars were determined by Sweeley et al. [193] as being 10 mg of carbohydrate, 1.0 cm³ of anhydrous pyridine, 0.2 cm³ hexamethylsilazane, and 0.1 cm³ trimethylchlorosilane. The reaction is complete in a few minutes, at room temperature, and the derivatives are stable for several days when stored in stoppered vials. N,O-bis-trimethylsilylacetamide [194] is another very useful trimethylsilylating agent that has been ranked 50 times more powerful than many of the monosubstituted amides [29, 195, 196].

Although the O-dimethylsilyl ether derivatives have not been used as much as O-trimethylsilyl ether derivatives, they possess the advantage of being much more volatile and the retention times are reduced by a factor of two from the corresponding trimethyl derivatives [188, 197].

O-Trifluoroacetyl ester derivatives of the mono-, di-, and trisaccharides have been prepared and their successful separation by GC has been reported [198]. Using this method, a variety of carbohydrates have been separated on a 1.0 m × 0.1 cm column containing 20% silicone gum rubber on a 40/60 mesh support. Retention times for hexoses ranged from 3.5 to 11 min. Prefractionation of biological samples is necessary before these derivative preparations are used. Following this procedure allows the samples to be divided into acidic, neutral, and basic sugars. This initial separation is best accomplished by ion-exchange chromatography. This type of technique, in combination with GC, has enabled the determination of the mucopolysaccharides [199-201].

A method to destroy the anomeric center of aldoses involves conversion of the aldehyde group (anomeric center of aldoses) to the acid with alkaline bromine or iodine [202]. The center may also be removed by forming the alditol by sodium borohydride reduction [203]. Thus, excellent methods of carbohydrate determination are available on complex mixtures providing that fractionation of the sugars into classes is employed prior to quantitative GC analysis.

The hexosamines have long been a great topic of interest to soil and plant scientists and excellent GC methods are now available for their determination. The first GC separation was accomplished by Jones et al. [204], who resolved the anomeric acetates of D-glucosamine and D-galatosamine by placing a column packed with 5% neopentylglycol sebacate on glass beads at the top of a column containing 1% SE-30 on glass beads at 214°C.

Successful quantitative gas chromatographic separation of the hexosamines also has been achieved using the N-acetyl and tetra-O-trimethylsilyl ether derivatives [205-207]. Several methods of forming the N-acetyl derivative are available and the preferred ones appear to be those described by White [208], Roseman and Ludpweig [209], and Kobata and Ziro [210].

Perry [205] used a 120 × 0.5 cm column of 10% neopentylglycol sebacate on 80/100 mesh Chromosorb W for quantitative separation of the N-acetyl and tetra-O-trimethylsilyl derivatives of D-glucosamine and D-galactosamine. Other columns used have been 11% neopentylglycol adipate on 80/100 mesh Chromosorb W [210] and 3% SE-30 on 100/120 mesh silanized support [207]. Quantitative separation of the hexosaminitols has been achieved as the N-acetyl and penta-O-trimethylsilyl ethers [211].

Hexamethyldisilazane and trimethylchlorosilane derivatives have been used for the analysis of D-glucosamine and D-galatosamine of mucopolysaccharides [199, 212]. The separation of the hexosamines was achieved on a Dowex 50 column (H⁺ form) [213] and then they were trimethylsilylated [193].

The gas chromatographic analysis was performed on a column of 3% SE-30 or 1% QF-1 on 100/140 mesh siliconized Gas Chrom P at 140°C. D-glucosamine separated into two anomeric forms on both stationary phases, as did D-galactosamine on QF-1, byt the D-galactosamine gave only a single peak on SE-30.

Presently, GC techniques are available for both qualitative and quantitative analysis of acidic carbohydrates. The O-trimethylsilyl or methyl ester O-trimethylsilyl ether derivatives allow sufficient thermal stability and volatility for the determination of the uronic acids, aldonic acids, and sugar phosphates. Each hexuronic acid can produce four isomeric glycoside derivatives as well as lactone glycoside derivatives. This multiplicity of GC peaks can be eliminated by converting the acid to its barium salt and subsequently reducing it with sodium borohydride [204]. Each of the hexuronic acids is converted to a single derivative and resolution of mixtures can be made of a column of 10% neopentylglycol sebacate on 80/100 mesh Chromosorb W.

Wood pulp aldonic acids have been analyzed as the alditol acetates on a column consisting of 1.5% ethylene glycol succinate polyester and 1.5% XF-1150 cyanoethyl silicone polymer on Gas Chrom P [214]. The authors were not able to separate O-trimethylsilyl ethers of D-galactonic and D-gluconic acids; however, by converting these compounds to their lactones and reducing them with sodium borohydride, they achieved a satisfactory separation. Racemic mixtures of hexonic acids and other carbohydrates also have been resolved by Pollock and Jermany [215, 216]. Other types of carbohydrates that have been successfully analyzed by GC include mucopolysaccharides, glycoproteins, glycolipids, and milk nucleotides [217].

A cursory listing of other determinations of biochemicals would be:

1. Free fatty acid homologs from C_2 through C_{18} [218].
2. Development of GC methods for the analysis of phospholipids [219].
3. Organic acids in ensiled forages [220].
4. Determination of arysulfonic acids and their salts following alkali fusion. The phenol products are determined by GC and the sulfites by titrimetry [221].
5. Sugars in soils have been determined by a flame ionization detector technique using myoinositol or quebrachetol as the internal standard [222].
6. GC methods for the determination of vitamin A and its isomers [223], vitamin C [224], vitamin D [225], and vitamin E in pharmaceuticals [226].
7. Radio gas chromatographic analysis of fatty acids and steroids using a proportional flow-through counter for ^{14}C and ^{3}H following cracking of the GC separated substances [227, 228].
8. Monosaccharides in soils have been reduced to alditols, acetylated, and then separated by temperature programmed GC from 125° to 195° C [229].

9. Nitrogen compounds have been determined in oil shale after separation on an Apiezon L column [230, 340, 341].

D. Organic Compounds

Little research on the characterization of the humic and fulvic materials of soils has been accomplished by gas chromatography. There is, however, a very large body of literature available for paper, thin-layer, and liquid chromatography of humus materials (see Chaps. 7-10).

Nagar [231] attempted to characterize several soil humic acids together with a lignin and a "humic acid" isolated from an Aspergillus niger culture by pyrolysis gas chromatography. The best separation of the pyrolyzed materials was achieved on a 4% XE-60 nitrile silicone polymer on 60/80 mesh siliconized Diatoport. Nineteen major and 37 minor components were found as pyrolysis products from the lignin. The humic acids portions furnished similar patterns. Several peaks were common to the pyrolysis products of the lignin, soil humic acids, and the Aspergillus niger "humic acid." However, insufficient data were available to correlate quantitative relationships among the various materials.

Some of the commonly occurring phenolic carboxylic acids in natural products have been identified [232]. These same products may also be found as the degradation products of the lignins and humic and fulvic acids. Two stationary phases, i.e., SE-30 (10%) and diethylene glycol succinate (10%) on 60/80 mesh Diatoport S, provide excellent resolution of p-hydroxybenzoic acid, vanillic acid, o-coumaric acid, ferulic acid, syringic acid, and m-hydroxybenzoic acid. The resolution of p-coumaric acid from protocatechuic acid proved difficult. Temperature programming was used for:

1. An SE-30 column in which the temperature was controlled isothermally for 9 min at 100° C and then programmed, at 25°C/min, to 250°C. Retention times ranged from 2.5 min for p-hydroxybenzoic acid to 15 min for m-hydroxybenzoic acid.
2. A diethylene glycol succinate column in which the program was set to hold isothermally for 3 min at 100°C and then to increase at 25°C/min up to 275°C. Retention times, under these conditions, ranged from 2.5 min for p-hydroxybenzoic acid to 7 min for syringic acid.
3. A diethylene glycol succinate column run isothermally for 3 min at 100°C and then temperature programmed to 200°C at a rate of 25°C/min. These conditions permitted the separation of protocatechuic acid from p-coumaric acid but not that of ferulic and syringic acids.

Alkaline potassium permanganate [212, 233, 234] and nitric acid [234] may be used as the oxidizing agents for investigating the oxidation products of soil organic matter. Schnitzer and Desjardins [233] oxidized the organic matter from the Bh horizon of a Podzol with alkaline $KMnO_4$. The portion that was soluble in organic solvents could then be methylated and identified by gas chromatographic techniques. Twelve benzene carboxylic acids, as their methyl esters, were identified using a column of 10% SE-30 on 80/100 mesh Chromosorb W-HMDS and a FID. A similar study of Danish illuvial organic matter showed that steam distillation of the alkaline permanganate oxidation product yielded 1.6% aliphatic monocarboxylic acids [233]. The gas chromatographic column was 5% neopentylglycol succinate on 40/60 mesh Gas Pack F. An FID was used for the detection. Acetic acid was present in the greatest yield (73% of the sum of the monocarboxylic acids), while the remaining 27% consisted of n- and iso-acids containing up to eight carbon atoms. Furthermore, methylation of the organic solvent extractable oxidation products gave 12 benzene carboxylic acids, which were identified. This accounted for 1.6% of the original organic matter. Peat humic acids and brown and weathered coals were oxidized with alkaline $KMnO_4$ [212] and identified by gas chromatography as di-, tri-, tetra-, penta-, and hexabenzene carboxylic acids with a predominance of tri-, and tetra- substituted acids. On the basis of this type of information some postulates concerning the organic matter have been made [212, 235].

Hydrogenolysis has also been examined as means of degrading organic matter fractions from a mulch soil into specific low molecular weight fractions. The n-alkanes produced by this treatment were characterized by gas chromatography [236]. These individual alkanes were separated on a 5% SE-30 crushed firebrick column utilizing an FID and temperature pro- gramming between 100° and 325° C at 6° C/min. By this technique, 25 n-alkanes (C_{11}-C_{35}) were detected as hydrogenolysis products with the greatest concentrations in the C_{22}-C_{32} range.

Organic compounds in the Orgueil and Murray carbonaceous chondrites have been examined by a combination of GC and MS [237]. A hydrocarbon distribution similar to that found in the chondrites was synthesized in the presence of iron meteorite powder. This reaction yielded a metastable distribution of normal and slightly branched alkanes and alkenes, including isoprenoid hydrocarbons from C_9 to C_{12}. Sustained reheating caused partial transformation of aliphatics to aromatics. Metal carbonyls were decomposed at 200°C after any unreacted CO and H_2 had been removed. The condensable organic compounds were then vacuum distilled into a stainless steel capillary column coated with Apiezon L. The column was then eluted with helium at 1 atm pressure and the entire effluent passed into a mass spectrometer source. The mass spectrum can then be continuously displayed on an oscilloscope.

The determination of methoxy groups in lignins and humic materials has always been a problem of interest to the lignin and soil organic matter chemists. The method of Zeisel [32] is very time consuming and requires a high degree of skill for accurate reproducible analyses. The final deter- mination necessitated nonaqueous methods of titrimetry, which gave poor end points. Kikuchi and Miki [238] have described a method for the gas chromatographic determination of the alkyl halides formed by the Zeisel reaction. The sample was first digested with Hl and the alkyl halides formed were adsorbed on 30/60 mesh silica gel coated with 10% Apiezon L. The adsorption tube was heated to about 155° C and the desorbed alkyl halides were introduced into a gas chromatograph. The use of direct calibration curves gave standard deviations ranging from 0.3 to 0.7 for $-OCH_3$ to $-OC_3H_7$.

These determinations and characterizations will provide the reader with some idea of the range and scope of gas chromatographic usefulness in attempts to determine structures of lignins and humic and fulvic acids. Many variations and refinements of these techniques will present themselves in the near future. Table 5 lists some of the assorted organic-type compounds that have been investigated in soil samples.

E. Gases

Gas chromatography is easily adapted to the analysis of various stable gases provided suitable preinjection apparatus is available for gas collection. Injection of the gaseous sample directly or absorption of the gases by a suitable solvent can be used. Cigarette smoke was absorbed in cold NaOH prior to the determination of phenolic constituents by GC [280]. Application of direct analysis for the separation of natural gas from sewer gas and for the determination of gasoline spillage in drill core soil samples has been reported [281]. The procedure allows detection of natural gas leaks at very low concentrations; separation of the trace hydrocarbons has been achieved on a 5% squalane liquid-phase column.

Gas chromatography has been very useful in studies of nitrogen transformation in soils and lake sediments. Greene and Pust [282] described a method for determining NO_2 with a molecular sieve 5A column. Nitrogen dioxide was converted to NO by reaction with the water (placed at the head of the column) and the NO produced determined by GC. In 1960 Smith et al. [283] modified the procedure to allow NO_2 analysis in the presence of O_2. Nitrogen dioxide was frozen out with an acetone-dry ice mixture and then vaporized following removal of O_2 from the system. Gibson et al. [284] analyzed silicate samples of rocks and meteorites for nitrogen content. A molecular sieve 5A column and TC detector was used. Steyn and Delwiche [285] followed the rate of nitrogen fixation in California soils, utilizing a Porapak R column and FID. Some early experiments on the adaptation of GC to soil nitrogen studies were done by Smith and Clark [286].

Separations of N_2, N_2O, NO, NO_2, NH_3, O_2, and CO_2 were achieved on columns of molecular sieve 5A, silica gel, silica gel plus ascarite, coconut charcoal, and polyethylene glycol on firebrick. No one column was capable of separating all gases; however, of these evaluated, the molecular sieve allowed separations of O_2, N_2, NO, and NO_2. Burford [287] determined N_2 N_2O, CO_2, and O_2 in soil samples also, using molecular sieves 5A and 13 X

TABLE 5

Analysis of Organic-type Compounds in Soil

Sample	Analysis	Column	Detector	Reference
Nixon sandy loam soil	Mixed chloroazobenzenes	5% SE-30 on 60/80 mesh Chromosorb W	FID	239
Oil shale	Triterpene alcohol, isoarborinol	3% SE-30 on 80/100 mesh Chrom Q, 2% OV-17 on 100/120 mesh Gas Chrom Q		240
Pueblito de Allende meteorite	Organic constituents	Apiezon L	FID	241
Lunar samples	Organic and organogenic matter	200 mesh Porapak Q	FID	242
Lunar samples	Organic Compounds	Polyphenyl ether capillary column	MS	243
Meteorites	Material of biological origin	10% SE-96 Ipegal 990 on capillary column	Ion current	244
Soil	Four isomers of benzene hexachloride	0.5% Apiezon L on 60/80 mesh Chromosorb P	Electron affinity	245
Soil	Aromatic hydrocarbons	6% Tricresyl phosphate on 60/80 mesh Chromosorb W	Dual FID	246
Soil	1,2-Dibromo-2-chloro-propane	5% DC-11 60/80 mesh Chromosorb W-AW	ECD	247

TABLE 5. —Continued

Sample	Analysis	Column	Detector	Reference
Sediments from Saanich Inlet	Amino acids	UCON 75H-90-000, Carbowax 20M, and XE-60 coated capillary columns	FID	248
Marine sediments	Sterols	3% JXR on 60/80 mesh Gas Chrom Q	FID	249
Lunar samples	Carbon compounds	Polyphenyl ether	FID	250
Soil	2,4-D free acid	11% OV-17 on 80/100 mesh Gas Chrom Q, 10% DC-200 on 100/120 mesh Gas Chrom Q	ECD	251
Soil fulvic acid	Fatty acids	2.5% SE-30 on 60/80 mesh Chromosorb W-HMDS	FID	252
Water, biological materials, and soil	Fluoroacetamide residues	20% Polyoxyethylene glycol on Chromosorb W-AW, DMCS, 20% dinonyl phthalate on silanized Chromosorb	FID	253
Soil	Organic acids	33% Dioctylsebacinate + 15% sebacic acid on 60/80 mesh Chromosorb W-AW	FID	254
Soil	cis-9-Fatty acid hydratase	20% SE-30 on 80/100 mesh Chromosorb W	FID	255

Sample	Compound	Column	Detector	Reference
Soil	Higher fatty acids and triglycerides	15% Polyethylene glycol succinate	FID	256
Sediments	Fatty alcohols	10% FFAP on 60/80 mesh Chromosorb G	FID	257
Soil	1,2-Dibromo-3-chloro-propane	5% DC-11 on 60/80 mesh Chromosorb W		258
Soil	Chlorinated hydrocarbons	9.8% DC-200 + 15.8% QF-1 on 90–100 mesh Anakrom ABS	Alkali flame AFD	259
Lunar material	Organic compounds	SCOT, SF-96, 1000cs, and Igepal CO-880, SCOT coated with Apiezon L	FID	260
Messel oil shale	Isoarborinal, a penta-cyclic alcohol	3% SE-30 on 80/100 mesh Gas Chrom Q, 2% OV-17 on 100/120 mesh Gas Chrom Q		261
Soil, water	Propylamine, dipropylamine, and propanol	8–9% Amine 220 on 80/100 mesh Chromosorb W	FID	262
Mars	Life	15% SE-30 on 60/80 mesh Chromosorb W	FID	263
Soil	1,3-Dichloropropene	10% Carbowax 20M on 70/80 mesh Anakrom ABS	ECD-tritium	111
Field weathered soil	[^{14}C]isopropyl-4,4'-dibromobenzilate	5% DOW 11 on 70/80 mesh Chromosorb W	ECD or P det.	264
Soil	Nitrite reduction products	5% SE-30 on 80/100 mesh Chromosorb W	ECD	265

TABLE 5. — Continued

Sample	Analysis	Column	Detector	Reference
Soils	2,6-Dichlorobenzo-nitrile	10% DC-200 on 80/100 mesh Gas Chrom Q	ECD	162
Green River shale	Acidic compounds	3% OV-17 on Gas Chrom Q		266
Green River shale	Steroids and triterpenes	Apiezon L		267
Soils	Ethylene dibromide	20% Castorwax	FID	268
Green River formation kerogen	Acids	3% SE-30 on Chromosorb	MS	269, 270
Solid samples	Volatiles	10% Carbowax 20 M on 60/80 mesh Gas Chrom Q	Dual FID	271
Onverwacht sediments	Organic matter	Carbowax coated open tubular columns	FID	272
Sediments	Isoprenoids acids	0.8% Apiezon L on silicone-treated, acid-washed Chromosorb W; 25% FFAP (same support)		273
Soil	cis- and trans-dichloro-propene residues	5% Carbowax 20M on 70/80 mesh Anakrom SD	ECD	274
Sedimentary rocks	n- and branched-chain alkanes	Apiezon L on SCOT	FID	275

Lunar dust	Organic compounds	1.5% OV-17 on 80/100 mesh Chromosorb G and 150 mesh Porapak Q		276
Precambrian rocks	Aliphatic hydrocarbons	Polysev coated capillary columns	FID	277
Soil	Metabolic pathway where _Pseudomonas_ converts alcohols to ions	Porapak Q, Carbowax 20M + 3% silver nitrate	FID	278
Rubber, rubber dust, tar, baseboard rubber, gasoline, etc.	Something that may interfere with GC analysis of extracts	10% DC-200 on 90/100 mesh Anakrom ABS	ECD	279, 342

plus a TC detector. Overrein [288] used gas chromatographic analysis of nitrogen gases to investigate urea transformation in soil. Acid-washed chromatographic grade coconut charcoal gave good resolution of H_2, N_2, CO, CO_2, N_2O, and O_2 [289]. Gases produced by reaction of nitrite with lignins, humic substances, and phenolic constituents were investigated using the same technique for determination of nitrogen gases [290]. Alkyl nitrites were determined by the procedure of Harrison and Stevenson [291].

Greene et al. [292] achieved separation of H_2, O_2, N_2, CO, CH_4, CO_2, C_2H_2, and C_2H_4 with a 3-m charcoal column heated to 170° C. Heated alumina columns (150° C) have been used efficiently to separate hydrocarbon gases [292]; on these columns the order of elution is in the order of increasing boiling points. Oxygen, N_2, NO, and CO have been satisfactorily separated with 40/60 mesh silica gel at the temperature of acetone-dry ice [293]. This analysis requires almost 1 hr for complete elution and this lengthy time produces somewhat diffuse peaks. The separation of N_2, N_2O, NO_2, CO, and CO_2 is improved if a column of I_2O_5 is attached to the end of the silica gel column. Complete separation of the gas mixture was obtained in less than 15 min using this technique. Van Cleemput [294], Tackett [295], Brazhnikov [296], and Smith and Dowdell [297, 343, 344] have published information concerning the gases in soil atmospheres. In all three cases more than one column was necessary for the analyses.

Ritchie [209] developed a gas chromatographic technique to study soil aeration. The technique allowed rapid, quantitative analysis of CO_2, O_2, and N_2 with a dual-column arrangement of silica gel at 100°C and molecular sieve at ambient temperature.

Berck and Gunther [298] studied the sorption affinity of soils for phosphine; the final measurement was made with a flame photometric detector (FPD). Bell [299, 300, 345-347] analyzed gas mixtures (H_2, N_2, NO, CH_4, CO_2, and N_2O) collected above the water on flooded soils. A TC detector was employed for the measurement.

F. Inorganic Compounds

Although few applications of GC have been utilized in soil chemistry for the analysis of inorganic constituents, especially metals, there is little doubt that many of the metal analyses by GC can be performed on soil samples. I am assuming that proper extraction of the inorganic species has been achieved and interferences removed or converted to chemically inactive components. GLC determinations for metal complexes possess very good sensitivity and adaptability (see Chapter 10 for the use of ion exchange in the analysis of soil samples).

Stevenson and Mendez [301] investigated soil humic acid for inorganic components. They used TLC for cleanup and the GC separation was performed with 10% SE-30, 10% DEGS, and 10% Carbowax on Diatoport S.

Flame ionization was the means of detection. Higashikuze and Kidaka [302] determined hydroscopic water content of clay minerals on a 1.5 m × 3 mm column of Porapak Q and a thermal conductivity detector.

I should like to discuss some of the GC applications of inorganic samples so that the reader may be aware of the potentiality of this technique. With proper sampling and cleanup many of the following techniques and/or determinations could easily be adapted to soil samples.

The earliest use of GLC for determining metal complexes is the one described for the separation of $SnCl_4$ and $TiCl_4$ [303]. It was shown that, on a column of 31% n-hexadecane on Chromosorb, the ratio of vapor pressures of $SnCl_4$ and $TiCl_4$ was very close to the inverse ratio of their retention times. A partial resolution of the optical isomers of Cr(III) hexafluoroacetylacetonate was first achieved on a 80/120 mesh dextro quartz column [304]. Later the same research group investigated the gas chromatographic separations of metal chelates of acetylacetone, trifluoroacetylacetone, and hexafluoroacetylacetone [305]. Separation of multicomponent mixtures was carried out quickly and efficiently. The use of temperature programming and a thermionic emission ionization detector greatly improved the resolution of the mixture. Hill and Gesser [306] developed a column that separated the β-diketone chelates of Be, Al, and Cr (7.5% SE-30 on 40/60 mesh firebrick). The technique gave good resolution and reasonable retention times with plots of sample size vs peak area being linear.

Fluorine-containing metal chelates can be detected by electron capture [307]; as little as 2×10^{-12} g of the metal can be accurately determined.

Some very impressive investigations of metal chelates have been performed in Japan [308-311]. Yamakawa et al. [311] were able to obtain a correlation between the character of metal acetylacetonates and their retention volumes. Quantitative analysis of Cr(III) hexafluoroacetylacetonate has been described in the concentration range of 10^{-3} to 10^{-8} g/cm^3. If 2.0 μl samples are utilized, this amounts to 2×10^{-6} to 2×10^{-11} g of the metal [312]. The complete separations of the Al, Ga, In, Be, and Th trifluoroacetylacetonates also have been described [313]. Thermal conductivity was the means of detection, giving a relative mean error of 2% for the Al, Ga, In, and Be complexes. Interferences from Sc, Cr, Cu, Mn, Zr, Hf, Zn, and Th were discussed.

Separation of ferrocene derivatives has been achieved on a 2.5% Apiezon L-Chromosorb W column using a TCD system [309]. Juvet and Fisher [314] introduced a rapid, direct GLC method for quantitative analysis of the volatile fluorides of metals in alloys, metal oxides, carbides, sulfides, acetates, and nitrates by reaction with F_2 in a specially designed reactor-injector system. Beryllium can be analyzed in airborne dusts [315] and biological fluids [316] by conversion to the trifluoroacetylacetonate.

Synthesis, characterization, and chromatographic properties of the 1,1,1,2,2,3,3-heptafluoro-7,7-dimethyl-4,6-octanedione chelates of Fe, Ni, Pd, Cr, Cu, Al, and Be have been described [317]. All the complexes were eluted without apparent decomposition and the total sample required was < 0.1 mg. Subnanogram amounts of Co(III) (4×10^{-11}) have also been determined using the same ligand [318].

Metals in lunar soil and rocks are receiving a great deal of attention [319, 320]. Beryllium and Cr concentrations have been determined on Apollo 12 lunar samples as their trifluoroacetylacetonates [319]. Lunar sample analyses for Be, Fe, Co, Ni, and Cr are being conducted by GLC-MS analysis, using trifluoroacetylacetone and heptafluorodimethyloctanedione chelates [320]. The lunar samples are first fused with sodium carbonate, dissolved in perchloric acid buffered with sodium acetate, and then subjected to solvent extraction with benzene-trifluoroacetylacetone. Beryllium (in lunar soils) can be accurately determined in amounts of 6×10^{-14} g by this technique.

Mercury, in the 10^{-5}-g range, as mercury alkyls has been determined by gas chromatography [321, 322]. Alkyltin [323] and alkyllead [324, 325] have been determined in gasolines by the gas chromatographic technique, also. A unique application of GLC has been developed for the determination of magnesium metal in the presence of magnesium compounds. The H_2 evolved during the reaction of the sample with sulfuric acid is fed directly into the carrier gas stream of a gas chromatograph and determined directly [326]. Micromoles of hydrogen can be detected and can readily be translated to amounts of free metallic Mg in fuels and other sample types.

Additional information regarding the use of GC for the separation and/or analysis of metal chelates may be found in the review articles by Sievers [327] and Ross and Sievers [328]. A metal chelate complex can be determined by GC if it possesses (1) high volatility, (2) thermal stability, and (3) resistance to degradation reactions. Nearly 50 elements have been investigated by the use of GC techniques. A great percentage of the elements are becoming more important in soils because of their use as plant nutrients or environmental pollutants.

REFERENCES

1. A. Turk, J. Turk, and J. T. Wittes, Ecology, Pollution, Environment, W. B. Saunders, Philadelphia, Pa., 1972, Chaps. 2 and 3, pp. 26-54.

2. C. A. Edwards, Residue Rev., 13, 83 (1966).

3. H. E. Morrison, H. H. Crowell, S. E. Crumb, and R. W. Lauderdale, J. Econ. Entomol., 41, 374 (1948).

4. A. C. Foster, USDA Circ., 41, 862 (1951).

5. E. B. Dennis and C. A. Edwards, Plant Pathol., 13, 173 (1964).

6. W. D. Guenzi and W. E. Beard, Proc. Soil Sci. Amer., 32, 522 (1968).

7. C. R. Harris, J. Econ. Entomol., 57, 946 (1964).

8. D. A. Greatz, G. Chesters, T. C. Daniel, L. W. Newland, and G. B. Lee, J. Water Poll. Contr. Fed., Part 2, R76-R94 (1970).

9. J. G. Konrad and G. Chesters, J. Agr. Food Chem., 17, 226 (1969).

10. J. G. Konrad, D. E. Armstrong, and G. Chesters, Agron. J., 59, 591 (1967).

11. J. G. Konrad, G. Chesters, and D. E. Armstrong, Soil Sci. Soc. Amer. Proc., 33, 259 (1969).

12. N. Sethunathan, Seminar-International Rice Research Institute, Manila, Philippines, 1970, pp. 1-24.

13. M. G. Merkle, R. W. Bovey, and R. Hall, Weeds, 14, 161(1966).

14. J. G. Iyer, G. Chesters, and S. A. Wilde, Weed. Res., 9, 53 (1969).

15. E. J. Bonelli, H. Hartman, and K. P. Dimick, J. Agr. Food Chem., 12, 333 (1964).

16. M. Beroza and M. C. Bowman, Proc. University of Missouri's 3rd Annual Conference on Trace Substances in Environmental Health, University of Missouri, Columbia, Mo., 1969.

17. K. I. Beynon and K. E. Elgar, Analyst, 91, 143 (1966).

18. H. F. Beckman and D. G. Crosby, Food Tech., 17, 32(1963).

19. K. P. Dimick and H. Hartman, Wilkens Instrument and Research, Inc. Bull., W-106, 1963.

20. J. A. Burke, J. Assoc. Off. Agr. Chem., 48, 1037 (1965).

21. A. Bevenue, Analytical Methods for Pesticides, Plant Growth Regulators and Food Additives, Vol. 1, (G. Zweig, ed.), Academic Press, New York, pp. 189-222.

22. M. J. de Faubert Maunder, H. Egan, and J. Roburn, Analyst, 89, 157 (1964).

23. J. A. Burke and L. Giuffrida, J. Assoc. Off. Agr. Chem., 47, 326 (1964).

24. R. P. W. Scott and D. W. Grant, Analyst, 89, 179 (1964).

25. J. A. Gaul, J. Assoc. Off. Agr. Chem., 49, 389 (1966).

26. E. S. Goodwin, R. Goulden, and J. G. Reynolds, Analyst, 86, 697 (1961).

27. W. A. Bosin, Anal. Chem., 35, 833 (1963).

28. H. P. Burchfield and D. E. Johnson, Guide to the Analysis of Pesticide Residues, Vol. 2, U.S. Govt. Printing Office, Washington, D.C., 1965.

29. J. F. Klebe, J. Amer. Chem. Soc., 86, 3399 (1964).

30. H. Egan, E. W. Hammond, and J. Thomson, Analyst, 89, 175 (1964).

31. D. L. Petitjean and C. D. Lantz, J. Gas Chromatog., 1, 23 (1963).

32. S. Zeisel, Monatsch., 6, 989 (1885).

33. W. H. Gutemann and D. J. Lisk, J. Agr. Food Chem., 11, 470 (1963).

34. L. Giuffrida, N. F. Ives, and D. C. Bostwick, J. Assoc. Off. Agr. Chem., 49, 8 (1966).

35. D. Oaks, K. P. Dimick, and H. C. Hartman, Varian Aerograph Reprint, Walnut Creek, Calif., No. 122, 1966.

36. G. Chesters, J. G. Konrad, B. D. Schrag, and L. Everett, Instrumental Methods for Analysis of Soils and Plant Tissue, SSSA, 677 South Segoe Road, Madison, Wisc., 1971.

37. H. Beckman and W. O. Gauer, Bull. Env. Contam. Toxicol., 1, 149 (1966).

38. S. S. Brody and J. E. Chaney, J. Gas Chromatog., 4, 42(1966).

39. R. S. Juvet and R. P. Durbin, Anal. Chem., 38, 565 (1966).

40. R. Goulden, E. S. Goodwin, and L. Davies, Analyst, 88, 941 (1963).

41. H. Shuman and J. R. Collie, J. Assoc. Off. Agr. Chem., 46, 992 (1963).

42. J. A. Burke, J. Assoc. Off. Agr. Chem., 46, 198 (1963).

43. R. Goulden, Qual. Plant. Mater. Veg., 11, 381 (1964).

44. H. Beckman and A. Bevenue, J. Chromatog., 10, 231 (1963).

45. D. M. Coulson, Technical Report No. II, Report No. 5, Stanford Research Institute, Menlo Park, Calif., SRI Project No. P-3198, 1961, p. 7.

46. F. A. Gunther, R. C. Blinn, and G. K. Kohn, Nature (London), 193, 573 (1962).

47. D. D. Phillips, G. E. Pollard, and S. B. Soloway, J. Agr. Food Chem., 10, 217 (1962).

48. H. Curtius, Z. Klin. Chem., Klin. Biochem., 6, 122 (1968).

49. J. Normand and R. Lombre, Method. Phys. Anal., 4, 61 (1968).

50. R. Teranishi, R. E. Lundin and J. R. Scherer, in Chemistry and Physiology of Flavors (H. W. Schultz, E. A. Day, and L. M. Libbey, eds.), Avr., Westport, Conn., 1969, Chapter 7.

51. C. F. Ellis, R. F. Kendall, and B. H. Ecoleston, Anal. Chem.,
    37, 511 (1965).

52. R. Teranishi, J. Dairy Sci., 52, 816 (1969).

53. W. S. Bowers, H. M. Fales, M. J. Thompson, and E. C. Uebel,
    Science, 154, 1020 (1966).

54. C. Hammar, B. Holmstedt, and R. Ryhage, Anal. Biochem., 25,
    532 (1968).

55. C. G. Hammar, I. Hanin, B. Holmstedt, R. J. Kitz, D. J. Jenden,
    and B. Karlen, Nature (London), 220, 915 (1968).

56. J. Han and M. Calvin, Geochim. Cosmochim. Acta, 33, 733 (1969).

57. R. A. Hites and K. Biemann, Anal. Chem., 40, 1217 (1968).

58. B. Holmstedt and R. Tham, J. Chromatog., 19, 286 (1965).

59. R. Ryhage and E. von Sydow, Acta Chem. Scand., 17, 2025 (1963).

60. J. Sjovall and R. Vihko, Steroids, 7, 447 (1966).

61. F. Vane and M. G. Horning, Anal. Letters, 2, 357 (1969).

62. M. H. Studier and R. Hayatsu, Anal. Chem., 40, 1011 (1968).

63. T. Furuya, H. Kojima, and H. Sato, Chem. Pharm. Bull., 15,
    1362 (1967).

64. A. Barlow, R. S. Lehrle, and J. C. Robb, Polymer, 8, 523 (1967).

65. K. Ettre and P. F. Varadi, Anal. Chem., 34, 752 (1962).

66. C. E. R. Jones and A. F. Moyles, Nature (London), 189, 222 (1961).

67. R. Kaiser, Chem. In Britain, 5, 54 (1969).

68. C. E. R. Jones and G. E. J. Reynolds, Soc. Chem. Ind. Monograph,
    26, 260 (1967).

69. R. S. Lehrle, Lab. Practice, 17, 696 (1968).

70. P. G. Simmonds, G. P. Shulman, and C. H. Stempridge, J. Chroma-
    tog. Sci., 7, 36 (1969).

71. B. Kolb, G. Kemmner, F. H. Schleser, and E. Wiedeking, Z.
    Anal. Chem., 220, 166 (1966).

72. B. Aliprandi, F. Cacace, and G. Ciranni, Anal. Chem., 36, 2445
    (1964).

73. F. Cacace and Inam-Ui-Haq, Science, 131, 732 (1960).

74. F. Cacace and G. Perez, J. Labelled Compounds, 2, 102 (1966).

75. F. Cacace, R. Cipollini, G. Perez, and E. Possagno, Gazz. Chim. Ital. , 91, 804 (1961).

76. A. Karmen, Packard Tech. Bull. , 14, (1965).

77. F. Cacace and Inam-Ul-Haq, La Ricerca Scientifica, 30, 3 (1960).

78. S. P. Cram and J. L. Brownlee, Jr. , J. Gas Chromatog. , 5, 353 (1967).

79. S. P. Cram and J. L. Brownlee, Jr. , J. Gas Chromatog. , 6, 305 (1968).

80. S. P. Cram and J. L. Brownlee, Jr. , J. Gas Chromatog. , 6, 313 (1968).

81. W. E. Westlake and F. A. Gunther, Residue Revs. , 18, 176 (1967).

82. J. H. Hamence, P. S. Hall, and D. J. Caverly, Analyst, 90, 649 (1965).

83. A. K. Klein and J. O. Watts, J. Assoc. Off. Agr. Chem. , 47, 311 (1964).

84. H. B. Pionke, G. Chesters, and D. E. Armstrong, Analyst, 94, 900 (1969).

85. W. W. Sans, J. Agr. Food Chem. , 15, 192 (1967).

86. B. E. Frazier, G. Chesters, and G. B. Lee, Pest. Monit. J. , 4, 67 (1970).

87. W. L. Lamar, D. F. Goerlitz, and L. M. Low, Geol. Survey Water-Supply Paper, 1818-B, 1-12 (1965).

88. W. L. Trautman, G. Chesters, and H. B. Pionke, Pest. Monit. J. , 2, 93 (1968).

89. T. W. Duke, J. I. Lowe, and A. J. Wilson, Bull. Env. Contam. Toxicol. , 5, 171 (1970).

90. J. J. Hickey and D. W. Anderson, Science, 162, 271 (1968).

91. D. C. Holmes, J. H. Simmons, and J. O'G. Tatton, Nature (London), 216, 227 (1967).

92. D. A. Ratcliffe, Nature (London), 215, 208 (1967).

93. R. W. Risebrough, D. B. Menzel, D. J. Martin, and H. S. Olcott, Nature (London), 216, 589 (2967).

94. G. Widmark, J. Assoc. Off. Anal. Chem. , 50, 1069 (1967).

95. L. W. Newland, G. Chesters, and G. B. Lee, J. Water Poll. Contr. Fed. , Part 2, R174-R188 (1969).

96. K. D. Berlin, T. H. Autin, M. Nagabhushanam, M. Peterson, J. Calvert, L. A. Wilson, and D. Hopper, J. Gas Chromatog., 3, 256 (1965).

97. M. C. Bowman and M. Beroza, J. Assoc. Off. Anal. Chem., 53, 499 (1970).

98. J. G. Konrad, H. B. Pionke, and G. Chesters, Analyst, 94, 490 (1969).

99. B. Krakow, Anal. Chem., 41, 815 (1969).

100. G. E. Pollock and L. H. Frommhagen, Anal. Biochem., 24, 18 (1968).

101. A. Bevenue, J. N. Ogata, and H. Beckman, J. Chromatog., 35, 17 (1968).

102. R. L. King, N. A. Clark, and R. W. Hemken, J. Agr. Food Chem., 14, 62 (1966).

103. R. R. Schnobrus and W. F. Phillips, J. Agr. Food Chem., 15, 661 (1967).

104. E. P. Lichtenstein, T. W. Fuhremann, and K. R. Schultz, J. Agr. Food Chem., 16, 348 (1968).

105. F. L. Carter, C. A. Stringer, and D. Heinzelman, Bull. Env. Contam. Toxicol., 6, 249 (1971).

106. P. C. Kearney, E. A. Woolson, J. R. Plimmer, and A. B. Irensee, in Residue Reviews, Vol. 29, (F. A. Gunther, ed.), Springer-Verlag, New York, 1969, pp. 137-49.

107. J. Hurter, H. Zurrer, and E. Reuthinger, Z. Lebensmittelunters, u.-Forsch., 130, 20 (1966).

108. I. H. Williams, R. Kore, and D. G. Finlayson, J. Agr. Food Chem., 19, 456 (1971).

109. R. I. Asai, W. E. Westlake, and F. A. Gunther, Bull. Env. Contam. Toxicol., 4, 278 (1969).

110. D. A. Graetz, G. Chesters, T. C. Daniel, G. B. Lee, and L. W. Newland, J. Water Poll. Contr. Fed., 42, R-76 (1970).

111. M. Leistra, J. Agr. Food Chem., 18, 1124 (1970).

112. J. G. Saha, J. Assoc. Off. Anal. Chem., 54, 170 (1971).

113. D. K. R. Stewart, D. Chisolm, and M. T. H. Ragab, Nature (London), 229, 47 (1971).

114. A. R. Swoboda, G. W. Thomas, F. B. Cady, R. W. Baird, and W. G. Knisel, Env. Sci. Tech., 5, 141 (1971).

115. Y. O. Shin, J. J. Chodan, and A. R. Wolcott, J. Agr. Food Chem.,
     18, 1129 (1970).

116. D. O. Eberle and D. Novak, J. Assoc. Off. Anal. Chem., 52, 1067
     (1969).

117. C. R. Malone, A. G. Winnet, and K. Helrich, Bull. Env. Contam.
     Toxicol., 2, 83 (1967).

118. A. R. Swoboda and G. W. Thomas, J. Agr. Food Chem., 16, 923
     (1968).

119. J. F. Parr, G. H. Willis, and S. Smith, Soil Sci., 110, 306 (1970).

120. J. H. Caro, J. Agr. Food Chem., 19, 78 (1971).

121. M. Chiba, W. N. Yule, and H. V. Morley, Bull. Env. Contam.
     Toxicol., 5, 263 (1970).

122. T. P. Methratta, R. W. Montagna, and W. P. Griffith, J. Agr.
     Food Chem., 15, 648 (1967).

123. J. G. Saha and W. W. A. Steward, Can. J. Plant Sci., 47, 79 (1967).

124. W. H. Gutemann and D. J. Lisk, J. Gas Chromatog., 6, 124 (1968).

125. A. J. B. Powell, T. Stevens, and K. A. McCully, J. Agr. Food
     Chem., 18, 224 (1970).

126. S. Voerman and A. F. H. Besemer, J. Agr. Food Chem., 18, 717
     (1970).

127. J. E. Fahey, J. W. Butcher, and M. E. Turner, Pest. Monit. J.,
     1, 30 (1968).

128. D. O. Eberle and D. Novak, J. Assoc. Off. Anal. Chem., 52, 1067
     (1969).

129. K. I. Beynon, M. J. Edwards, K. Elgar, and A. N. Wright, J. Sci.
     Food Agri., 19, 302 (1968).

130. W. N. Yule, M. Chiba, and H. V. Morley, J. Agr. Food Chem., 15,
     1000 (1967).

131. F. L. Carter and C. A. Stringer, Bull. Env. Contam. Toxicol., 5,
     422 (1970).

132. J. G. Saha and H. McDonald, J. Agr. Food Chem., 15, 205 (1967).

133. J. R. Duffy and N. Wong, J. Agr. Food Chem., 15, 457 (1967).

134. J. G. Saha, C. H. Craig, and W. K. Janzen, J. Agr. Food Chem.,
     16, 617 (1968).

135. E. P. Lichtenstein, K. R. Schultz, R. F. Skrentny, and P. A. Scott,
     J. Econ. Entomol., 58, 742 (1965).

136. A. M. Gardner, J. N. Damico, E. A. Hansen, E. Lustig, and R. W. Storherr, J. Agr. Food Chem., 17, 1181 (1969).

137. H. B. Pionke, G. Chesters, and D. E. Armstrong, Analyst, 94, 900 (1969).

138. L. W. Getzin and I. Rosenfield, J. Agr. Food Chem., 16, 598 (1968).

139. M. G. Merkle and F. S. Davis, USDA-ARS-Corps Res. Div. and Texas Agr. Exp. Sta., College Station, Texas 77843, Mimeo reprint.

140. H. L. Morton, F. S. Davis, and M. G. Merkle, Weed Sci., 16, 88 (1968).

141. R. E. Wildung, G. Chesters, and D. E. Armstrong, Weed Res., 8, 213 (1968).

142. A. E. Smith, J. Agr. Food Chem., 17, 1052 (1969).

143. C. E. McKone and R. J. Hance, J. Agr. Food Chem., 15, 935 (1967).

144. C. E. McKone, J. Chromatog., 44, 60 (1969).

145. C. A. Benfield and E. D. Chilwell, Analyst, 89, 475 (1964).

146. C. I. Hannon, J. Angelini, and R. Wolford, J. Gas Chromatog., 1, 27 (1963).

147. J. J. Jurinak and T. S. Inouye, Soil Sci. Soc. Amer. Proc., 27, 602 (1963).

148. D. H. Smith and R. S. Shigenaga, Soil Sci. Soc. Amer. Proc., 25, 160 (1961).

149. S. E. Katz and R. F. Strusz, Bull. Env. Contam. Toxicol., 3, 258 (1968).

150. K. I. Beynon and A. N. Wright, J. Sci. Food Agr., 19, 718 (1968).

151. R. Bartha and D. Pramer, Science, 156, 1617 (1967).

152. R. Bartha, J. Agr. Food Chem., 16, 602 (1968).

153. A. M. Mattson, R. A. Kahrs, and J. Schneller, J. Agr. Food Chem., 13, 120 (1965).

154. K. J. Meulemans and E. T. Upton, J. Assoc. Off. Anal. Chem., 49, 976 (1966).

155. R. F. Moseman and W. A. Aue, J. Chromatog., 49, 432 (1970).

156. K. Ishikawa, R. Shinohara, and K. Akasaki, Agr. Biol. Chem., 35, 1161 (1971).

157. C. E. Mendoza and J. B. Shields, J. Assoc. Off. Anal. Chem., 54, 986 (1971).

158. W. H. Gutemann and D. J. Lisk, J. Assoc. Off. Anal. Chem., 51, 688 (1968).

159. P. J. Burcar, R. L. Wershaw, M. C. Goldberg, and L. Kahn, Analysis Instrumentation, Vol. 4 (Fowler, Harmon and Rowe, eds.), Plenum, New York, 1967, pp. 215-224.

160. D. O. Eberle and W. D. Horman, J. Assoc. Off. Anal. Chem., 54, 150 (1971).

161. C. I. Harris, J. Agr. Food Chem., 17, 80 (1969).

162. G. C. Briggs and J. E. Dawson, J. Agr. Food Chem., 18, 97 (1970).

163. H. H. Cheng, Bull. Env. Contam. Toxicol., 6, 28 (1971).

164. H. L. Pease, J. Agr. Food Chem., 15, 917 (1967).

165. V. A. Jolliffe, B. E. Day, L. S. Jordan, and J. W. Mann, J. Agr. Food Chem., 15, 174 (1967).

166. W. H. Gutemann and D. J. Lisk, J. Assoc. Off. Anal. Chem., 51, 688 (1968).

167. M. L. Beall, E. A. Woolson, T. J. Sheets, and C. I. Harris, J. Agr. Food Chem., 15, 208 (1967).

168. I. J. Belasco and H. L. Pease, J. Agr. Food Chem., 17, 1414 (1969).

169. H. P. Hermanson, M. Siewierski, and K. Helrich, J. Assoc. Off. Anal. Chem., 52, 175 (1969).

170. K. I. Beynon, L. Davies, K. Elgar, and A. N. Wright, J. Sci. Food Agr., 17. 151 (1966).

171. P. J. Burcar, R. L. Wershaw, M. C. Goldberg, and L. Kahn, Analysis Instrumentation, Vol. 4 (Fowler, Harmon, and Rowe, eds.), Plenum, New York, 1967, pp. 215-224.

172. J. S. Thornton and C. A. Anderso, J. Agr. Food Chem., 13, 509 (1965).

173. J. O'G. Tatton and P. J. Wagstaffe, J. Chromatog., 44, 284 (1969).

174. J. C. Casely, Bull. Env. Contam. Toxicol., 3, 180 (1968).

175. K. Groves and K. S. Chough, J. Agr. Food Chem., 18, 1127 (1970).

176. A. Starcl, J. Agr. Food Chem., 17, 871 (1969).

177. C. Zomzely, G. Marco, and E. Emery, Anal. Chem., 34, 1414 (1962).

178. H. A. Saroff, A. Karmen, and J. W. Healy, J. Chromatog., 9, 122 (1962).

179. S. Makisumi, C. H. Nicholls and H. A. Saroff, J. Chromatog., 12, 106 (1963).

180. S. Makisumi and H. A. Saroff, J. Gas Chromatog., 3, 21 (1965).

181. G. E. Pollock and A. H. Kawauchi, Anal. Chem., 40, 1356 (1968).

182. G. E. Pollock and V. I. Oyama, J. Gas Chromatog., 4, 126 (1966).

183. G. E. Pollock, V. I. Oyama, and R. D. Johnson, J. Gas Chromatog., 3, 174 (1965).

184. J. W. Berry, Advan. Chromatog., 2, 271 (1962).

185. C. T. Bishop, in Methods of Biomedical Analysis, Vol. 10 (D. Glick, ed.), Interscience, New York.

186. C. T. Bishop, Adv. Carbohyd. Chem., 19, 95 (1964).

187. H. P. Burchfield and E. E. Storrs, Biomedical Applications of Gas Chromatography, Academic, New York, 1962.

188. W. W. Wells, C. C. Sweeley, and R. Bentley, in Biomedical Applications of Gas Chromatography (H. A. Szymanski, ed.), Plenum, New York, 1963.

189. E. J. Hedgley and W. G. Overend, Chem. Ind.(London),1960, 378 (1960).

190. E. Bayer, Gas Chromatography, Elsevier, Amsterdam, The Netherlands, 1961.

191. R. J. Ferrier, Tetrahedron, 18, 1149 (1962).

192. R. J. Ferrier and M. F. Singleton, Tetrahedron, 18, 1143 (1962).

193. C. C. Sweeley, R. Bentley, M. Makita, and W. W. Wells, J. Amer. Chem. Soc., 85, 2497 (1963).

194. L. Birkofer, A. Ritter, and W. Giesler, Angew. Chem., 75, 93 (1963).

195. J. F. Klebe, J. B. Bush, Jr., and J. E. Lyons, J. Amer. Chem. Soc., 86, 4400 (1964).

196. J. F. Klebe, H. Finkbeiner, and D. M. White, J. Amer. Chem. Soc., 88, 3390 (1966).

197. W. R. Supina, R. F. Kruppa, and R. S. Henly, J. Amer. Oil Chem. Soc., 44, 74 (1967).

198. M. Vilkas, H. Jan, G. Boussac, and M. C. Bonnard, Tetrahedron Letters, 1441 (1966).

199. J. Kärkkäinen, A. Lehtonen, and T. Nikkari, J. Chromatog., 20, 457 (1965).

200. J. H. Kim, B. Shome, T. Liao, and J. G. Pierce, Anal. Biochem., 20, 258 (1967).

201. A. Lehtonen, J. Kärkkäinen, and E. Haahti, J. Chromatog., 24, 179 (1966).

202. I. M. Morrison and M. B. Perry, Can. J. Biochem., 44, 1115 (1966).

203. M. Abdel-Akher, J. K. Hamilton, and F. Smith, J. Amer. Chem. Soc., 73, 4691 (1951).

204. H. G. Jones, J. K. N. Jones, and M. B. Perry, Can. J. Chem., 40, 1559 (1962).

205. M. B. Perry, Can. J. Biochem., 42, 451 (1964).

206. J. M. Richey, H. G. Richey, Jr., and R. Schraer, Anal. Biochem., 9, 272 (1964).

207. C. C. Sweeley and B. Walker, Anal. Chem., 36, 1461 (1964).

208. T. White, J. Chem. Soc., 1940, 428.

209. S. Roseman and J. Ludowieg, J. Amer. Chem. Soc., 76, 301 (1954).

210. A. Kobata and S. Zirô, Biochim. Biophys. Acta, 107, 405 (1965).

211. M. I. Horowitz and M. R. Delman, J. Chromatog., 21, 300 (1966).

212. T. A. Kukharenko, V. I. Belikova, and L. V. Motovilova, Pochvoved., No. 2, 65 (1969).

213. N. F. Boas, J. Biol. Chem., 204, 553 (1953).

214. E. Sjöström, P. Haglund, and J. Janson, Acta Chem. Scand., 20, 1718 (1966).

215. G. E. Pollock and D. A. Jermany, J. Gas Chromatog., 6, 412 (1968).

216. G. E. Pollock and D. A. Jermany, J. Gas Chromatog., 8, 296 (1970).

217. J. H. Sloneker, Biomed. Appl. Gas Chromatog., 2, 87 (1968).

218. J. G. Nikelly, Anal. Chem., 36, 2244 (1964).

219. M. G. Horning, G. Casparrini, and E. C. Horning, J. Gas Chromatog., 7, 267 (1969).

220. D. A. Shearer and W. E. Cordukes, Can. J. Plant Sci., 42, 686 (1962).

221. S. Siggia, L. R. Whitlock, and J. C. Tao, Anal. Chem., 41, 1387 (1969).

222. J. M. Oades, M. A. Kirkman, and G. H. Wagner, Soil Sci. Soc. Amer. Proc., 34, 30 (1970).

223. M. Vecchi, W. Vetter, W. Walther, S. J. Jermstad, G. W. Schutt, Helv. Chim. Acta, 50, 1243 (1967).

224. M. Vecchi and K. Kaiser, J. Chromatog., 26, 22 (1967).

225. L. V. Aviola and S. W. Lee, Anal. Biochem., 16, 193 (1966).

226. H. C. Pillsbury, A. J. Sheppard, and D. A. Libby, J. Assoc. Offic. Anal. Chem., 50, 809 (1967).

227. B. Kolb and E. Wiedeking, Z. Anal. Chem., 243, 129 (1968).

228. B. Kolb and E. Wiedeking, Angew. Gas Chromatog., Heft 13E, 11p. (1969).

229. M. A. Kirkman, Dissertation Abstr., 30, 1544B (Oct. 1969).

230. B. R. Simoneit, A. L. Burlingame, and H. K. Schnoes, Nature (London), 226, 75 (1970).

231. B. R. Nagar, Nature (London), 199, 1213 (1963).

232. J. Mendez and F. J. Stevenson, J. Gas Chromatog., 4, 483 (1966).

233. E. H. Hansen and M. Schnitzer, Soil Sci. Soc. Amer. Proc., 30, 745 (1966).

234. M. Schnitzer and J. G. Desjardins, Can. J. Soil Sci., 44, 272 (1964).

235. E. H. Hansen and M. Schnitzer, Soil Sci. Soc. Amer. Proc., 31, 79 (1967).

236. G. T. Felbeck, Jr., Trans. Comm. II and IV, Int. Soc. Soil Sci., Aberdeen, Scotland, 1966, 11.

237. M. H. Studier, R. Hayatsu, and E. Anders, Geochim. Cosmochim. Acta, 32, 151 (1968).

238. N. Kikuchi and T. Miki, Bunseki Kagaku, 17, 1102 (1968).

239. P. C. Kearney, J. R. Plimmer, and F. B. Guardia, J. Agr. Food Chem., 17, 1418 (1969).

240. P. Albrecht and G. Ourisson, Science, 163, 1192 (1969).

241. J. Han, B. R. Simoneit, A. L. Burlingame and M. Calvin, Nature (London), 222, 364 (1969).

242. J. Oró, J. Gilbert, W. Updegrave, J. McReynolds, J. Ibanez, E. Gil-Av, D. Flory, and A. Zlatkis, J. Chromatog. Sci., 8, 297 (1970).

243. B. Nagy, C. M. Drew, P. B. Hamilton, V. E. Modzeleski, M. E. Murphy, W. M. Scott, H. C. Urey, and M. Young, Science, 167, 770 (1970).

244. D. G. Simmonds, G. P. Shulman, and C. H. Stempridge, J. Chromatog. Sci., 7, 36 (1969).

245. S. Pennington and C. E. Meloan, J. Chromatog., 27, 250 (1967).

246. P. Simonart and L. Batistic, Nature (London), 212, 1461 (1966).

247. D. E. Johnson and B. Lear, J. Chromatog. Sci., 7, 384 (1969).

248. E. Peterson, K. A. Kvenvolden, and F. S. Brown, Science, 169, 1079 (1970).

249. D. Attaway and P. L. Parker, Science, 169, 674 (1970).

250. S. Nagy, W. M. Scott, V. Modzeleski, L. A. Nagy, C. M. Drew, W. S. McEwan, T. F. Thomas, P. B. Hamilton, and H. C. Urey, Nature (London), 225, 1028 (1970).

251. D. W. Woodham, W. G. Mitchell, C. D. Loftis, C. W. Collier, J. Agr. Food Chem., 19, 186 (1971).

252. M. Schnitzer and G. Ogner, Israel J. Chem., 8, 505 (1970).

253. R. Sawyer, B. G. Cox, E. J. Dixon, and J. Thompson, J. Sci. Food Agr., 18, 287 (1967).

254. T. S. C. Wang, S. Y. Cheng, and H. Tung, Soil Sci., 103, 360 (1967).

255. E. N. Davis, L. L. Wallen, J. C. Goodwin, W. K. Rohwedder, and R. A. Rhodes, Lipids, 4, 356 (1969).

256. T. W. G. Want, Y. Liang, and W. Shen, Soil Sci., 107, 181 (1969).

257. J. Sever and P. L. Parker, Science, 164, 1052 (1969).

258. D. E. Johnson and B. Lear, Soil Sci., 106, 31 (1968).

259. S. Lakota and W. A. Aue, J. Chromatog., 44, 472 (1969).

260. S. R. Lipsky, R. J. Cushley, C. G. Horvath, and W. J. McMurray, Science, 167, 778 (1970).

261. P. Albrecht and G. Ourisson, Science, 163, 1192 (1969).

262. H. P. Hermanson, K. Helrich, and W. F. Carey, Anal. Letters, 1, 941 (1968).

263. V. I. Oyama, Nature (London), 200, 1058 (1963).

264. R. D. Cannizzaro, T. E. Cullen, and R. T. Murphy, J. Agr. Food Chem., 18, 728 (1970).

265. L. A. Bulla, C. M. Gilmour, and W. B. Bollen, Nature (London), 225, 664 (1970).

266. R. C. Murphy, M. V. Djuricic, S. P. Markey and L. Biemann, Science, 165, 695 (1969).

267. W. Henderson, V. Wollrab and G. Eglinton, Chem. Comm., 1968, 710.

268. J. J. Jurinak and T. S. Nonye, Soil Sci. Soc. Amer. Proc., 27, 602 (1963).

269. A. L. Burlingame and B. R. Simoneit, Nature (London), 222, 741 (1969).

270. A. L. Burlingame and B. R. Simoneit, Science, 160, 531 (1968).

271. R. J. Levins and R. M. Ikeda, J. Gas Chromatog., 6, 331 (1968).

272. B. Nagy and L. A. Nagy, Nature (London), 223, 1226 (1969).

273. M. Blumer and W. J. Cooper, Science, 158, 1463 (1967).

274. I. H. Williams, J. Econ. Entomol., 61, 1432 (1968).

275. V. E. Modzeleski, W. D. MacLeod, Jr., and B. Nagy, Anal. Chem., 40, 987 (1968).

276. C. Ponnamperuma, K. Kvenvolden, S. Chang, R. Johnson, G. Pollock, D. Philpott, I. Kaplan, J. Smith, J. W. Schopf, C. Gerke, H. Hodgson, I. Breger, B. Halpern, A. Duffield, K. Krauskopf, E. Barghoorn, H. Holland, and K. Keil, Science, 167, 760 (1970).

277. J. Oró, J. Gilbert W. Updegrave, J. McReynolds, J. Ibanez, E. Gil, D. Flory, and A. Zlatkis, J. Chromatog. Sci., 8, 297 (1970).

278. H. O. Belser and C. E. Castro, J. Agr. Food Chem., 19, 23 (1970).

279. K. H. Deubert, Bull. Env. Contam. Toxicol., 5, 379 (1970).

280. R. H. Crouse, J. W. Garner, and H. O'Neill, J. Gas Chromatog., 1, 18 (1963).

281. M. E. Bednas and D. S. Russell, J. Gas Chromatog., 5, 592 (1967).

282. S. A. Greene and H. Pust, Anal. Chem., 30, 1039 (1958).

283. D. H. Smith, F. S. Nakayama, and F. E. Clark, Soil Sci. Soc. Amer. Proc., 24, 145 (1960).

284. E. K. Gibson, Jr. and C. B. Moore, Anal. Chem., 42, 461 (1970).

285. P. L. Steyn and C. C. Delwiche, Env. Sci. Tech., 4, 1122 (1970).

286. D. H. Smith and F. E. Clark, Soil Sci. Soc. Amer. Proc., 24, 111 (1960).

287. J. R. Burford, J. Chromatog. Sci., 7, 760 (1969).

288. L. Overrein, Ph.D. Thesis, Purdue University Library, Lafayette, Indiana, 1963.

289. F. J. Stevenson and R. M. Harrison, Soil Sci. Soc. Amer. Proc., 30, 609 (1966).

290. F. J. Stevenson, R. M. Harrison, R. Wetselaar, and R. A. Leeper, Soil Sci. Soc. Amer. Proc., 34, 430 (1970).

291. R. M. Harrison and F. J. Stevenson, J. Gas Chromatog., 3, 240 (1965).

292. S. A. Greene, M. Z. Moberg and E. M. Wilson, Anal. Chem., 28, 1369 (1956).

293. D. H. Szulczewski and T. Higuchi, Anal. Chem., 29, 1541 (1957).

294. O. van Cleemput, J. Chromatog., 45, 315 (1969).

295. J. L. Tackett, Soil Sci. Soc. Amer. Proc., 32, 346 (1968).

296. V. V. Brazhnikov, L. M. Muhin, V. A. Otrozchenko, R. I. Fedorova, and K. I. Sakodynsky, Chromatographia, 3, 552 (1971).

297. K. A. Smith and R. J. Dowdell, J. Chromatog. Sci., 11, 655 (1973).

298. B. Berck and F. A. Gunther, J. Agr. Food Chem., 18, 148 (1970).

299. R. G. Bell, Soil Sci., 106, 408 (1968).

300. R. G. Bell, Soil Sci., 105, 78 (1968).

301. R. J. Stevenson and J. Mendez, Soil Sci., 103, 383 (1967).

302. H. Higashikuze and Y. Kidaka, Bunseki Kagaku, 19, 554 (1970).

303. H. Freiser, Anal. Chem., 31, 1440 (1959).

304. R. E. Sievers, R. W. Moshier and M. L. Morris, Inorg. Chem., 1, 966 (1962).

305. R. E. Sievers, B. W. Ponder, M. L. Morris, and R. W. Moshier, Inorg. Chem., 2, 693 (1962).

306. R. D. Hill and H. Gesser, J. Gas Chromatog., 1, 11 (1963).

307. W. D. Ross, Anal. Chem., 35, 1596 (1963).

308. K. Arakawa and K. Tanikawa, Bunseki Kagaku, 16, 812 (1967).

309. K. Tanikawa and K. Arakawa, Chem. Pharm. Bull., 13, 926 (1965).

310. K. Tanikawa, K. Hirano, and K. Arakawa, Chem. Pharm. Bull., 15, 915 (1967).

311. K. Yamakawa, K. Tanikawa, and K. Arakawa, Chem. Pharm. Bull., 11, 1405 (1963).

312. W. D. Ross and G. Wheeler, Jr., Anal. Chem., 36, 266 (1964).

313. J. E. Schwarberg, R. W. Moshier, and J. H. Walsh, Talanta, 11, 1213 (1964).

314. R. S. Juvet, Jr. and R. L. Fisher, Anal. Chem., 38, 1860 (1966).

315. M. H. Noweir and J. Cholak, Env. Sci. Tech., 3, 927 (1969).

316. M. L. Taylor, E. L. Arnold, and R. E. Sievers, Anal. Letters, 1, 735 (1968).

317. R. E. Sievers, J. W. Connolly, and W. D. Ross, J. Gas Chromatog., 5, 241 (1967).

318. W. D. Ross, W. G. Scribner, and R. E. Sievers, Proc. 8th Int. Symp. Chromatog., Dublin, Eire, Sept. 1970.

319. R. E. Sievers, K. J. Eisentraut, M. F. Richardson, and W. R. Wolf, Abst. of Lecture at IUPAC 13th Int. Conf. Coord. Chem., Zakopane, Poland, Sept. 1970.

320. R. E. Sievers, K. J. Eisentraut, C. W. Harris, A. S. Hilton, M. F. Richardson, W. R. Wolf, and T. L. Isenhour, Abst. from IUPAC 6th Int. Symp. Microtechniques, Graz, Austria, Sept. 1970.

321. S. Nishi and Y. Horimoto, Bunseki Kagaku, 17, 75 (1968).

322. S. Nishi and Y. Horimoto, Bunseki Kagaku, 17, 1247 (1968).

323. Y. Jitsu, N. Kudo, K. Sato, and T. Teshima, Bunseki Kagaku, 18, 169 (1969).

324. E. J. Bonelli and H. Hartman, Anal. Chem., 35, 1980 (1963).

325. H. J. Dawson, Jr., Anal. Chem., 35, 542 (1963).

326. V. D. Hogan and F. R. Taylor, Anal. Chem., 40, 1387 (1968).

327. R. E. Sievers, Coord. Chem., 1969, 270-288.

328. W. D. Ross and R. E. Sievers, Devel. Appl. Spectrosc., 8, 181 (1970).

329. R. J. Whiteoak, M. Crofts, and R. J. Harris, Pestic. Sci., 4(3), 319 (1973).

330. A. M. Mattson, R. A. Kahrs, and R. T. Murphy, Residue Rev., 32, 371 (1970).

331. A. Bevenue and J. N. Ogata, J. Chromatog., 46(1), 110 (1970).

332. H. Y. Young and A. Chu, J. Agr. Food Chem., 21(4), 711 (1973).

333. R. Purkayas, J. Agr. Food Chem., 22(3), 453 (1974).

334. S. U. Khan, J. Agr. Food Chem., 22(5), 863 (1974).

335. J. P. Rouchaud and J. R. Decallonne, J. Agr. Food Chem., 22(2), 259 (1974).

336. K. Nishi, M. Onishi, M. Yasumatsu, and M. Hattori, Takedo Kenkyusho Ho, 31(2), 163 (1972).

337. C. E. McKone and R. J. Hance, J. Chromatog., 69(1), 204 (1972).

338. D. E. Ott, G. Formica, G. F. Liebig Jr., D. O. Eberle, and F. A. Gunther, J. A. O. A. C., 54(6), 1388 (1971).

339. C. D. Ercegovich, S. Witkonton, and K. B. Steen, J. Agr. Food Chem., 20(3), 734 (1972).

340. M. Wurst, V. Vancura, and L. Kalachova, J. Chromatog., 91, 469 (1974).

341. P. G. Brisbane, M. Amato, and J. N. Ladd, Soil Biol. Biochem., 4(1), 51 (1972).

342. W. S. Updegrove and J. Oro, Res. Phys. Chem., Proc. Lunar Int. Lab. Symp., 3rd 53-74, Pergamon Press, Oxford, England, 1969.

343. K. J. Eisentraut, D. J. Griest, and R. E. Sievers, Anal. Chem., 43(14), 2003 (1971).

344. C. J. Soderquist, D. G. Crosby, and J. B. Bowers, Anal. Chem., 46(1), 155 (1974).

345. A. M. Blackmer, J. H. Baker, and M. E. Weeks, Soil Sci. Soc. Amer. Proc., 38(4), 689 (1974).

346. K. A. Smith and R. J. Dowdell, J. Chromatog. Sci., 11(12), 655 (1973).

347. J. H. Taylor and G. R. Shoemake, J. Chromatog. Sci., 10(1), 48 (1972).

Chapter 7

LIQUID CHROMATOGRAPHIC ANALYSIS IN SOIL CHEMISTRY

Gordon Chesters and D. A. Graetz*

Water Resources Center
Department of Soil Science and
Water Chemistry Program
University of Wisconsin — Madison
Madison, Wisconsin

*Present Address: Soil Science Department
                 Institute of Food and Agricultural Sciences
                 University of Florida
                 Gainesville, Florida

I. INTRODUCTION

The applications of liquid chromatographic techniques to soils have been limited. In the field of characterization of the humic materials of soils attention has been focused principally on separation of soil extracts based on molecular fractionation by Sephadex gel chromatography. Ratios of the amounts of humic components present in different molecular weight fraction ranges have been used to qualitatively characterize humic and fulvic acids from different Great Soil groups.

For pesticide analysis, liquid chromatography has been used almost exclusively as a means to clean up soil and sediment extracts prior to some final method of quantitative determination. Furthermore, using a series of eluting solvents, some attempts have been made to fractionate pesticidal residues to improve the specificity of their identification. The value of liquid chromatography as a cleanup procedure over that of paper chromatography, for example, is based on the high carrying capacity of the support.

The lack of quantitative procedures based on liquid chromatography stems largely from the poor resolution that can be achieved by this method and up to this time no satisfactory detection systems for the column eluent have been devised.

Much information is available on soil sampling, and techniques used are very similar irrespective of the types of analysis to be conducted. For further information on the techniques of sampling and the design of sampling on a statistical basis, the reader is referred to review articles [1-4].

Since the major use of liquid chromatographic procedures in soils is as a cleanup technique, a discussion of separation of interferences and concentration techniques runs throughout the chapter and particularly in the section dealing with pesticide analysis.

II. METHODS AND USES OF LIQUID CHROMATOGRAPHIC ANALYSIS IN SOIL CHEMISTRY

A. Humic and Fulvic Acids

In order to adequately discuss the chromatography of humic material, it is necessary to describe the methods of isolation and the nomenclature of the isolated components. This can best be done with a simple flow sheet (Fig. 1).

The terms to be used most frequently in this discussion are "humic acid" (alkali soluble, acid insoluble) and "fulvic acid" (alkali soluble, acid soluble), portions of soil organic matter. Detailed reviews of the isolation of these fractions have been prepared [5-9].

The vast majority of investigations using liquid chromatographic techniques for the characterization of soil humic materials have centered on the use of Sephadex gel filtration columns to provide a fractionation based on molecular weight. The Sephadex gels available and their properties are described in Table 1 [10].

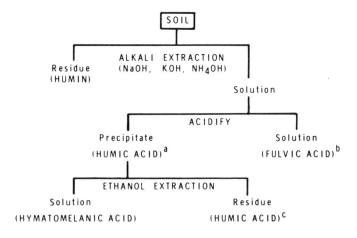

Figure 1. Method of isolation and nomenclature of soil humic materials. (a) American terminology; (b) sometimes termed "crenic" and "apocrenic" acids; (c) German terminology.

TABLE 1

Properties of Sephadex Dextran Gels[a]

Gel grade	Molecular weight fractionation range[b]	Water regain (g water/g dry gel)	Bed volume (ml/g dry gel)
G-10	<700	1.0 ± 0.1	2
G-15	<1500	1.5 ± 0.1	3
G-25	100–5000	2.5 ± 0.2	5
G-50	500–10,000	5.0 ± 0.3	10
G-75	1000–50,000	7.5 ± 0.5	12–15
G-100	1000–100,000	10.0 ± 1.0	15–20
G-150	1000–150,000	15.0 ± 1.5	20–30
G-200	1000–200,000	20.0 ± 2.0	30–40

[a]Taken from ref. 10.

[b]Determined for polysaccharides.

Sephadex G gel filtration media are all crosslinked dextrans and consist of three-dimensional networks of polysaccharide chains. The lowest porosity gel is G-10, which excludes molecules of molecular weight >700, while G-200 allows the passage of molecules up to a molecular weight of 200,000. All Sephadex G gels are delivered as beads that swell in water and other liquids. Because G-10 has the lowest porosity and greatest number of crosslinkages, it swells to the least extent, which can be seen from the water regained value and bed volume shown in Table 1. All of the gels except G-25 and G-50 are supplied with a particle size range of 40-120 μm. Sephadex G-25 and G-50 are supplied in fine and coarse mesh sizes of 20-80 μm and 100-300 μm. Superfine preparations for thin-layer gel filtration are supplied at a particle size of 10-40 μm [10].

A series of gel filtration media termed "Sepharoses" is prepared from agarose — a linear polysaccharide consisting of alternating residues of D-galactose and 3,6-anhydro-L-galactose units [11]. Sepharoses are supplied as a dense suspension of swollen beads in distilled water containing 0.02% sodium azide added as a bacteriostatic agent. Two types are available, Sepharose 4B and Sepharose 2B. Sepharose 4B is supplied as a 4% agarose concentration and has a particle size range in the swollen state of 40-190 μm and a molecular weight fractionation range of 300,000-3,000,000. Sepharose 2B is available as a 2% suspension with particles of 60-250 μm and a molecular weight fractionation range of 2,000,000-25,000,000 [11].

For best results from gel filtration experiments, care is necessary in the choice of column dimensions [12]. For analytical purposes, columns of 2.5-cm diameter have proved to be satisfactory; narrower columns give wall effects and larger columns, although satisfactory, are not usually recommended because of the proportionally larger sample size required. With large columns it is difficult to obtain uniform application across the sample bed, and hence it is difficult to obtain an even distribution of the flow rate over the column cross-sectional area.

Very large dimension columns can be used for preparative purposes and the mechanical properties of the Sephadex gels allow the preparation of stable beds of very large dimensions. Beds of as much as 1000 liters have been used in continuous operations when the dense gels are used as bed materials.

The height of the gel bed depends on the type of separation to be accomplished. For such processes as desalting, relatively short beds (50 cm or less) are satisfactory. For more critical separations, longer beds should be used.

Gel filtration chromatography has been applied to a wide diversity of materials and its rapid acceptance indicates that the technique is relatively simple to operate and is highly effective. Its effectiveness is marked most clearly when the technique is applied to naturally occurring materials displaying a wide range of molecular weights. It therefore lends itself to a fractionation of components based on molecular weights and also to determination of molecular weights [13]. With these considerations in mind, it appears that gel filtration techniques are ideally suited to fractionation and characterization of the humic materials of soils — a highly polydispersed material.

Using a variety of grades of Sephadex, Dell'Agnola and his colleages [14-16] have fractionated humic acids into molecular weight classes. Humic acids were extracted by four different reagents: (1) 0.5N NaOH, (2) 0.1M NaHCO$_3$ and 0.1M Na$_2$CO$_3$ at a ratio of 2:1, (3) 0.1M Na$_4$P$_2$O$_7$, and (4) 0.1M NaF [14]. These fractions displayed a wide variation in carbon-nitrogen ratios, namely, 6.4, 8.8, 10.1, and 13.0 for the NaOH, NaHCO$_3$-Na$_2$CO$_3$, Na$_4$P$_2$O$_7$, and NaF extracts, respectively. The Sephadex columns used for gel filtration were 50 cm long and 2.7 cm in diameter. The Sephadex gel grades used were G-25, G-50, and G-100, and the columns were eluted with 0.02M Na$_2$B$_4$O$_7$. Eluted fractions were divided into four molecular weight classes with ranges of <4000, 4000-9000, 9000-100,000, and >100,000. It was found that the Na$_4$P$_2$O$_7$-extracted humic acid displayed different characteristics from the other extracts. It was suggested that the structural characteristics of the humic acid might have been determined largely by the alkalinity of the extractants.

To investigate the biochemical properties of humiclike substances synthesized by microorganisms, a molecular weight comparison based on Sephadex gel chromatography was made between the Na$_4$P$_2$O$_7$ (0.1M at pH 7) extracts of a mor and a mull soil layer, a Rendzina soil, two horizons of a sphagnum peat bog, and the humiclike material synthesized by a strain of Azotobacter chroococcum [17]. Each extract could be divided into two to five fractions and the mean molecular weights of each fraction were established on a column of Sephadex G-50 or G-75. Column dimensions were 100 cm long × 1.5 cm diameter. The hexose (by the anthrone method [18]) and amino acid (by a combination of the Rosen method [19] and thin-layer chromatography) contents of the fractions were determined. During the fractionation procedure the sample collector was fitted with a Spectronic 20 colorimeter and the absorbance of each fraction was measured at 450 nm. For the mor, mull, Rendzina, and two peat bog humic acid extracts, an inverse relationship was shown between the molecular weight of the fractions and their absorbance at 450 nm. Conversely, a direct relationship was found for the bacterially synthesized humiclike extract. Similarly, a direct relationship was found between molecular weight and the hexose and amino acid contents of the fractions derived from the mor, mull, and Rendzina extracts, while an inverse relationship was found for the peat bog samples. No definitive relationships were established for the bacterial extract [17].

In a related vein, Rashid and King [20, 21] fractionated solutions of humic acids on a molecular weight basis using a series of Sephadex gels, namely, G-50, G-100, and G-200, to give molecular weight exclusion at 10,000, 100,000, and 200,000. The data were used to determine average molecular weights of the fractions and these were related to the elemental composition and empirical formulas of the fractions.

Swift and Posner [13] have conducted a detailed investigation of the nature of the interactions between gels and solutes, with particular reference to the known interactions between Sephadex-type gels and compounds that are aromatic, phenolic, heterocyclic, or charged. These observations are discussed in terms of their relevancy to humic acid chemistry. The gels investigated include Sephadex G and Sepharose B (manufactured by Pharmacia Fine Chemicals, Uppsala, Sweden), Bio-gel A and P (manufactured by Bio-Rad, Richmond, Calif.), Sagarose and Agarose (Serevac, Maidenhead, England), and Ionagar (Oxford Manufacturing Co., London, England). The gels were prepared according to the specifications of the manufacturers except for the Ionagar, which was prepared by the method of Andrews [22]. Column dimensions were 30 cm long × 2 cm diameter and the eluting solvents were distilled water, 0.05N NaCl, 0.5N NaCl, and a series of buffer solutions based on carbonate, bicarbonate, and tris [2-amino-2-(hydroxymethyl)-1,3-propanediol].

The humic acid samples, which were isolated initially in the acid form, were solubilized in 0.05N NaOH to a pH of approximately 7 [13]. The samples, containing 10 mg of humic acid in a 2-ml volume, were applied to the column. Effluent was collected as 2-ml volumes and the elution curves were plotted by measuring the absorbance of the fractions at 400 nm. Because of interactions between the gels and humic acids, e.g., adsorption and charge effects, it was concluded that other gel types were to be preferred over the Sephadex types for investigations with humic acids. However, in some alkaline buffer systems, significant levels of adsorption of humic acids were found on Sagarose and Bio-gel P. This effect was particularly evident in the carbonate-bicarbonate buffer systems. Tris buffer appeared to have marked advantages in preventing adsorption of humic acids during gel filtration chromatography. Similar methods have been used to determine molecular weight distribution of humic acids in a soil from Spain [23, 24].

Mehta et al. [25] have determined the average molecular weight and the molecular weight range of various humic fractions (including one humic and three fulvic acid extracts) isolated from the B horizons of podzol soils by the gel filtration technique on Sephadex of different pore sizes. Average molecular weights were between 3000 and 25,000 with a range of molecular weights of 2000–100,000. The method was found to be suitable for detection of differences in humic components isolated by different techniques. Drying the samples under high vacuum at 100°C had no effect on average molecular weight. However, treating the samples with 1N HCl at room temperature for 1 month served to increase average molecular weight, while alkali treatment at room temperature appeared to partially decrease polymerization of the humic colloids, thereby decreasing the molecular weight.

The molecular weight distribution of humic and fulvic acids isolated from marine clays on the Scotian shelf showed an extremely wide range of molecular weight types [21]. Using these molecular weight fractions, the stimulatory effect of humic substances on marine dinoflagellates was determined.

By fractionating fulvic acids on Sephadex G-25 and G-50 and conducting functional group analysis on the isolated fractions, it has been shown that differences in molecular weight can be related to structural differences of the fractions [27]. All of the fractions showed the presence of phenolic hydroxy groups but were distinguished by differences in their mutual positions and their positions relative to other ring substituents. High molecular weight fulvic acid fractions showed the presence of keto structures but did not contain phenolic hydroxy groups in the ortho and para positions nor were any strong-acid groups found. A note of caution with regard to analysis of gel filtered fulvic acid fractions has been sounded by Wildenhain and Henseke [28], who have shown that rapid autooxidation of these fractions may occur.

A system has been developed for the separation of fulvic acids on Sephadex gels to provide chromatographically homogenous fractions that differed from each other with respect to average molecular weight, nitrogen content, numbers of carboxylic acid and phenolic hydroxy groups, as well as infrared and nuclear magnetic resonance spectra [29]. The low molecular weight fractions had properties that resembled those of the so-called "green humic acids." Molecular weights of fulvic acid fractions based on Sephadex gel chromatography were two to ten times higher than molecular weights determined by vapor pressure osmometry [29]. A critical discussion of the relationship between actual molecular weights and nominal molecular weights assigned to Sephadex gels has been presented by Ladd [30].

Spectroscopic investigations of soil humic acids in the visible range have normally shown featureless spectra. However, investigations of humic acids from some Japanese soils have shown the widespread distribution of a "P-type" humic acid that is commonly present in podzolic soils and alpine grassland soils and to a lesser extent in forest brown earth and yellow earth soils [31, 32]. The "P-type" humic acids are commonly referred to as "green humic acids" and show absorption bands near 620, 570, and 450 nm. Humic acids extracted from soils with 0.1N NaOH were subjected to gel filtration chromatography on Sephadex G-25 and three distinct fractions were eluted [33]. The first two fractions were dark brown in color, while the third, slower moving component showed a strong emerald green color in alkaline solution. Although dark brown in color, the second fraction had an absorption spectrum very similar to that of the "green humic acid." It was shown that the humic acid fraction of some Japanese soils possessed in excess of 10% of "green humic acid." Although not important in the surface soils of podzols from Great Britain, the "green humic acids" assume greater significance in the A_2 and B horizons. Kumada and Hurst [33] have suggested that metabolites from the sclerotia formed by the fungus Cenococcum graniforme might be the source of the green humic acids.

By gel filtration [34] on a large (48 cm long × 14 cm diameter) 12% agar (Oxoid "Ionagar" No. 2) column, a cold NaOH-extracted humic acid was fractionated into 50 fractions. The column was prepared as described by

Andrews [22]. The humic acid was converted to its ammonium salt by
neutralization to pH 7 with 0.5N NH_4OH and dissolved in 150 ml of a buffer
solution containing KCl, $NaHCO_3$, and ethylenediamine tetraacetic acid.
This solution was eluted through the column and 50 × 200 ml fractions of
the colored eluent were collected. Fractions 1-25 were acidified to pH 1
with 5N HCl and the precipitated humic acids were washed with water until
free of chloride ions. Fractions 26-29, 30-35, 36-41, and 41-50 were
combined and concentrated under reduced pressure at $60°C$ to volumes of
200 ml. These combined fractions were treated in a manner similar to
that described for fractions 1-25. By spectral analysis, differences in the
humic acid fractions were recorded and related to the average molecular
weight of the fraction.

A neutralized humic acid dissolved in 0.2M NaCl and containing 3.33
mg/ml of each preparation was eluted through Sephadex gel columns with
distilled water. The columns, 4.1 cm in diameter, were 32 cm (G-25),
42 cm (G-50), and 52 cm (G-100) long [30]. Discrete separation into two
major brown colored fractions, a and b, was achieved in all cases. With
the G-25 gel a third, minor, yellow-brown component could be distinguished.
Fractions a and b from replicate runs of a given preparation were concentrated to 20-30 ml at $<45°C$. The fractions were acidified with 0.2-0.3 ml
of 5N HCl and centrifuged, and the precipitates were washed twice with 10
ml of 0.1N HCl. The humic acid components were dried in vacuo and stored
at room temperature. Samples of the dried and ground humic fractions were
dissolved in NaOH, the solutions were adjusted to pH 7.0, and their absorption spectra were recorded. Extinction coefficients were calculated for
solutions containing 1.0 mg/ml, from which data information on the
molecular weights of the fractions was obtained. Similar kinds of investigations, using salt solutions as eluent, have been conducted by Lindqvist [35],
Dell'Agnola and Maggioni [36], Koter et al. [37], and Passera and Maggioni
[38]. Although Sephadex and other types of gel filtration have been used in
the fractionation and characterization of complex naturally occurring organic
polymers, it seems safe to say that more investigations of more efficient
systems for fractionation of humic acids are needed before this method can
become capable of shedding a great deal of light on the chemical characterizations of soil humic and fulvic colloids.

Other types of column chromatography have been used to develop methods
of fractionation of soil humic colloids. Gradient elution studies have been
conducted on sodium pyrophosphate extracts of soils adsorbed on Celite 535
(Johns Mansville) [39]. The humic acids were isolated from a peat by the
methods described by Bremner [40] and Lemieux et al. [41]. The Celite
columns (10 g in columns 2.5 cm in diameter and 7.5 cm long) were prepared
using a slurry in acetone [41]. Elution was started with (a) 500 ml 100%
acetone by forming a gradient with distilled water until the eluting solvent
was 60% acetone and with (b) 500 ml 90% acetone diluted on a gradient with
0.001M $Na_4P_2O_7$ until the eluting solution became 50% with respect to

acetone. Fractions were collected and their absorbances were determined at 534 nm. Some fractionation of the humic acids was achieved, with the early fractions traveling in clear bands while the later bands became continually more diffuse.

A method of adsorption chromatography of humic acids on different kinds of alumina has been described by Scholz [42] and Scheffer et al. [43]. It was possible to separate a solution containing a mixture of humic acids into seven components with different types of alumina. The column consisted of different types of Woelm aluminas with the acid alumina at the base of the column, neutral alumina in the middle, and basic alumina at the top. Bands of the humic acid fractions were located at different parts of the column and could only be released from the alumina with concentrated formic acid or dimethylsulfoxide. Differences between the fractions were shown by their visible and ultraviolet spectra. It was suggested that the types of ring systems present in the fractions differed and that the types of substituents on the rings were different. A changing relationship of quinoid and phenolic configurations was readily evident in the infrared region and the spectra indicated that the fractions were in a high state of purity.

Following hydrogenolysis of a muck soil, the saturated hydrocarbons were separated from other compounds in the hydrogenation mixture on a 4 × 40 cm magnesia-silica gel (Florisil) adsorbent [44]. After 330 g of 60/100 mesh Florisil (activated at 600°C for 6 hr) was poured into 600 ml of hexane in the column, a known aliquot of the hydrogenation product was placed on top of the column and eluted with hexane until a colored fraction appeared. This procedure allowed the satisfactory separation of n-alkanes by gas-liquid chromatography with little or no interference.

Humic acids, extracted from soils by the neutral pyrophosphate method [45] with the ethanol and benzene treatments omitted [40], were fractionated on a 0.5-g Celite 535 column [46]. The column was eluted with isopropanol-water (6:4, 9:1, or 15:1) at a rate of three drops per minute. After 75 ml of eluent were collected, the column was allowed to run dry and the adsorbent was extruded from the column. Three humic acid fractions were separated by this procedure using any one of the solvent ratios; however, isopropanol-water ratios of 6:4 and 9:1 caused more streaking than the 15:1 ratios. The fractions isolated were (a) a residual humic material at the column surface, (b) a dark band about 2 cm from the top, and (c) a yellowish-brown band collected in the eluate. These fractions were hydrolyzed with 6N HCl, desalted electrolytically, and examined for amino acids by two-dimensional paper chromatography [47]. Although the amino acid content of the three fractions differed somewhat, the Celite 535-isopropanol-water fractionation procedure did not achieve the desired degree of fractionation and, in addition, the results were not easily reproducible.

Because only limited resolution can be achieved by present methods of column chromatography and because detection capability is lacking, the application of this method to humic acid chemistry is unlikely to be satisfactory unless more efficient systems are developed.

B. Phosphorus Compounds

Phosphorus compounds in soils are most often fractionated by paper or
anion-exchange chromatographic methods; however, column chromatographic methods have been used occasionally in conjunction to separate
organic phosphorus from excess salts prior to quantitative estimation of
ribonucleic acid phosphorus in soils by anion-exchange chromatography.
Organophosphorus compounds were extracted from soils with 1N NaOH.
Humic acids were precipitated from the alkaline extract by acidification
and soluble organophosphorus in the acidified extract was separated from
salts by adsorption on activated alumina prior to anion-exchange chromatography. Activated alumina (40 g), prewashed to remove colloidal material,
was shaken with the acid extract for 1 hr and the suspension was transferred
to a chromatographic column. Salts were removed with distilled water and
organophosphorus compounds were eluted with 300 ml of 3N NH_4OH. The
eluate was concentrated, particulate alumina was removed by centrifugation
(and washed twice with 3N NH_4OH), and NH_3 was removed by vacuum distillation. Almost complete adsorption of nucleotides was accomplished with
alumina, and elution of the nucleotides with 3N NH_4OH was essentially
quantitative.

Cellulose column chromatography has been used to obtain pure samples
of free inositols in a hydrolyzate of a soil phytin preparation [49]. The procedure used was that of Angyal et al. [50]. The sample was added to a
2.5 × 40 cm column containing 25 cm of cellulose powder. Ten-milliliter
fractions of acetone-water (4:1) eluate were collected. Fractions were
evaporated to dryness, dissolved in 0.2 ml of water, and examined by paper
chromatography using the above solvent [51].

Activated carbon columns have been used to separate inorganic and
organic phosphorus in soil extracts. Soil phosphorus was extracted with
NH_4OH; NaOH extracts failed to give quantitative separation on the carbon
column [52]. Use of Norit-A carbon was suggested because of its low
phosphorus content. The carbon, pretreated by heating overnight in concentrated HCl, was washed with water, oven dried, and finally washed with
successive volumes of 95% ethanol, 6N NH_4OH, and concentrated HCl. The
column was prepared by adding Celite (0.25 g) to a 1.6 × 40 cm glass column,
followed by 4 g of carbon. The soil extract, 0.5N with respect to HCl, was
added to the column followed by 0.25 g of Celite to flocculate suspended
organic matter and to facilitate infiltration through the column. Inorganic
phosphorus was eluted with 0.5N HCl; 100-300 ml appeared to be adequate
when inorganic phosphorus contents were <1000 μg. For higher phosphorus
contents a control containing inorganic phosphorus in amounts equal to that
in the soil extract was suggested.

Sorbed organic phosphorus was determined by extruding the carbon,
adding 5 ml of 6N NH_4OH, and 1 ml of 10% $Mg(NO_3)_2$, evaporating to dryness
on a steam plate, and igniting at $500°C$ in a muffle furnace. The ignited
material was treated with 20 ml of 1N HCl and heated for 10 min on a steam

plate. The acid solution was diluted to volume with water and analyzed for phosphorus according to the method of Mehta et al. [53]. Retention of organic phosphorus by the carbon against elution by water was >95%.

C. Carbohydrates

Hydrolyzates of soil extracts containing carbohydrates have been purified and fractionated by a variety of liquid chromatographic procedures employing cellulose, Celite, carbon-Celite, carbon, and Sephadex columns. Liquid chromatography has been used to obtain samples of various components of carbohydrate extracts for determination of physical constants and for derivative formation [54, 55] as well as to aid in separation and identification of carbohydrates by paper chromatographic procedures. The extraction of carbohydrates and methods of hydrolysis from soils are discussed in Chapter 8, Section V, B.

Liquid chromatographic procedures that have been used in the analysis of soil extracts containing carbohydrates are presented in Table 2.

The use of cellulose column chromatography for the analysis of carbohydrates has been discussed briefly by Whistler and BeMiller [73]. Included in the discussion are columns, column packing, eluting solvent, temperature effects, sample application, collection of samples, and detection of components. A column height to diameter ratio of 9:1 has been found most satisfactory. Column packing is a critical operation and is most often accomplished by the slurry method. A slurry is made by blending cellulose powder with a solvent, usually acetone. A medium slurry is to be preferred; if the slurry is too thin the column tends to pack too tightly. The slurry is poured into the chromatographic tube and is stirred slowly with a glass rod just above the packing line to prevent settling and channeling. Acetone is allowed to drain during packing. After packing, the column is washed for several days with the eluting solvent. As the eluting solvent moves through the column, water is adsorbed by the cellulose, causing the column to tighten. If the column becomes too tight, a particle size separation of the cellulose may occur. If this becomes a problem, a more viscous solvent, such as n-butanol or the eluting solvent, may be used to replace acetone in making the packing slurry.

The eluting solvent of choice can be determined by paper chromatographic studies. All solvents should be distilled before use. Formic acid is not recommended as a solvent because it changes the cellulose to such an extent that it cannot be reused. When solvents contain an acid or base, these should be removed prior to any concentration step conducted on the eluent, thereby preventing alteration of the sample components.

Flow rates can be increased by application of a positive pressure to the column head; suction on the effluent is not recommended since it may cause the lower portion of the column to be freed of solvent. Temperature can also be raised to increase flow rates; however, temperature changes during elution should be avoided.

TABLE 2

Liquid Chromatographic Methods Used for the Analysis of Carbohydrates in Soils

Column material	Eluting solvent	Application	Reference
Cellulose, 3.2 × 30 cm	Ammonia–water–saturated n–butanol (1:99)	Analysis of constituent sugars in hydrolyzate of a pine forest soil	56
Cellulose, 4 × 6 cm	n–butanol–ethanol–water–NH$_4$OH (40:10:49:1)	Separation of monosaccharides in a soil polysaccharide hydrolyzate	57
Cellulose, 1.2 × 40 cm	Acetone–n–butanol:water (7:2:1)	Separation of sugars in forest litter hydrolyzates	58–60
Cellulose, 2.5 × 110 cm, heated at 32° ± 2°C	Water–saturated redistilled butanol	Separation of sugars in a soil polysaccharide hydrolyzate	61
Cellulose, 2 × 20 cm	Water saturated collidine	Separation of sugars in a soil polysaccharide hydrolyzate	62
Cellulose, 3.5 × 37 cm	Ethyl acetate–pyridine–water (8:2:1)	Separation of sugars in a high molecular weight polysaccharide hydrolyzate	63
Carbon–Celite (1:1), 4.6 × 50 cm	A. 300 ml water B. Linear gradient of 0.20% ethanol	Separation of sugars in a soil polysaccharide hydrolyzate	64–66
Celite	Water–saturated butanol–light petroleum (80:20)	Further separation of sugars as described in Refs. 67–69	64–66

Charcoal	0.1N HCl, 90% aqueous acetone, water, followed by 0.5N NaOH in sequence	Fractionation of fulvic acids	67–70
Alkaline activated Wofatit E	Water, acetone, dimethylformamide, followed by 0.5% NaOH in sequence	Separation of carbohydrates	71
Sephadex G–100, 5.5 × 100 cm	Distilled water	Separation of colored components from high molecular weight poly-saccharides in a 0.6N H_2SO_4 soil extract	63, 72

Celite 535, a low-cost, inert, nonadsorbing material, is often used as a support for the aqueous phase in column partition chromatography of sugars and sugar derivatives [74]. Celite columns provide better resolution than cellulose columns and gravity flow rates through Celite columns are often ten times greater than flow rates through cellulose columns. Carbon-Celite mixtures also have found wide application. These columns are particularly useful when large amounts of material must be separated; however, gradient elution is often necessary for the separation of complex mixtures [75].

Duff [64] used the carbon-Celite column for the separation of constituent sugars contained in peat and soil hydrolyzates. Water-extracted polysaccharides were hydrolyzed with 1.5 liters of 1N H_2SO_4 at 100°C for 5 hr. The hydrolyzate was neutralized with $BaCO_3$ and, following removal of barium salts, the solution was evaporated to a small volume. The solution was transferred to an ultrasorb (vegetable charcoal)-Celite 535 (220:220 g) column that had been pretreated with 5% CH_3COOH and washed with water until neutral. Concentrated HCl also has been used in place of acetic acid [75]. After addition of the sample, 300 ml of water was eluted and discarded. Sugars were eluted with a linear gradient of 0-20% ethanol and the eluate was collected in 5-ml fractions. For soils a carbon-celite column (22:22g) was used and eluted with 0-2% ethanol. The fractions were examined by paper chromatographic methods and similar fractions were combined and evaporated to a syrupy consistency.

Combined fractions containing several sugars were further fractionated on Celite 535 columns using water saturated butanol-light petroleum (80:20) as solvent. Celite 535 was treated overnight with concentrated HCl at room temperature, washed free of acid with water, and dried at 110°C. The column was prepared by adding water (1:1, v/w) followed by addition of sufficient nonaqueous solvent to obtain a free flowing slurry [41, 74]. After addition of the slurry the nonaqueous solvent is recycled through the column until the Celite forms a firm bed. The column is further firmed by placing a filter disk on top of the Celite and compressing the column with a plunger.

Bouhours and Cheshire [65] used the methods described by Duff [64] and Lemieux [74] in conjunction with gas-liquid chromatography to determine 2-0-methylxylose and 3-0-methylxylose in peat hydrolyzates. However, in an investigation of the transformation of [^{14}C]glucose and [^{14}C]starch in soil, Cheshire et al. [66] concluded that sugars in soil hydrolyzates were best separated as borate complexes by anion-exchange chromatography [76].

Carbon liquid chromatography has been used to fractionate fulvic acids [67-70]. The fulvic acid extract was adjusted to pH 2.5-3.0 and filtered in a 15-cm Buchner funnel containing a 5-cm bed of animal charcoal washed with 0.1N HCl. All of the color and most of the organic matter was removed by the charcoal. After the fulvic extract had been eluted, the column was washed with 500 ml of 0.1N HCl. The combined eluate and 0.1N HCl fraction contained simple organic materials, such as amino acids, purine bases, and sugars. The charcoal was extracted with 90% aqueous acetone until the eluate was colorless. This fraction contained phenolic glycosides and a carbohydrate material. A subsequent elution with water until the extract no longer gave a precipitate on addition of excess acetone yielded a polysaccharide containing glucose and glucuronic acid groups. The organic matter remaining on the charcoal was eluted with 0.5N NaOH.

Wildenhain and Henseke [71] used activated alkaline Wofatit E with water, acetone, dimethyl formamide (DMF), and 0.5% NaOH as successive solvents to fractionate an ethanol extract of a peat fulvic acid. The water eluted fraction contained amino acids and carbohydrates, while the acetone fraction contained aromatic aldehydes and ketones. Dimethyl formamide eluted materials with properties similar to humic acids containing carbonyl and phenolic hydroxy groups, and NaOH eluted a large number of organic acids.

Gel chromatography also has found wide application for the fractionation of a variety of carbohydrates. Polyacrylamide gels, hydrophilized polystyrene, and porous glass or silica beads are best suited for gel chromatographic analysis of carbohydrates [77]. Dextran and agarose gels also have been widely used for this purpose. However, since the latter gels consist of carbohydrates, they may cause sample contamination and are being replaced by noncarbohydrate materials [77].

Barker et al. [63, 72] were able to separate colored components of a 0.6N H_2SO_4 soil extract from high molecular weight polysaccharides with Sephadex G-100. Only polysaccharides extracted with dilute acids were separated from colored material on Sephadex and it was concluded that acid was necessary to liberate polysaccharide materials from colored soil organic components.

After acid extraction of the soil, the filtrate was neutralized with $NaHCO_3$, centrifuged at 2000 rpm for 30 min, and filtered through a bacterial filter pad to remove all suspended materials. The filtrate was dialyzed against water and the nondialyzable material was freeze dried to yield a crude polysaccharide mixture. The mixture was dissolved in water and eluted from a Sephadex G-100 column (5.5 × 100 cm) with distilled water. Most of the polysaccharide materials were eluted in the 400-1400 ml eluate, while the colored lower molecular weight materials were eluted in the 1400-2300 ml eluate fraction. Further fractionation and identification were accomplished by ion-exchange and paper chromatography.

D. Waxy Components

Morrison and Bick [78] have investigated the organic material extracted from peat and mineral soil by a benzene-ethanol solvent, using column chromatography with alumina and magnesium trisilicate. The peat or soil samples were extracted with benzene-ethanol (2:1) in glass soxhlet extractors to obtain a crude wax fraction, which, after removal of the solvent, was extracted with light petroleum at 60-80°C in soxhlet extractors, followed by refluxing for 1 hr with isopropanol-methanol (1:1) mixture. A ratio of 1 g of wax to 10 ml of solvent was employed. The hot solvent was decanted and the undissolved components were re-extracted with the solvent. A precipitate, the refined wax, was formed on cooling. This material was redissolved in benzene (15 ml/g), neutralized with 0.5N alcoholic KOH, treated with excess $CaCl_2$ in ethanol, and heated. The acids were precipitated as calcium salts, while nonacids remained in solution. Acids were converted to methyl esters [79] prior to column chromatographic analysis. Nonacids were refluxed with benzene (10 ml/g) and 1N ethanolic KOH (5 ml/g) for 6 hr. Esterified acids formed by this treatment were precipitated with $CaCl_2$.

Free and esterified acids were fractionated as methyl esters on magnesium trisilicate (100 g) as described by Downing et al. [80]. Fractions (5 ml) were collected and examined by infrared spectroscopy. Simple methyl esters were eluted with benzene-light petroleum (20:80), hydroxylated methyl esters with chloroform, and aromatic methyl esters with ethanol-chloroform (5:95).

Neutral alumina, activity II (2.0 × 40 cm columns), at 50°C was used to fractionate the unsaponifiable material. The sample, obtained as a precipitate in the saponification procedure, was redissolved in a small volume of hot light petroleum for application to the column. The column was eluted with light petroleum, benzene, and ethanol-benzene (1:9), respectively. Fractions (5 ml) were collected and examined by infrared spectroscopy. Fractions of similar chemical nature were combined. The first fraction eluted by petroleum ether contained a mixture of n-alkanes. A second fraction, also eluted by petroleum ether, contained methyl ketones, while a third fraction, eluted by benzene, was composed primarily of long-chain primary alcohols. The ethanol-benzene eluate contained long-chain alcohols, carbonyl-containing compounds, and other unidentified components. Fractions were purified further, prior to gas chromatographic analysis, on smaller columns of activity I alumina using n-hexane or benzene as solvent.

E. Pesticides

Liquid chromatography finds wide application in the analysis of pesticides primarily as a cleanup procedure prior to qualitative and quantitative analysis by infrared spectroscopy; polarography; and paper, thin-layer, and gas chromatography. Because of its large carrying capacity, liquid chromatography is particularly well suited for cleanup of highly contaminated materials, such as soil extracts [81].

Liquid chromatographic procedures have been used extensively for cleanup of biological and soil extracts containing organochlorine insecticides [81, 82] and organophosphorus insecticides and herbicides [83]. Many of the liquid chromatographic techniques used as cleanup procedures were developed initially for plant tissue extracts, and their applicability to soil extracts can only be inferred. The trade and chemical names of the pesticides discussed in this section are given in Table 3.

TABLE 3

Trade and Chemical Names of Pesticides

Trade name(s)	Chemical name
	Organochlorine Insecticides
Heptachlor	1,4,5,6,7,8,8-Heptachloro-3a,4,7,7a-tetrahydro-4,7-endo-methanoindene
Heptachlor epoxide	2,3-Epoxide of heptachlor
Aldrin	1,2,3,4,10,10-Hexachloro-1,4,4a,5,8,8a-hexahydro-1,4-endo-exo-5,8-dimethanonaphthalene
Dieldrin	6,7-Epoxide of aldrin
Endrin	1,4-Endo-endo isomer of dieldrin
γ-Chlordane	1,2,3,5,6,7,8,8-Octachloro-2,3,3a,4,7,7a-hexahydro-4,7-methanoindene
p,p'-DDE	2,2-Bis(p-chlorophenyl)-1,1-dichloroethylene
o,p'-DDE	o-Chlorophenyl isomer of p,p'-DDE
p,p'-DDD or TDE or Rhothane	2,2-Bis(p-chlorophenyl)-1,1-dichloroethane
p,p'-DDT	2,2-Bis(p-chlorophenyl)-1,1,1-trichloroethane
o,p'-DDT	o-Chlorophenyl isomer of p,p'-DDT
p,p'-Methoxychlor	2,2-Bis(p-methoxyphenyl)-1,1,1-trichloroethane
γ-BHC or Lindane	γ-Isomer of 1,2,3,4,5,6-hexachlorocyclohexane
Toxaphene	Chlorinated camphene containing 67-69% chlorine
	Organophosphorus Insecticides
Disulfoton or Disyston	O,O-Diethyl-S-[2-(ethylthio)ethyl] phosphorodithioate
Nemacide or VC-13 nemacide	0-2,4-Dichlorophenyl-0,0-diethyl phosphorothioate
Carbophenothion or Trithion	0,0-Diethyl phosphorodithioate
Malathion	S-1,2-Bis(ethoxycarbonyl)ethyl-0,0-dimethyl phosphorodithioate

TABLE 3. — Continued

Trade name(s)	Chemical name
Parathion	0,0-Diethyl-0-p-nitrophenyl phosphorothioate
Diazinon	0,0-Diethyl-0-(2-isopropyl-4-methyl-6-pyrimidyl)-phosphorothioate
Azinphosmethyl or Guthion	0,0-Diethyl-S-4-oxo-1,2,3-benzotriazin-3-(4H)-ylmethyl phosphorodithioate

Triazine Herbicides

Propazine	2-Chloro-4,6-bis(isopropylamino)-s-triazine
Ametryne	2-Methylthio-4-ethylamino-6-isopropylamino-s-triazine
Prometryne	2-Methylthio-4,6-bis(isopropylamino)-s-triazine
Atrazine or Aatrex	2-Chloro-4-ethylamino-6-isopropylamino-s-triazine
Simazine	2-Chloro-4,6-bis(ethylamino)-s-triazine
Prometone	2-Methoxy-4,6-bis(isopropylamino)-s-triazine

Miscellaneous

	2,6-Dichlorobenzonitrile
Isobenzan or Telodrin	1,3,4,5,6,7,8,8-Octachloro-1,3,3a,4,7,7a-hexahydro-4,7-methanoisobenzofuran

Liquid chromatographic techniques are divided into two types: adsorption chromatography, in which the stationary phase (adsorbent) is a solid, and partition chromatography, in which the stationary phase consists of a liquid adsorbed on a solid. Adsorption chromatography with florisil, alumina, silica gel, and carbon as the adsorbents has been used most extensively for cleanup of soil extracts prior to insecticidal determination [81, 82]. Activity of the adsorbent is varied by the addition of water and is related inversely to the water content of the adsorbent. Hexane or benzene containing small quantities of polar solvents, such as ether or acetone, are used most frequently as the eluting solvent [82]. The solvent polarity needed to elute a particular pesticide is dependent on the stationary phase. The commonly used adsorbents have been classified by Beynon and Elgar [82] and are listed in Chap. 1, Table 7, Sec. III,B,5, in the order of their increasing adsorptivity.

The commonly used eluting solvents, in order of their increasing polarity, are hexane, cyclohexane, benzene, methylene chloride, diethyl ether, ethyl acetate, acetone, and alcohols.

Although liquid chromatography has found wide application for the clean-up of soil extracts, no universally recommended procedure has been described. Until recently, most procedures were those developed to meet the needs of a specific problem, i.e., determination of one or two pesticides extracted from soils with a particular solvent. However, in 1967, Duffy and Wong [84] employed a Florisil column that allowed separation of ten organochlorine insecticides from coextracted indigenous soil components that interfered with determination of the insecticides following soil extraction with a hexane-2-propanol (3:1) solvent. Prior to chromatographic cleanup, 100 ml of petroleum ether were added to the soil extract and the 2-propanol was removed by washing with 500 ml of distilled water containing 10 ml of a saturated solution of NaCl followed by two successive washings with 500 ml of distilled water. The hexane-petroleum ether mixture was dried over anhydrous Na_2SO_4 and flash evaporated to a volume of 5-10 ml. The concentrated extract was chromatographed on a column 300 mm long and 16 mm in internal diameter. The adsorbent consisted of Florisil, which had been activated by heating at $650°C$, stored at $130°C$, and aged at room temperature for 3 days prior to use. The adsorbent was added to the column by means of a slurry technique. The insecticides were eluted with 75 ml of a mixture containing 120 ml of diethyl ether diluted to 1000 ml with redistilled petroleum ether followed by 250 ml of a mixture of 50 ml of diethyl ether and 4 ml of 1,4-dioxane diluted to 1000 ml with petroleum ether. Each eluent fraction was collected separately, concentrated to a small volume, and diluted with hexane to a volume of 5 ml. The initial eluent fraction contained heptachlor, aldrin, γ-chlordane, p,p'-DDE, o,p'-DDT, p,p'-DDD, and p,p'-DDT, while the second fraction contained heptachlor epoxide, 1-hydroxychlordene, and dieldrin. King et al. [85] used a similar column for cleanup of heptachlor and heptachlor epoxide residues extracted with acetone and partitioned into petroleum ether. Insecticides were eluted with 200 ml of 6% (v/v) diethyl ether in petroleum ether. In both investigations, detection and quantitative determination were achieved by electron capture gas chromatography.

Nine organochlorine insecticides contained in hexane-acetone (49:51) extracts of soils and sediments were separated from coextracted interfering components on a Florisil column [86]. Glass columns (500 mm × 20 mm) were packed with 10 g of 60/80 mesh Florisil stored at $125°C$ prior to use. A 10-mm layer of Na_2SO_4 (heated to $400°C$ for 8 hr) was placed on top of the Florisil and the column was prewashed with 50 ml of hexane. After a 5-ml aliquot of the sample was added to the column, it was eluted with 100 ml of hexane at a rate of 5 ml/min and the eluent was collected as fraction 1. The column was further eluted with 150 ml of hexane-diethyl ether (85:15) and collected as fraction 2. Heptachlor and aldrin were recovered in fraction 1 and γ-BHC, heptachlor epoxide, dieldrin, endrin, p,p'-DDD, p,p'-DDT, and p,p'-methoxychlor in fraction 2. Recovery of each of the insecticides was essentially quantitative, i.e., in excess of 95%. The procedure was recommended for cleanup of lake and stream sediment extracts, which contain large amounts of interfering components with gas chromatographic retention time values of <10 min.

Extracts of lake sediments containing residues of toxaphene were purified sufficiently for microcoulometric gas chromatographic analysis on a H_2SO_4-Celite column [87]. The column was prepared by the addition of 6 ml of 3:1 fuming concentrated H_2SO_4 to 10 g of Celite 454 and adding this mixture to a bed of 3 g of dry Celite 454. Toxaphene was eluted from the column with 100 ml of purified hexane.

A method for cleanup of organochlorine pesticide residues in acetonitrile extracts of vegetables [88] was used for the cleanup of hexane-acetone (9:1) soil extracts prior to analysis by microcoulometric gas chromatography [89] and infrared spectroscopy [90]. Florisil preactivated at 650° C was heated at 135°C for 5 hr and stored in glass-stoppered bottles at 135°C prior to use. Column packing consisted of a 100-mm bed of Florisil contained in a 300 × 20 mm column. Preparatory to addition of the sample, the column was prewetted with a small amount of petroleum ether. Pesticides were eluted initially with 250 ml of 6% diethyl ether in petroleum ether, followed by 250 ml of 15% diethyl ether in petroleum ether at a rate of 5 ml/min. The success of this method as a cleanup procedure for soil extracts suggested that many methods used for the cleanup of plant extracts can, with minor modifications, be suitably applied to soil extracts.

Liquid chromatography employing Florisil also has been used to separate propylene carbonate (used as the extractant) from soil and plant pesticide residue [91]. Propylene carbonate causes peak suppression when extracts are analyzed by electron capture gas chromatography and must be removed so that injected solutions contain ≤0.1% propylene carbonate. The column was prewashed with 50 ml of petroleum ether and an aliquot of extract equivalent to 2.5 g of sample was transferred to the column. Aldrin, chlordane, and p,p'-DDT and its analogs were eluted with 200 ml of petroleum ether. Other organochlorine insecticides were eluted with a subsequent 200-ml portion of 7% diethyl ether in petroleum ether and, finally, organothiophosphorus compounds were eluted with 200 ml of 25% diethyl ether in petroleum ether.

Recently, polychlorinated biphenyls were found to interfere with the analysis of organochlorine insecticides by electron capture gas chromatographic techniques. Polychlorinated biphenyls are contained in a wide variety of products and may eventually become soil contaminants. These compounds possess structures and properties similar to the DDT group of pesticides and pose a particular problem when soils are extracted and analyzed for DDT and its analogs and degradation products. Reynolds [92] has shown that polychlorinated biphenyls can be separated from many of the organochlorine insecticides by means of cleanup on a Florisil column. Essentially quantitative removal of polychlorinated biphenyls was achieved on a column consisting of 19 g of 60/100 mesh Florisil (stored at $130°$ C prior to use). Elution with 200 ml of hexane separated heptachlor, aldrin, and p, p'-DDE. Clean separation of γ-BHC, heptachlor epoxide, dieldrin, p, p'-DDD, and p, p'-DDT from the polychlorinated biphenyls was obtained by elution with 250 ml of 20% diethyl ether in hexane. Although the method was not evaluated for soil extracts, it was highly efficient for cleanup of extracts of animal tissues.

Column chromatography with Florisil as the adsorbent was used for fractionation of pesticides as a method of cleanup and to improve pesticide identification [93]. The following procedure, in conjunction with gas chromatographic retention times and derivative preparation, was used to identify pesticides in extracts of soils and vegetable crops.

Thirty grams of 60/100 mesh Florisil (preheated at $660°$C) mixed with 2 g of anhydrous Na_2SO_4 were added to a chromatographic column and an additional 5 g of Na_2SO_4 was placed on top of this bed. The column was prewashed with 50 ml of benzene and two successive 25-ml portions of petroleum ether. If the activity of the Florisil was found to be too high, 0.5 ml of absolute ethanol was included in the prewash procedure to provide the desired activity. Pesticides in 5-10 ml of petroleum ether together with the washings from the flask containing the extract were added to the column. The pesticides were eluted with a series of four solvents, namely, (1) 200 ml of petroleum ether, (2) 200 ml of 5:1 benzene-petroleum ether, (3) 200 ml of chloroform, and (4) 150 ml of acetone. (The numbers indicating the solvents correspond with the fraction numbers used below.) The first 40 ml of eluent was discarded. A flow rate of 6 ml/min was used throughout and fractions 3 and 4 were evaporated to dryness and dissolved in petroleum ether prior to analysis.

In the initial experiments, 19 organochlorine and organophosphorus insecticides were fractionated. Heptachlor, aldrin, isobenzan, p, p'-DDE, o, p'-DDT, and p, p'-DDT were eluted in fraction 1. Fraction 2 contained γ-BHC, disulfoton, nemacide, heptachlor epoxide, γ-chlordane, dieldrin, endrin, p, p'-DDD, carbophenothion, and p, p'methoxychlor. Fraction 3 contained diazinon, malathion, and parathion, while fraction 4 contained metabolites of the organophosphorus insecticides azinphosmethyl and diazinon. It was suggested that other metabolites of the organophosphorus insecticides would most likely elute in fraction 4.

Cleanup of hexane-acetone (49:51) extracts of highly organic lake sediments was evaluated using column chromatographic techniques with acid, neutral, and basic alumina; MgO:Celite (4:1); and Florisil as supports [94]. Preliminary investigations showed that the MgO-Celite column converted γ-BHC, p, p'-DDD, and p, p'-DDT to other compounds that were not identified. This eliminated MgO-Celite from further consideration as an adsorbent for cleanup of a variety of pesticides. Of the three alumina-type adsorbents, highest recoveries of pesticides were attained with neutral and basic aluminas. However, the basic alumina invariably contained a contaminant that co-chromatographed with heptachlor epoxide and that could not be

removed from the alumina without markedly changing its chemical properties. Neutral alumina possessed similar elution characteristics to
Florisil but was chosen for more intensive investigation because of its
superior carrying capacity.

Best resolution of insecticides from sediment extract interferences was
provided by neutral alumina containing 12.6% water by weight. Generally,
the alumina column is prewashed with a solvent, such as diethyl ether,
but this step was eliminated because no significant interferences were
removed and the activity of the alumina was decreased. The insecticides
were eluted from 10 g of the alumina with hexane and the eluent was
collected on a fraction collector. The numbers and volumes of the fractions
collected were related to the concentration range of the insecticides. With
insecticide concentrations ranging from 52 ppb for γ-BHC to 1305 ppb for
p,p$'$-methoxychlor (concentrations are chosen to give similar gas chromatographic peak heights for each insecticide) in the lake sediment, hexane
fractions of 5, 10, and 65 ml were collected. For a concentration range
fivefold lower, hexane fractions of 5, 10, 5, 20, and 40 ml were collected.
Insecticide recovery ranged from 90 to 100%. Application of this method
to hexane-acetone extracts of sediments allowed cleanup of only six of the
ten organochlorine insecticides, namely, γ-BHC, dieldrin, o,p$'$-DDT,
p,p$'$-DDT, p,p$'$-DDD, and p,p$'$-methoxychlor. Heptachlor, heptachlor
epoxide, aldrin, and p,p$'$-DDE were not resolved from the interfering
indigenous sediment components. It was concluded that complete resolution
of the ten insecticides from interferences was unlikely to be achieved with
a single combination of neutral alumina activity and size of hexane fractions.

An alumina column (2 g of alumina type H obtained from Laporta, Ind.)
was used for cleanup of the bottom sediments of ponds prior to dieldrin
analysis by electron capture gas chromatography [95]. Acetone (5% by
volume) in hexane was used as the solvent but the volume of solvent was not
specified.

Alumina also has been used for cleanup of soil extracts prior to herbicide
analysis [96, 97]. A routine method for the analysis of triazine herbicides
in soils was described by Mattson et al. [96]. Herbicide extraction was
achieved by refluxing with an acetonitrile-water mixture (10:1); cleanup
was accomplished by use of an alumina column prior to final determination
of the herbicides gas chromatographically, using a chloride sensitive titration cell as detector. After extraction, the solvent was filtered and an
aliquot shaken with a dilute aqueous Na_2SO_4 solution. Herbicides were
extracted from the aqueous phase with methylene chloride. The methylene
chloride containing the herbicides was evaporated to dryness and the residue
was dissolved in CCl_4 preparatory to its addition to a 12.5-g alumina
column. Activity IV alumina (10% water) was used for separation of propazine, ametryne, and prometryne and activity V alumina (15% water) was
used to fractionate atrazine, simazine, and prometone. After the solvent
containing the sample had penetrated the alumina, an additional 65 ml of
CCl_4 penetrated the column, 50 ml of 5% diethyl ether in methylene chloride

was added. The eluent containing the herbicides was evaporated to dryness and the residue dissolved in benzene prior to gas chromatographic analysis.

Soils containing the herbicide 2,6-dichlorobenzonitrile were extracted with 20% acetone in hexane (v/v) [97]. Cleanup of the extracts sufficient to allow determination by electron capture gas chromatography was achieved on an alumina type H column. Following extraction, the acetone was partitioned into water and an aliquot of the hexane phase was added to the column, which had been prewetted with 1% acetone in hexane. The initial 3 ml of solvent was discarded and the herbicide was collected in the subsequent 9 ml of eluent.

Combinations of several adsorbents have been used in cleanup procedures. Soil extracts (pentane-isopropanol, 3:1) containing heptachlor and heptachlor epoxide were purified using a Florisil-carbon column [98]. Isopropanol was removed from the mixture by partitioning into distilled water. When residue contents of the soil were low, it was necessary to extract a greater amount of soil, and a prior treatment with fuming H_2SO_4 was essential (2 × 35 ml portions). The column was made up of the following sequence of adsorbents: 10 g of Florisil, 10 g of a Florisil-carbon (15:2) mixture, and 5 g of Florisil. A layer of Na_2SO_4 was placed on top of the mixed column. The column was prewashed with 100 ml of 5% ether in pentane followed by 100 ml of pentane, and samples were introduced and washed onto the column with 2 × 20 ml portions of 2% benzene in pentane. Heptachlor was eluted quantitatively from the column with 100 ml of 2% benzene in pentane, and 5% ether in pentane (225 ml) was used to elute the heptachlor epoxide.

Wilkinson et al. [99] used a mixture of Florisil and Celite 545 (5:1, w/w) containing 4.5% H_2O for the purification of acetone-petroleum ether (1:9, v/v) extracts of soil containing aldrin, dieldrin, heptachlor, and heptachlor epoxide. Acetone was removed from the soil extract by partitioning into H_2O prior to addition to 10 g of the adsorbent mixture. Aldrin and heptachlor were eluted with 40 ml of petroleum ether. Subsequent elution with 80 ml of 25% benzene in petroleum ether removed heptachlor epoxide and dieldrin from the column.

Magnesia-Celite columns were used to purify soil extracts containing a number of organochlorine insecticides [100-102]. Cleanup of hexane-isopropanol (1:1 by volume) extracts of soil containing heptachlor [100] and aldrin and endrin [101] was achieved on this type of column. The residues were partitioned into petroleum ether and chromatographed on a 1:1 magnesia-Celite column. Insecticides were eluted with 250 ml of benzene-hexane (15:85 by volume). Hendrick et al. [102] used a 4:1 magnesia-Celite column for purification of hexane-isopropanol soil extract containing aldrin and dieldrin. Isopropanol was removed by partitioning into water prior to adding the sample to the column. No information was given regarding the eluting solvent; however, it may have been the benzene-hexane (15:85) mixture used by Saha and McDonald [101] for eluting extracts of

TABLE 4

Liquid Chromatographic Systems for Cleanup of Soil Extracts Containing Pesticide Residues

Adsorbent[a]	Eluting solvent[b]	Pesticides	Reference
	Florisil		
Florisil heated to 650°C, reactivated at 130°C	A. 75 ml of diethyl ether-petroleum ether (12:88) B. 250 ml of diethyl ether-1,4-dioxane-petroleum ether (50:4:946)	Organochlorine insecticides	84
Florisil heated to 650°C, reactivated at 125°C (10 g)	A. 100 ml of hexane B. 150 ml of hexane-diethyl ether (85:15)	Organochlorine insecticides	86
Florisil heated to 650°C, reactivated at 135°C (8 g)	A. 250 ml of diethyl ether-petroleum ether (6:94) B. 250 ml of diethyl ether-petroleum ether (15:85)	Organochlorine insecticides	89, 90
Florisil heated to 650°C, reactivated at 130°C	200 ml of diethyl ether-petroleum ether (6:94)	Heptachlor and heptachlor epoxide	85
Florisil, 5% H_2O (30 g)	A. 200 ml of petroleum ether B. 200 ml of diethyl ether-petroleum ether (7:93) C. 200 ml of diethyl ether-petroleum ether (25:75)	To separate propylene carbonate from organochlorine and organophosphorus insecticides	91
Florisil heated at 650°C, reactivated at 130°C (19 g)	A. 200 ml of hexane B. 250 ml of diethyl ether-hexane (20:80)	Separation of organochlorine insecticide from polychlorinated biphenyl	92

Florisil heated to 660°C (30 g)	A. 200 ml of petroleum ether B. 200 ml of benzene–petroleum ether (5:1) C. 200 ml of chloroform D. 150 ml of acetone	Fractionation of organochlorine and organophosphorus insecticides as an aid to identification	93
Alumina			
Neutral alumina, 12.6% H_2O (10 g)	Hexane — sequence of amounts dependent on insecticide concentration A. For high concentrations — successive portions of 5, 10, and 65 ml B. For low concentrations — successive portions of 5, 10, 5, 20, and 40 ml	Fractionation of organochlorine insecticides in lake sediments	94
Alumina type H (2 g)	Acetone–hexane (5:95)	Dieldrin in sediments	95
Alumina of activity IV (12.5 g)	A. 65 ml of carbon tetrachloride B. 50 ml of diethyl ether–methylene chloride (5:95)	The triazine herbicides, propazine, ametryne, and prometryne	96
Alumina of activity V (12.5 g)	A. 65 ml of carbon tetrachloride B. 50 ml of diethyl ether–methylene chloride (5:95)	The triazine herbicides, atrazine, simazine, and prometone	96
Alumina type H	12 ml of acetone–hexane (1:99)	2,6-dichlorobenzonitrile	97

TABLE 4. — Continued

Adsorbent[a]	Eluting solvent[b]	Pesticides	Reference
	Mixed Adsorbents		
1. Florisil (10 g) 2. Florisil–carbon (15:2) (10 g) 3. Florisil (5 g)	A. 100 ml of benzene–pentane (2:98) B. 225 of diethyl ether–pentane (5:95)	Heptachlor and heptachlor epoxide	98
Florisil–Celite 545 (5:1), 4.5% H_2O (10 g)	A. 40 ml of petroleum ether B. 80 ml of benzene–petroleum ether (25:75)	Aldrin, heptachlor, heptachlor epoxide, dieldrin	99
Magnesia–Celite (1:1) (15 g)	250 ml of benzene–hexane (15:85)	Organochlorine insecticides	100, 101
Magnesia–Celite (4:1)	Not given	Aldrin, dieldrin	102
1. Celite (3 g) 2. 3:1 fuming H_2SO_4 (6 ml): Celite (10 g)	100 ml of hexane	Toxaphene in sediments	87

[a]Column consists of a series of layers with layer 1 at the bottom of the column.
[b]Letters indicate sequence of eluting solvents.

rutubaga from a 4:1 magnesia-Celite column. A summary of liquid chromatographic systems used for cleanup of soil extracts containing pesticide residues is given in Table 4.

It is apparent from the above discussion that no single liquid chromatographic cleanup technique is universally applicable to all pesticides. Florisil and alumina appear to be used most often in cleanup of soil extracts. Many methods have been evaluated for a single purpose and, while they may be suitably applied to other applications, each application should be evaluated carefully. Once the method has been determined adequate for a given purpose, standard compounds must be carried through the procedure since "standard" adsorbents show considerable variation between batches and between manufacturers.

REFERENCES

1. R. G. Petersen and L. D. Calvin, in Methods of Soil Analysis (C. A. Black, D. D. Evans, J. L. White, L. E. Ensminger, and F. E. Clark, eds.), American Society of Agronomy, Madison, Wisc., 1965, pp. 54-72.

2. M. G. Cline, Soil Sci., 58, 275 (1944).

3. W. G. Cochran, Sampling Techniques, Wiley, New York, 1953.

4. M. L. Jackson, Soil Chemical Analysis, Prentice-Hall, Englewood Cliffs, N. J., 1958, pp. 10-53.

5. G. T. Felbeck, Advan. Agron., 17, 327 (1965).

6. J. L. Mortensen and F. L. Himes, in Chemistry of the Soil, 2nd ed. (F. E. Bear, ed.), Reinhold, New York, 1964, pp. 206-241.

7. P. Dubach and N. C. Mehta, Soils Fert., 26, 293 (1963).

8. M. M. Kononova, Soil Organic Matter, Pergamon, London, 1961.

9. J. M. Bremner, J. Soil Sci., 5, 214 (1954).

10. Sephadex and Other Separation Products, Pharmacia Chemicals, Division of A B Pharmacia, Uppsala, Sweden, undated, p. 4.

11. Pharmacia Fine Chemicals, Inc., Sepharose Agarose Gels in Bead Form, Beckman Hansson AB/Svegea, 1967, 15p.

12. Pharmacia Fine Chemicals, Inc., Sephadex — Gel Filtration in Theory and Practice, Beckman Hansson AB/Eklunds & Vasatryck, 1966, 56p.

13. R. S. Swift and A. M. Posner, J. Soil Sci., 22, 237 (1971).

14. G. Dell'Agnola, A. Maggioni, and G. Ferrari, Agrochim., 9, 224 (1965).

15. G. Dell'Agnola and A. Maggioni, Estratto Dal Bollettino Scientifico Della Facottà di Chimica Industrale-Bologna, 23, 333 (1965).

16. G. Dell'Agnola, G. Ferrari, and A. Maggioni, Ricerca Scientifica, 34(II-B), 347 (1964).

17. M. Robert-Géro, C. Hardisson, L. Leborgne, and G. Pignaud, Ann. Pasteur, 111, 750 (1966).

18. J. R. Helbert and K. D. Brown, Anal. Chem., 29, 1464 (1957).

19. H. Rosen, Arch. Biochem. Biophys., 67, 10 (1957).

20. M. A. Rashid and L. H. King, Chem. Geol., 7, 37 (1971).

21. M. A. Rashid and L. H. King, Geochim. Cosmochim. Acta, 33, 147 (1969).

22. P. Andrews, Nature (London), 196, 36 (1962).

23. E. Dorado Bernal, Anal. Edafol. Agrobiol., 28, 269 (1969).

24. E. Dorado and J. DelRio, Anal. Edafol. Agrobiol., 28, 869 (1969).

25. N. C. Mehta, P. Dubach, and H. Deuel, Z. Pflanzenernaehr. Dueng. Bodenk., 102, 128 (1963).

26. A. Prakash and M. A. Rashid, Limnol. Oceanog., 13, 598 (1968).

27. W. Wildenhain and G. Henseke, Z. Anal. Chem., 229, 271 (1967).

28. W. Wildenhain and G. Henseke, Thaer-Archiv, 12, 679 (1968).

29. M. Schnitzer and S. I. M. Skinner, Gel Filtration of Fulvic Acid, A Soil Humic Compound, in Isotopes and Radiation in Soil Organic Matter Studies, International Atomic Energy Agency, Vienna, 1968, pp. 41-55.

30. J. N. Ladd, Soil Sci., 107, 303 (1969).

31. K. Kumada and O. Sato, J. Soc. Soil Manure (Japan), 36, 373 (1965).

32. K. Kumada and O. Sato, Soil Plant Nutrit. (Japan), 8, 31 (1962).

33. K. Kumada and H. M. Hurst, Nature (London), 214, 631 (1967).

34. R. S. Swift, B. K. Thornton, and A. M. Posner, Soil Sci., 110, 93 (1970).

35. I. Lindqvist, Acta Chem. Scand., 21, 2564 (1967).

36. G. Dell'Agnola and A. Maggioni, Boll. Sci. Fac. Chim. Ind. Bologna, 23, 321 (1965).

37. M. Koter, B. Chodan, J. Chodan, and H. Panak, Zezyty Nauk Wyzszej Szkoly Rol. Wroclaw. Melioracja, 15, 343 (1963).

38. C. Passera and A. Maggioni, Ric. Sci. Suppl., 36, 652 (1966).

39. A. R. Bromfield, C. B. Coulson, and R. I. Davies, Chem. Ind., 1959, 601 (1959).

40. J. M. Bremner, J. Agr. Sci., 46, 247 (1955).

41. R. U. Lemieux, C. T. Bishop, and G. E. Pelletier, Can. J. Chem., 34, 1365 (1956).

42. H. Scholz, Z. Pflanzenernaehr. Dueng. Bodenk., 84, 159 (1959).

43. F. Scheffer, W. Ziechmann, and H. Scholz, Z. Pflanzenernaehr. Dueng. Bodenk., 85, 50 (1959).

44. G. T. Felbeck, Jr., Trans. Int. Cong. Soil Sci., Comm. II and IV, Aberdeen, 1966, 11 (1966).

45. J. M. Bremner and H. Lees, J. Agr. Sci., 39, 274 (1949).

46. C. B. Coulson, R. I. Davies, and E. J. A. Khan, J. Soil Sci., 10, 271 (1959).

47. C. B. Coulson, R. I. Davies, and E. J. A. Khan, J. Sci. Food Agr., 10, 209 (1959).

48. A. P. Adams, W. V. Bartholemew, and F. E. Clark, Soil Sci. Soc. Amer. Proc., 18, 40 (1954).

49. D. J. Cosgrove, Australian J. Soil Res., 1, 203 (1963).

50. S. J. Angyal, P. T. Gilham, and C. G. MacDonald, J. Chem. Soc., 1957, 1417 (1957).

51. S. J. Angyal, D. J. McHugh, and P. T. Gilham, J. Chem. Soc., 1957, 1432 (1957).

52. C. A. Anderson and C. A. Black, Soil Sci. Soc. Amer. Proc., 29, 255 (1965).

53. N. C. Mehta, J. O. Legg, C. A. I. Goring, and C. A. Black, Soil Sci. Soc. Amer. Proc., 18, 443 (1954).

54. L. Hough, J. K. N. Jones, and W. H. Wadman, Nature (London), 162, 448, (1948).

55. L. Hough, J. K. N. Jones, and W. H. Wadman, J. Chem. Soc., 1949, 2511 (1949).

56. E. Alvsaker and K. Michelsen, Acta Chem. Scand., 11, 1794 (1957).

57. R. L. Whistler and K. W. Kirby, J. Amer. Chem. Soc., 78, 1755 (1956).

58. F. J. Sowden and K. C. Ivarson, Plant Soil, 16, 389 (1962).

59. F. J. Sowden and K. C. Ivarson, Soil Sci., 94, 340 (1962).

60. U. C. Gupta, F. J. Sowden, and P. C. Stobbe, Soil Sci. Soc. Amer. Proc., 27, 380 (1963).

61. R. B. Duff, J. Sci. Food Agr., 3, 140 (1952).

62. W. G. C. Forsyth, Biochem. J., 46, 141 (1950).

63. S. A. Barker, M. H. B. Hayes, R. G. Simmonds, and M. Stacey, Carbohyd. Res., 5, 13 (1967).

64. R. B. Duff, J. Sci. Food Agr., 12, 826 (1961).

65. J. F. Bouhours and M. V. Cheshire, Soil Biol. Biochem., 1, 185 (1969).

66. M. V. Cheshire, C. M. Mundie, and H. Shepard, Soil Biol. Biochem., 1, 117 (1969).

67. W. G. C. Forsyth, Biochem. J., 41, 176 (1947).

68. E. Schlichting, Z. Pflanzenernaehr. Dueng. Bodenk., 61, 97 (1953).

69. R. F. Keefer, Diss. Abstr., 24, 4329 (1964).

70. R. F. Keefer, F. L. Himes, and J. L. Mortensen, Soil Sci. Soc. Amer. Proc., 30, 415 (1966).

71. W. Wildenhain and G. Henseke, Z. Chem., 5, 457 (1965).

72. S. A. Barker, P. Finch, M. H. B. Hayes, R. G. Simmonds, and M. Stacey, Nature (London), 205, 68 (1965).

73. R. L. Whistler and J. N. BeMiller, in Methods in Carbohydrate Chemistry, (R. L. Whistler and M. L. Wolfrom, eds.), Academic, New York, 1962, pp. 47-50.

74. R. U. Lemieux, in Methods in Carbohydrate Chemistry, (R. L. Whistler and M. L. Wolfrom, eds.), Academic, New York, 1962, pp. 45-47.

75. R. L. Whistler and J. N. BeMiller, in Methods in Carbohydrate Chemistry, (R. L. Whistler and M. L. Wolfrom, eds.), Academic, New York, 1962, pp. 42-44.

76. G. N. Catravas, in Automation in Analytical Chemistry, Technicon Symposia, Vol. 1, Mcdiad, White Plains, N.Y., 1966, pp. 397-400.

77. S. H. Churms, in Advances in Carbohydrate Chemistry and Biochemistry, (R. S. Tipson, ed.), Academic, New York, 1970, pp. 13-51.

78. R. I. Morrison and W. Bick, J. Sci. Food Agr., 18, 351 (1967).

79. D. T. Downing, Z. H. Kranz, and K. E. Murray, Australian J. Chem., 13, 82 (1960).

80. D. T. Downing, Z. H. Kranz, J. A. Lamberton, K. E. Murray, and A. H. Redcliffe, Australian J. Chem., 14, 253 (1961).

81. H. B. Pionke and G. Chesters, Soil Sci. Soc. Amer. Proc., 32, 749 (1968).

82. K. I. Beynon and K. E. Elgar, Analyst, 91, 143 (1966).

83. H. P. Burchfield and D. E. Johnson, Guide to the Analysis of Pesticide Residues, U.S. Govt. Printing Office, Washington, D.C., 1965, 2 vols.

84. J. R. Duffy and N. Wong, J. Agr. Food Chem., 15, 457 (1967).

85. R. L. King, N. A. Clark, and R. W. Hemken, J. Agr. Food Chem., 14, 62 (1966).

86. W. L. Trautman, Ph.D. Thesis, Univ. Of Wisc. Library, Madison, Wisc., 1970.

87. L. C. Terriere, U. Kügemagi, A. R. Gerlach, and R. L. Borovicka, J. Agr. Food Chem., 14, 66 (1966).

88. L. Johnson, J. Assoc. Off. Anal. Chem., 45, 348 (1962).

89. J. I. Teasley and W. S. Cox, J. Agr. Food Chem., 14, 519 (1966).

90. W. R. Payne, Jr., and W. S. Cox, J. Assoc. Offic. Anal. Chem., 49, 989 (1966).

91. R. R. Schnorbus and W. F. Phillips, J. Agr. Food Chem., 15, 661 (1967).

92. L. M. Reynolds, Bull. Env. Contam. Toxicol., 4, 128 (1969).

93. W. W. Sans, J. Agr. Food Chem., 15, 192 (1967).

94. M. G. Browman, Ph.D. Thesis, Univ. of Wisc. Library, Madison, Wisc., 1970.

95. C. A. Edwards, A. R. Thompson, K. I. Beynon, and M. J. Edwards, Pest. Sci., 1, 169 (1970).

96. A. M. Mattson, R. A. Kahrs, and R. T. Murphy, Residue Revs., 32, 371 (1970).

97. K. I. Beynon, L. Davies, K. Elgar, and A. N. Wright, J. Sci. Food Agr., 17, 151 (1966).

98. R. T. Murphy and W. F. Barthel, J. Agr. Food Chem., 8, 442 (1960).

99. A. T. S. Wilkinson, D. G. Finlayson, and H. V. Morley, Science, 143, 681 (1964).

100. J. G. Saha and W. W. A. Stewart, Can. J. Plant Sci., 47, 79 (1967).

101. J. G. Saha and H. McDonald, J. Agr. Food Chem., 15, 205 (1967).

102. R. D. Hendrick, F. L. Bonner, T. R. Everett, and J. E. Fahey,
      J. Econ. Entomol., 59, 1388 (1966).

Chapter 8

PAPER CHROMATOGRAPHIC ANALYSIS IN SOIL CHEMISTRY

Gordon Chesters and D. A. Graetz*

Water Resources Center
Department of Soil Science and
Water Chemistry Program
University of Wisconsin — Madison
Madison, Wisconsin

*Present Address: Soil Science Department
                   Institute of Food and Agricultural Sciences
                   University of Florida
                   Gainesville, Florida

I. INTRODUCTION

Paper chromatographic analyses have received extensive usage in soil
chemistry. However, because of difficulty in quantitative extraction of
soil components, particularly polymeric materials, and because of the
large amounts and multiplicity of coextractives, this chromatographic
method has been used largely for qualitative identification of components
present in soil hydrolyzates. For analyses of indigenous soil components,
perhaps the most extensive quantitative use of paper chromatography is for
determination of amino acids in soil hydrolyzates, but this form of analysis
has been superceded largely by column ion-exchange chromatography, which
is more accurate, sensitive, and reliable and which possesses greater
carrying capacity.

For pesticide analyses of soil and sediment extracts, paper chromatography has served several functions, including cleanup of samples prior to
determination by some other method of analysis; semiquantitative determination of pesticides in soil and sediment extracts; identification of intact
pesticides and, more importantly, identification of unknown metabolites and
other degradation products of pesticides; and screening tests for possible
pesticidal contamination of soils and sediments. The usefulness of paper
chromatography as a cleanup procedure has declined in recent years because
of its low carrying capacity and almost all cleanup procedures for soil,
sediment, and plant extracts presently involve a column chromatographic
technique. Furthermore, the method possesses the disadvantage of being
too tedious and time consuming for routine quantitative analyses, and gas
chromatographic methods of analysis have proved to be more efficient.
Paper chromatography still possesses utility for separating multiple-
component systems and for identifying unknown components, particularly
when it is supplemented by some other methods. The identification of unknown
components cannot be made on the basis of a single R_f value in a single
solvent system because the coextraction of other indigenous soil components
as well as chromatographic conditions may alter R_f values. Nevertheless,
as a supplementary technique to other techniques of identification, paper
chromatography serves, and will continue to serve, a useful purpose in soil
chemistry, not only in the field of pesticide analysis, but also in characterization of the monomer components of soil organic polymers. Because the
present status of the chemistry of soil organic fractions is such that a large
amount of qualitative analysis is still essential, paper chromatographic
methods are likely to retain this function.

In the determination of inorganic components of soils, developments in
paper chromatographic methodology continue to be made; however, widespread environmental use of this technique for determining toxic elements
is unlikely, particularly for those metal ions capable of forming volatile
coordination complexes. This type of analysis can be performed less
tediously and with greater accuracy and sensitivity by gas chromatographic
methods.

Thus, paper chromatography does not possess sufficient carrying capacity to be an effective cleanup technique for use on soil, sediment, or plant extracts, nor is it sufficiently adaptable to routine analysis to serve an important function in quantitative soil chemical analysis, but it is likely to retain its utility for identification of indigenous and added soil components.

The discussion of methodology and uses of paper chromatographic techniques in soil chemistry has been divided into six sections based on types of materials to be investigated. Some of the divisions between sections have, of necessity, been chosen arbitrarily. For example, although proteins and carbohydrates comprise a major portion of soil humic and fulvic fractions, they have been discussed separately, largely because investigators examining the carbohydrate or proteinaceous compositions of humic fractions have chosen to confine themselves to a single grouping of chemical entities. Thus, the discussion of humic materials has been confined almost entirely to the types of aromatic constituents isolated from humic materials subjected to some form of oxidative or reductive degradation procedure. Other types of soil chemical analyses discussed include soil organophosphates, inorganic components of soils and fertilizers, and organochlorine and organophosphorus insecticides, and, to a lesser degree, herbicides and fungicides.

II. ACCURACY AND PRECISION

Paper chromatography has been used in soil chemical analyses largely for qualitative identification, and information pertaining to accuracy, precision, and limits of detection is rather sparse. In the area of pesticide analyses amounts to 1 μg or less have been determined. Lower limits of detection for metal ions are 0.005 μg for most others.

III. SAMPLING

Much information is available on soil sampling, and techniques used are very similar irrespective of the type of analysis to be conducted; the reader is referred to the original references for further information [1-4]. It should perhaps be pointed out that because of the qualitative nature of much of the soil analysis it is not nearly so important to get homogeneous samples from exactly the same depth in the soil profile as it would be for quantitative analysis.

IV. SEPARATIONS AND CONCENTRATION TECHNIQUES

Separation of interferences and concentration techniques used in paper chromatography are discussed for the particular analysis to be performed in the following sections.

V. METHODS AND USES OF PAPER CHROMATOGRAPHIC ANALYSIS IN SOIL CHEMISTRY

A. Humic Materials and Their Degradation Products

A brief discussion of methods of extraction and terminology used to identify isolated soil organic matter components is given in Chap. 7, Sec. II, A.

Very little characterization of intact humic and fulvic acids has been completed, and the bulk of this section deals with the methodology and usage of paper chromatographic techniques for characterization of humic degradation products. Although carbohydrates and amino acids constitute a portion of the humic fraction of soils, for the purpose of this discussion they are considered separately.

As early as 1959, Coulson et al. [5] attempted to chromatograph intact humic acids and plant lignins using a paper-strip technique similar to that described by Coulson [6]. Paper (Whatman No. 4) in 18 × 2 in. strips was placed in 20 × 3 in. glass tubes fitted with rubber stoppers and glass rods hooked at one end to accomodate up to three filter strips in a vertical position. The humic acid (0.1 mg in 0.1 ml of 0.1N NaOH) was applied to the paper strip at a point 1.5 in. from the lower end and the paper dipped into 10-15 ml of solvent. The solvent was allowed to move 9-12 in. and the papers were dried and examined under ordinary and ultraviolet light. Some 50 different solvents were used, as described by Khan [7]. The three most promising solvents were isopropanol-H_2O (6:4), tetrahydrofuran-H_2O (9:1), and dioxane-H_2O (9:1). These solvents used in an ascending manner produced a slight separation of components. Many others produced either a streaky separation or failed to move the acids at all. A complex formed between the humic acids and cetyl pyridinium chloride [8] did not fractionate by paper chromatography.

Fractionation of humic and fulvic materials extracted from soils was attempted by paper chromatographic methods [9]. The techniques included ascending, descending, circular, and two-dimensional chromatographic methods using acidic, basic, and neutral solvents. Among these techniques the circular method was the most suitable and versatile. In acidic solvents, such as n-butanol-acetic acid-water, n-butanol-formic acid-water, n-propanol-formic acid-water, and isopropanol-formic acid-water, the separated individual components were characterized. The neutral solvent systems, namely, methyl ethyl ketone-ethanol-water, benzene-acetone-water, dioxane-acetone-water, and acetone-methanol-water, were examined in detail. Only the methyl ethyl ketone-ethanol-water (6:3:1) achieved some separation, but even this solvent system was not satisfactory. A few basic solvent systems were tried because of their known specificity for indolic and phenolic acids and related compounds. The most successful of these solvent systems was isopropanol-concentrated ammonium hydroxide-water (20:1:2). Whatman No. 1 chromatography paper was used in all techniques and all experiments were carried out at room temperature in tanks saturated with solvent vapors.

Visualization of components was achieved using 0.1% Bromocresol Green in 99.5% ethanol. In the case of basic solvents, the pH of the indicator was adjusted to 7.2 and for acid solvents to pH 7.8. Visualization also was accomplished by the use of ultraviolet light. The ultraviolet examination revealed the presence of a fluorescent material in all but one of the humic acids examined.

Sinha and Shukla [10] have made a detailed investigation of the products of $KMnO_4$ oxidation of humic acids [11, 12]. Oxidation of the humic acids was carried out in conical flasks at $55°C$ for 15 hr and separations were studied by ascending, descending, and circular paper chromatographic methods. A variety of compounds were identified. Solvents used included n-butanol-acetic acid-water (12:3:5), n-butanol-pyridine-water (1:1:1), methanol-pyridine water (16:0.8:4) and ethanol-concentrated ammonium hydroxide-water (18:1:1). Reagents for locating spots on the papers were ninhydrin, diazotized sulfanilic acid, nitrosonaphthol, Bromocresol Green, and 2,6-dichloroquinone chloroimide. Among the compounds identified as degradation products were several amino acids, anthranalic acid, amino benzoic acid, S-benzyl penicillamine, phenyl ethylamine, and imidazole carboxylic acid.

In a comparison of the alkaline extracts of Aspergillus niger tissue with soil humic acids, Kang and Felbeck [13] have shown that the amino acid and simple sugar components of the hydrolyzates of these materials are very similar. A two-way ascending paper chromatographic technique was used for identification and semiquantitative estimation of the amino acids and sugars. The paper support was Whatman No 1 paper (50 × 50 cm) and the solvents used for amino acids consisted of n-butanol-glacial acetic acid-water (4:1:5) for the first dimension and phenol-ethanol-water (4:1:1) for the second dimension. Prior to use of the second solvent, a 0.3% NH_4OH solution was equilibrated with the solvent by placing 50 ml of the solution in a beaker in the bottom of the developing tank. For chromatography of sugars the solvents used were n-butanol-pyridine-water (2:2:1) for the first dimension and isopropanol-water (4:1) for the second dimension. The procedures for preparation of acid hydrolyzates, color development, and identification of the chromatographic spots followed those described by Crossan and Lynch [14].

Methods have been described for the fractionation of low molecular weight materials extracted with dilute acid from the B horizon of a podzol soil [15]. An initial fractionation could be achieved by precipitation with increments of Pb(II), or by the addition of an excess of Ba(II) followed by an excess of Cu(II), and finally by an excess of Pb(II). Metal ions were removed by passing the soil extract through a column of Dowex 50. Further fractionation was achieved on cellulose column. For paper chromatographic analysis of these materials, a large number of solvent systems were tried, including neutral, basic, acidic, and buffered solvents. Buffered and unbuffered papers were used. The most useful solvent systems were methyl-ethyl ketone-acetone formic acid-water [16] and those based on isopropanol-water made either acid with CH_3COOH or alkaline with NH_4OH or buffered. The proportion of isopropanol used varied from 40 to 60%. For the higher molecular weight darker materials, the more aqueous solvents were preferred. Acidic, basic, or buffered solvents gave more compact spots with less tailing than did the neutral solvents. Best indication of separation of the materials was obtained by viewing under near and far ultraviolet light. The paper chromatograms showed that with each method of fractionation a series of materials was obtained with no clear differenced between them [15]. The first fractions to be eluted from the columns and the last materials to be precipitated by the metals had the lightest color, had the highest R_f values on the paper chromatograms, and gave the most characteristic reactions with sprays used to locate the materials on the chromatograms. Thus, the first fractions eluted from the cellulose columns were strongly fluorescent in ultraviolet light and gave characteristic reactions with the Folin-Ciocalteu reagent.[17], α-nitroso-β-naphthol, and diazotized sulfanilic acid, which are used as tests for phenols [18]; aniline-xylose, used as a test for acids; and dinitrophenylhydrazine, used as a test for carbonyls.

In an examination of lignoprotein components isolated from compost, Jenkinson and Tinsley [19] identified the amino acid composition by paper chromatography of hydrolyzates. Ten milliliters of the hydrolyzates containing approximately 1 mg of nitrogen were evaporated to dryness under reduced pressure, dissolved in 5 ml of H_2O, and again evaporated to dryness. The residue was taken up in 1.5 ml of H_2O and centrifuged and 0.04 ml of the clear supernatant liquid was spotted on Whatman No. 1 chromatography paper. The chromatograms were developed using the solvent systems described by Wolfe [20]. The amino acids were made visible by dipping the papers in a 0.2% solution of ninhydrin in acetone. The following amino acids were identified: cysteic acid, aspartic acid, glutamic acid, lysine, arginine, glycine, histidine, serine, alanine, tyrosine, proline, valine, threonine, isoleucine, leucine, and phenylalanine. Methionine is not detected because it is not separated from glycine in these solvent systems, nor is methionine sulfone resolved from valine. Chromatograms dipped in 0.1% isatin in acetone confirmed the presence of proline and the absence of hydroxyproline. Detection and estimation of peptide-bound lysine was made with 2,4-dinitrofluorobenzene [21]. This reaction proceeds because the ε-amino groups of lysene are free to react with the reagent and the α-amino groups are not. The method developed by Blackburn and Lowther [22] was used for separating the dinitrofluorobenzene derivatives of amino acids.

Phenolic acids have been obtained as degradation products of humic acids [23-25] and from decomposing straw [26]. However, reviews of the chemistry of soil organic matter indicate that only p-hydroxybenzoic acid had been found in the free state [27-29]. In 1964, Whitehead [30] was able to identify p-hydroxybenzoic, vanillic, p-coumaric, and ferulic acids in the free state in soils by paper chromatographic methods. Extraction of the phenolic acids from the soils was effected by shaking 250 g of soil with 2.5 g of CaO and 250 ml of H_2O. The filtrate was acidified with H_2SO_4 to pH 2 and treated with freshly precipitated zinc ferrocyanide to remove fatty materials [31]. The solution was readjusted to pH 2 and extracted with an equal volume of peroxide-free diethyl ether as described by Hamence [31]. The ethereal solution was washed with H_2O, dried over anhydrous Na_2SO_4, and evaporated to a small volume. Phenolic acids were isolated using three successive solvent systems by the descending technique. The solvent systems were isopropanol-concentrated NH_4OH–H_2O (10:1:1) for 40 hr, n-butanol-pyridine–H_2O (14:3:3) for 16 hr, and 2% CH_3COOH for 3.5 hr. Whatman No. 3MM paper was used in all cases. Confirmatory evidence of identification of the phenolic acids was provided by the colors obtained on staining with diazotized p-nitroaniline and diazotized sulfanilic acid [32]. Each phenolic zone was extracted with absolute ethanol and the absorbance of each was determined in the wavelength range 220-400 nm. Absorption curves were prepared by plotting absorbance against wavelength as described by Bradfield and Flood [33]. Total amounts of phenolic acids obtained from the soils ranged from 0.004 to 0.009% of the total organic matter content of the soils. These small quantities are to be expected in view of the ready susceptibility of these compounds to microbial attack [34-36].

In the mid-1960's, reports of high yields of phenols and phenolic acids obtained by the reductive cleavage of humic acids with sodium amalgam were made [25, 37]. These findings could not be confirmed by Mendez and Stevenson [38]. These authors reported that the ether-soluble fraction obtained by reduction of humic acids with sodium amalgam consisted of a complex mixture of aliphatic and aromatic compounds, with the former predominating. Further information on these components was obtained by paper and other chromatographic methods [39]. Paper chromatography was used to separate phenolic and phenolic acid constituents. On Whatman No. 1 paper, chromatograms were developed using the following solvent systems: acetic acid-chloroform (1:10), isopropanol-concentrated ammonium hydroxide-water (10:1:1), glacial acetic acid-concentrated hydrochloric acid-water (30:3:10), ethyl acetate-benzene (4:5), toluene-acetic acid-water (4:1:5), and solvents B (chloroform-methanol-formic acid-water) and C (benzene-ethyl methyl ketone-formic acid-water) described by Reio [16]. Visualization of the spots is achieved with diazotized sulfanilic acid or 2,6-dichloroquinone chloroimide. Negative results obtained using $FeCl_3$ and $K_3Fe(CN)_6$, phospho-molybdic acid, ammonium metavanadate, and benzidine for locating spots. Additional identification was achieved by a gas-liquid chromatographic technique on materials eluted from excised spots.

Oxidative degradation of humic acids has been accomplished by several methods [40-44]. Among the variety of compounds obtained are phenols, phenolic aldehydes, and phenolic acids, all of which can be obtained from plant lignins by the same chemical methods. Greene and Steelink [23] used the CuO method for oxidation of humic acids as described by Pearl and Beyer [45, 46]. Humic acids extracted from a podzol B horizon were hydrolyzed with 2N HCl to remove carbohydrates and mineral material. The residue from this treatment was oxidized with a mixture of NaOH and CuO in a high-pressure stirring autoclave at $170°C$ for 3 hr. Paper chromatographic analysis revealed the presence of p-hydroxybenzaldehyde, syringaldehyde, vanillic acid, p-hydroxybenzoic acid, m-hydroxybenzoic acid, and 3,5-dihydroxybenzoic acid. Details of the chromatographic techniques used were not described. In a related fashion, Mehta et al. [47] have examined the degradation products of humic materials oxidized with $NaClO_2$, H_2O_2, and $NaIO_4$. Following oxidation, the material was extracted with diethyl ether and subjected to paper chromatographic analysis. Whatman Nos. 1, 3, and 4 papers were used in all experiments. Development of chromatograms for the organic acids was attained with four solvents, namely, n-butanol-pyridine-water (1:1:1), n-butanol-glacial acetic acid-water (12:3:5), ethanol-concentrated ammonium hydroxide-water (16:7:5), and ethanol-concentrated ammonium hydroxide (70:30) as described by Germain et al. [48]. For chromatography of phenolic compounds chloroform-methanol-water-formic acid (1000:100:96:4), as described by Reio [16], and n-butanol-pyridine-water (14:3:3), as described by Smith [18] were used. With these techniques several organic acids and phenols were identified.

Alkaline nitrobenzene oxidation of lignins [49-55] and humic acids [24, 43, 49, 55-58] has been used as a basis for characterizing humic acids and to determine whether lignins are a principal source material for the synthesis of humic acids. The major products of oxidation are phenolic aldehydes, namely, p-hydroxybenzaldehyde, vanillin, syringaldehyde, and their carboxylic acid analogs. Paper chromatography has been used commonly to identify and quantify these compounds [24, 43, 49]. The phenolic aldehydes in ethanol solution were separated on Whatman No. 52 paper buffered at pH 10 and dried at room temperature [48]. The solvent consisted of t-amyl alcohol saturated with the pH 10 buffer and the chromatograms were developed by the descending technique described by Consden et al. [59]. The papers were dried at room temperature and the aldehydic spots were located by their characteristic fluorescence in ultraviolet light. After sections containing the isolated aldehydic spots were cut from the chromatograms and extracted with 0.2N NaOH, the absorbance of the solutions was determined at the appropriate ultraviolet wavelength.

In an investigation of the decomposition of humic acid by fungi, Hurst et al. [60] used paper chromatographic methods for separating phenols. It was possible to separate m-hydroxybenzoic acid from its aldehyde and alcohol analogs on Whatman No. 4 paper using n-butanol-concentrated NH_4OH (4:1) or the faster running "Desaga" chromatostrips using Kieselgel G (Merck, Darmstadt) with $CHCl_3$-10% CH_3COOH (1:1) as solvent. Phenolic compounds were located by spraying with diazotized sulfanilic acid followed by 10% Na_2CO_3. This reagent reveals m-hydroxybenzoic acid and its alcohol as strong orange spots, while m-hydroxybenzaldehyde forms a much weaker brown spot. The aldehyde is more easily detected by its strong green fluorescence in ultraviolet light after it is sprayed with diazotized sulfanilic acid or by its reaction with 2,4-dinitrophenylhydrazine.

By drastic extraction methods, several investigators have broken down soil organic matter and identified aromatic products. Forsyth [61] and Drosdova [62] claim to have extracted phenolic glycosides from fulvic acids, while Schlichting [63] repeated these investigations and found no evidence for such compounds. In an attempt to find evidence for the presence of loosely bound phenolic groupings in soil organic matter extracts, Keefer et al. [64] purified low molecular weight materials from a NaOH extract of a peat soil. The components were separated paper chromatographically using as solvents distilled H_2O, butanone-acetone-formic acid-water (40:2:1:6) as described by Reio [16], n-C_4H_9OH-CH_3COOH-H_2O (4:1:5), and isopropanol-pyridine-acetic acid-water (8:8:4:1) as described by Gordon et al. [65]. One area on each of the chromatograms gave color reactions that were similar to the flavanoid compounds catechin and phloridzin. By the use of diazotized spray reagents some eight components were identified, which suggested the presence of phenols, aromatic and aliphatic amines, β-diketo compounds, and imidazoles [64].

It is well known that certain microorganisms can synthesize humiclike polymers from aromatic compounds [12, 66, 67]. The alkali fusion products of these polymers were reported by Robert-Géro et al. [68]. The methods used for paper chromatographic analysis were those discussed earlier by Whitehead [30]. Among the compounds identified were p-hydroxybenzalde-hyde; catechol; resorcinol; m- and p-hydroxybenzoic acids; 2,3-, 2,4-, 2,5-, 2,6-, 3,4-, and 3,5-dihydroxybenzoic acids; vanillin; 3-hydroxy-5-methoxy-, 4-hydroxy-3-methoxy-, and 4-hydroxy-3,5-dimethoxybenzoic acids, cinnamic acid; and o-hydroxy-, 3,4-dihydroxy-, and 4-hydroxy-3-methoxycinnamic acids.

As a perusal of the above data shows, most paper chromatographic investigations of humic materials and their degradation products are largely qualitative and are used to obtain a picture of the constitution of humic colloids. Paper chromatography is particularly well suited to this type of analysis. However, as the detailed chemical characterization of humic colloids progresses it is anticipated that more rigorous quantitative analyses can be made by gas chromatographic and other techniques.

B. Carbohydrates

Almost without exception, carbohydrate investigations have involved hydrolysis of the soil organic matter to produce the constituent sugars followed by some method of identification. Paper chromatography has been, along with column chromatographic techniques, a popular method of analysis.

Almost invariably, the hydrolysis of either intact soil or isolated poly-saccharides has been carried out with H_2SO_4, although the concentration of H_2SO_4 and the length of treatment have varied considerably.

A common pretreatment has involved 72% H_2SO_4. It is important to insure that the temperature does not rise significantly during the 2-hr duration of the pretreatment [69-73]. The H_2SO_4 solution is the most frequently diluted to 1N and refluxed for approximately 8 hr. At the termination of refluxing and after filtering, sufficient powdered $Ba(OH)_2$ is added slowly with stirring to the supernatant solution to neutralize approximately three-quarters of the acid, followed by addition of a saturated $Ba(OH)_2$ solution to bring the pH to approximately 4. The precipitated $BaSO_4$ is filtered and washed with H_2O. If a 10-g soil sample has been used, the combined filtrate is concentrated to 20-30 ml on a rotary evaporator. The pH of the solution is adjusted to 6.0-6.5 with a $Ba(OH)_2$ solution, the $BaSO_4$ is removed by centrifugation, and the precipitate is washed with H_2O. The supernatant is evaporated to dryness and the material is taken up in 2 ml of H_2O. Many methods of hydrolysis do not include a pretreatment with 72% H_2SO_4 but merely an 8-hr refluxing with H_2SO_4. Concentrations used include 0.5N [70], 1.0N [70, 71, 74], 2.0N [70, 75], 3.0N [76, 77], and 6.0N [78]. The methods of hydrolysis have included heating on a boiling water bath for times ranging from 1.5 to 18 hr [74-76, 78], refluxing usually for 8 hr [69-73], and heating in an autoclave at 121°C for 1 hr [70].

A unique method of hydrolysis introduced by Parsons and Tinsley [79] involved hydrolysis of 2 mg of sample with 3 ml of 98% HCOOH for 12 hr at 100°C followed by 3 ml of 1.0N H_2SO_4 for 6 hr at 100°C in sealed tubes. This method was found optimum for release of sugars from soils and composts. The HCOOH was removed as the ethyl ester by gently warming with absolute ethanol. The H_2SO_4 was neutralized with $BaCO_3$ and the filtrate was diluted to a known volume.

Paper chromatographic systems that have been used for the identification of carbohydrates are presented in Table 1. Many chromogenic reagents have been used for visualization of sugars on paper chromatograms (Table 2).

Paper chromatography has proved qualitatively useful for identification of the sugar constituents of soils and other biological materials, and several systems have proved highly successful. Perhaps the biggest drawback to quantitative analysis has been the difficulty in developing methods for complete hydrolysis of soil organic matter.

C. Organic Phosphates

1. Phospholipids

Separations of glycolipids and phospholipids from animal tissue [113-119] and plant tissue [120-122] have been made chromatographically on silicic acid-impregnated paper.

Most investigations of soil phospholipids have been concerned only with determination of phosphate soluble in common fat solvents [123-127]. Choline has been isolated from soils [128, 129], but its source is uncertain

TABLE 1

Paper Chromatographic Systems Used for Identification of Sugars in Hydrolyzates of Soils and Other Biological Materials

Compounds to be determined	Paper support	Type of chromatography	Solvent system		Original reference to solvents	References in soils
			Solvent	Composition		
Sugars in isolated polysaccharide and intact soil hydrolyzates	Whatman No. 3	Descending	Ethyl acetate–pyridine–H_2O	8:2:1	80	76, 79
	Whatman No. 1	Circular	Ethyl acetate–CH_3COOH–H_2O	9:2:2	80	76, 79
			n–Butanol–pyridine–H_2O	Not stated	81	82–84
			Ethyl acetate–pyridine–H_2O	Not stated	85	82–84
			Ethyl acetate–CH_3COOH–H_2O	Not stated	86	82–84
Sugars in column fractionated materials	Whatman No. 1	Ascending, developed twice for 16 and 8 hr	n–Butanol–pyridine–benzene–H_2O	5:3:1:3	87	82–84, 88
Sugars in intact soil hydrolyzate	Whatman No. 1 buffered in 0.1M phosphate	Ascending, developed three times for 16, 8, and 16 hr	Acetone–n–butanol–H_2O	5:4:1	87	82–84, 88
Sugars in hydrolyzates of isolated polysaccharides	Whatman No. 1	Circular	Ethyl acetate–pyridine–H_2O	40:11:6	82	75
			n–Butanol–CH_3COOH–H_2O	4:1:5 (upper layer)	89	75
			n–Butanol–pyridine–H_2O	6:4:3	81, 90	75

TABLE 1.— Continued

Compounds to be determined	Paper support	Type of chromatography	Solvent system		Original reference to solvents	References in soils
			Solvent	Composition		
			Ethyl acetate–CH_3COOH–H_2O	3:1:3 (upper layer)	86	75
			Ethyl acetate–pyridine–CH_3COOH–H_2O	5:5:1:3	91	75
	Not stated	Not stated	H_2O–saturated phenol		89	75
			H_2O–saturated–n–butanol		74	74
			H_2O–Saturated phenol		89	74
			H_2O–Saturated s–collidine		74	74
			H_2O–Saturated amyl alcohol		74	74
Sugars in fulvic acid hydrolyzates	Whatman No. 4	Ascending	n–Butanol–pyridine–H_2O	6:4:3	90	77,90
			n–Butanol–CH_3COOH–H_2O	4:1:5	90	77,92
Sugars in hydroly- zates of humic and lignin materials and "humic" materials synthesized by microorganisms	Whatman No. 1	Not stated	n–Butanol–ethanol–H_2O	40:11:19	93	94–96
			n–Butanol–CH_3COOH–H_2O	2:1:1	93	94–96
			Ethyl acetate–pyridine–H_2O	2:1:2	97	94–96
			Ethyl acetate–n–propanol–H_2O	1:6:3	98	94–96

Study	Paper	Method	Solvent	Ratio	Ref.	Ref.
Cellulose from soil organic matter	Chromatographic methods are probably similar to those used in [69].					73
Relationship between mycostasis and free sugars in soils	Whatman No. 1	Not stated	n-Butanol-pyridine-H_2O	6:4:3	81,90	99,100
Identification of 2-0-methylxylose and 3-0-methyl-xylose in peats	Whatman No. 1	Descending	n-Butanol-CH_3COOH-H_2O	4:5:1		101
			H_2O-Saturated n-Butanol		74	101
			Ethyl acetate-pyridine-H_2O	12:5:4		101
Transformations of [^{14}C]glucose in soil	Whatman No. 1	Descending	H_2O-Saturated n-butanol		74	102

TABLE 2

Reagents Used for Location of Carbohydrate Spots
on Paper Chromatograms

Reagent	Reference
Aniline-diphenylamine	75,103
Triphenyltetrazolium bromide	75,83,104
Aniline hydrogenphthalate	75,77,83 105,106
α-Naphthylamine-phosphoric acid	75,81,83
p-Anisidine-phosphoric acid	75,107
Periodate oxidation followed by nitroprusside-piperazine	75,108
Acetylacetone-p-dimethylamino benzaldehyde	74,75,89,109
p-Anisidine-HCl in n-butanol saturated with H_2O	75,76,101
2N NH_4OH, sodium thiosulfate/sodium metabisulfite solution	99,100
2-Amino biphenyl	79,110
Resorcinol	74,111
Phenol-H_2SO_4	71,112

since it occurs in plants as the free base and in such compounds as acetyl choline as well as lipids. The only investigation dealing directly with phospholipids in soils is an attempt by Hance and Anderson [130] to identify characteristic hydrolysis products of phospholipids from soils. Lipids were extracted from the soil using a variety of organic solvents after the soil was pretreated with HF and HCl. A portion of the lipid fraction precipitated following evaporation of the combined extracts. The insoluble lipids were extracted with light petroleum followed by a mixture of chloroform and methanol. The chloroform-methanol extract contained 84% of the extracted lipid phosphorus.

Glycerophosphate was released from the crude lipid extract by alkaline hydrolysis and was identified paper chromatographically as follows: 200 g of air-dried soil were treated with 300 ml of a mixture containing equal volumes of 2.5% HF and 2.5% HCl. After 48 hr, the mixture was filtered through Whatman No. 42 paper and the soil was washed with H_2O until free of acid. An amount of the residue equivalent to 50 g of soil was extracted

with successive 100-ml portions of acetone, light petroleum (b. p. 40-60°C),
an ethanol-benzene mixture (1:4), and a methanol-chloroform mixture (1:1).
The combined extracts were evaporated under reduced pressure at a temperature below 30°C to a final volume of 50 ml. The precipitate was centrifuged and the supernatant liquid retained (solution A). The residue was
stirred with 50 ml of light petroleum (b. p. 40-60°C) for 5 min and the
supernatant liquid (solution B) was decanted. The residue was treated with
50 ml of a chloroform-methanol mixture (4:1), in which almost all the
material was soluble (solution C). The residue was discarded. Solution C
was utilized to characterize lipid hydrolysis products. A 5-ml quantity of
solution C was evaporated to dryness and transferred to a sealed glass tube
with 2.5 ml of 0.5N ethanolic NaOH. Hydrolysis was effected in a boiling
water bath for 3 hr. The contents of the tube were transferred to a separatory funnel with 3 ml of H_2O and acidified with 0.5N HCl; a precipitate
formed, which was removed by shaking with 5 ml of diethyl ether. The
aqueous layer in a tapered centrifuge tube was made alkaline with dilute
NH_4OH and 1 ml of 10% Ba $(CH_3COO)_2$ and 5 ml of ethanol were added. On
standing, a small amount of precipitate was formed, which was centrifuged
and treated with Amberlite IR-120 cation-exchange resin in the H form and
50 μl of H_2O; the mixture was stirred until the precipitate dissolved. This
solution was spotted on Whatman No. 1 chromatography paper treated
previously with aqueous disodium ethylenediamine tetraacetic acid (Na_2EDTA)
to remove heavy metal ions and the chromatogram was developed by the
ascending method of Bandurski and Axelrod [131], using methanol-NH_4OH-
H_2O (6:1:3) as solvent for 16 hr. Phosphates on the chromatograms were
detected by the $FeCl_3$-sulfosalicyclic acid reagent of Wade and Morgan
[132]. Two of the compounds were glycerophosphate and orthophosphate,
and the former co-chromatographed with authentic glycerophosphate.

Choline and ethanolamine were liberated from the crude lipid fraction by
acid hydrolysis, but prior to paper chromatographic identification it was
necessary to remove nonlipid nitrogen from the hydrolyzate by the method
of McKibbin and Taylor [133]. Paper chromatography of choline was
achieved on Whatman No. 1 chromatography paper using the ascending
technique of Huennekens et al. [134] for 16 hr with a mixture of n-propanol-
glacial CH_3COOH-H_2O (8:1:1 by volume) as solvent. The paper was dried
and suspended in a chamber containing I_2 crystals until brown spots appeared
at the sites of nitrogen-containing compounds [135].

Paper chromatographic conditions for identification of ethanolamine were
those described by Block and Bolling [136]. The chromatogram on Whatman
No. 1 chromatography paper was developed by the ascending method for 16
hr in 77% aqueous ethanol. Visualization of ethanolamine was achieved by
spraying the dried paper with 0.4% ninhydrin in acetone [137] and heating
at 80°C for 10 min.

As this brief discussion confirms, the paper chromatographic identification
and determination of phospholipids in soils have been very sparse, likely
because these compounds comprise only a very small percentage of the
organophosphorous content of soils.

2. Nucleic Acids

For many years, speculation has existed regarding the importance of nucleic acids and nucleoproteins in soils. The presence of substances having the properties of nucleic acids has been reported [127, 129, 138, 139]; however, more recent attempts to confirm their presence have proved inconclusive [140]. Best evidence now suggests that they comprise only 2 or 3% of the total organophosphorus of soils [141].

In an investigation of soil organic phosphates, Anderson [142, 143] has identified purine and pyrimidine bases in preparations of three mineral soils from northeastern Scotland. These bases were found in the alkali-soluble, acid-insoluble portion of the soil organic matter fraction but were not present in the free state because they could only be released on $HClO_4$ hydrolysis. The bases were separated from the remainder of the hydrolyzate by adsorption on an Amberlite IR-120 cation-exchange column, from which they were eluted with 5N HCl. They were identified by paper partition chromatography using the solvents butanol-NH_4OH [144] and isopropanol-HCl-H_2O [145]. No other details of the chromatographic procedure were given. However, guanine, adenine, cytosine, and thymine were identified in about equal amounts; uracil was present in much smaller quantities. The solvent described by MacNutt [144] for separation of purine and pyrimidine bases was n-butanol-H_2O-NH_4OH, which was prepared by addition of 1 ml of 15N NH_4OH to n-butanol saturated with H_2O. After the mixture was thoroughly shaken, the clear supernatant liquid was used for paper chromatographic identification. The isopropanol-HCl-H_2O solvent described by Wyatt [145] was prepared as follows: to 65 ml of absolute isopropanol (or 68 ml of 95% isopropanol) at $16°C$, sufficient concentrated HCl was added to give 0.2 g mol of HCl and H_2O was added to bring to a volume of 100 ml. It is believed that these two solvents were the ones used by Anderson [142, 143] to identify guanine, adenine, cytosine, thymine, and uracil in the $HClO_4$ hydrolysis product of soil humic acid.

3. Inositol Phosphates

Of the organophosphate compounds identified in soils, by far the greatest amount (more than 50% of total organic phosphorus) exists in the form of inositol phosphates or phytates [141]. As early as 1940, Dyer et al. [146] showed that a material having properties similar to those of inositol hexa-phosphate could be isolated from NaOH extracts of soils. Isolation of the hexaphosphate usually involved precipitation of humic acids by acidification of NaOH extracts of soil to provide a fulvic acid solution that was oxidized with hypobromite followed by precipitation of inositol hexaphosphates as their FE(III) salts [138, 146-149]. Anderson [150] merely precipitated inositol hexaphosphates in fulvic acid fractions as the barium salt. A paper chromatographic method was used to separate inositol hexaphosphate components from other organic phosphates in the crude "phytate" preparation [150]. The chromatographic conditions were as follows: Whatman No. 1 paper was used which had previously been washed in 2N HCl, H_2O, and 0.05% aqueous solution of EDTA to which a slight excess of aqueous ammonia had been added. The papers were dried and placed in open tanks in very dilute aqueous ammonia in such a way that the ammonia ascended along the paper. This treatment carried any basic impurities to the top of the paper and thin strips containing the impurities were cut from the top of the paper. After it had dried, the paper was ready for use. When a range of solvents was used, it was shown that acid solvents did not differentiate between the various inositol phosphates and mixtures of these compounds moved in a compact spot. An alkaline solvent consisting of 0.5N NH_4OH (7:3) resolved mixtures of inositol phosphates into (1) inositol hexaphosphate ($R_f = 0.06$), (2) an unresolved mixture of inositol tetra- and triphosphate ($R_f = 0.17$), (3) inositol diphosphate ($R_f = 0.32$), and (4) inositol monophosphate ($R_f = 0.48$). The chromatograms were developed in an ascending manner at $23°C$ until the solvent front had moved 33 cm. After they had dried, the spots were visualized by spraying with either $FeCl_3$-sulfosalicylic acid [132] or molybdate-$HClO_4$ [151]. The molybdate-$HClO_4$ method proved to be more sensitive but after partial air drying it was found that the papers had to be heated at $90°C$ for 30 min between glass plates when trace quantities of the phosphates were chromatographed, and this made the papers very brittle and difficult to handle. When inositol hexaphosphate was hydrolyzed enzymatically, the chromatographed hydrolyzate produced orthophosphate spots that masked the presence of inositol monophosphate ($R_f = 0.48$).

Mixtures containing other phosphate esters in addition to inositol phosphates could be chromatographed by a two-dimensional method [150]. The mixture was applied in a single spot and ascending development with an acid solvent (1:1, acetone-30% acetic acid) moved the inositol phosphates in a compact spot but resolved other phosphate esters. A second development of the chromatogram at right angles to the first using the methanol-aqueous ammonia solvent described above separated the inositol phosphates in a straight line, while other phosphates were located at other points on the paper.

For soil samples, 500 g of air-dried soil was leached, in the cold, with 0.2N HCl until no precipitate was obtained when the filtrate was reacted with ammonium oxalate [150]. The soil was washed with water and extracted with 2 liters of 1N NaOH at $60°C$ for 4 hr. After filtering, 1200 ml of the extract was treated with 70 ml of glacial CH_3COOH, followed by HCl, until a pH of 0.5 was achieved. The precipitated humic acid was centrifuged and washed with a small amount of H_2O. The supernatant and washings were made alkaline with NH_4OH and the bulky precipitate was removed by centrifugation, washed with H_2O and ethanol, and dried. An excess of 10% $Ba(CH_3COO)_2$ was added to the supernatant solution and allowed to stand overnight. The precipitate was centrifuged, washed with H_2O and ethanol, and dried under vacuum. This precipitate contained the bulk of the soil inositol phosphates.

Prior to paper chromatographic analysis, the precipitates described above were dissolved in dilute HCl and centrifuged to remove insoluble material. Barium was removed either as $BaSO_4$ or by passing the extract through a cation-exchange resin. One-dimensional chromatography with methanol-aqueous ammonia was used to develop the chromatograms. Total inositol phosphates present in a variety of soils averaged 50 mg of P_2O_5 per 100 g of soil. In a related vein, a method of paper electrophoresis has been described for the determination of inositol phosphates [152].

In 1962, Cosgrove [153-155] used an anion-exchange resin method to separate phytic acid constituents of soil including the commonly found hexaphosphate of myoinositol and the hexaphosphates of DL-inositol and scylloinositol. Extraction of phytin from soil was achieved in a manner similar to that described by Anderson [150]. After phytic material was extracted, a suitable aliquot was used for paper chromatography as follows: for inositol phosphates, a descending chromatographic method was used on Whatman No. 1 paper for 18 hr. The solvent consisted of n-propanol-18N NH_4OH-H_2O (5:4:1) as described by Desjobert and Petek [156]. The chromatographic paper was treated by successive washings with 2N HCl, H_2O, 1% Na_2 EDTA adjusted to pH 9.0 with NH_4OH, and finally H_2O. Phosphate-containing zones on the chromatograms were visualized by the molybdate-$HClO_4$ method of Hanes and Isherwood [151] as modified by Harrap [157]. Similar paper chromatographic methods were used by Greaves et al. [158] in an examination of the phytase activity of Aerobacter aerogenes.

In a paper chromatographic method described for characterization of phytin in peas, Fowler [159] used Whatman No. 541 filter paper with a solvent consisting of 40 ml of isopropanol, 1 g of picric acid, and 20 ml of H_2O. The technique used was that of Longenecker [160] as modified by Fowler [161]. Other solvents that provided resolution of phytin from pyrophosphate and orthophosphate were t-butanol (80 ml)-picric acid (3.5 g)-H_2O (25 ml), t-butanol (80 ml)-trichloracetic acid (4 g)-H_2O (20 ml) and t-butanol (60 ml)-98% formic acid (20 ml)-H_2O (30 ml).

At this time, it is convenient to mention the paper chromatographic methods used for the separation and determination of inositols. Cosgrove [154] used the chromatographic methods developed by Angyal et al. [162], who has described methods for determination of a wide variety of cyclitols, including inositols, quercitols, and inositol methyl ethers. Recommended solvents include acetone-water (4:1), phenol-water (4:1), butanol-acetic acid-water (4:1:1), and ethanol-water-concentrated ammonium hydroxide (20:4:1). The best method of visualizing cyclitols on paper is the $AgNO_3$ method described by Trevelyan et al. [163] as modified by Anet and Reynolds [164]. Fixation by thiosulfate provides black spots, which can be preserved as a permanent record of the chromatograms. The Scherer reagent (10 g HgO + 10 g HNO_3 in 200 ml water) is specific for inositols but lacks sensitivity [165], while the borate-Phenol Red reagent is useful because it can be used with high sensitivity for those cyclitols that form complexes with borate [166].

Thus, this reagent can be used to visualize very small amounts of cis-inositol in the presence of larger amounts of myoinositol. The borate-Phenol Red reagent also is a useful reagent when it is necessary to preserve the cyclitols, since the other reagents described are destructive of the cyclitols.

The method used by Cosgrove [154] for determination of inositols in soil hydrolyzates was essentially that described above [162]. Specific inositol fractions were concentrated to 5-10 ml by vacuum evaporation. The HCl content was adjusted to 5N and the sample was heated in a sealed tube at 110°C for 40 hr. The contents of the tubes were vacuum evaporated at 50°C and the residue dissolved in H_2O, decolorized with charcoal, and deionized with Zeo Carb 225 in the H form and DeAcidite FF in the OH form. After filtration, the solution was concentrated and a suitable aliquot used for paper chromatography. Paper chromatographic methods, although not widely used in soil phosphate chemistry may eventually prove very useful for characterizing organophosphorous components.

D. Amino Acids

Recently, two excellent reviews have been written on soil nitrogenous compounds, covering the amino acid fraction in great detail [167, 168]. The nitrogen content of soils ranges from 0.02% in some subsoils to as much as 2.5% in highly organic soils. Of the total nitrogen, some 95-99% exists in the form of organic compounds. Approximately one-half of the organic nitrogen in soil hydrolyzates has been characterized and the largest single fraction is in the form of amino acids [167, 168]. Methods used for the hydrolysis of proteins in biological systems other than soils have been adapted to the hydrolysis of amino acid-containing soil polymers. Bremner [169] found that the use of 6N HCl with refluxing for 12 hr provided maximal release of amino acids from soils. Thus, the most commonly used procedure of soil hydrolysis involves refluxing with 6N HCl for 12-24 hr. It has been reported that in arid soils 72 hr of refluxing is often necessary to release maximal amounts of amino acids [170].

Determinations of α-amino nitrogen liberated by hydrolysis of surface soils show contents ranging from 20 to 50% of the total nitrogen present [169, 171-188]; the amount of α-amino nitrogen decreases with depth in the profile.

The methods of paper chromatography that have been used in soil investigations are diversified and include ascending, descending, circular, and one- and two-dimensional techniques. It is now clear that soils, irrespective of their source, show only slight qualitative variation from one to another [172, 175, 177, 180, 184, 185, 187-197]; however, there are indications that the quantitative amino acid composition of soils is influenced by climate, cultivation, and fertilization [170, 172, 175, 177, 187, 188, 194, 195]. For the sake of brevity, the types of paper chromatographic analysis for amino acids to which soil hydrolyzates have been subjected are shown in

TABLE 3

Paper Chromatographic Systems Used for Identification of Free and Combined Amino Acids in Soils

Compounds to be determined	Paper support	Type of chromatography	Solvent system		Original reference to solvents	References in soils
			Solvent	Composition		
Water-soluble amino acids and amino acids in soil hydrolyzates	Whatman No. 1	Circular, ascending, descending	n-Butanol-CH₃COOH-H₂O	4:1:1	198-200	196, 201-204
			H₂O-saturated phenol		59	196, 203, 204, 205
			Ethyl acetate-pyridine-H₂O	14:3:3	206	196, 203, 204
			Phenol-n-butanol-CH₃COOH-H₂O	5:5:2:10	207	204
			t-Butanol-ethyl acetate-H₂O-HCOOH	160:160:39:1	208	204
		Ascending	Phenol-H₂O	3:1	200, 209	187
			n-Butanol-CH₃COOH-H₂O		200, 209	187
			s-Collidine-2,4-lutidine-H₂O	1:1:1	200, 209	187
			H₂O-Saturated s-collidine			205
			H₂O-Saturated amyl alcohol			205
			H₂O-Saturated o-cresol			205
	Whatman No. 1 saturated with pH 8.4 buffer	Circular	m-Cresol saturated with pH 8.4 buffer		207	204

Sample	Paper	Development	Solvent system	Ratio	Ref.	Ref.
	Whatman No. 1	Two-dimensional	t-Amyl alcohol–phenol s-Collidine–phenol	Not stated		205
	Whatman No. 4	Two-dimensional	H_2O-Saturated phenol in an atmosphere of NH_3		109,210	
Free amino acids in humus	Whatman No. 4	Two-dimensional	n-Butanol–CH_3COOH–H_2O	12:3:5	109	211
	Whatman No. 4	Two-dimensional	n-Butanol–CH_3COOH–H_2O	Not stated	196	212
			Phenol–concentrated NH_4OH–H_2O		213,214	212
Amino acid composition of fulvic and humic acid isolates	Whatman No. 1	Not stated	n-Butanol–CH_3COOH–H_2O	10:3:7		215
	Whatman No. 4	Two-dimensional	Phenol in an atmosphere of NH_3 and HCN		216	189,190, 217–222
			s-Collidine–2,4-lutidine	1:1		
Amino acids in upland peat	Not stated	Not stated	Phenol–H_2O in an NH_3 atmosphere	4:1	222	223
			n-Butanol–CH_3COOH–H_2O	31:6:13	222	223
Amino acids in a peat bog	Not stated	Not stated	n-Butanol–CH_3COOH–H_2O	4:1:5 (upper layer)		224
			n-Propanol–H_2O	7:3	225	224
			n-Propanol–concentrated NH_4OH	4:1	225	224
			Ethanol–n-butanol–H_2O–C_2H_5COOH	10:10:5:2	225	224
			n-Butanol–HCOOH–H_2O	6:1:2	225	224

Table 3.— Continued

Compounds to be determined	Paper support	Type of chromatography	Solvent system Solvent	Composition	Original reference to solvents	References in soils
Separation of methionine, cystine and leucine in soil hydrolyzates	Whatman No. 4	Not stated	t-Amyl alcohol		226–228	190
Separation of histidine and tyrosine in soil hydrolyzates	Whatman No. 4	Not stated	n–Butanol–CH_3COOH–H_2O	12:3:5	109	190
Amino acids and peptides produced by partial hydrolysis of soil	Not stated	Not stated	n–Heptane–pyridine	7:3	32	229
			n–Heptane–ethylene chloride–75% HCOOH	12:6:1	230	229
			n–Heptane–ethylene chloride–75% HCOOH	6:12:1	230	229
		Two–dimensional	n–Heptane–pyridine	7:3	32	
			n–Heptane–ethylene chloride–75% HCOOH	6:12:1	230	229
Free amino acids in Hibiscus esculentus L.	Whatman No. 1	Two–dimensional	Phenol n–Butanol–CH_3COOH–H_2O	4:1:5 (upper layer)		231–233
Tryptophane as a growth promoting substance in plants	Not stated	Descending	Isopropanol–H_2O	4:1	234	234

Table 3. Many types of chromogenic reagents have been used for locating amino acid spots on the chromatograms (Table 4). However, ninhydrin has been used almost universally because it reacts with all of the amino acids isolated from soil samples. The other visualizing agents have been used to provide definitive characterization for a particular amino acid.

E. Pesticides

The majority of paper chromatographic techniques developed for pesticide residue analysis were introduced between 1950 and 1960, and several reviews dealing with techniques of paper chromatographic analysis of pesticides are available [106, 247, 248]. Mitchell [249] provides an excellent discussion of ascending paper chromatographic techniques developed for the analysis of pesticides in foods.

TABLE 4

Reagents Used for Location of Amino Acid Spots
on Paper Chromatograms

Reagent	Reference
Ninhydrin	234, 235
Ninhydrin-$Cu(NO_3)_2$	236
Isatin	187, 198-200, 237
Chloramine T in Ehrlich's reagent	238
Nessler's reagent, HIO_4	239
Phenol, NaOCl	206
p-Anisidine, amylnitrite, ammonia	240
HIO_4, γ-collidine, $CH_3COO\,NH_4$, CH_3COOH, acetylacetone	241
HIO_4, γ-collidine, sodium nitroprusside piperidine	241
Isatin in Ehrlich's reagent	242, 243
Potassium iodoplatinate	190, 199, 200, 243, 244
$CuCO_3$	245
Alloxan	199, 200, 246

Most of the paper chromatographic procedures used in pesticide residue analysis were developed for use on food and biological samples. However, if the necessary cleanup procedures can be devised, the methods should be equally applicable to plant and soil extracts. For this reason some of these procedures are described here as well as those techniques developed specifically for analyzing soil extracts.

Paper chromatography is most often used to separate residues in extracts that have been prepurified [250]. However, occasionally it is used as a cleanup procedure for biological or plant material. Menn et al. [251] reported a paper chromatographic cleanup procedure used on biological samples. An acetone extract of the sample is transferred to the center of a Whatman No. 4 chromatographic filter paper, 2.5 × 1 in., and tapered to a point at one end. One microliter of a 0.5% solution of N,N'-dimethyl-p-1-naphthylazoaniline (NDA) was applied to the spot. The NDA serves as a visual indicator of insecticide movement. An ascending or descending technique can be used in the washing procedure. In the former, the wide end of the paper is dipped into a beaker containing 5-10 ml of acetonitrile and allowed to develop. Most of the lipoid material remains at the original spot while the insecticide and reference dye (NDA) move with the solvent. The solvent is transferred from the tip of the paper to another chromatographic paper, forming a spot on the latter. The spot is dried and more solvent is transferred; this process is repeated until the dye and insecticide transfer is complete. In the downward washing technique, the spotted paper is hung in a cup containing acetonitrile. The solvent is collected in a receiving cup for 30 min, concentrated, and transferred to another chromatographic paper for development.

A paper chromatographic-acetonitrile cleanup was also used for extracts of fruits and vegetables by Major and Barry [252]. The crude extract was applied to Whatman No. 1 chromatographic paper (8 × 8 in.) 1 in. from the lower edge. The paper was developed using acetonitrile as the mobile solvent. When the solvent reached 1 in. from the top of the paper, the paper was removed and allowed to dry and the process was repeated. At this point the insecticides are close to the top of the paper, while most of the fatty and waxy materials remain near the origin. The paper is rotated 180° so that the isolated insecticides are at the bottom of the paper. Standards are spotted on the line and the chromatograms are developed by the procedure of Mills [253].

Paper chromatographic cleanup of soil extracts has not been used most likely because of its tediousness and low carrying capacity [254].

1. Organochlorine Pesticides

Applications of paper chromatographic techniques have found wide use for the separation, detection, and identification of chlorinated pesticides. Little progress was made with paper chromatographic systems using mobile solvents alone until it was found that an immobile solvent must be applied to the chromatographic paper to achieve satisfactory pesticide migration [249] at which time numerous methods of analysis were introduced.

A large number of solvent systems have been used for separating chlorinated pesticides. Most pesticide mixtures can be separated using a combination of aqueous and nonaqueous solvent systems. Aqueous solvent systems most often use refined soybean oil mixed with various percentages of diethyl ether as the immobile phase. However, any refined vegetable oil, animal oil, or mineral oil should also be adequate. For the mobile solvent, water mixed with acetic acid, acetone, acetonitrile, 1,4-dioxane, ethanol, methanol, 2-methoxyethanol, or pyridine may be used [255]. Nonaqueous solvent systems used include diethyl ether (or mixtures of the ether with ethanol, or acetone) in combination with acetic anhydride, dimethyl acetamide, dimethylcyanamide, dimethylformamide, dimethylfuran, dimethylhydrogen phosphite, dimethylphthalate, 1,4-dioxane, fenchone, formamide, glycerol, glycols, 2,2-iminodiethanol, isopropylbenzene, isopropylformate, 1-methoxy-2-propanol, 1-methylnaphthalene, 2-phenoxyethanol, N-isopropylbenzene-sulfonamide, or trimethylphosphate. Mobile solvents used most frequently are the volatile paraffins, namely, petroleum ethers, lead-free gasoline, hexanes, heptanes, octanes, kerosenes, and the light mineral oils [255].

In a series of reports, Mitchell and his co-workers [255-268] have described several paper chromatographic procedures for the separation of mixtures of pure and technical grade organochlorine pesticides, which with adequate cleanup techniques may be applicable to the analysis of plant and soil extracts. The R_f values on paper chromatograms of 113 pesticide chemicals, mainly of the organochlorine type, using an aqueous and a nonaqueous solvent system for each pesticide have been tabulated [255]. The preferred solvent systems were soybean oil as the immobile phase and 2-methoxyethanol-H_2O (75 : 25) as the mobile phase for the aqueous solvent; 2-phenoxyethanol was used as the immobile phase and 2,2,4-trimethylpentane as the mobile phase for the nonaqueous solvent. For chromatography, Whatman No. 1 paper, washed in water and air dried, was used. The immobile solvent is applied by dipping the paper in a tank containing the solvent. After it has dried, the paper is developed in a tank containing the mobile solvent for approximately 4.25 hr. For the nonaqueous system, paper is dipped in the immobile phase and transferred immediately to the chromatographic tank for development. Development time is approximately 1.25 hr. After the solvent front ascended close to the top of the paper, the paper was removed from the tank and the solvent front marked; the papers were hung in a hood to dry, usually overnight.

The chromogenic agent used for locating the spots consisted of a solution of $AgNO_3$ and 2-phenoxyethanol prepared by dissolving 1.7 g $AgNO_3$ in 5 ml of H_2O, adding 2-phenoxyethanol (10 ml for nonaqueous systems and 20 ml for aqueous systems), and diluting to 200 ml with acetone. A few drops of 30% H_2O_2 were added if the solution darkened. The chromatograms were sprayed or dipped into the above solution and immediately exposed to strong ultraviolet light, first from the back of the paper and then from the front.

Chlorinated pesticides appear as darkened spots. The chromatograms can be preserved if excess silver is washed from the paper with water [255] or hyposulfite and water [248].

McKinley and Mahon [269] have applied the above techniques to residue analysis of plant extracts. Three solvent systems were used: an immobile phase of 14% 2-phenoxyethanol in diethyl ether with a mobile phase of pure grade 2,2,4-trimethylpentane, an immobile phase of 4% liquid paraffin in diethyl ether with a mobile phase of 40% aqueous pyridine, and an immobile phase of 2% liquid paraffin in diethyl ether with a mobile phase of 70% aqueous acetone.

McKinley and Mahon [269] extracted 900-1100 g of plant material with benzene by the procedure of Gunther and Blinn [270]. Prior to chromatography, the dried benzene extract was partitioned between acetonitrile and n-hexane. Twenty milliliters of n-hexane saturated with acetonitrile was added to the dry benzene extract, stirred, allowed to stand 10-15 min, restirred, and transferred to a 125-ml separatory funnel. The n-hexane was extracted with 4 ×10 ml portions of acetonitrile. The acetonitrile extracts were combined, reextracted with n-hexane, and reduced in volume to 2-3 ml at 60°C and evaporated to dryness at room temperature. The residue was dissolved in 1 ml of ethyl acetate and transferred to a 5-ml tube. The ethyl acetate was removed with a stream of N_2 gas. The tube was sealed and at the time of analysis 0.1 ml of ethyl acetate was added and a 2-μl portion was applied to the chromatography paper.

Whatman No. 1 paper was used for the residue analysis. The paper was washed by dipping briefly in 2% aqueous $AgNO_3$ and rinsed with running water for 20 min. Silver chloride, which may be deposited during the washing procedure, was removed by dipping the paper into 28% NH_4OH. If Cl-free water is available, the NH_4OH treatment is unnecessary. The immobile phase was applied to the paper by pouring it over the paper. The immobile phase that runs off the paper should not be reused because of the rapid evaporation of the ether.

Plant extracts remaining from the acetonitrile-n-hexane cleanup were chromatographed using a solvent system of 14% 2-phenoxyethanol in diethyl ether as the immobile phase and 2,2,4-trimethylpentane as the mobile phase. The developed chromatogram was sprayed with the $AgNO_3$ reagent and exposed to strong ultraviolet light until each of the standards was revealed as a dark spot. Samples containing pesticides were rechromatographed using the second and third solvent systems described earlier [269]. R_x values (the distance moved by the unknown compound divided by the distance moved by a reference standard) rather than R_f values were used to designate movement of the pesticides.

Lindane (γ-BHC) was identified in plant and soil extracts by San Antonio [271] using a modification of the method described by Mitchell [261, 266]. No cleanup procedure was used, and in the presence of large quantities of plant oils and pigments only the nonaqueous system could be used. The solvent pair of n-hexane and N,N-dimethylformamide (DMF) was used with

Whatman No. 17 paper as support. The heavier paper takes up a considerable amount of the DMF immobile phase that is necessary to provide resolution of lindane from plant pigments. The chromatographic procedure was similar to that described above [265]. Five to 10 μg of lindane were necessary to develop visible spots with the AgNO$_3$ reagent; somewhat greater sensitivity was obtained using Whatman No. 1 paper. However, as much as 3 mg of lindane can be chromatographed without overloading the paper.

This method was further modified by San Antonio [271, 272] to increase its value as a quantitative procedure for determining chlorinated insecticides in soils. The same solvent system was used as described above [272]; however, the amount of DMF in the immobile-phase solution was varied to obtain R$_f$ values of approximately 0.5. For slow-moving substances, such as lindane, 25% DMF was used, while 40% was used for heptachlor, p,p$'$-DDT, and dieldrin and 60% for fast-moving substances, such as aldrin. Another modification was to include the AgNO$_3$ reagent in the immobile phase. This is a time-saving step and increases sensitivity and precision by providing a more consistent and uniform distribution of the indicator on the paper. Details of the procedure are: after addition of the freshly prepared immobile phase, the paper is dried and clamped in a spotting press. This consists essentially of a chamber designed to reduce the volatilization of DMF during the spotting process. The chamber is comprised of three metal sheets: a flat bottom sheet, a 0.075-in. thick middle sheet cut out to fit the paper, and a top sheet containing 21 spotting holes, which are covered with aluminum squares when not in use. Samples are spotted 4 in. above the lower edge of the paper because the solvent front movement is sufficiently slow to allow development of circular spots. The entire sample solution (3 μl) is spotted at one time and should contain 5-25 μg of insecticide. Soil extracts were applied without prior cleanup.

Immediately after it is spotted, the paper is formed into a cylinder and transferred to a chromatographic tank at a temperature of 14-17°C. The solvent front reaches the top of the paper in 30 min, the paper is removed from the solvent, and the spots are allowed to diffuse for an additional 4.5 hr to obtain spots of satisfactory area and color density. The paper is removed from the chamber and allowed to dry overnight. Prior to irradiation the paper is exposed to an H$_2$O vapor-saturated atmosphere at room temperature for 20 min to insure adequate darkening of the substances. Irradiation consists of exposure to ultraviolet light for 5 min to the back and 35 min to the front side of the paper. Quantitative analysis was obtained by weighing (nearest milligram) cutout spots and plotting weights against the logs of known quantities of standards. This relationship was linear and was used to determine the unknowns.

Presently, most of the quantitative and qualitative determinations of organochlorine pesticide residues are made by gas chromatographic methods. Paper chromatography is useful for screening tests, confirmation of gas chromatographic results, and determination of degradation products not determinable gas chromatographically.

2. Organophosphorus Insecticides

Paper chromatographic methods used for the analysis of organophosphorus residues are similar to those described for organochlorine compounds. However, organophosphorus residue analysis involves the determination of a variety of metabolites and oxidation products.

The method described by McKinley and Mahon [269] was used to detect a number of organophosphorus insecticides employing bromosuccinimide and fluorescein reagents as described by Cook [273]. Bates [274] described a general method involving extraction, cleanup, and analysis by paper chromatographic methods. Acetone was used as the extractant for plant materials. Fats and waxes were removed from the extracted plant material by freezing from acetone at -70°C [275]. Further cleanup was accomplished by column chromatography. Nonpolar compounds were treated using a 1.5-cm column containing a slurry of 0.5 g of Nuchar carbon, 2.0 g of MgO, and 1.5 g of powdered cellulose in a benzene-$CHCl_3$ (1:1) mixture. Insecticides were eluted with 150 ml of the above solvent, which was evaporated to near dryness and the residue was made up to 10 ml with $CHCl_3$ for chromatographic analysis. Polar compounds were eluted in a similar column containing a slurry of 3 g of MgO in $CHCl_3$. Two fractions of eluate were collected, the first consisting of 150 ml of $CHCl_3$ and the second of 100 ml of methanol. Between collection of the two fractions, acetone (50 ml) was used to wash the column. The acetone wash was discarded and both fractions were evaporated to near dryness and the residue dissolved in $CHCl_3$ for analysis.

Two-dimensional paper chromatography was used for separating the nonpolar compounds. A 5-ml aliquot of the cleaned extract, evaporated to a small volume, was spotted on a washed Whatman No. 20 paper (10 × 10 in.). The most satisfactory working range for pure compounds is 5-10 μg of material; however, to detect minor components in other extracts larger amounts generally are needed. Whatman No. 20 paper was used because it is slower running than Whatman Nos. 1 and 3MM and gives more compact spots and better resolution. However, either of the other two papers may be used successfully. Two marker spots are used in such a way that one is developed in each direction without interfering with the analysis of the sample. The paper is dipped into a 30% solution of DMF in acetone (immobile phase) and developed in one direction (15 cm) with hexane. After development, the paper is dried (30 min) and dipped in a solution of 5% liquid paraffin in diethyl ether and the second dimension (15 cm) is developed in DMF-H_2O (1:1) and the papers are dried at 50°C.

Polar compounds are resolved using 20% formamide in acetone as the immobile phase. The paper is developed twice with hexane and then turned through 90° and redeveloped in a benzene-$CHCl_3$ (9:1) mixture.

The insecticides or metabolites were detected by ultraviolet light causing quenched or darkened areas. Most of the compounds used could be located at the 1-μg level. Several spray reagents were used to detect the insecticides at the 1-μg level. Organophosphorus compounds containing sulfur were detected by the yellow or red coloration formed with 2,6-dibromo-N-chloro-p-quinonimine or the blue-purple color formed with $AgNO_3$-Bromophenol Blue. Most compounds were also detected by exposing the paper to Br_2 fumes and spraying with dibromo-4-methylumbelliferone. Quantitative analysis was achieved by a wet digestion technique [276] used to convert organophosphorus compounds to orthophosphates, which were determined spectrophotometrically as the Molybdenum Blue complex [277].

A similar chromatographic procedure was used by Mitchell [278] for the separation of eleven organophosphorus insecticides. A number of solvent systems were investigated. Best resolution for an aqueous system was obtained using 10% mineral oil as the immobile phase and 50% aqueous DMF as the mobile phase. The most satisfactory nonaqueous system was 20% DMF and 2,2,4-trimethylpentane. Both one- and two-dimensional procedures were satisfactory.

Coffin and McKinley [279] were able to separate parathion, methyl parathion, EPN (O-ethyl-O-p-nitrophenyl phenylphosphonothioate) and their oxons using two aqueous systems, namely 5% silicone 550 in diethyl ether as the immobile phase and acetone-absolute ethanol-H_2O (1:1:2) as the mobile phase, and 5% silicone 550 in diethyl ether and acetone-H_2O (1:9). The first system separated the thionophosphate esters and the second separated the phosphate esters. Insecticides were detected as yellow spots after bromination (vapor-phase treatment with 10% Br_2 in $CHCl_3$ for 1 min) followed by spraying with 5% KOH. Two-dimensional methods can be used with the above systems; however, two one-dimensional separations were more convenient and gave better separations.

Solvent systems similar to those described by Coffin and McKinley [279] were useful for detecting most organophosphorus insecticides and their metabolites [280]. Acetylated paper (40-50%) was used with acetone-H_2O (1:9). The detection technique was the enzymatic method developed by Cook [281] as modified by Getz and Friedman [282] for determining enzyme-inhibiting pesticides. An enzyme-indicator solution consisting of 20 ml of pooled human blood serum, 60 ml of H_2O, 2 ml of 0.1N NaOH, and 3.25 ml of Bromthymol Blue indicator (0.15 g of Bromthymol Blue in 25 ml of 0.1N NaOH) was sprayed on the dry chromatograms and incubated for 30 min at room temperature in a tank satuarated with H_2O vapor. The paper was sprayed with a substrate consisting of 2 g acetylcholine bromide in 100 ml of H_2O. Within 10 min, the background of the paper turned yellow, while the spots containing the enzyme-inhibiting pesticide remained blue. Certain immobile solvent systems were found to contain enzyme inhibitors that interfered with pesticide detection by turning the paper blue; also acids in the immobile phase were capable of turning the paper yellow. In these cases, a method of transferring the chromatographic spots from one paper to another was developed. The "clean paper" sprayed with the enzyme-indicator solution was placed on the sample chromatogram between stainless steel plates and incubated for 15 min. The paper containing the enzyme indicator after transfer of the chromatographic spots was sprayed with the substrate solution.

Some insecticides require activation in order to inhibit the enzyme. Activation can be accomplished by a bromination method as follows: 0.1 ml of Br_2 is placed in a tank and after the vapors saturate the tank the paper is placed in the container for 30 sec. It is then placed in a forced air cabinet at 50-60°C for 15 min.

MacRae and McKinley [283] developed a detection method for organo-phosphorus insecticides based on the method described by Wade and Morgan [284] for phosphate esters. Insecticides were separated using Whatman No. 1 paper with 8% mineral oil as the immobile phase and methanol-H_2O-0.5% NH_4OH (19:1:1) as the mobile phase or 100% acetylated paper with 2% mineral oil as the immobile phase and 70% aqueous acetone as the mobile phase. Developed chromatograms were sprayed with 0.1% $FeCl_3 \cdot 6H_2O$ in 80% ethyl alcohol followed by 1% salicylsulfonic acid in 80% ethyl alcohol. Detection depends on the fixation of Fe(III) ions by phosphate esters. Free Fe(III) ions react with salicylsulfonic acid to produce a background color. With Whatman No. 1 paper, the chromatograms were exposed to Br_2 vapor to oxidize the thiophosphate ester. Alternatively, bromination could be omitted for detection of phosphate esters. Adequate coloration occurs only when the residual moisture in the chromatogram is at a pH of 1.5-2.5.

On acetylated paper, organophosphate insecticides produced yellow spots when exposed to Br_2 vapors. When sprayed with $FeCl_3 \cdot 6H_2O$ and salicyl-sulfonic acid both thiophosphate and phosphate esters appeared as white spots. On drying, the thiophosphates appeared as intense yellow spots and the phosphates became invisible. Further exposure to air (about 12 hr) results in the phosphates appearing as white spots on a mauve background and the thiophosphates remaining as yellow spots. Thus, thiophosphate and phosphate esters can be detected on the same chromatogram.

Detection of 5 $\mu g/cm^2$ of most of the insecticides was possible on What-man No. 1 paper. Increased sensitivity was achieved on acetylated paper. Further advantages of the acetylated paper are the ease of spotting and the development of more compact spots.

Watts [285] developed a procedure capable of detecting 0.5 μg of thio- and nonthiophosphate pesticides using 4-(p-nitrobenzyl)pyridine (NBP) as a chromogenic reagent. Samples were spotted on Whatman No. 1 paper and the chromatogram was developed with 20% DMF in diethyl ether as the immobile phase and isooctane as the mobile phase. After development, the paper was air dried for 10 min and sprayed with 2% NBP in acetone followed by 5% diammonium citrate in H_2O; the latter reagent was used for diazinon. The paper is placed in a 110°C oven for 10 min, sprayed with 10% tetra-ethylenepentamine in acetone, and returned to the oven for 1 min. On cooling, the pesticide spots became blue with the exception of diazinon, which turned red. Color intensities of the spots were proportional to the quantity of pesti-cide applied, thereby allowing quantitative estimation of the sample.

As with organochlorine pesticides, gas chromatography is being used widely for the analysis of organophosphorus insecticides and paper chromatography is limited to the role of screening tests, confirmation work, and detection of degradation products. With regard to the latter, paper chromatographic determinations can be quite useful considering the large number of degradation products encountered with organophosphorus insecticides.

3. Herbicides

Some methods have been described for the paper chromatographic determination of herbicides and fungicides, but far fewer than for insecticides. Several of the chlorinated herbicides have been determined by the method described by Mitchell [286]; however, several chlorinated herbicides could not be resolved by this method. The commonly used herbicides 2,4-D (2,4-dichlorophenoxyacetic acid) and 2,4,5-T (2,4,5-trichlorophenoxyacetic acid) were resolved on Whatman No. 1 chromatography paper using a mobile solvent consisting of t-butanol-concentrated NH_4OH-H_2O-2-butanone (9:1:20:70) [286]. Location of spots on the chromatograms was achieved with a chromogenic agent of $AgNO_3$ and 2-phenoxyethanol in acetone followed by ultraviolet light.

Yip [287] described a procedure for detecting 2,4-D, MCP (2-methyl-4-chlorophenoxyacetic acid), 2,4-DB [4-(2,4-dichlorophenoxy)butyric acid], 2,4,5-T, and silvex or 2,4,5-TP [2-(2,4,5-trichlorophenoxy)propionic acid] and several esters of 2,4-D extracted from wheat. The acids were methylated prior to chromatography by dissolving them in 2 ml of diazomethane and shaking occasionally for 10 min. The methylated acids were chromatographed on Whatman No. 1 paper using two solvent systems, namely 35% DMF in diethyl ether as the immobile phase and 2,2,4-trimethylpentane as the mobile phase; and 10% mineral oil in diethyl ether as the immobile phase and 50% acetonitrile in H_2O as the mobile phase. The first system was preferred. Esters of 2,4-D were separated using 10% mineral oil in diethyl ether as the immobile phase and 65% acetonitrile in H_2O as the mobile phase. This system did not resolve the butoxyethanol ester from the isopropyl ester or the butyl ester from the propylene glycol butyl ether. The first pair was resolved using 10% phenoxyethanol in diethyl ether (immobile phase) and 2,2,4-trimethylpentane (mobile phase), while the second pair was resolved by 50% DMF in diethyl ether (immobile phase) and 2,2,4-trimethylpentane (mobile phase). Silver nitrate and ultraviolet light were used for locating and detecting the spots.

Mitchell [288] developed a chromatographic procedure that would separate several substituted urea herbicides and their trichloroacetate (TCA) salts. These included fenuron (3-phenyl-1,1-dimethylurea), fenuron TCA, monuron [3-(4-chlorophenyl)-1,1-dimethylurea], monuron TCA, diuron [3-(3,4-dichlorophenyl)-1,1-dimethylurea], diuron TCA, 3-(2-chlorophenyl)-

1-methylurea, linuron [3-(3,4-dichlorophenyl)-1-methoxy-1-methylurea],
and neburon [3-(3,4-dichlorophenyl)-1-n-butyl-1-methylurea]. Trichloro-
acetic acid was also detected in these systems. An aqueous solvent system
consisting of 10% USP grade heavy mineral oil in diethyl ether as the
immobile phase and 25% tetrahydrofuran in H_2O as the mobile phase and a
nonaqueous system of 25% glacial CH_3COOH in 2,2,4-trimethylpentane as
the mobile phase (no immobile phase was required) were used with Whatman
No. 1 paper as support. The chromogenic agent consisted of 0.85 g of
$AgNO_3$ and 10 ml of 2-phenoxyethanol diluted to 200 ml with acetone. All
compounds except TCA were detected under ultraviolet light and all but the
fenurons were detected by the chromogenic reagent in combination with
ultraviolet light.

Pentachlorophenol, used as a herbicide or fungicide, was determined
paper chromatographically on Whatman No. 1 paper using 10% mineral oil
in diethyl ether as the immobile phase and a 70:30 mixture of acetone-H_2O
as the mobile phase [289]. Silver nitrate in 2-phenoxyethanol was used for
detection. Pentachlorophenol and pentachlorophenol ether were resolved
and identified by this procedure. Lower limits of detection were 0.4 μg and
0.2 μg, respectively.

Mitchell [290] was able to resolve ATA (3-amino-1,2,4-triazole) from
impurities in the technical grade product using a mobile phase of 20 ml of
H_2O diluted to 100 ml with 1,4-dioxane. Spraying with ammoniacal $AgNO_3$
followed by heating at 130°C for 10-15 min provided a sensitivity of 0.5 μg
of ATA.

Simazine [2-chloro-4,6-bis(ethylamino)-s-triazine] (^{14}C labeled) and
hydroxysimazine were determined in soils by Harris [291]. Soils were first
extracted with 100 ml methanol for 2 hr in a soxhlet apparatus and shaken
with 25 ml of 0.5N NaOH for 2 hr. The methanol and neutralized NaOH
extracts were chromatographed on Whatman No. 1 paper using two solvent
systems, 30% CH_3COOH in H_2O on kerosene-soaked paper and isoamyl
alcohol saturated with 0.1N HCl on Whatman No. 1 paper. Radioactivity on
the paper strips was detected with a 4π strip scanner.

Two additional 2-chloro-s-triazines (atrazine and propazine) and their
hydroxy analogs extracted from soils [291] were chromatographed in a
similar system with the exception that the isoamyl alcohol was saturated
with 0.3N HCl [292]. Sodium hydroxide extracts described above required
a cleanup step prior to chromatography. The extracts were acidified to pH
4.0 and the precipitated material was removed by centrifugation. After the
pH of the supernatant solution was adjusted to 7.0, the precipitate was again
removed by centrifugation. The triazines were removed from the supernatant solution by adsorption on activated charcoal; the charcoal was filtered
and the triazines were eluted from the charcoal with hot methanol and
concentrated for analysis.

[14]C-Hydroxyatrazine and degradation products extracted from soils and
sediments with a series of solvents (HCl, methyl cyanide-H_2O, benzene-
methanol) were analyzed paper chromatographically using Whatman No. 1

paper with a solvent consisting of $180:1:20$ isopropanol-glacial $CH_3 COOH$-20% aqueous $CH_3 COONH_4$ [293]. Developed chromatograms were scanned with a radiochromatogram scanner for ^{14}C detection.

Paper chromatographic analysis of herbicides apparently has not reached the same degree of sophistication that has been achieved with insecticides. However, there appears to be considerable potential for use of paper chromatographic methods since other chromatographic techniques, particularly gas chromatography, have not been particularly adaptable to herbicide analysis.

As can be seen from the above discussion, a wide variety of paper chromatographic methods are available for the separation and identification of pesticides. When applied to the determination of pesticides in plant and soil extracts, considerable cleanup is required prior to analysis. In the relatively few cases where paper chromatographic techniques have been applied to plant and soil extracts, adequate cleanup procedures have been devised and it is likely that other procedures not developed specifically for plant and soil extracts can be modified for such use. However, other chromatographic methods have largely replaced paper chromatographic methods. Paper chromatography is used most frequently where other chromatographic methods are unsuitable or unavailable and is of particular value where several unknown degradation products are to be identified.

F. Inorganic Components

Paper chromatographic techniques have been used to determine a variety of inorganic components in soil and plant extracts. Methods have been developed for many metal cations and for a number of anions, particularly for phosphates.

A large number of solvent mixtures has been used to separate inorganic components. Ketone solvent systems often are used for separating mixtures of metals in soil and plant extracts [294-296]. Hunt et al. [294] investigated two solvent systems, isobutylmethylketone containing HCl and ethylmethylketone mixtures with HCl and H_2O, for the separation of Cu(II), Co(II), and Ni(II) from Fe(III) and other elements found in soils. The most satisfactory system was ethylmethylketone-HCl-H_2O ($15:3:2$). For chromatographic analysis, a 0.01-ml aliquot of a bisulfate fusion of a soil sample dissolved in dilute ($1:1$) HCl containing 5% HNO_3 was spotted on Whatman No. 1 filter paper. To obtain consistent results, humidity, temperature, and drying time were controlled during drying of the sample spot. These conditions were obtained by placing a beaker containing the paper in a boiling H_2O bath. After 3 min, the paper was placed in the solvent chamber and allowed to develop from 30-40 min. After development, the acid remaining on the paper was neutralized by a 2-min exposure to NH_3 vapor and the metals were detected by spraying the paper (front and back) with 0.1% solution of rubeanic acid in ethanol-H_2O ($3:2$). Nickel(II) appeared as a blue-purple band

(R_f = 0.10), Co(II) as an orange-yellow band (R_f = 0.45), Cu(II) as an olive green band (R_f = 0.65), and Fe(III) as a brown hydroxide located at the solvent front. Quantitative data were obtained by comparing color intensity with known standards. Approximately 20 ppm (lower detection limit of 0.05 μg) of each metal could be detected in the soil.

Duffield [295] used this method for the determination of Cu(II), Co(II) and Ni(II) in plant material. Zinc(II) also was separated in this solvent system and was visualized by spraying with a dilute solution of dithizone in $CHCl_3$. The color produced by Zn(II) was not of uniform intensity and was not useful for quantitative analysis. A mixture of ten metal cations found commonly in soils was separated on Whatman No. 1 paper using a descending technique with ethylmethylketone-amyl alcohol-HCl (3:5:2) [296]. The metals were located using 1% alizarin in ethanol followed by exposure to NH_3. The following metals were separated: Al(III) (R_f value = 0.05), Mg(II) (0.08), Mn(II) (0.13), Cu(II) (0.50), Mo(VI) (0.60), B(III) (0.69), Ca(II) (0.78), Zn(II) (0.89), and Fe(III) (0.96). The same cations were separated by a solvent mixture of n-butanol-ethanol-CH_3COOH-H_2O (4:4:1:1) using a circular technique on Whatman No. 1 paper [296]. However, Fe(III) and Zn(II), and Mn(II) and Al(III) were not separated distinctly and Ca(II) could not be detected. The R_f values for the elements were Mg(II), 0.22; Mo(VI), 0.28; Cu(II), 0.48; Mn(II), 0.52; Al(III), 0.58; Zn(II), 0.68; Fe(III), 0.70; and B(III), 0.78. Approximately 0.5 mg of each metal could be detected by the descending and circular techniques.

Similarly, a solvent mixture of n-butanol-HCl-H_2O (100:23:17) was used to separate Fe(III), Mn(II), Zn(II), and Cu(II) in plant material [297]. Materials that interfered in the chromatographic separation, namely K(I), Ca(II), Mg(II), and phosphate, were removed by ion-exchange chromatography. Pyrophosphates were hydrolyzed by boiling plant tissue in 1.0N HCl for 10 min. After it dried, the residue was dissolved in a mixture of acetone-HCl-H_2O (6:4:1) and applied to a column (13 cm long) containing 100 to 200 mesh Dowex 1-X8 resin washed with the solvent mixture. The resin was washed with three volumes of the same solvent mixture to remove K(I), Ca(II), and Mg(II) and the trace elements were eluted with four volumes of H_2O.

The eluate from the exchange column was evaporated to dryness in a tared tube and redissolved in dilute HCl (1:1). Reduced iron was oxidized with one drop of H_2O_2. For quantitative analysis, the specimen tube was weighed with and without the sample. Prior to sample application, Whatman No. 1 paper was washed by soaking in 2N HCl for 30 min, rinsed with H_2O, and dried. After the spotted paper was allowed to equilibrate with the solvent for 1 hr, it was developed (30 cm) by a descending technique.

The separated compounds in a standard mixture were located either by using known R_f values or by the use of 0.5% 2-nitroso-1-naphthol-4-sulfonic acid in 50% ethanol containing 4% anhydrous CH_3COONa as a color-forming reagent. Manganese did not form a colored complex with this reagent but when Co(II) was added to the standard mixture [Co(II) and Mn(II) move together], Mn(II) was located. The developed standard mixture was cut from the paper, neutralized with NH_3 vapor, and sprayed. Colors and R_f values were Mn(II) and Co(II), red, $R_f = 0.16$; Cu(II), brown, $R_f = 0.29$; Fe(III), green, $R_f = 0.84$; and Zn(II), orange, $R_f = 0.96$. Sample compounds developed simultaneously with the standard mixture were located by comparison with a standard mixture.

Alternatively, the metals were located directly and more precisely on the sample chromatogram without affecting the later quantitative analysis. Iron(III) develops a faint yellow color on drying. Zinc(II) is located within 3 cm of the solvent front. The areas occupied by these metals are cut from the paper and analyzed. Copper(II) and manganese(II) are located as brown areas by spraying with formaldoxime reagent (20 g of paraformaldehyde and 55 g of hydroxylammonium sulfate in 100 ml of H_2O) and exposing to NH_3 vapor.

Colorimetric methods were used to determine the metals in the 5-40 μg range after elution from the paper. Lesser amounts were determined directly on the paper by measuring the reflectance of the colored complexes.

Acetone-ethyl acetate-6N HCl (9 : 9 : 2) was used by Coulson et al. [298] for the separation of Cu(II), Co(II), Ni(II), and Zn(II) in plant materials and soils. The chromatogram was developed by an ascending procedure [6] for 6 hr, air dried, and exposed to NH_3 vapor to neutralize the HCl. A 0.1% ethanolic solution of rubeanic acid was used for detecting Ni(II), Co(II), and Cu(II) and an aqueous alkaline solution of 1-nitroso-2-naphthol (0.5 mg/ml) was used for Co(II) detection. Copper(II), Co(II), and Ni(II) were determined quantitatively by reflectance densitometry [299] and Zn(II) was determined colorimetrically at 578 nm.

Iron(III) was determined in soil extracts by circular chromatography on Whatman No. 1 paper using CH_3COOH-methanol (3 : 1) as solvent [300]. Iron(III) was visualized as a purple compound formed with an alcoholic solution of sodium alizarin sulfonate after exposure to NH_3 vapors. The Fe(III) was cut from the paper, extracted with 0.1N HCl, and determined colorimetrically. Acetone-8N HCl (9 : 1) also has been used as a solvent for separating Fe(III) from soil extracts [301].

A variety of solvent systems containing HNO_3 were used to separate uranium from othermetals [302]. Hunt et al. [294] has used a solvent system consisting of 10% HNO_3 and 5% H_2O in ethyl acetate for paper chromatographic analysis. Similar solvent systems were used by Szekely [303] and Thompson and Lakin [304]. Purushottam [305] used a solvent mixture consisting of 0.2 g of citric acid, 18 ml of acetone, 1 ml of CH_3COOH, and 1 ml of 1 : 3 HNO_3. Location of uranium on paper strips was achieved by $K_4Fe(CN)_6$ [294, 303-305]. Approximately 0.1 μg of uranium could be detected as a brown spot [294].

Lead(II) was separated from other metals in dilute HNO_3 soil extracts using Whatman No. 1 paper, a solvent mixture of dry methanol containing 5% HCl, and dithizone as a chromogenic agent. A 0.002% dithizone solution was used to detect small amounts of Pb(II) with a detection limit of 0.1 μg.

When several micrograms of Pb(II) were present it was necessary to use a
0.02% dithizone solution. Quantitative determinations were made by comparing band widths to known standards.

A circular paper chromatographic method also has been used for
determination of Pb(II) [306]. Heavy metals were concentrated using 20%
dithizone in $CHCl_3$ at pH 10.5. After destruction of the dithizonates, the
metals were separated using a solvent mixture of diethyl ether-methanol-
H_2O-HNO_3 (25:15:15:1). Lead(II) was located using a mixture of rubeanic
acid-alizarin-salicylaldoxime-dithizone (1:1:1:0.5). The Pb(II) was
determined colorimetrically after the paper was ashed, using a solution of
0.273 g of 4-(2-pyridylazo)resorcinol in 100 ml of H_2O for color development.

Molybdenum was separated paper chromatographically from plant
extracts using n-butanol saturated with 10% HCl [295]. Separation was
improved by conditioning the paper in a H_2O-saturated atmosphere. A
dithiol reagent (0.1% toluene-3,4-dithiol in 0.1% thioglycolic acid in 1.0N
NaOH) was used for color development. Molybdenum forms a green band
and 0.5-5.0 μg was determined. Sensitivity is increased by allowing n-butyl
acetate to diffuse up the strips. This technique concentrated the colored band
and allowed detection of 0.005 μg of Mo and visual comparison of standards
in the 0.02-0.2 μg range.

Bönig [307] used circular paper chromatography to separate Mo from a
mixture of metals that had been concentrated using dithiol. Ethanol-HCl-H_2O
(4:1:5) was used as the solvent and Mo was determined colorimetrically
with dithiol.

Ba(II), Sr(II), Ca(II) and Mg(II) extracted from dolomite and limestone
were determined quantitatively by paper chromatographic analysis [308]. A
descending method was used with a solvent consisting of 50 ml of methanol,
30 ml of isopropanol, 2 ml of HCOOH, 15 ml of H_2O, and 2.5 g of $HCOONH_4$.
For qualitative analysis a descending technique was used for 6 hr; the solvent moved approximately 30 cm. For quantitative analysis, the chromatograms were developed for 14-16 hr. The solvent front moved beyond the end
of the chromatogram, thereby increasing resolution. Two identical samples
were spotted when quantitative analysis was required. One was used for
sample location and the other for quantitative analysis.

Ba(II) and Sr(II) were located on unwashed paper with 0.1% aqueous sodium
rhodizonate; Ca(II) was located with pyrogallol-4-carboxylic acid and Mg(II)
with 0.5% 8-hydroxyquinoline in methylated spirits-H_2O (3:2). On acid-
washed paper 8-hydroxyquinoline was sensitive for all the metals when
exposed to NH_3 and ultraviolet light.

Extraction of the metals from the paper is necessary for quantitative
analysis. Sr(II), Ca(II), and Mg(II) are extracted with 1% HCOOH. After 2 hr
the HCOOH is decanted and the paper is washed with 3 x 15 ml portions of
H_2O. The combined HCOOH and H_2O extracts are evaporated to dryness in a
Pt crucible under infrared light. Organic matter is removed from the residue
by gentle heating and the residue is dissolved in 0.01N HCl. The quantity of

acid used is chosen to provide a 10% solution when the sample is diluted to volume. Alternatively, the paper can be burned directly, without HCOOH treatment, and the residue dissolved in 0.01N HCl In addition to this method's being more rapid, improved recovery of Sr(II) is obtained. It should be noted that following extraction of Ba(II) with HCOOH, the paper should not be burned because the high temperature is likely to promote formation of insoluble barium residues. The paper is removed from the extract by filtration and the filtered extract is evaporated to a small volume at low temperature, after which it is evaporated to dryness under an infra-red lamp and the residue is dissolved in H_2O. The extracted metals are determined spectrophotometrically using o-cresolphthalein complexone for Ba(II), Sr(II), and Ca(II) and Eriochrome Black T for Mg(II).

A solvent system consisting of pyridine-ethanol-1.5N CH_3COOH (2:2:1) has been used to separate Na(I), K(I), Ca(II), and Mg(II) from plant extracts [309]. After separation, Na(I) and K(I) were determined by polarographic methods and Ca(II) and Mg(II) by colorimetric methods.

Paper chromatographic methods also have been adapted for the separation of anions in soil and plant extracts. Pyridine-1.5N NH_4OH-n-butanol (2:2:1) and pyridine-ethanol-2N CH_3COOH (2:2:1) were used to separate anions extracted from soil and plant material by boiling H_2O. The anions were located using a solution of Methyl Red, Phenol Red, and Bromthymol Blue. PO_4^{3-}, SO_4^{2-}, NO_3^-, NO_2^-, Cl^-, and $B_4O_7^{2-}$ were separated and detected as pink, green, green, gray-blue, pink, and brown spots, respectively. Prior to the separation, organic impurities were removed by ion-exchange techniques.

Separation of several inorganic phosphate compounds has been accomplished by paper chromatographic methods. Pyrophosphate was identified by paper chromatography in soil extracts obtained by pretreating the soil with dilute HCl and extracting with hot NaOH [310]. Phosphorous compounds were isolated by anion-exchange chromatography using either the Cl^- or the $HCOO^-$ form of Dowex 1. The phosphate fractions, obtained were freed from the eluting salt by gel filtration on Sephadex G-10. Paper chromatographic analysis was conducted on the acid-labile fraction after separation by a series of ion-exchange fractionations. Solvent systems used were those described by Hanes and Isherwood [151] for the separation of phosphate esters. The acid-labile phosphate isolated by ion-exchange chromatography was separated with a solvent consisting of n-propanol-NH_4OH-H_2O (6:3:1) by a descending technique for 48 hr. Three compounds were detected by spraying with an $FeCl_3$-salicylsulfonic acid spray reagent [132]; one of the compounds was found to be an acid-labile phosphate. The R_{pp} value (distance moved relative to pyrophosphate) of the acid-labile phosphate suggested that it was a diphosphate, possibly pyrophosphate. The other two compounds were an unidentified phosphate and a non-phosphorus-containing complexing agent, probably ammonium oxalate. The ammonium oxalate in the original solution was removed by freezing and the solution chromatographed for 5 days using the same solvent. The acid-labile phosphate fraction was extracted from the paper and rechromatographed using a solvent system consisting of isopropyl ether-90% HCOOH (3 : 2) with descending development for 8 hr. A single spot was obtained which corresponded to pyrophosphate.

Water- and citrate-soluble phosphates have been separated paper chromatographically by fixing the citrate-soluble phosphates with Na_2CO_3 and allowing the water-soluble components to develop with the solvent mixture [311]. Whatman No. 1 paper was used for the separations. Ascending, descending, circular, and two dimensional techniques were investigated. Circular chromatography with Whatman No. 1 paper and a chromogenic agent of $K_4Fe(CN)_6$ was the preferred method of analysis.

Nine solvent systems gave satisfactory separation of H_2O-soluble phosphate from citrate-soluble phosphate, namely, n-propanol-H_2O-acetone (6 : 1 : 3), n-butanol-acetone-H_2O (5 : 2 : 3), n-amyl alcohol-acetone-H_2O (3 : 5 : 2), isobutanol-acetone-H_2O (6 : 3 : 1), ethyl acetate-isopropanol-H_2O (3 : 3 : 4), ethyl acetate-ethanol-H_2O (3 : 3 : 4), ethylmethylketone-ethanol-H_2O (2 : 3 : 5), n-amyl alcohol-n-butanol-methanol-H_2O (3 : 2 : 2 : 3), and n-amyl alcohol-n-propanol-H_2O-acetone (3 : 3 : 3 : 2). In each of the above systems the citrate-soluble phosphates remained close to the center of the paper provided that 5% Na_2CO_3 was added to the sample spot prior to development. Water-soluble phosphates displayed R_f values of 0.28-0.98.

Paper chromatographic methods for the analysis of inorganic components in soil and plant extracts have been supplanted largely by spectrometric methods. However, paper chromatographic methods are used commonly where specialized apparatus for spectrometric analysis is not available. Analysis of anions, particularly phosphates, by paper chromatographic methods continues to receive attention.

It is perhaps safe to conclude that paper chromatographic analysis is not the powerful tool it was in the years between the mid-1940's and the early 1960's because for many uses it has been superceded by more accurate, precise, and sensitive methods. It retains its effectiveness as a means of separating unknown multicomponent mixtures, and in laboratories where expensive gas chromatographic and spectrophotometric equipment is unavailable it is still possible to conduct important soil chemistry research in such areas as determining degradation pathways of organic compounds in soils, early screening investigations of pesticide residues in soils, and separating and determining trace nutrient elements in soil and plant extracts. Furthermore, paper chromatography is valuable for introducing students to the theory of chromatography using equipment that is readily accessible in almost every college teaching laboratory in the world.

REFERENCES

1. R. G. Petersen and L. D. Calvin, in Methods of Soil Analysis (C. A. Black, D. D. Evans, J. L. White, L. E. Ensminger, and F. E. Clark, eds.), American Society of Agronomy, Madison, Wisc., 1965, pp. 54-72.

2. M. G. Cline, Soil Sci., 58, 275 (1944).

3. W. G. Cochran, Sampling Techniques, Wiley, New York, 1953.

4. M. L. Jackson, Soil Chemical Analysis, Prentice-Hall, Englewood Cliffs, N.J., 1958, pp. 10-53.

5. C. B. Coulson, R. I. Davies, and E. J. A. Khan, J. Soil Sci., 10, 271 (1959).

6. C. B. Coulson, J. Sci. Food Agr., 9, 281 (1958).

7. E. J. A. Khan, M.Sc. Thesis, Univ. of Wales, 1958.

8. R. I. Davies, C. B. Coulson, and C. Luna, Chem. Ind., 1957, 1544 (1957).

9. A. Sinha and R. N. Shukla, Technology, 2, 19 (1965).

10. A. Sinha and R. N. Shukla, Technology, 3, 17 (1966).

11. J. R. Wright and M. Schnitzer, Can. J. Soil Sci., 39, 44 (1959).

12. K. Kumada, A. Suzuki, and A. Aizawa, Nature (London), 191, 415 (1961).

13. K. S. Kang and G. T. Felbeck, Jr., Soil Sci., 99, 175 (1964).

14. D. F. Crossan and D. L. Lynch, Phytopathology, 48, 55 (1958).

15. F. J. Sowden and H. Deuel, Soil Sci., 91, 44 (1961).

16. L. Reio, J. Chromatog., 1, 338 (1958).

17. I. M. Hais and K. Macek, Handbuch der Paierchromatographie, Band I, G. Fisher, Jena, 1958.

18. I. Smith, Chromatographic Techniques, Heinemann, London, 1958.

19. D. S. Jenkinson and J. Tinsley, J. Soil Sci., 10, 245 (1959).

20. M. Wolfe, Biochim. Biophys. Acta, 23, 186 (1957).

21. F. Sanger, Biochem. J., 39, 507 (1945).

22. S. Blackburn and A. G. Lowther, Biochem. J., 48, 126 (1951).

23. G. Greene and C. Steelink, J. Org. Chem., 27, 170 (1962).

24. R. I. Morrison, J. Soil Sci., 14, 201 (1963).

25. A. Burges, H. M. Hurst, S. B. Walkden, F. M. Dean, and M. Hirst, Nature (London), 199, 696 (1963).

26. W. Flaig, Maataloustieteellinen Aikakouskirja, 33, 1 (1961).

27. S. A. Waksman, Humus, Baillière, Tindall and Cox, London, 1936.

28. M. M. Kononova, Soil Organic Matter (T. Z. Nowakowski and G. A. Greenwood, Transl.), Pergamon, London, 1961

29. W. Flaig, J. C. Salfeld, and K. Haider, Landw. Forsch., 16, 85 (1963).

30. D. C. Whitehead, Nature (London), 202, 417 (1964).

31. J. C. Hamence, Analyst, 69, 229 (1944).

32. I. Smith, Chromatographic and Electrophoretic Techniques, Vol. 1, 2nd ed., Heinemann, London, 1960.

33. A. E. Bradfield and A. E. Flood, J. Chem. Soc., 1952, 4740 (1952).

34. N. E. K. Henderson, in The Ecology of Soil Fungi (D. Parkinson and J. S. Waid, eds.), Liverpool Univ. Press, Liverpool, 1960.

35. W. C. Evans, Ann. Rep. Chem. Soc. (London), 23, 279 (1956).

36. H. M. Hurst, in Enzyme Chemistry of Phenolic Compounds (J. Pridham, ed.), Pergamon, London, 1963.

37. N. A. Burges, H. M. Hurst, and S. B. Walkden, Geochim. Cosmochim. Acta, 28, 1547 (1964).

38. J. Mendez and F. J. Stevenson, Soil. Sci., 102, 85 (1966).

39. F. J. Stevenson and J. Mendez, Soil Sci., 103, 383 (1967).

40. J. M. Bremner, J. Soil Sci., 5, 214 (1954).

41. S. S. Dragunov, N. N. Zhelokhovtseva, and E. I. Strelkeve, Pochvoved., 1948, 409 (1948); through Chem. Abstr., 44, 6995 (1949).

42. G. C. Esh and S. S. Guha-Sircar, J. Indian Chem. Soc., 17, 326 (1940).

43. R. I. Morrison, J. Soil Sci., 9, 130 (1958).

44. P. Dubach and N. C. Mehta, Soils Fert., 26, 293 (1963).

45. I. A. Pearl and D. L. Beyer, J. Amer. Chem. Soc., 76, 6106 (1954).

46. I. A. Pearl and E. E. Dickey, J. Amer. Chem. Soc., 74, 614 (1952).

47. N. C. Mehta, P. Dubach, and H. Deuel, Z. Pflernaehr. Dueng, Bodenk., 101, 147 (1964).

48. J. Germain, J. Montreuil, and P. Koukos, Bull. Soc. Chim., 1959, 115 (1959).

49. R. E. Wildung, G. Chesters, and D. E. Behmer, Plant Soil, 32, 221 (1970).

50. W. J. Brickman and C. V. Purvis, J. Amer. Chem. Soc., 75, 4336 (1953).

51. K. Freudenberg, W. Lautsch, and K. Engler, Berichte, 73, 167 (1940).

52. K. R. Kavanagh and J. M. Pepper, Can. J. Chem., 33, 24 (1955).

53. J. C. Pew, J. Amer. Chem. Soc., 77, 2831 (1955).

54. F. E. Roadhouse and D. MacDougall, Biochem. J., 63, 33 (1956).

55. J. E. Stone and M. J. Blundell, Anal. Chem., 23, 771 (1951).

56. J. M. Bremner, Z. Pflernaehr. Dueng. Bodenk., 69, 32 (1955).

57. S. Gottlieb and S. B. Hendricks, Soil Sci. Soc. Amer. Proc., 10, 117 (1945).

58. H. W. Scharpenseel, Z. Pflernaehr. Dueng. Bodenk., 88, 95 (1960).

59. R. Consden, A. H. Gordon, and A. J. P. Martin, Biochem. J., 38, 224 (1944).

60. H. M. Hurst, A. Burges, and P. Latter, Phytochemistry, 1, 227 (1962).

61. W. G. C. Forsyth, Biochem. J., 41, 176 (1947).

62. T. V. Drozdova, Pochvoved., 1955, 83 (1959); Reprinted in Soils Fert., 19, 27 (1956).

63. E. Schlichting, Z. Pflernaehr. Dueng. Bodenk., 61, 97 (1953).

64. R. F. Keefer, F. L. Himes, and J. L. Mortensen, Soil Sci. Soc. Amer. Proc., 30, 415 (1966).

65. H. T. Gordon, W. W. Thornburg, and L. N. Werum, J. Chromatog., 9, 44 (1962).

66. A. Capriotti, Nature (London), 190, 464 (1961).

67. K. T. Wieringa, Plant Soil, 21, 333 (1964).

68. M. Robert-Géro, G. Vidal, C. Hardisson, L. Leborgne, and J. Pochon, Trans. Int. Symp. Humus et Planta IV, Prague, 1967, pp. 39-41.

69. F. J. Sowden and K. C. Ivarson, Soil Sci., 94, 340 (1962).

70. U. C. Gupta, F. J. Sowden, and P. C. Stobbe, Soil Sci. Soc. Amer. Proc., 27, 380 (1963).

71. U. C. Gupta and F. J. Sowden, Can. J. Soil Sci., 45, 237 (1965).

72. U. C. Gupta and F. J. Sowden, Soil Sci., 96, 217 (1963).

73. U. C. Gupta and F. J. Sowden, Soil Sci., 97, 328 (1964).

74. W. G. C. Forsyth, Biochem. J., 46, 141 (1950).

75. A. Grov, Acta Chem. Scand., 17, 2301 (1963).

76. R. L. Whistler and K. W. Kirby, J. Amer. Chem. Soc., 78, 1755 (1956).

77. J. F. Dormaar, Soil Sci., 103, 417 (1967).

78. D. L. Lynch, E. E. Hearns, and L. J. Cotnoir, Jr., Soil Sci. Soc. Amer. Proc., 21, 160 (1957).

79. J. W. Parsons and J. Tinsley, Soil Sci., 92, 46 (1961).

80. L. Hough, Methods Biochem. Anal., 1, 205 (1954).

81 K. V. Giri and V. N. Nigam, J. Indian Inst. Sci., Sect. A, 36, 49 (1954).

82. E. Alvsaker and K. Michelsen, Acta Chem. Scand., 11, 1794 (1957).

83. R. B. Duff, J. Sci. Food Agr., 3, 140 (1952).

84. R. B. Duff, Chem. Ind., 1952, 1104 (1952).

85. F. G. Fischer and H. Z. Dörfel, Physiol. Chem., 297, 164 (1954).

86. M. A. Jermyn and F. A. Isherwood, Biochem. J., 44, 402 (1949).

87. F. Petrelli, N. Siliprandi, and L. Sartorelli, J. Chromatog., 3, 377 (1960).

88. D. L. Lynch, L. M. Wright, and H. O. Olney, Soil Sci., 84, 405 (1957).

89. S. M. Partridge and R. G. Westall, Biochem. J., 42, 238 (1948).

90. H. H. Johnston, Soil Sci. Soc. Amer. Proc., 25, 415 (1961).

91. M. Gee and R. M. McCready, Anal. Chem., 29, 257 (1957).

92. B. R. Nagar, Nature (London), 194, 896 (1962).

93. L. Hough, J. K. N. Jones, and W. H. Wadman, J. Chem. Soc., 1950, 1702 (1950).

94. M. Robert, Ann. Inst. Pasteur, 106, 801 (1964).

95. M. Robert-Géro, C. Hardisson, L. Leborgne, and G. Vidal, Ann. Inst. Pasteur, 113, 903 (1967).

96. M. Robert-Géro, G. Vidal, C. Hardisson, L. Leborgne, and J. Pochon, Ann. Inst. Pasteur, 113, 911 (1967).

97. F. A. Isherwood and M. A. Jermyn, Biochem. J., 48, 515 (1951).

98. N. Albon and D. Gross, Analyst, 77, 410 (1952).

99. C. G. Dobbs and D. A. Griffiths, Rept. Forest Res., 1960, 86 (1961).

100. D. A. Griffiths and C. G. Dobbs, Nature (London), 199, 408 (1963).

101. J.-F. Bouhours and M. V. Cheshire, Soil Biol. Biochem., 1, 185 (1969).

102. M. V. Cheshire, C. M. Mundie, and H. Shepard, Soil Biol. Biochem., 1, 117 (1969).

103. J. L. Buchan and R. I. Savage, Analyst, 77, 401 (1952).

104. K. Wallenfels, Naturwiss., 37, 491 (1950).

105. S. M. Partridge, Nature (London), 164, 443 (1949).

106. R. J. Block, E. L. Durrum, and G. Zweig, A Manual of Paper Chromatography and Paper Electrophoresis, Academic, New York, 1958.

107. S. Mukherjee and H. C. Srivastava, Nature (London), 169, 330 (1952).

108. J. T. Edward and D. M. Waldron, J. Chem. Soc., 1952, 3631 (1952).

109 S. M. Partridge, Nature (London), 158, 270 (1947).

110. H. T. Gordon, W. W. Thornburg, and L. N. Werum, Anal. Chem., 28, 849 (1956).

111. W. G. C. Forsyth, Nature (London), 161, 239 (1948).

112. M. Dubois, K. A. Gilles, J. K. Hamilton, P. A. Rebers, and F. Smith, Anal. Chem., 28, 350 (1956).

113. L. Ruzicka, A. Eschenmoser, and H. Heusser, Experimentia, 9, 357 (1953).

114. E. Zavarin and A. B. Anderson, J. Org. Chem., 21, 332 (1956).

115. A. di Prima and T. Storto, Rev. Ital. Essenz-Profumi, Piante Offic., 42, 283 (1960).

116. G. Rath, Parfuem. Kosmetik, 41, 307 (1960).

117. S. Dietze, Parfuem. Kosmetik, 42, 43 (1961).

118. D. Holness, J. Soc. Cosmetic Chemists, 12, 370 (1961).

119. P. Teisseire, Bull. Soc. Chim., 2, 384 (1963).

120. B. H. Kingston, Am. Perfumer Cosmetics, 79, 50 (1964).

121. P. Teisseire, Recherches (Paris), 14, 12 (1964).

122. L. Domange and S. Longueville, Compt. Rend., 247, 209 (1958).

123. G. Nannes, J. Landwirtsch., 47, 45 (1899).

124. D. F. Sokolov, Pochvoved., 1948, 502 (1948).

125. J. Stoklasa, Centbl. Bakteriol. Parasitenk., 29, 385 (1911).

126. V. Vincent, Compt. Rend. 17th Cong. Chim. Ind. (Paris), 2, 861 (1937).

127. C. L. Wrenshall and R. R. McKibbin, Can. J. Res., 15B, 475 (1937).

128. K. Aso, Tokyo Imperial Univ. Coll. Agr. Bull., 6, 277 (1905).

129. E. C. Shorey, U.S. Dept. Agr. Bur. Soils Bull., 88, 1 (1913).

130. R. J. Hance and G. Anderson, Soil Sci. , 96, 157 (1962).

131. R. S. Bandurski and B. Axelrod, J. Biol. Chem. , 193, 405 (1951).

132. H. E. Wade and D. M. Morgan, Biochem. J. , 60, 264 (1955).

133. J. M. McKibben and W. E. Taylor, J. Biol. Chem. , 178, 17 (1949).

134. F. M. Huennekens, D. J. Hanahan, and M. Uziel, J. Biol. Chem. , 206, 443 (1954).

135. R. Munier, Bull. Soc. Chim. Biol. , 33, 862 (1951).

136. R. J. Block and D. Bolling, The Amino Acid Composition of Proteins and Foods, Charles C Thomas, Springfield, Ill. , 1951.

137. G. Toennies and J. J. Kolb, Anal. Chem. , 23, 823 (1951).

138. C. L. Wrenshall and W. J. Dyer, Soil Sci. , 51, 235 (1941).

139. C. A. Bower, Iowa Agr. Exp. Sta. Res. Bull. , 362, 1 (1949).

140. A. P. Adams, W. V. Bartholomew, and F. E. Clark, Soil Sci. Soc. Amer. Proc. , 18, 40 (1954).

141. G. Anderson, in Soil Biochemistry (A. D. McLaren and G. H. Peterson, eds.), Marcel Dekker, New York, 1967, pp. 67-90.

142. G. Anderson, Nature (London), 180, 287 (1957).

143. G. Anderson, Soil Sci. , 86, 169 (1958).

144. W. S. MacNutt, Biochem. J. , 50, 384 (1952).

145. G. R. Wyatt, Biochem. J. , 48, 584 (1951).

146. W. J. Dyer, C. L. Wrenshall, and G. R. Smith, Science, 91, 319 (1940)

147. C. A. Bower, Soil Sci. , 59, 277 (1945).

148. D. H. Smith and F. E. Clark, Soil Sci. , 72, 353 (1951).

149. E. J. N. Pedersen, Plant Soil, 4, 252 (1953).

150. G. Anderson, J. Sci. Food Agr. , 7, 437 (1956).

151. C. S. Hanes and F. A. Isherwood, Nature (London), 164, 1107 (1949).

152. P. W. Arnold, Biochim. Biophys. Acta, 19, 552 (1956).

153. D. J. Cosgrove, Nature (London), 194, 1265 (1962).

154. D. J. Cosgrove, Australian J. Soil Res. , 1, 203 (1963).

155 D. J. Cosgrove, Soil Sci. , 102, 42 (1966).

156. A. Desjobert and F. Petek, Bull. Soc. Chim. Biol. , 38, 871 (1956).

157. F. E. G. Harrap, Analyst, 85, 452 (1960).

158. M. P. Greaves, G. Anderson, and D. M. Webley, Biochim. Biophys. Acta, 132, 412 (1967).

159. H. D. Fowler, J. Sci. Food Agr., 7, 381 (1956).

160. W. H. Longenecker, Anal. Chem., 21, 1402 (1949).

161. H. D. Fowler, M.Sc. Thesis, Bristol Univ., Bristol, England, 1953.

162. S. J. Angyal, D. J. McHugh, and P. T. Gilham, J. Chem. Soc., 1957, 1432 (1957).

163. W. E. Trevelyan, D. P. Procter, and J. S. Harrison, Nature (London), 166, 444 (1950).

164. E. F. L. J. Anet and T. M. Reynolds, Nature (London), 174, 930 (1954).

165. P. Fleury, J. E. Courtois, and P. Malangeau, Bull. Soc. Chim. Biol., 35, 537 (1953).

166. S. J. Angyal and D. J. McHugh, J. Chem. Soc., 1957, 1423 (1957).

167. J. M. Bremner, in Soil Nitrogen (W. V. Bartholomew and F. E. Clark, eds.), American Society of Agronomy, Madison, Wisc., 1965, pp. 93-149.

168. J. M. Bremner, in Soil Biochemistry (A. D. McLaren and G. H. Peterson, eds.), Marcel Dekker, New York, 1967, pp. 19-66.

169. J. M. Bremner, J. Agr. Sci., 39, 183 (1949).

170. S. Singh and G. S. Bhandari, J. Indian Soc. Soil Sci., 11, 1 (1963).

171. J. M. Bremner, J. Agr. Sci., 52, 147 (1959).

172. S. C. Wang, T. K. Yang, and S. Y. Chen, Rep. Taiwan Sugar Exp. Sta., 34, 57 (1964).

173. D. R. Keeney and J. M. Bremner, Soil Sci. Soc. Amer. Proc., 28, 653 (1964).

174. R. T. Kojima, Soil Sci., 64, 157 (1947).

175. F. J. Stevenson, Soil Sci. Soc. Amer. Proc., 20, 204 (1956).

176. H. H. Cheng and J. van Hove, Pedologie, 14, 8 (1964).

177. F. J. Stevenson, Soil Sci. Soc. Amer. Proc., 18, 373 (1954).

178. K. Kobo, R. Tatsukawa, and Y. Oba, Rapports 6th Int. Cong. Sci. Sol. (Paris), B, 485 (1965).

179. V. V. Rendig, Soil Sci., 71, 253 (1951).

180. D. I. Parker, F. J. Sowden, and H. T. Atkinson, Sci. Agr., 32, 163 (1952).

181. W. Wittich, Schriftenreihe Forstl. Fak. Univ. Goettingen, 4, 1 (1952).

182. W. Laatsch, E. Schlichting, Z. Pflernaehr. Dueng. Bodenk., 62, 50 (1953).

183. E. Schlichting, Z. Pflernaehr. Dueng. Bodenk., 61, 1 (1953).

184. F. J. Sowden, Soil Sci., 80, 181 (1955).

185. F. J. Sowden, Soil Sci., 82, 491 (1956).

186. A. H. Cornfield, J. Sci. Food Agr., 8, 509 (1957).

187. J. L. Young and J. L. Mortensen, Ohio Agr. Exp. Sta. Res. Circ., 61, 1 (1958).

188. T. Yamashita and T. Akiya, J. Sci. Soil Manure (Japan), 34, 255 (1963).

189. J. M. Bremner, Nature (London), 165, 367 (1950).

190. J. M. Bremner, Biochem. J., 47, 538 (1950).

191 T. D. Biswas and N. B. Das, J. Indian Soc. Soil Sci., 5, 31 (1957).

192. J. Carles, L. Soubiés, and R. Gadet, Compt. Rend., 247, 1229 (1958).

193. F. J. Sowden and K. C. Ivarson, Plant Soil, 11, 249 (1959).

194. J. Carles and J. Decau, Sci. Proc. Roy. Dublin Soc., A1, 177 (1960).

195. J. Carles and J. Decau, Ann. Agron., 11, 557 (1960).

196. A. Grov, Acta Chem. Scand., 17, 2319 (1963).

197. C. C. Wang, S. Y. Chen, and T. K. Yang, Rep. Taiwan Sugar Exp. Sta., 34, 93 (1964).

198. H. F. Linskens, Papierchromatographie in der Botanik, Springer-Verlag, Berlin, 1959, p. 109.

199. R. J. Block, R. LeStrange, and G. Zweig, Paper Chromatography, Academic, New York, 1952.

200. R. J. Block, Anal. Chem., 22, 1327 (1950).

201. F. J. Sowden, Can. J. Soil Sci., 37, 143 (1957).

202. F. J. Sowden, Soil Sci., 102, 202 (1966).

203. A. Grov, Acta Chem. Scand., 17, 2316 (1963).

204. A. Grov and E. Alvsaker, Acta Chem. Scand., 17, 2307 (1963).

205. D. I. Davidson, F. J. Sowden, and H. J. Atkinson, Soil Sci., 71, 347 (1951).

206. F. Turba, Chromatographische Metoden in der Protein-Chemie, Springer-Verlag, Berlin, 1954, p. 170.

207. K. Krishnamurthy and M. J. Swaminathan, J. Sci. Indian Res., 14C, 79 (1955).

208. R. J. Block, Trans. Int. 2nd Cong. Biochim., 1952, 171 (1952).

209. A. Saifer and I. Oreskes, Anal. Chem., 28, 501 (1956).

210. L. Fowden, Biochem. J., 48, 327 (1951).

211. C. C. Dadd, L. Fowden, and W. H. Pearsall, J. Soil Sci., 4, 69 (1953).

212. P. Simonart and F. Peeters, Trans. 5th Int. Cong. Soil Sci., 3, 132 (1954).

213. A. J. Woiwood, J. Gen. Microbiol., 3, 312 (1949).

214. A. J. Woiwood and F. V. Linggood, Nature (London), 163, 218 (1949).

215. C. Petronici, Agrochime, 11, 360 (1967).

216. C. E. Dent, Biochem. J., 43, 169 (1948).

217. J. M. Bremner, J. Agr. Sci., 45, 469 (1955).

218. J. M. Bremner, J. Agr. Sci., 46, 247 (1955).

219. R. I. Davies, C. B. Coulson, and C. Luna, Chem. Ind., 1957, 1544 (1957).

220. J. M. Bremner, Z. Pflernaehr. Dueng. Bodenk., 69, 32 (1955).

221. J. M. Bremner, Z. Pflernaehr. Dueng. Bodenk., 71, 63 (1955).

222. C. B. Coulson, J. Sci. Food Agr., 6, 674 (1955).

223. C. B. Coulson, R. I. Davies, and E. J. A. Khan, J. Sci. Food Agr., 10, 209 (1959).

224. L. Leborgne, S. Junque, G. Vidal, C. Hardisson, M. Robert-Géro, and J. Pochon, Rev. Ecol. Biol. Sol., 4, 553 (1957).

225. P. Wollenweber, J. Chromatog., 9, 369 (1962).

226. E. Work, Biochim. Biophys. Acta, 3, 400 (1949).

227. E. Work, Bull. Soc. Chim. Biol., 31, 138 (1949).

228. E. Work, Nature (London), 165, 74 (1950).

229. F. J. Sowden, Soil Sci., 102, 264 (1966).

230. J. Sjöquist, Biochim. Biophys. Acta, 41, 20 (1960).

231. S. D. Démétriadès, Nature (London), 177, 95 (1956).

232. S. D. Démétriadès and P. T. Constantinon, Compt. Rend., 242, 2384 (1956).

233. S. D. Démétriadès, Ann. Inst. Phytopathol. Benaki, 9, 10 (1953).

234. S. R. Wilkinson and A. J. Ohlrogge, Nature (London), 204, 902 (1964).

235. R. T. Williams and R. L. M. Synge, Partition Chromatography, Biochem. Soc. 3rd Symp., Cambridge Univ. Press, Cambridge, England, 1949.

236. E. D. Moffat and R. J. Lytle, Anal. Chem., 31, 926 (1959).

237. R. Acher, C. Fromageot, and M. Jutisz, Biochim. Biophys. Acta, 5, 81 (1950).

238. H. Stegemann and H. F. Griffen, Naturwiss., 45, 263 (1958).

239. R. J. Block, Food Ind., 22, 824 (1950).

240. F. Sanger and H. Tuppy, Biochem. J., 49, 463 (1951).

241. D. P. Schwartz, Anal. Chem., 30, 1855 (1958).

242. J. B. Jepson and I. Smith, Nature (London), 172, 1100 (1953).

243. R. Consden, A. H. Gordon, and A. J. P. Martin, Biochem. J., 40, 33 (1946).

244. H. M. Winegard, G. Toennies, and R. J. Block, Science, 108, 506 (1948).

245. H. R. Crumpler and C. E. Dent, Nature (London), 164, 441 (1949).

246. S. Rosebeek, Chem. Weekblad, 46, 813 (1950).

247. W. P. McKinley, in Analytical Methods for Pesticides, Plant Growth Regulators, and Food Additives, Vol. 1 (G. Zweig, ed.), Academic, New York, 1963, pp. 227-252.

248. G. Zweig, in Chromatographic Reviews, Vol. 6 (M. Lederer, ed.), Elsevier, Amsterdam, 1964, pp. 110-128.

249. L. C. Mitchell, J. Assoc. Off. Agr. Chem., 40, 999 (1957).

250. W. W. Thornburg, in Analytical Methods for Pesticides, Plant Growth Regulators, and Food Additives, Vol. 1 (G. Zweig, ed.), Academic, New York, 1963, pp. 87-108.

251. J. J. Menn, M. E. Eldefrawi, and H. T. Gordon, J. Agr. Food Chem., 8, 41 (1960).

252. A. Major, Jr., and H. C. Barry, J. Assoc. Off. Agr. Chem., 44, 202 (1960).

253. P. A. Mills, J. Assoc. Off. Agr. Chem., 42, 734 (1959).

254. H. B. Pionke and G. Chesters, Soil Sci. Soc. Amer. Proc., 32, 749 (1968).

255. L. C. Mitchell, J. Assoc. Off. Agr. Chem., 41, 781 (1958).

256. L. C. Mitchell, J. Assoc. Off. Agr. Chem., 35, 920 (1952).

257. L. C. Mitchell and W. I. Patterson, J. Assoc. Off. Agr. Chem., 36, 553 (1953).

258. L. C. Mitchell, J. Assoc. Off. Agr. Chem., 36, 1183 (1953).

259. L. C. Mitchell, J. Assoc. Off. Agr. Chem., 37, 216 (1954).

260. L. C. Mitchell, J. Assoc. Off. Agr. Chem., 37, 530 (1954).

261. L. C. Mitchell, J. Assoc. Off. Agr. Chem., 37, 996 (1954).

262. L. C. Mitchell, J. Assoc. Off. Agr. Chem., 39, 484 (1956).

263. L. C. Mitchell, J. Assoc. Off. Agr. Chem., 39, 891 (1956).

264. L. C. Mitchell, J. Assoc. Off. Agr. Chem., 39, 980 (1956).

265. L. C. Mitchell, J. Assoc. Off. Agr. Chem., 39, 985 (1956).

266. L. C. Mitchell, J. Assoc. Off. Agr. Chem., 40, 294 (1957).

267. L. C. Mitchell, J. Assoc. Off. Agr. Chem., 45, 682 (1962).

268. L. C. Mitchell, J. Assoc. Off. Agr. Chem., 46, 988 (1963).

269. W. P. McKinley and J. H. Mahon, J. Assoc. Off. Agr. Chem., 42, 725 (1959).

270. F. A. Gunther and R. C. Blinn, Analysis of Insecticides and Acaricides, Interscience, New York, 1955.

271. J. P. San Antonio, J. Agr. Food Chem., 7, 322 (1959).

272. J. P. San Antonio, J. Assoc. Off. Agr. Chem., 43, 721 (1960).

273. J. W. Cook, J. Assoc. Off. Agr. Chem., 37, 987 (1954).

274. J. A. R. Bates, Analyst, 90, 453 (1965).

275. C. Anglin and W. P. McKinley, J. Agr. Food Chem., 8, 186 (1960).

276. Joint Demeton-Methyl Residues Panel, Analyst, 87, 485 (1962).

277. D. Caverly and P. S. Hall, Analyst, 86, 478 (1961).

278. L. C. Mitchell, J. Assoc. Off. Agr. Chem., 43, 810 (1960).

279. D. E. Coffin and W. P. McKinley, J. Assoc. Off. Agr. Chem., 46, 223 (1963).

280. H. P. Burchfield and D. E. Johnson, Guide to the Analysis of Pesticide Residues, U.S. Govt. Printing Office, Washington, D.C., 2 vols., 1965.

281. J. W. Cook, J. Assoc. Off. Agr. Chem., 38, 150 (1955).

282. M. E. Getz and S. J. Friedman, J. Assoc. Off. Agr. Chem., 46, 707 (1963).

283. H. F. MacRae and W. P. McKinley, J. Assoc. Off. Agr. Chem., 44, 207 (1961).

284. H. E. Wade and D. M. Morgan, Nature (London), 171, 529 (1953).

285. R. R. Watts, J. Assoc. Off. Agr. Chem., 48, 1161 (1965).

286. L. C. Mitchell, J. Assoc. Off. Agr. Chem., 44, 720 (1961).

287. G. Yip, J. Assoc. Off. Agr. Chem., 47, 343 (1964).

288. L. C. Mitchell, J. Assoc. Off. Anal. Chem., 49, 1163 (1966).

289. A. Bevenue and H. Beckman, Residue Rev., 19, 83 (1967).

290. L. C. Mitchell, J. Assoc. Off. Agr. Chem., 43, 87 (1960).

291. C. I. Harris, Weed Res., 5, 275 (1965).

292. C. I. Harris, J. Agr. Food Chem., 15, 157 (1967).

293. R. J. Hance and G. Chesters, Soil Biol. Biochem., 1, 309 (1969).

294. E. C. Hunt, A. A. North, and R. A. Wells, Analyst, 80, 172 (1955).

295. W. D. Duffield, Analyst, 83, 503 (1958).

296. A. Sinha and S. K. Ghosh, Technology, 2, 197 (1965).

297. R. A. Webb, D. G. Hallas, and H. M. Stevens, Analyst, 94, 794 (1969).

298. C. B. Coulson, R. I. Davies, and C. Luna, Analyst, 85, 203 (1960).

299. C. W. Ayers, Mikrochim. Acta, 1956, 1333 (1956).

300. N. Kamalam and A. Mariakulandai, J. Indian Soc. Soil Sci., 6, 169 (1958).

301. G. Stefanovic, Anal. Abstr., 3, 919 (1956).

302. T. V. Arden, F. H. Burstall, and R. P. Linstead, J. Chem. Soc., 1950, s311 (1950).

303. A. Szekely, Agrikem. Talajt., 9, 381 (1960).

304. C. E. Thompson and H. W. Lakin, U.S. Geol. Survey Bull., 1036-L, 209 (1957).

305. D. Purushottam, J. Sci. Ind. Res., 19B, 449 (1960).

306. G. Bönig and H. Heigener, Landw. Forsch., 19, 117 (1966).

307. G. Bönig, Landw. Forsch., 9, 101 (1956).

308. F. H. Pollard, J. F. W. McOmie, and J. V. Martin, Analyst, 81, 353 (1956).

309. A. E. Petrov-Spiridonov, Izvest. Timiryazev. Sal'skokhoz. Akad., 1959, 229 (1959): through Chem. Abstr., 54, 6385 (1960).

310. G. Anderson and J. D. Russel, J. Sci. Food Agr., 20, 78 (1969).

311. A. Sinha and S. K. Ghosh, Technology, 2, 76 (1965).

Chapter 9

THIN-LAYER CHROMATOGRAPHIC ANALYSIS IN SOIL CHEMISTRY

Wayne Thornburg

Del Monte Corporation Research Center
Walnut Creek, California

I. INTRODUCTION

Soil is no doubt one of the most complex mixtures existing in nature. The task of identifying and quantitating all of the compounds present is formidable. Some of the organic compounds and trace elements are present at very low levels, but it cannot be assumed that a compound or element is of no consequence if it is present in low concentrations.

The analysis of trace elements or organic compounds in complex mixtures is very difficult unless the material for which the analysis is being made is first separated in a relatively pure form. Thin-layer chromatography (TLC) is an outstanding technique for the separation of organic compounds and inorganic ions and is a useful semiquantitative analytical procedure.

After the soil constituents are separated on the TLC plate they can often be eluted and identified and quantitated by modern instrumental analytical techniques.

II. THIN-LAYER CHROMATOGRAPHY FOR SOIL ANALYSIS

A. Pesticide Residues and Trace Elements

The continued use of organic crop control agents has resulted in the upper layers of soil in some heavily cultivated areas becoming a repository for low levels of persistent pesticides and their metabolites. These chemicals may have an adverse effect on plant and animal life and can be leached from the soil into waterways, where they enter the marine food chain.

Thin-layer chromatography is an inexpensive and efficient tool for separating, identifying, and semiquantitating pesticidal residues. It is also useful as confirmatory evidence for other methods of detection; often the coating containing the separated pesticides can be removed from the TLC plate and extracted, and the pesticide quantitated by another technique.

Many of the analytical techniques used for trace element analysis in soil have, in the past, been crude. While TLC has not found extensive use in soil analysis, it has been used for the analysis of inorganic ions. Thus, by combining proved soil extraction procedures with TLC, rapid qualitative knowledge of the trace elements available for plant growth could be gained. If quantitation were required, the elements of interest could then be quantitated by atomic absorption or a colorimetric technique.

B. Accuracy and Precision

The quantitation of some TLC separations has become possible within the accuracy and precision required for many analyses by the use of scanning photometers and spectrophotometers. However, TLC itself should not be considered a good quantitative tool. Usually, accurate quantitation without removal of the material from the plate is very difficult, and the accuracy that can be achieved by direct measurement techniques is often somewhat unsatisfactory.

The absolute limit of detection on a TLC plate depends to a large extent on the technique of the operator, and the sensitivity improves as the operator gains experience. The sensitivity of pesticides to reagents containing silver nitrate is from 0.05 to 1 μg for cleaned up samples, and up to tenfold less for those not put through a cleanup procedure. Other techniques have sensitivities of about 1.5-5 μg. The sensitivity varies with reactivity of the compound being tested. It is important to remember that the use of quality plates, highly purified solvents, and cleaned up extracts produces marked improvement in the sensitivity of any TLC analysis. However, for the greatest accuracy and reliability, the pesticide or soil constituent should be quantitated by colorimetric or instrumental techniques after removal from the plate.

Improvements are constantly being made in detection procedures, and the literature should be consulted before an extensive TLC analytical program is undertaken.

III. PESTICIDE RESIDUE ANALYSIS

A. General Considerations

Pesticides that accumulate in the soil are degraded to a series of metabolites by the action of sunlight, moisture, and bacteria. Some types of pesticides, such as organophosphate and carbamate insecticides, have a short half-life, while some chlorinated insecticides may remain in the soil as the original compound or stable metabolites for several years.

Thin-layer chromatography is an effective way to separate, and, in conjunction with chemical or gas chromatographic techniques, to quantitate these chemicals. Pionke and Chesters [1] reviewed the determination of organochlorine insecticides in soils and waters. This review is divided into three parts; i.e., extraction of organochlorine insecticides, cleanup of the extracts, and analysis of the purified extracts. A large portion of final section on analysis is given over to GLC techniques because of the preponderance of its use for pesticide determination, but TLC and other methods are evaluated.

B. Extraction Procedures

The first step in soil pesticide analysis is to separate the pesticide from the soil. It is necessary to recover as much pesticide and metabolites as possible from the soil, and to do this in reasonable time.

The majority of pesticides in common use are soluble in nonpolar organic solvents and thus can be extracted from the soil by a solvent extraction procedure. Water-soluble herbicides and metabolites usually call for special extraction procedures. Generally, the soil sample is received in a moist condition, and if the soil is dried prior to analysis the pesticide may be lost by volatilization. It is generally better to extract the soil on an as-is condition and determine the percent moisture on a subsample. The pesticide level on a dried soil basis can then be calculated.

A technique developed by the Shell Chemical Company [2] for the removal of pesticides from soil is to blend 100 g of soil with 125 ml of water and 200 ml of benzene in an explosion-proof Waring Blender for 5 min. The slurry is then transferred to a gallon jar and rotated for 20 min. The benzene layer is separated by centrifugation and clarified by filtration through filter paper.

Lichtenstein and co-workers [3] studied the degradation of aldrin and heptachlor in field soils during a 10-year period. Soils were extracted in a homogenizer with redistilled acetonitrile (2 ml/g of wet soil) followed by concentration of the extract to a low known volume at 25°C in a flash evaporator. The extract was then analyzed by TLC.

Saha [4] compared several methods of extracting C^{14} dieldrin from soil. The best solvent system was an exhaustive 8-hr soxhlet extraction with methanol-chloroform (1 : 1). I found the addition of water to soil helps desorb organochlorine pesticides from soil.

Chiba and co-workers [5] studied the suitability of TLC, GLC, and bio-assay for the determination of chlorinated pesticide residues in soil without cleanup using five solvent systems. Background interference varied in TLC for the different systems and soil types and was directly related to the amount of coextractives. Their work emphasized the importance of extract cleanup prior to final quantitation.

C. Cleanup and Concentration Procedures

After the pesticide is removed from the bulk of the soil by solvent extraction, concentration and cleanup of the extract is usually necessary prior to TLC analysis. The use of uncleaned up extracts will often lead to incorrect R_f values, and often a streak is produced on the TLC plate rather than discrete spots for the pesticides.

A number of techniques have been developed to clean up extracts prior to TLC analysis. It is important to remember that most pesticides are volatile and can be lost unless care is taken in the evaporation of extracts. In any analysis, standards, blanks, and recovery samples, where known amounts of pesticides are added to check soil, should be put through the complete procedure. The use of a three-ball Snyder column or a reduced pressure evaporator, such as the Rinco Evaporator, is recommended for the concentration of extracts. Some pesticides can be evaporated in a water bath at temperatures not over 50°C with a gentle stream of dry air if enough organic material is present to prevent crystallization and evaporation of the pesticide. As the amount of organic material is reduced by the cleanup procedure there is an increased possibility of loss of the pesticide.

Extracts are usually cleaned up by solvent partition, column chromatography, or a selective adsorption technique or a combination of these. If the extract contains waxy or fatty material, solvent partition between hexane and acetonitrile is often useful.

Florisil, PR grade, is very effective for the cleanup of extracts. A Shell chromatographic column is filled with Florisil to a depth of 4.5 in. A half-inch layer of anhydrous sodium sulfate is added. The residue from the evaporated soil extract is dissolved in a small volume of hexane and poured on to the column. A 200 ml quantity of hexane containing 6% ethyl ether is prepared and poured through the column. The eluate is collected and 200 ml of hexane containing 15% ethyl ether is added to the column. The eluate is collected, 200 ml of hexane containing 50% ethyl ether is added, and the eluate again collected. The eluates are evaporated in a boiling water bath using a three-ball Snyder column. The extracts are taken to a convenient volume and are ready for TLC analysis. The use of the three-eluate system provides considerable fractionation of the pesticides. This technique was studied in detail by Johnson and Stansbury [6], by Beckman and co-workers [7], and by Sans [8].

Another effective cleanup procedure is the use of Nuchar-Attaclay. The Nuchar-Attaclay is added to the benzene extract, the mixture is shaken and allowed to settle, and the Nuchar-Attaclay is washed five times with benzene saturated with water. Sissons and co-workers [9] found water-saturated toluene containing 50% hexane to be a suitable solvent for use with Nuchar-Attaclay.

Silica gel and the aluminas have also been used for cleanup procedure for special applications.

D. TLC Analysis of Extracts

1. Plate Preparation and Manipulation

TLC plates of excellent quality and reproducibility are available from a number of suppliers and their use is highly recommended for pesticide analysis. For most pesticide analyses silica gel or alumina coated plates or sheets have been found to be the most satisfactory.

The sensitivity of TLC procedures can be increased by washing the TLC plate prior to use to remove inorganic ions. The solvent used by Gordon and Hewel [10] to wash filter paper prior to paper chromatographic analysis can be used on some commercial TLC plates. This solvent consists of a mixture of water, pyridine, and acetic acid in the ratio of 80:15:5. The washing solvent mix is allowed to ascend the plate, carrying the inorganic ions to the solvent front. The plate is allowed to dry at room temperature and then re-activated at $100°C$ for about 30 min.

The next step in the analysis of pesticide residues is to spot the extract on the TLC plate at several levels along with a graded series of standards to be run on the same plate. It is often convenient to use a standard solution containing a mixture of common pesticides if the R_f values are known for each pesticide under standard conditions using a given type of TLC plate and solvent mix.

Various mixtures of nonpolar solvents, sometimes with the addition of a small amount of polar solvent, are used to develop the TLC plate. Powell and co-workers [11] used 10% benzene in hexane to develop silica gel GHR plates in their analysis of aldrin and dieldrin.

Lichtenstein and co-workers [3], in their study of the degradation of aldrin and heptachlor, separated soil extracts on aluminum oxide G-coated glass plates. Chromatograms from aldrin-treated soils were developed with isooctane-pyridine, 7:3, or with isooctane-diethyl ether, 7:3, while the chromatograms from five heptachlor-treated soil extracts were developed with isooctane or with cyclohexane-ethyl acetate, 7:3. If the R_f values on the separation of the pesticides and their metabolites are not satisfactory, a slight change in the composition of the developing solvent system can often achieve a satisfactory separation.

In analyzing for metabolites of organophosphorus pesticides a more elaborate solvent system may be necessary to effect separation. Mücke and co-workers [12] used Silica Gel G (Merck) plates to separate and isolate the metabolites of diazinon. The following solvent systems were used:

a. Ethyl acetate-ethanol-ammonia, 80:15:5
b. Toluene-acetic acid-H_2O, 60:60:6
c. Benzene-ethanol, 9:1
d. Dioxane-ethanol-H_2O, 5:3:1
e. Ethanol-ethyl acetate-acetic acid, 3:4:2
f. Acetonitrile-ammonia-H_2O, 40:9:1
g. Benzene-chloroform-ethyl acetate-propanol, 4:2:4:2

These solvent mixtures should be satisfactory for organophosphorus pesticidal compounds that the analyst may encounter.

2. Detection Techniques

After the pesticides have been separated on the thin-layer plate they are detected. Chlorine-containing compounds on thin-layer plates can best be visualized by spraying with the reagents as described by Mitchell [13] and subsequent exposure to short-wave ultraviolet light for 10 min. This chromogenic reagent is prepared in the following manner: 1.7 g of $AgNO_3$ are dissolved in water, 10 ml of 2-phenoxy ethanol and 50 ml of ethanol are added and the solution is diluted with H_2O to 200 ml. (Should temperature changes cause separation of the 2-phenoxyethanol, ethanol is added in 1-ml increments with mixing until it is redissolved.)

In order to quantitate the pesticide, two accurately measured portions of the purified and concentrated extract can be spotted side by side. After development of the chromatogram one-half of the plate is covered and the other half is visualized by the Mitchell technique. The unsprayed portion of the plate corresponding to each of the different spots observed on the sprayed side of the plate is then scraped off and extracted with acetone. The acetone extract is then evaporated at room temperature, and the residue is redissolved in hexane. The pesticides are then quantitated by electron capture GLC analysis or by microcoulometry.

Many systems have been developed to visualize organophosphorus pesticides on thin-layer plates. Unfortunately none of these are completely satisfactory for all compounds.

Sethunathan and Yoshida [14] located compounds of diazinon and two of its potential analogs on TLC plates by spraying with 0.1% Rhodamine B in 95% ethanol and locating the compounds by their characteristic fluorescence on exposure to short-wave ultraviolet light.

Mendoza and co-workers [15] used liver esterases and 5-bromoindoxyl acetate to detect cholinesterase inhibitory organophosphorus pesticides on TLC plates. The areas of inhibition appear as white spots on an intense blue background.

Getz and Wheeler [16] used three selective chromogenic sprays to detect organophosphorus insecticides. These were prepared and used as follows:

a. Silver nitrate-Bromcresol Green was used for detecting the thiophosphoryl configuration. Five milliliters of 1% solution of Bromcresol Green was mixed with 95 ml of a 2% $AgNO_3$ in aqueous methanol ($AgNO_3$ is dissolved in just enough water and added to the methanol). The TLC plate was sprayed uniformly with this solution and dried in a forced draft oven at no higher than $100°C$. The plate was then cooled and sprayed with pH 4.4 citrate or acetate buffer. The plate was again dried at $100°C$. The background turned yellow and a bluish-black or magenta spot indicated the presence of an insecticide with a thiophosphoryl configuration.

b. 4-(p-Nitrobenzyl)pyridine was used for general detection. A 2% solution in acetone and a 10% solution of tetraethylenepentamine in acetone was prepared. The plate was sprayed with the 4-(p-nitrobenzyl)pyridine solution and heated at $100°C$ for 10 min, then sprayed with the solution of tetraethylenepentamine. The insecticides appeared as purple-blue spots on a white background.

c. A serum cholinesterase inhibition procedure was used to detect the P=O configuration   Twenty milliliters of horse serum (Difco) and 15 ml of indicator (1.2% Bromothymol Blue in 0.1N NaOH) were mixed with 2 ml of 0.1N NaOH and 60 ml of water. A 20 x 20 cm sheet of Whatman chromatographic paper was sprayed with the solution of enzyme-dye until the paper was saturated. The sheet was placed in contact with the TLC plate, and a glass plate was placed on top as a weight. The enzyme was allowed to incubate for 30 min. The TLC plate was then sprayed with a 2% solution of acetylcholine chloride. The inhibiting insecticide was detected as a blue spot on a yellow background. The reagents for this procedure should be prepared just prior to use.

As in the case of chlorinated pesticides, organophosphorus pesticides can best be quantitated by scraping the spots from the TLC plate, extracting the pesticide from the TLC material with methanol, and quantitating the organo-phosphorus pesticide by gas chromatography. Phosphorus-containing materials can be quantitated by using a thermionic detector or the flame photometric detector in the phosphorus mode.

If sufficient sample is available and the compound in question contains sulfur, quantitation may be by microcoulometry or flame photometry in the sulfur mode.

Abbott and co-workers [17] used TLC for the separation and identification of residues of urea and carbamate herbicides. They found that most of these compounds could be detected on TLC plates by spraying with a 2% solution of p-dimethylaminobenzaldehyde in ethanol followed by 2N aqueous sulfuric acid and heating at 150°C for 20 min.

Geissbuehler and Gross [18] developed a specific detection of urea herbi-cide residues by separating their amines on cellulose TLC plates. The phenylurea herbicides were hydrolyzed with alkali and steam distilled, and the amines were extracted into hexane. The amines were then extracted into 0.5N HCl and diazotized. The diazotized amines were reacted with N-ethyl-1-naphthylamine to form colored compounds. The dyes were extracted into ethyl ether and spotted on plates coated with a 300 μm layer of cellulose powder along with appropriately prepared standards. The plates were developed for 15 min with an HCONMe-0.5N HCl-EtOH (60:20:20) in a solvent-saturated chamber. The authors recorded colors and R_f values for most of the commonly used phenylurea herbicides.

3. Quantitation Techniques

If the soil extracts containing the pesticides have been thoroughly cleaned up using a procedure such as the Florisil column, they can be used for confirm-atory analysis and quantitation by GLC, using selective detectors. Thin-layer analysis is a semiquantitative technique, so it is usually best to use a second procedure for quantitation. If the sample is analyzed by TLC, GLC, and by removing the spot from the TLC plate, dissolving the pesticide in a solvent and analyzing this material by GLC, the pesticide can be identified and quantitated with reasonable accuracy. The spot removed from the TLC plate for GLC analysis must not have been reacted with any detecting reagent nor contain any chemical that interferes with GLC analysis. The extracts from several spots can be combined for GLC analysis if the level of the pesticide is low.

IV. INORGANIC ION ANALYSIS

The trace element content of a soil is dependent almost entirely on that of the rocks from which the soil parent material was derived and on the processes of weathering to which the soil-forming materials have been subjected. The uptake of trace elements by plants is the obvious criterion of their availability, but this availability of a trace element in a soil, as shown by the capacity of a plant to take it up, is as much a function of the plant as of the soil. By appropriate choice of soil extractant, some indication can be obtained regarding the form in which trace elements occur.

A. Extraction and Cleanup Procedure

The collection, preparation, and storage of the sample represent important steps in the analytical operation. For samples on which determinations of trace element content are to be made, great care should be exercised to avoid contamination during and after sampling. Since soils are very heterogeneous, the analysis of replicate samples should be considered. Blank determination should be made simultaneously with each group of samples. The blanks should include all reagents used in the determination.

For the determination of adsorbed cations, the choice of extractant must be guided by the degree of extraction required. Furthermore, the extractant must be such that it can be removed prior to TLC analysis. Ammonium acetate at pH 7 and acetic acid (2. 5%) at pH 2. 5 are such extractants. Procedures utilizing chromic acid digestion or sodium carbonate fusion cannot be used since these techniques add chemicals that interfere with the TLC analysis.

After the soil is extracted, some of the organic materials can be removed by solvent extraction of the aqueous layer. The ammonium acetate and acetic acid can be removed by heating the extract to near dryness and redissolving any solid material for TLC analysis.

The use of organic complexing agents producing solvent-soluble complexes with metallic ions should also be considered as a preliminary separation and cleanup technique.

B. TLC Analysis of Extracts

The solvent systems used for inorganic TLC analysis are in general extensions of those used for paper chromatography. These systems are generally mixtures of organic solvents together with aqueous acid, base, or buffers.

Many TLC materials contain such metallic ions as iron, alkali metals, and alkaline earth ions. These can often be removed by acid treatment of the adsorbent prior to coating the TLC plate. Commercial plates can be washed prior to use.

1. Plate Preparation and Manipulation

Bark and co-workers [19] used reversed-phase TLC of metal ions with tributyl phosphate (TBP) at various concentrations of hydrochloric acid. These authors prepared their own TLC plates in the following manner: 15 g of cellulose powder was mixed with a solution of purified tributyl phosphate in carbon tetrachloride (70 ml of a 5% solution) to give a homogeneous slurry that was spread as an even layer 0.3 mm thick for five 20 x 20 cm glass plates. The layers were allowed to air dry in an air oven at room temperature for 1 hr to remove the carbon tetrachloride. This operation should be carried out in a fume hood. The samples in water were spotted on the plate in 1-μl aliquot and allowed to dry for a standard time.

The plates were developed using various concentrations of HCl (0.1-9M). The solvent was allowed to move a fixed distance on the plates. The development took from 1 to 3 hr. The plates were then removed from the development chamber and heated for 20 min at 100° ± 2°C to remove the acid and most of the TBP. It is necessary to remove most of the TBP as indicated to allow the chromogenic indicators to form complexes with some of the "free" metal ions.

2. Detection Techniques

After the plates are developed visual identification of the metal ions is made by spraying the plate with the chromogenic indicators. The chromogenic reagents are prepared as follows:

a. 3,5,7,2',4'-Pentahydroxyflavone (morin), 0.1% in ethanol
b. 8-Hydroxyquinoline, 2% in chloroform

TABLE 1

Inorganic Detection Techniques for TLC in Soil Chemistry

Treatment	Appearance	Metal
Sprayed with morin, exposed to NH$_3$, and held under ultra-violet light	Yellow fluorescent spots	Alkali metals
Sprayed with 8-hydroxyquino-line, exposed to NH$_3$, and held under ultraviolet light	Yellow fluorescent spots	Alkaline earths and Al
Sprayed with a 1 + 1 mixture of PAN and p-dimethylamino benzylidenerhodanine and viewed in visible light	Red spots Grey spots	Cr, Mn, Ni, Cu, Zn, ZrO^2+, Cd, In, Sn, Hg, Tl Fe(II), Fe(III)

c. 1-(2-Pyridylazo)-2-naphthol (PAN), 0.1% in ethanol

d. p-Dimethylaminobenzylidenerhodanine, 0.1% in ethanol

Table 1 lists the chromogenic reagent, the appearance of the plate, and the ion detected.

3. Quantitation Techniques

The ions can be quantitated on another portion of the extracts by an instrumental technique, such as atomic absorption spectrophotometry. Another technique for the TLC analysis of metal inorganic ions is to form the dithiazone derivative by conventional procedures and to chromatograph this derivative on Silica Gel G using a solvent mixture of benzene and methylene chloride (50 : 10 v/v). As metal dithiazonates are differently and characteristically colored, no spray reagent is needed for detection. The spots can be scraped from the plate and the metal dithiazonates dissolved in a suitable solvent and quantitated by spectrophotometry.

V. ANALYSIS OF THE ORGANIC CONSTITUENTS OF SOIL

The organic fraction of the soil is a complex mixture of substances the composition of which is determined by the plant and animal residues added to the soil and by the biological, physical, and chemical transformations of these substances; therefore, the soil contains a vast array of compounds. No systematic identification of these compounds has been undertaken. However, a beginning has been made on the study of these chemicals in soil. Ritchie [20] reviewed the recent advances in the chromatographic analysis of geologic materials with special emphasis on the TLC analysis of rocks, minerals, and soils.

Gupta and Sowden [21] determined sugars and uronic acids in soils. The soil was treated with 72% sulfuric acid prior to hydrolysis with 1N sulfuric acid. The carbohydrates were then separated by TLC and detected by spraying the plates with phenol-sulfuric acid.

Humus-containing soils contain fatty acids, triglycerides, and steroids. Tsunoda and Oba [22] identified triglycerigdes, cholesteryl stearate, squalene, cholesterol, and higher fatty acids in soil extracts by gel filtration followed by TLC analysis.

Wang and co-workers [23] made an extensive study of soil lipids. The soils were treated for 48 hr with 1.25% HF-1.25% HCl (1 : 1) and extracted with $CHCl_3$-MeOH (2 : 1) and then with methanolic NaOH (pH 11). The soil lipids obtained by this procedure were resolved into classes by TLC. The higher fatty acids were purified by methylation and rechromatographed on silica gel containing 10% Na_2CO_3. Further analysis and identification was then made by GLC analysis.

Polycyclic aromatic hydrocarbons can be extracted from soils high in organic matter. Hansen and Schnitzer [24] extracted humic and fulvic acid from soil and subjected these acids to zinc dust distillation and fusion. The reaction products were purified by vacuum sublimation and then separated by a combination of preparative TLC on silica gel and on cellulose. The polycyclic aromatic hydrocarbons were identified by UV spectroscopy and spectrophotofluorometry.

Kunte [25] determined ten polycyclic aromatic hydrocarbons in water, dust, soil, and plant extracts using TLC and located and identified the separated materials by fluorescence.

Thin-layer chromatography can also be used to detect mineral and tar oils in soil. The oils can either be naturally occurring or a result of leaks or pollution. Kreiger [26] described a rather elaborate procedure for determining oil in soil. With the increased effort toward the prevention of oil pollution these techniques offer a rapid procedure for the detection of leaks.

Preparative TLC offers a rapid method for the isolation of large amounts of materials from soil which can then be quantitated by other techniques. Since soil contains such a complex mixture of materials TLC should always be considered for the first purification step in the isolation of a chemical of interest. The isolated chemical can then be quantitated by other techniques.

REFERENCES

1. H. B. Pionke and G. Chesters, Soil Sci. Soc. Amer. Proc., 32, 749 (1968).

2. Shell Development Co., Agricultural Division, Modesta, California, 1963. Extraction of Halogenated Hydrocarbon Pesticide Residues from Crops, Soils and Animal Products. Analytical Methods MNS-1/63.

3. E. P. Lichtenstein, K. R. Schultz, T. W. Fuhremann, and T. T. Liang, J. Agr. Food Chem., 18, 100 (1970).

4. J. G. Saha, Bull. Env. Contam. Toxicol., 3, 26 (1968).

5. M. Chiba, W. N. Yule, and H. V. Morley, Bull. Env. Contam. Toxicol. 5, 263 (1970).

6. D. P. Johnson, H. A. Stansburg, Jr., J. Assoc. Off. Agr. Chem., 49, 399 (1966).

7. H. Beckman, K. Carroll, W. Thornburg, R. I. Vetro, R. D. Smith, A. Mittler, M. B. Keckes, A. Bevenue, and A. Kawano, J. Assoc. Off. Agr. Chem., 50, 1251 (1967).

8. W. W. Sans, J. Agr. Food Chem., 15, 192 (1967).

9. D. J. Sissons, G. M. Telling, and C. D. Usher, J. Chromatog., 33, 435 (1968).

10. H. T. Gordon and C. A. Hewel, Anal. Chem., 27, 1471 (1955).

11. A. J. B. Powell, T. Stevens, and K. A. McCully, J. Agr. Food Chem., 18, 224 (1970).

12. W. Mücke, K. O. Alt, and H. O. Essers, J. Agr. Food Chem., 18, 208 (1970).

13. L. C. Mitchell, J. Assoc. Off. Agr. Chem., 40, 999 (1957).

14. N. Sethunathan and T. Toshida, J. Agr. Food Chem., 17, 1192 (1969).

15. C. E. Mendoza, P. J. Wales, D. L. Grant, and K. A. McCully, J. Agr. Food Chem., 17, 1196 (1969).

16. M. E. Getz and H. G. Wheeler, J. Assoc. Off. Agr. Chem., 51, 1101 (1968).

17. D. C. Abbott, K. W. Blake, K. R. Tarrant, and J. Thomson, J. Chromatog., 30, 136 (1967).

18. H. Geissbuehler and D. Gross, J. Chromatog., 27, 296 (1967).

19. L. S. Bark, G. Duncan, and R. J. T. Graham, Analyst, 92, 347 (1967).

20. A. S. Ritchie, Quart. Colo. School Mines, 64, 427 (1969).

21. U. C. Gupta and F. J. Sowden, Can. J. Soil Sci., 45, 237 (1965).

22. T. Tsunoda and Y. Oba, Nippon Kagaku Zasshi, 89, 831 (1968).

23. T. S. C. Wang, Yu-C. Liang and Wei-C. Shen, Soil Sci., 107, 181 (1969).

24. E. H. Hansen and M. Schnitzer, Soil Sci. Soc. Amer. Prod., 33, 29 (1969).

25. H. Kunte, Arch. Hyg. Bakteriol., 151, 193 (1967).

26. H. Krieger, Gas-Wasserfach, 104, 695 (1963).

Chapter 10

ION-EXCHANGE ANALYSIS IN SOIL CHEMISTRY

Harold F. Walton

University of Colorado
Department of Chemistry
Boulder, Colorado

I. INTRODUCTION

It was in soil that ion exchange was first recognized by the English agricultural chemists H. S. Thompson [1] and J. T. Way [2] in 1850. The ion-exchanging power of soil is the main reason for "The Power of Soil to Absorb Manure," to quote the title of Way's classic paper, and it is the subject of many publications today. However, the use of ion-exchange chromatography have appeared on this subject in recent years. Most soil analyses must be rapid and simple and need not be highly precise. Ion-exchange chromatography is generally a slow process and one that requires a certain amount of skill.

To analyze the mineral constituents of soil, in the sense of determining the chemical elements present, is the same thing as analyzing a rock. Much use has been made of ion exchange in rock analysis in cases where there are special requirements for high precision and where small amounts of trace constituents must be separated and concentrated. Scanning the pages of a journal, such as <u>Geochimica et Cosmochimica Acta</u>, the careful reader will note many examples of the use of ion exchange to separate and concentrate trace constituents in rocks, usually in the chemical treatments following neutron activation analysis. Ion exchange has been used for this purpose in the analysis of lunar rocks. A full discussion of these topics is outside the scope of this book. The reader is referred to a recent cooperative text on geochemical analysis [3] and to recent publications by Strelow and collaborators [4-7]. The point is that in spite of the selectivity and sensitivity of such techniques as atomic absorption spectroscopy, X-ray spectroscopy, ultraviolet emission spectroscopy, and activation analysis, there is no substitute for careful and effective chemical separations where really precise and accurate elemental analysis is required.

I cite a recent reference to the analysis of rocks and soils [8] because of the significance of lead and molybdenum as pollutants. It describes the ion-exchange concentration of trace elements from rocks and soils as a prelude to spectrochemical analysis. The sample is ashed at 450° C (if a soil) and then dissolved in a mixture of concentrated hydrochloric and hydrofluoric acids. The hydrofluoric acid and silicon fluoride are removed by evaporation; then the solution is made about 1.5M in hydrochloric acid and passed into a column of strong-base anion-exchange resin that has been previously washed with 1.5M HCl. Hydrochloric acid of this concentration is passed. Many elements, including the alkali and alkaline earth elements and aluminum, are washed out of the column immediately, followed by cobalt. Then molybdenum is eluted, followed by lead. The elements Ag, Bi, Sb, Sn, Zn, and Cd are held more strongly and are removed by washing with decreasing concentrations of hydrochloric acid. They are recovered in concentrated form, as solutions that can easily be evaporated and analyzed by emission spectroscopy without interference from large backgrounds of major constituents.

This book is concerned with man's contamination of his natural environment, and for the remainder of this section I shall restrict myself to this topic.

II. ACCURACY AND PRECISION

As pointed out in Chapter 1, ion exchange is the least useful chromatographic technique for quantitative or qualitative analysis. Stating any values pertaining to accuracy and/or precision would thus be meaningless. However, information can be given regarding selectivity and resolution. "Selectivity" refers to the preference of the ion exchanger to accept one ion over another. It is possible to draw some generalities with regards to one exchange as compared to another An ion exchanger prefers (has the tendency towards):

a. The counterion of higher oxidation state
b. The counterion possessing the greatest polarizability
c. The counterion requiring the least solvated equivalent volume
d. The counterion interacting the most with the resin's fixed ionic groups
e. The counterion that participates in complex formation, to the lesser extent, with the co-ion

Increasing the degree of crosslinking or decreasing the temperature and solution concentration of counterions generally enhances selectivity of an ion exchanger. In light of the above discussion a selectivity sequence of the most commonly encountered anions and cations can be put forth [9]:

a. For anion-exchange resins:
   Citrate > sulfate > oxalate > iodide > nitrate > chromate > bromide > thiocyanate > chloride > formate > acetate > fluoride.

Hydroxide falls before fluoride on strong-base resins and moves to the left as the base strength decreases.

b. For cation-exchange resins:
   $Ba(II) > Pb(II) > Sr(II) > Ca(II) > Ni(II) > Cd(II) > Cu(II) > Co(II) > Zn(II) >$
   $Mg(II) > UO_2(II) > Tl(I) > Ag(I) > Cs(I) > Rb(I) > K(I) > NH_4(I) > Na(I) > Li(I)$

The hydrogen ion varies in position depending on the strength of the resin; e.g., for a strong-acid resin, hydrogen comes after Na(I) and moves to the left as the acid strength of the resin decreases.

III. CONTAMINANTS FROM FUNGICIDE SPRAYS

It is common practice to spray fruit trees and ornamental shrubs and trees with fungicides and disinfectants containing copper. Other elements, including zinc, arsenic, and lead, are also used. The leaves of the plants and the soil under the leaves must be analyzed from time to time, and occasionally ion exchange has been used to aid in this analysis. Fourcy [10] described a systematic analysis of the metal content of vine leaves, grapes, wine, and the soil from whence they came. Organic material was ashed by heating the samples to $450°C$, and then the ashes were extracted with hydrochloric acid. Copper was separated at this point by electrolysis at a controlled cathode potential. The solution, free of copper, was passed through the strong-base anion-exchange resin Dowex-1 x 10, 200/400 mesh, in a column 6 mm in diameter with a 12-cm depth of resin.

Phosphoric acid was eluted with 8N HCl, then cobalt with 5N HCl, iron with 0.5N HCl, and finally zinc with 0.005N HCl. Portions of 15-30 ml of each concentration were used. Potassium and sodium ions passed through the anion-exchange resin at the beginning. They were separated from phosphoric acid by passing them through a small column of the cation-exchange resin Dowex 50 x 12; phosphoric acid continued through the column, while sodium and potassium ions were retained. Sodium was eluted with 0.2N HCl and potassium with 0.5N HCl.

In the work cited, the elements were determined by neutron activation; the ash, free from organic matter, was irradiated in a nuclear reactor before the separations were performed.

Zinc was determined in soils and plants by a similar method without neutron activation [11]. The plant material or soil was ashed and then extracted with concentrated hydrochloric acid to get the "total" zinc. This is contrasted with the "extractable" zinc, dissolved out of the soil by hydrochloric acid at pH 3.0 without previous ashing. The distinction between "total" and "extractable" constituents of soil is very important. Sometimes it is necessary to know the total amount of a particular constituent, such as phosphate or a trace metal, but more often it is necessary to know the amounts available as nutrients to plants; this is the "extractable" concentration, measured after extracting the soil with a mild reagent in a prescribed manner.

In the work mentioned [11], the extracts were passed into a column of the strong-base anion-exchange resin Dowex 1 x 8 in the chloride form. Zinc was absorbed as its chloride complex. From hydrochloric acid solutions more concentrated than 0.1M the absorption of zinc as the chloride complex $ZnCl_4^{2-}$ is very strong. After the column was washed to remove unabsorbed metallic constituents, the zinc was eluted (so the abstract says) with 1M KCl. (A better choice might have been 0.01M KNO_3.) It was determined spectrophotometrically with zincon.

The zinc content of soils is commonly 20 ppm or greater. Zinc is therefore a poor indicator of industrial pollution. Cadmium, the chemical behavior of which is similar to zinc, is much less common and much more toxic than zinc. Its presence in amounts exceeding 1 ppm usually does indicate pollution. Cadmium is more strongly held than zinc on an anion-exchange resin at hydrochloric acid concentrations below 0.1M.

A good way to separate cadmium from zinc and most other metals is to absorb cadmium and zinc together from 0.1M hydrochloric acid on a resin, such as Dowex-1, and then elute with 0.01M hydrochloric acid. Zinc comes out first and cadmium later. Separation is improved by adding 10-25% of methanol [12]. Once the zinc is out of the column, the cadmium can be eluted more quickly by changing to 0.001M hydrochloric acid.

Another approach is to use the fact that the iodide complex of cadmium is much more stable than that of zinc. The solution containing zinc and cadmium is made 0.3M in potassium iodide and 0.4M in sulfuric acid and then passed through a small column of anion-exchange resin, about 5 cm long, in the chloride form. Cadmium is absorbed along with some of the zinc. The absorbed zinc is removed by passing more 0.3M KI-0.4M H_2SO_4; then, when all the zinc is removed, cadmium is eluted with 2M nitric acid [13].

Although these procedures were designed for ores and metallurgical samples, they should be applicable to soils. Presumably, the final measurement would be made by atomic absorption spectroscopy. If so, it would not be necessary to separate cadmium completely from zinc, only to recover it without loss and free from most of the accompanying elements.

IV. RADIOACTIVE CONTAMINANTS

Artificial radioactive elements could not be present naturally in soil, and their presence ipso facto indicates contamination. The most serious contaminant, and the one to which most attention has been given, is plutonium. In locations where nuclear fuel is processed, the determination of traces of plutonium in soil is a matter of urgency.

A. Plutonium

Plutonium is an α emitter and is determined by counting its α ray emission. Because α rays are easily absorbed the counting must be done on a sample that is essentially pure. Plutonium metal is electroplated from solution onto a disk of stainless steel, platinum, or tantalum, and the emission from the electroplated deposit is then counted. To compensate for possible losses during chemical preparation an internal isotopic standard, ^{236}Pu is added. This is a short-lived isotope of plutonium, the common isotope being the long-lived ^{239}Pu. The α rays from the two isotopes have different energies and can be compared by a multichannel counter.

An early method for separating plutonium from iron and other elements by ion exchange [14] depended on the relative stabilities of the EDTA complexes. One hundred grams of soil was ashed at 450 °C and then extracted overnight with 6M HCl. The extract was diluted to 1M and made 0.1M in EDTA and was then passed through a column of strong-acid cation-exchange resin, Dowex 50 x 8. Iron and thorium form anionic or uncharged EDTA complexes and pass through the resin column, while plutonium, the complexes of which are less stable, is retained. The column is washed with 1N HCl and then with 3N HCl, which strips off the plutonium as Pu(III) and Pu(IV). A second purification is now performed. The plutonium is absorbed on a strong-base anion-exchange resin from 9N HCl and then eluted by a solution 6N in HCl and 0.2N in HF [15]. It is electrodeposited from this solution, which contains plutonium only.

The absorption on an anion-exchange resin of the anionic nitrate complex of Pu(IV) was used by Chu [16] and Wong [17] to separate plutonium from other metals. In the method of Chu, soil (100 g) was leached with concentrated nitric and hydrochloric acids for a few hours after adding the ^{236}Pu tracer. The extract was evaporated and taken up in 8M HNO_3. The solution was heated and 2 ml of 5% sodium nitrate were added to oxidize the plutonium to Pu(IV). Five milliliters of a slurry of strong-base anion-exchange resin was added and stirred to start the absorption of the plutonium; then the solution and suspended resin were poured onto a column of resin 3 cm in diameter and 30 cm long. The resin was 4% crosslinked, instead of the customary 8%, to give faster absorption. The column retained the plutonium. After it was washed with 8M nitric acid and 12M hydrochloric acid to remove accompanying elements, the plutonium was eluted with 100 ml of $0.4M$ HNO_3-$0.01M$ HF. In the more dilute acid the plutonium changes from the 4+ oxidation state to 3+, where it is easily eluted.

Wong's procedure [17] was similar in the sense that plutonium was absorbed by the anion-exchange resin from 8M nitric acid containing sodium nitrite plus hydrogen peroxide, but it was eluted with concentrated hydrochloric acid containing ammonium iodide. The iodide reduced plutonium to Pu(III). The method was used for sea water as well as for soils and sediments With sea water, iron was added and the plutonium coprecipitated with Fe_2O_3. The tracer [236] Pu was added in every case. This is necessary, as the total recovery, or chemical yield, was less than 100% and sometimes as low as 60-70%.

Talvitie [18] modified the procedure to absorb plutonium on an anion-exchange resin from 9M HCl. Hydrogen peroxide was added beforehand to insure the formation of Pu(IV), which forms a stable, strongly absorbed chloride complex. A resin of low crosslinking, 2%, was used to speed absorption and desorption. Iron was washed out of the column with 7.2M HNO_3; then plutonium was eluted with 1.2M HCl containing hydrogen peroxide. Again, the more dilute acid caused reduction of Pu(IV) to Pu(III), in which oxidation state it is not held by the resin.

Again, [236] Pu tracer was used, and the final measurement was made by alpha counting of an electrodeposited disk of plutonium metal. Chemical yields ranged from 83 to 102%, and the average standard deviation of the plutonium-239 determination was 4%. The method was applied to sea water, urine, bone ash, and biological samples, as well as to soils and sediments.

A much simpler, but rougher, method for determining plutonium in soils was described by Crisco and Lada [19]. It depends on the absorption of Pu(IV) from 7.5M nitric acid by paper impregnated with a strong-base ion-exchange resin.

B. Strontium-90

Radioactive strontium-90 was measured in soils and environmental samples by Ibbett [20]. The soil was extracted with hydrochloric acid, excess acid was evaporated, and the residue was taken up in a solution containing both EDTA and citrate. Under these conditions, but calcium and magnesium are more strongly complexed than strontium. The solution was passed through a column of strong-acid cation-exchange resin at pH 5.0. Calcium, magnesium the rare earths, iron, and aluminum all passed, while strontium was retained The column was washed with water to remove EDTA and citrate; then the strontium was eluted with 2M hydrochloric acid.

C. Uranium

Uranium may or may not be a contaminant. Traces of uranium are wide-spread in surface waters, ground waters, soils, and rocks. Uranium in its highest oxidation state (6+) forms soluble compounds that are easily transported by water. Moreover, it is not particularly toxic. Readers interested in the environmental chemistry of uranium and the analytical methods used in its study should consult the work of Korkisch [21, 22], who has developed highly efficient, selective ion-exchange methods to separate traces of this element from soils, water, and rocks. These methods combine ion exchange with solvent extraction. For example, uranium is extracted, together with iron, copper, and other metals, by tributyl phosphate from 6M hydrochloric acid. The extract is mixed with methyl glycol and dilute aqueous hydrochloric acid to give a homogeneous solution, which is then passed through a column of anion-exchange resin. Iron, copper, and other metals (which form uncharged ion pairs under these circumstances) are not absorbed by the resin, while uranium is absorbed; it is then eluted selectively with 1M hydrochloric acid.

V. GENERAL CONCLUSIONS

Although the number of references to the analysis of contaminated soils by ion-exchange chromatography is small, the methods used for water analysis (Chapter 15) can be adapted to soils with little modification. All that is needed is to bring the constituents into solution before using these methods. As was mentioned in Sec. III, there is a difference between the total concentration and the extractable concentration of an element. The extractable concentration is an arbitrary quantity, for it depends on the method of extraction, but it may yield more useful information than the total concentration, which includes elements present in grains and crystals of insoluble minerals.

REFERENCES

1. H. S. Thompson, J. Roy. Agr. Soc. England, 11, 68 (1850).

2. J. T. Way, J. Roy. Agr. Soc. England, 11, 313 (1850); 13, 123 (1852).

3. R. E. Wainerdi and E. A. Ukem (eds.), Modern Methods of Geochemical Analysis, Plenum, New York and London, 1971.

4. F. W.E. Strelow and C. H. S. W. Weinert, Talanta, 17, 1 (1970).

5. F. W. E. Strelow and M. D. Boshoff, Talanta, 18, 985 (1971).

6. F. W. E. Strelow, A. H. Victor, and E. Eloff, J. South African Chem. Inst., 24, 27 (1971).

7. F. W. E. Strelow, A. H. Victor, C. R. van Zyl, and E. Eloff, Anal. Chem., 43, 870 (1971).

8. H. H. LeRiche, Geochim. Cosmochim. Acta, 32, 791 (1968).

9. Dowex Ion Exchange, The Dow Chemical Company, Midland, Mich., 1959.

10. A. Fourcy, Compt. Rend., 261, 830 (1965).

11. A. Dartigues, Ann. Agron., 17, 75 (1966); Chem. Abstr., 65, 20487g (1966).

12. E. W. Berg and J. T. Truemper, Anal. Chem., 30, 1827 (1958).

13. S. Kallmann, G. Oberthin, and R. Lin, Anal. Chem., 30, 1846 (1958).

14. M. C. deBortoli, Anal. Chem., 39, 375 (1967).

15. L. Wish, Anal. Chem., 31, 326 (1959).

16. N. Y. Chu, Anal. Chem., 43, 449 (1971).

17. K. M. Wong, Anal. Chim. Acta, 56, 355 (1971).

18. N. A. Talvitie, Anal. Chem., 43, 1827 (1971).

19. C. Crisco and H. F. Lada, Rept. U.S. Army Nucl. Defense Lab., AD-665357 (1968).

20. R. D. Ibbett, Analyst, 92, 417 (1967).

21. J. Korkisch and I. Steffan, Mikrochim. Acta, 837 (1972).

22. J. Korkisch and W. Koch, Mikrochim. Acta, 865 (1973).

PART IV

WATER POLLUTION SECTION

Chapter 11

GAS CHROMATOGRAPHIC ANALYSIS IN WATER POLLUTION

Barbara S. Jacobson

Ross Laboratories
Columbus, Ohio

399

I. INTRODUCTION

Gas chromatography is an analytical tool well suited to the separation and quantitation of minute amounts of material from complex mixtures. Microgram to picogram levels of organic impurities in water samples can be determined. This makes it particularly useful and popular in water quality analyses. Identifying and quantitating impurities and pollutants are important steps in the control of water quality and the elimination of sources of pollution.

Any compound that can be volatilized without decomposition, or with reproducible decomposition, can be analyzed by gas chromatography with the proper choice of columns, detectors, and conditions. Preliminary elimination of compounds that may interfere can be accomplished by liquid-liquid extraction, liquid chromatography, thin-layer chromatography, or other means.

This chapter discusses many of the techniques and methods used in the gas chromatographic analysis of impurities in water samples. It provides examples of some of the problems that have been experienced and the approaches that have been taken to solve them. It is by no means a comprehensive study of all facets of the field but it is a basis on which a prospective analyst can build.

II. ACCURACY AND DETECTION LIMITS

The accuracy of a gas chromatographic method is determined by the degree of analyte recovery achieved in each phase of the analysis. The following points should be considered.

Samples collected must be representative of the waters being sampled. Collection techniques must not alter the concentration of the sample components. Precautions must be taken to prevent biological or chemical degradation of the sample before analysis. The sample cleanup or derivative preparation techniques must provide complete, or at least reproducible, recovery of the analyte.

The gas chromatographic columns must resolve the analyte from all other compounds. The proper selection of a detector can be as important as the selection of the column for separation. A detector that is sensitive to the analyte and relatively insensitive to interfering compounds will effectively "separate" the analyte simply by not measuring the other compounds. Injection port and column conditions must be selected so that the analyte retains its identity without suffering partial degradation, rearrangement, or reactions that would yield false results. Sample sizes must be selected so that the quantity of analyte reaching the detector is within its dynamic operating range.

Spiking the sample with known amounts of the analyte at various stages throughout the analysis can help pinpoint losses and help estimate the

accuracy. Use of an internal standard added at the beginning of the assay
can help compensate for any losses that do occur. To yield accurate
results, the internal standard must, of course, undergo exactly the same
concentration changes as the analyte and have the same sensitivity to the
detector used.

The detection limits are determined by the detector, the type and quantity
of interfering compounds present, and the degree of concentration achieved
in the sample preparation. In the absence of interfering peaks, the thermal
conductivity detector is capable of detecting quantities of material in the
microgram range; the coulometric detector in the nanogram range; the
flame ionization detector in the 10 pg range; and the electron capture and
thermionic (alkali flame) detectors in the picogram range [1].

III. SAMPLE ACQUISITION

Water samples for analysis by gas chromatography are generally collected
as grab samples or by means of one of the liquid chromatographic collection
techniques described in Chapter 12, Sec. II. Compounds of interest are
extracted from adsorption columns, and often from grab samples, into
organic solvents before analysis. Since many components of the samples
may be subject to biological or chemical degradation, care should be taken
to preserve their integrity. Containers used must not adsorb compounds
from the solution. The analysis should be performed as soon as possible
after collection of the sample. Sampling precautions are discussed by
Faust and Suffet [2]. Precautions that should be considered when handling
samples in the lab are discussed by Bevenue et al. [3].

Volatile components of a water pollution sample present special sampling
problems. Care must be taken to prevent loss of the analyte into the head
space of the sampling container. (Alternatively, conditions can be adjusted
specifically to drive all volatile components into the head space and then
the head space gas can be analyzed. The problem is then reduced to one of
gas-sample analysis, and the techniques discussed in Chapter 2 are more
applicable.) Special considerations for specific sample types are discussed
in Sec. V of this chapter.

IV. SEPARATION OF INTERFERENCES

While gas chromatography is a tool designed to be used in separating a mixture into its components, it cannot always be expected to completely separate
a mixture as complex and diverse as polluted water. In some cases it is
possible to inject a water sample directly into the gas chromatograph,
without preliminary cleanup, and produce meaningful results. More often it
is necessary to remove some of the unwanted components prior to gas chromatographic analysis. Occasionally it is advantageous to prepare a derivative

of the component of interest.   The amount of preliminary cleanup necessary
is dependent on the quantity and character of the component of interest and
on the type and amount of any interfering compounds.

A. Cleanup and Concentration

Many types of cleanup techniques are applied in the preliminary preparation
of water samples for gas chromatographic analysis. If the component of
interest is soluble in an organic solvent, it can be extracted from the water
sample into hexane [4-11], chloroform [12-15], ethyl ether [16], benzene
[17], combinations of these solvents [18-19], or other solvents [20-22].
By back extracting from the organic layer with aqueous phases of various pH,
various classes of compounds can be separated. For example, Rosen et al.
[23] analyzed water contaminated with petroleum products. They dissolved
the residue from the oil pollution sample in ethyl ether. The ether was then
extracted successively with water, acid, sodium bicarbonate, and sodium
hydroxide solutions to separate neutral fraction, bases, strong acids, and
weak acids.

Often a liquid chromatographic column is employed for further sample
cleanup. The theory and technique of liquid chromatography of water pollution
samples are discussed in Chapters 1 and 12. A comparison of several types
of columns used for cleanup in pesticide analysis has been made by Versino
et al. [24]. Table 1 lists some of the types of columns that are used with
various kinds of samples.

Thin-layer chromatography is also used as a cleanup step for gas chromatographic analyses [9, 34], although not nearly as frequently as liquid
chromatography. The theory and the technique of thin-layer chromatography
of water pollution samples are discussed in Chapters 1 and 14. Once the
separation by TLC is achieved, the spot containing the component of interest
is usually removed from the plate by scraping. The analyte is then extracted
from the coating material with a suitable solvent and the extract is analyzed
by gas chromatography.

Sample cleanup has also been accomplished by using chemical methods
to destroy interfering compounds. One example of this approach was reported
by Sodergren [35], who used fuming sulfuric acid or potassium hydroxide to
eliminate interferences in organochlorine pesticide analysis.

Concentration of samples, whether in the original aqueous solution or
after extraction into an organic solvent, can often be accomplished by evaporating the solvent in a concentrator or under vacuum. Care must be taken,
however, since volatile components may be lost by this technique. The
extraction of the analyte into an organic solvent, or a liquid or thin-layer
chromatographic step, may in itself serve to concentrate the sample.

Another approach used in separating trace organic contaminants from
water samples is the "freeze-out" technique. When the temperature of a
sample is reduced, the aqueous and organic fractions freeze at different

TABLE 1

Liquid Chromatographic Columns Used for Sample Cleanup

Type of sample	Type of column	Eluting solvent	Reference
Pesticides	Silica gel	Benzene	25
		Hexane–benzene	26
		Hexane, hexane–benzene, or hexane–ethyl ether	27
	Florisil	Benzene	28
		Hexane–ethyl ether	27, 29
	Alumina	Hexane, hexane–benzene	27
		Chloroform	30
	Silicic acid–Celite	Mixture of acetonitrile, hexane, and methylene chloride	31
Chlorinated naphthalenes	Silicic acid–Celite	Petroleum ether	32
Polychlorinated biphenyls	Florisil	Hexane–ethyl ether	19
	Silica gel	Hexane	4
	Silicic acid–Celite	Petroleum ether	32
Polynuclear aromatic hydrocarbons	Silica gel	Benzene	14
Phenols	Florisil	Ethyl ether	33

points. If the temperature is carefully controlled, the liquid organic fraction can be separated from the solid aqueous fraction by filtration. If an organic internal standard is added to the sample prior to the freezing step, complete recovery of the organic solutes is not necessary. This technique simultaneously separates and concentrates the organic components without loss of volatile substances. Phenols, acids, and guaiacol [36]; pesticides [29]; and other organic solutes [37] have been separated from water samples by this technique.

B. Derivative Preparation

There are three main reasons that the preparation of derivatives of sample components may be advantageous. First, the components of the sample may be difficult to separate chromatographically in their natural state and yet have derivatives that are readily separated. For example, the higher molecular weight fatty acids are highly polar compounds with low volatility. The methyl esters of these fatty acids, however, are compounds of low polarity and high volatility. Since gas chromatography depends on volatilizing compounds in a column, the advantage of more volatile compounds is obvious. The methyl esters of fatty acids are rather easily prepared using diazomethane or a BF_3-methanol reagent. Therefore, fatty acids are usually (although not always) chromatographed as the methyl esters.

Silylation is another technique used to decrease the polarity and increase the volatility of compounds. It can be applied to compounds containing active hydrogens, i.e., alcohols, acids, sterols, or amines. The main disadvantage of this technique is that the solvent and the reagents must be free of water. If some precautions are taken, silylation methods are usually rapid and simple.

The second reason for preparing a derivative of the analyte is to provide a molecule that is more readily detected, usually by an electron capture detector. Halogenated silylation agents are often used for this purpose [1], but other halogenated derivatives have also been prepared [12, 13].

The third reason for preparing derivatives is for purposes of component identification, which is discussed below in Sect. IV, C.

C. Peak Identification

Since the history of water samples is seldom completely known, it is important to positively identify the peaks of interest. Chromatography of the sample on two dissimilar columns to obtain retention data can verify peak identities. Other techniques have also been applied.

Treating the sample with reagents that destroy the component of interest can sometimes be helpful. For example, treating organomercuric compounds with metal sulfides will destroy them. If a peak tentatively identified from its retention time as an organomercuric compound vanishes after the solution is treated with sodium sulfide, this is further evidence to support the preliminary identification [38]. If the peak does not disappear, it probably does not correspond to an organomercuric compound.

Preparation and chromatography of derivatives of compounds under investigation can lend greater certainty to their identification. Many researchers have used this technique. Shafik et al. [39] hydrolyzed organophosphorus compounds to dialkyl phosphates and thiophosphates to help confirm their identities. Poinke [40] used potassium hydroxide in ethanol to prepare derivatives of p,p'-DDD, p,p'-DDT, and methoxychlor and hydrochloric acid in ethanol to prepare derivatives of endrin and dieldrin for verification. Greve and Wit [41] used ethanolic alkali to prepare derivatives to aid in the identification of endosulfan.

Many other techniques, such as IR, TLC, and mass spectroscopy, have been used in verification of gas chromatographic peaks. Gas chromatography is a valuable tool for separating and quantitating sample components, but retention time on one column alone cannot be used to positively identify a peak. Corroborating evidence must be provided.

V.   METHODS

A.   Pesticides

For purposes of analysis, organic pesticides can be classified as (1) chlorinated hydrocarbons, (2) organophosphorus compounds, or (3) other compounds. The first two categories are particularly well suited to analysis by gas chromatography because they include atoms that are readily detected by element selective detectors. The electron capture and microcoulombic detectors provide high sensitivity for the chlorinated pesticides. Thermionic (alkali flame) and flame photometric detectors are sensitive to phosphorus-containing compounds. Other detectors are also used, although they are not as specific or sensitive as these types. Some of the other pesticides have been analyzed by gas chromatography after preparation of halogenated derivatives [42]. More often, however, these are analyzed by thin-layer chromatography (Ref. 18 and Chapter 14), or by other means.

1. Chlorinated Pesticides

Lichtenberg et al. [43] analyzed samples of surface waters for 12 organochlorine pesticides and degradation products. The pesticides were extracted from the water samples into 15% ethyl ether in hexane. The extracts were dried over anhydrous sodium sulfate and concentrated to approximately 5 cm^3 in Knuderna-Danish evaporators. They were then carefully evaporated to 0.5 cm^3 in a warm-water bath. Up to 10 μl was injected into the gas chromatograph. If no response was detected, the extracts were further concentrated to 0.2 cm^3 and injected into the gas chromatograph again. The extracts were separated on a 6 ft $\times$ 1/4 in. o.d. coiled aluminum column packed with 5% OV-17 on 60/80 mesh Gas Chrom Q. The detector used was a nickel electron capture detector. The column was maintained at 205°C, the injection block at 250°C, and the detector at 360°C. The minimum detectable concentration was about 1-2 ng/liter for most pesticides.

This is typical of chlorinated pesticide analyses by gas chromatography. Samples for analysis are usually collected as grab samples or on activated charcoal filters. The pesticides are extracted from the water sample or the filter into an organic solvent. Chloroform [44], ethyl ether in hexane [18, 19], benzene [40], and hexane [5] have been found to be suitable solvents. Various liquid chromatographic columns have been used for cleanup of samples for pesticide analysis (See Table 1 and ref. [24]). Extracts can be concentrated, if necessary, by evaporation of the solvent, although some loss of volatile components has been reported [34].

One of the most troublesome sources of interference in organochlorine pesticide analysis has been polychlorinated biphenyls (PCB's). The gas chromatographic characteristics of PCB's and chlorinated pesticides are very similar. Several methods have been developed to avoid this interference problem. Armour and Burke [31] used liquid chromatography on a silicic acid-Celite column to separate PCB's from pesticides. Miles [28] converted DDT and its metabolites to dichlorobenzophenones prior to liquid chromatographic separations on a Florisil column. This derivative of DDT is more polar than the parent compound and easier to separate from the PCB's. The PCB's were not affected by the dehydrochlorination and oxidation steps. Dolan et al. [45] modified a Coulson electrolytic conductivity detector to allow precise control of furnace temperature. By adjustment of the temperature to 600°C and of the hydrogen flow rate to 1-2 cm^3/min, the sensitivity toward pesticides was increased while the sensitivity toward PCB's was decreased. This effectively reduced the amount of interference due to PCB's.

Some of the gas chromatographic columns used for chlorinated pesticide analysis are listed in Table 2. Shafer et al. [6] have compared a number of columns and conditions for such analyses.

2. Organophosphorus Pesticides

Sample preparation for organophosphorus pesticide analysis is very similar to that of the chlorinated pesticides. Identical sample preparations have been used for both types of samples [43, 53]. The main difference in the analysis is the detector used. Although the same detector (e.g., FID) can be used for both chlorinated and phosphorus-containing pesticides, the sensitivity of the assay is greatly increased if specific detectors are used. The flame photometric or thermionic detectors are generally used for organophosphorus pesticide residue analysis.

TABLE 2

Columns Used for Chlorinated Pesticide Analysis

Compounds	Columns	Detectors	Reference
Chlorinated pesticides	6 ft × 1/4 in. glass; 3% OV-1 on Chromosorb W-AW at 225°C	Coulson electrolytic conductivity	45
	6 ft × 1/4 in. aluminum; 5% OV-17 on Gas Chrom Q at 200°C	Electron capture	18
	6 ft × 4 mm glass; 8% QF-1 and 2% OV-17 on Gas Chrom Q at 200°C	Electron capture	46
	4 m × 4 mm glass; 10% DC-200 on Gas Chrom Q at 195°C	Electron capture	47
	1.8 m × 1/4 in. glass; 5% DC-200/7.5% QF-1 on Chromosorb WHP	Electron capture	24
Dieldrin, endrin, lindane, DDD, heptachlor epoxide, heptachlor, DDT, DDE, aldrin	4 ft × 1/4 in. glass; 5% DC-200 on Chromosorb P at 200°C and 150° C	Electron capture Microcoulombic	34

TABLE 2. — Continued

Compounds	Columns	Detectors	Reference
Heptachlor, heptachlor epoxide, cis-chlorodane, trans-chlorodane, dieldrin, endrin	5 ft × 1/8 in. glass; 4% DC–11 and 6% QF–1(1+1) on Chromosorb W–AW at 200°C	Electron capture	48
Aldrin, isodrin, dieldrin, endrin	4 ft × 4 mm stainless steel; 5% SE–30 on Chromosorb W–HMDS at 165°C	Electron capture	49
Chlorodane, heptachlor, heptachlor epoxide	4 ft × 1/4 in. glass; 3% silicone on Chromosorb W–AW–DMCS at 200°C	Electron capture	11
DDT, dieldrin	91 cm × 0.32 cm stainless steel; 5% Dow 11 on Chromosorb W programmed from 125°C to 200°C	Flame ionization	44
	6 ft × 4 mm Pyrex; 10% DC–200 on Anakrom ABS at 200°C	Electron capture	50
p,p'–DDT	5 ft × 1/8 in. Pyrex; 5% Dow 11 on Chromosorb W at 180°C	Electron capture	51

Chlordane	4 ft × 1/4 in. glass; 3% SE-30 on Chromosorb W-AW-DMCS at 190°C	Electron capture	52
Thiodan	1.5 m stainless steel; 3% SE-30 on Chromosorb W at 190°C	Electron capture	5
Endosulfan	5 ft × 1/8 in. Pyrex; 5% DC-200 on Aeropak 30 at 200°C	Electron capture	41

Hindin et al. [53] analyzed surface and ground waters for pesticides.
A measured amount of water was pumped through a column of activated
charcoal. The carbon was removed from the filter and air dried. The
pesticides were extracted from the carbon in a modified soxhlet extractor
using petroleum ether or petroleum naphtha. The extract was concentrated
to 10 cm^3. Extraneous organic matter was removed by liquid chromatography on an activated alumina column. The sample was separated on a
1.83 m × 0.63 cm stainless steel column packed with 10% Dow 11 on
Chromosorb W. The temperature was programmed from 160° to 260°C
at 260°C/min.

This is typical of procedures for organophosphorus pesticide analysis.
The samples are usually collected on activated carbon or other columns or
as grab samples. The pesticides are then extracted into an organic solvent.
The extracts are concentrated by evaporation as necessary. Liquid chromatography or other techniques are used, if needed, for sample cleanup.

Some of the gas chromatographic columns and detectors that have been
used for the analysis of organophosphorus pesticides are listed in Table 3.

3. Other Pesticides

Landrin has been determined by gas chromatographic analysis by Lau and
Maxmiller [42] in nanogram quantities. The pesticide was reacted with
trifluoroacetic anhydride and, after a liquid chromatographic cleanup step,
was analyzed on a gas chromatograph equipped with a 6 ft × 1/8 in. stainless
steel column packed with either 2% Reoplex 400 or 3% OV-17. The Reoplex
column temperature was 170°C and the OV-17 column temperature was
185°C. A tritium electron capture detector was used in both cases.

The sensitivity of this method is due to the preparation of a halogenated
derivative. Without such a derivative preparation, the carbamate insecticides
do not contain atoms that make them sensitive to any of the element selective
detectors, and the necessary sensitivity is lost. Other methods provide
better analyses for most of these compounds.

B. Polychlorinated Biphenyls (PCB's)

PCB's are chemically very similar to chlorinated pesticides. The techniques
used for sampling pesticides also apply to PCB's. It is also true, however,
that because they are similar, chlorinated pesticides can be a serious source
of interference in PCB analysis. Many procedures have been developed to
achieve the necessary sample cleanup.

Armour and Burke [31] separated PCB's from DDT and its analogs by
liquid chromatography on a silicic acid-Celite column prior to injection into
the gas chromatograph. The PCB's were eluted with petroleum ether prior

TABLE 3

Columns Used for Organophosphorus Pesticide Analysis

Compounds	Columns	Detectors	Reference
Organophosphorus pesticides	3 ft × 1/4 in. glass; 2% Reoplex 400 on Gas Chrom Q at 180°C	Flame photometric	18
	1.8 m × 1/4 in. o.d. stainless steel; 5% DC-200/7.5% QF-1 on Chromosorb WHP	CsBr thermionic	24
	2 m × 4 mm i.d. glass; 10% DC-200 on Chromosorb W at 200°C	KCl thermionic	47
	4 ft × 4 mm i.d. glass; 2% Reoplex 400 on Gas Chrom Q at 185°C	Flame photometric	43
	1.83 m × 0.63 cm stainless steel; 10% Dow 11 and 2% Epon 1001 on Chromosorb W programmed between 160° and 260°C at 20°/min	FID	53
Abate	6 ft × 1/8 in. stainless steel; 5% Dow 11 on Chromosorb W at 270°C	FID	15

TABLE 3. — Continued

Compounds	Columns	Detectors	Reference
Desmethyl derivatives of phosphate insecticides	3 ft glass; 2% Reoplex 400 and 10% QF–1 (1:1) on Gas Chrom Q at 145°C	CsBr thermionic	54
Malthion	5 ft × 1/4 in. o.d.; 5% DC–11 on Chromosorb W at 180°C	Electron capture (^{90}Sr)	55 106

to the elution of the pesticides with a combination of solvents. Beezhold and Stout [4] used silica gel columns for the separation of PCB's and DDT.

Miles [28] separated PCB's from DDT on Florisil columns after first converting the DDT to a dichlorobenzophenone. The PCB's were not affected by the chemical treatment.

Schmidt et al. [19] analyzed polychlorinated biphenyls in California coastal waters. Each grab sample was shaken with 15% ethyl ether in hexane immediately after collection to extract the PCB's and stop microbial growth. The organic layer was dried over anhydrous sodium sulfate and chromatographed on a 4-in. Florisil column to remove interferences. The eluent was chromatographed on columns of 3% QF-1 on 80/100 mesh acid-washed and DMCS-treated Chromosorb W. The injection port was at 230°C, the column at 190°C, and the electron capture detector at 250°C. Nitrogen was the carrier gas at 95 cm^3/min.

Tas and de Vos [56] characterized four main components of a commercial PCB mixture by using gas chromatography in conjunction with TLC, NMR, and IR spectroscopy. One gas chromatographic column was used in a preliminary cleanup and two additional columns were used in the final separation and confirmation. The first column was a 5 m × 6 mm i.d. aluminum column packed with 20% SF96 on Chromosorb W-AW. Nitrogen was the carrier gas at 75 cm^3/min. The column was maintained at 225°C.

The eluents from this column were collected and chromatographed again on each of two gas chromatographic columns. The first was a 1.9 m × 3 mm i.d. glass column packed with 3% OV-1 on 80/100 mesh Gas Chrom Q. The column was held at 190°C and the nitrogen carrier gas flow rate was 80 cm^3/ min. The second column was identical in size and operating conditions but was packed with 2.7% QF-1 on 80/100 mesh Gas Chrom Q.

Beezhold and Stout [4] have suggested the use of mixed standards for quantitating the results of a PCB analysis. The chromatogram of a PCB compound yields not one sharp peak but a series of peaks that are not completely resolved. The compound is identified by its "fingerprint" as well as its elution time. These authors claim that since detector response is not constant for variously chlorinated isomers, differences in quantitation can develop depending on which PCB is used as the standard. They suggest using a mixture of two PCB's (Archlor 1254 and 1260, for example) for standards. The two standards can be mixed to get a ratio of concentrations that yields a chromatogram resembling that of the unknown. Various concentrations of this mixture can then be used as standards to calculate the concentration of the unknown.

C. Herbicides

Chlorinated herbicides are well suited to analysis by gas chromatography in conjunction with an electron capture detector. Many investigators have analyzed herbicides using such a system.

Devine and Zweig [8] analyzed several chlorophenoxy herbicides and their esters (Dicamba, MCPA, 2,4-D, the isopropyl ester of 2,4-D, Silex, 2,4,5-T, and the n-butyl ester of 2,4-D). Three different columns were compared. The first, and most preferred by the analysts, was 5% SE-30 on 60/80 mesh Chromosorb W. The nitrogen flow rate was 175 cm^3/min and the column temperature was 175°C. The second column was 20% Carbowax 20M on 60/80 mesh acid-washed Chromosorb W. The nitrogen flow rate was 180 cm^3/min and the column temperature was 220°C. The third column was 2% QF-1 on 90/100 mesh Anakrom ABS. The nitrogen flow rate was 25 cm^3/min and the column temperature was 175°C. The inlet temperature and electron capture detector temperature were 230°C and 275°C, respectively, for all three columns. Aldrin was used as an internal standard. Detection limits were 0.01-0.05 ppb for all compounds tested except MCPA. Due to recovery problems experienced with MCPA, its detection limit was 2 ppb.

Diuron [N'-(3,4-dichlorophenyl)-N,N-dimethylurea] in surface waters has been analyzed by McKone and Hance [22]. A 100-cm^3 water sample was extracted twice by shaking for 1 min with 25 cm^3 portions of dichloromethane. The organic layers were combined and concentrated to 0.5 cm^3 under reduced pressure in a 35°C water bath. This residue was shaken with 5 cm^3 of saturated NaCl solution for 15 sec; then 5 cm^3 of 2,2,4-trimethylpentane was added and the mixture was shaken for 1 min. Aliquots of the upper (2,2,4-trimethylpentane) layer were analyzed on a 1.5 m × 3.5 mm o.d. stainless steel column packed with 5% E301 on 60/80 mesh Gas Chrom Q. The nitrogen carrier gas flow rate was 50 cm^3/min. The column, injection port, and electron capture detector temperatures were 155°C, 265°C, and 200°C, respectively.

Dalapon (2,2-dichloropropionic acid) in water samples was analyzed by Frank and Demint [16]. After the sample was nearly saturated with sodium chloride, the pH was adjusted to 1. The sample was then extracted with diethyl ether. The Dalapon was then extracted into 0.1N $NaHCO_3$, at pH 8, saturated with NaCl. The solution was brought to a pH of 1 by addition of 1:1 HCl and extracted again with ethyl ether. The ether fraction was reduced to a volume of 2 cm^3. Diazomethane was added to esterify the Dalapon. The ester was then chromatographed on 5 ft × 1/8 in. stainless steel columns packed with 10% FFAP on 60/80 mesh, HMDS-treated Chromosorb P. The nitrogen carrier gas flow rate was 30 cm^3/min and the column temperature was 140°C. The injection port and electron capture detector were at 215°C and 195°C, respectively.

Wilder [57] analyzed 2,4-dichlorphenoxyacetic acid dimethylamine salt in surface waters. The herbicide was extracted into methylene chloride, esterified with diazomethane, and chromatographed on 5% Dow 11 in a 5 ft × 1/8 in. stainless steel column. The column was maintained at 142°C, the injection port at 250°C, and the electron capture detector at 215°C. The carrier gas was nitrogen at 40 cm^3/min. As little as 10 ng of the herbicide per liter can be detected.

D. Dissolved Gases

The major problem involved in analyzing dissolved gases is separating the gases from the water. A number of approaches have been used.

Kilner and Ratcliff [58] used a precolumn of calcium sulfate, 10 in. × 1/2 in., to remove the water from the sample before it entered the column for separation of the gases. The gas separation column was 1 ft long, containing 44/60 mesh molecular sieve 5A. With argon as the carrier gas, 0.001 cm^3 of hydrogen in 0.2-0.3 cm^3 of H_2O could be determined with an accuracy of ±3%. Oxygen and N_2 could also be separated.

Navone and Fenninger [59] determined methane in much the same way. The sample passed through a 10-in. precolumn filled with 8 mesh Drierite to adsorb the water. A 40 mesh, 13× molecular sieve column (4 ft × 1/4 in.) was used for the final separation. A 0.05 cm^3 sample containing 0.02 mg methane per liter could be analyzed with precision and accuracy of ±5%.

Swinnerton et al. [60, 61] developed a method for stripping the gases from water samples. The liquid sample is injected into a chamber through which the carrier gas is passed. The carrier gas enters the sample chamber through a glass frit which divides it into very fine bubbles that effectively strip the dissolved gases from the liquid. The gas is then dried and sent through a 6 ft × 1/4 in. column packed with 30% HMPA (hexamethylphosphoramide) on 60/80 mesh Columnpak. This column separates CO_2 from O_2, N_2, CH_4, and CO. A second set of columns, one 4 ft × 1/4 in. packed with 60/80 mesh Columnpak followed by one 7 3/4 ft × 1/4 in. packed with 42/60 mesh 13× molecular sieve, separates O_2, N_2, CH_4, and CO. Carbon dioxide does not elute from the second set of columns and is presumably adsorbed.

Swinnerton and Linnenbom [62, 63] have also used this gas stripping chamber in conjunction with cold traps to concentrate the gases before they enter the separation column. The gases are released from the traps when sufficient sample has collected simply by increasing the trap temperature. Methane was separated from the other gaseous hydrocarbons on a column of silica gel, while the C_2-C_4 parafins and olefins were separated on an activated alumina column containing 10% Nujol. The higher molecular weight hydrocarbons were separated on a column of Chromosorb coated with 20% SF-96 or SF-30. Ethane (C_2H_6) and ethylene (C_2H_4) could be separated on an activated alumina column containing 10% silicone oil. Gaseous hydrocarbons have been detected down to one part in 10^{13} by this technique [63].

E. Organomercuric Compounds and Inorganic Substances

Nishi et al. [64] developed a method for the determination of methylmercuric compounds in aqueous solutions containing sulfide or other extraction-disturbing sulfur compounds. The sample was pretreated with 0.4% HCl, $HgCl_2$, and 0.4 NH_4OH and pre-extracted with hexane. Sufficient HCl was added to the sample to prepare a 4% acid solution. It was then shaken with 8 cm^3 of cystein solution (0.1 g of 1-cystein and 12 g of Na_2SO_4 in 100 cm^3 of H_2O) for 5 min. The cystein layer was separated, treated with 0.2 cm^3 of concentrated HCl and 1 cm^3 of hexane, and shaken again for 5 min. The hexane layer was then analyzed for methylmercuric compounds on a 1 m × 4 mm glass column packed with 5% polydiethylene glycol succinate on Chromosorb W. The column temperature was 130°C and an electron capture detector was used.

A method for the identification of organomercuric compounds was developed by Nishi and Horimoto [38]. The solution was first chromato-graphed as usual. It was then treated with an aqueous solution of thio compounds, such as glutathione; or with metal sulfides, such as NiS; or with thiosulfates, such as $Na_2S_2O_3$; or with metal powders, such as Fe, Al, Zn, or Pb, in the presence of water. Each of these treatments was sufficient to destroy the organomercuric compounds. The treated solution was then chromatographed again, and the peaks that had been eliminated from the chromatogram were identified as organomercuric compounds.

Chloride and bromide ions in aqueous solutions have been analyzed by Bergmann and Martin [65]. The aqueous sample was added to 80% H_2SO_4, which converted the ions to the hydrogen halides. The gases were swept out of the reaction chamber and onto a glass analysis column 160 cm × 8 mm. The liquid phase was 2.3% toluene and 5% n-heptane plus one drop of H_2SO_4 per gram of packing material. This was coated on granular Teflon. Using a thermal conductivity detector, 0.1 ppm of chloride and 0.2 ppm of bromide could be detected.

Bock and Semmler [66] developed a gas chromatographic analysis for fluoride ions. After addition of 20 cm^3 of 2N HCl to an 80-cm^3 sample, the solution was extracted for 35 min with 5 cm^3 of 0.05N Et_3SiCl in tetra-chloroethylene. An aliquot of the organic layer was chromatographed on a silicone oil/Sterchemol column at 120°C. Phenylchloride was used as an internal standard.

Phosphorus was determined in hexane or isooctane extracts of water by Addison and Ackman [67]. When a flame photometric detector was used with a 526-nm filter, 10^{-12} g of phosphorus could be detected. The columns used were 3% OV-1 or 3% SE-30 on Chromosorb W.

F. Organic Acids

Organic acids have been determined by direct gas chromatography of dilute aqueous solutions. Several authors [68-70] have experienced problems with ghosting or "memory peaks" in the gas chromatography of aqueous solutions. These problems have been solved in several different ways.

Ackman and Burgher [69] added formic acid to the carrier gas when analyzing fatty acids in aqueous solutions to suppress reversible adsorption and eliminate the appearance of ghost peaks. Using 6 ft × 1/8 in. stainless steel columns packed with DC550 silicone grease containing 5% stearic acid, Tween, or NPGA coated on Chromosorb W, as little as 0.01% fatty acid could be determined. Pentoxone (4-methoxy-4-methyl-2-pentone) was used as an internal standard.

Teflon solid support used in packing the gas chromatographic column helped eliminate ghosting problems experienced by Scher [70] and Lemoine et al. [71]. Scher has also cautioned that the syringe needle must be scrupulously cleaned between injections to eliminate apparent ghosting phenomena. Lemoine injected 2-10 μl samples of water between sample injections to eliminate ghosting.

Geddes and Gilmour [68] discovered that charred deposits at the injection port and the glass wool plugs at the ends of the columns were the sources of their ghosting problems in the analysis of C_2-C_5 acids. Replacing the glass wool with Teflon tape plugs and employing formic acid and frequent cleaning of the injection port remedied the ghosting problems.

Emery and Koerner [72] analyzed 0.1-0.01% each of acetic, propionic, isobutyric, n-butyric, isovaleric, and n-valeric acids directly by gas chromatography. A 1-m column packed with 20 wt% Tween 80 on acid-washed 60/80 mesh Chromosorb W was used. The column temperature was 110°C, and that of the injection block 210°C. No ghosting problems were discussed.

Fatty acids are often extracted into organic solvents prior to analysis by gas chromatography, thus eliminating the ghosting problems encountered with aqueous solutions. Derivatives of the acids are usually prepared.

Ackman and Hooper [21] subjected seawater samples to saponification, extracted the sample with petroleum ether, and prepared the methyl esters of the fatty acids with a methanol-BF_3 reagent. A capillary column coated with butanediolsuccinate polyester was used for separation.

Ocean water has also been analyzed for fatty acids by Slowley et al. [73]. The water samples were extracted with ethyl acetate. The solvent was removed and the methyl esters of the fatty acids were prepared. Capric, lauric, myristic, palmitic, steric, myristoleic, palmitoleic, oleic, linoleic, and linolenic acids were identified.

Lower molecular weight fatty acids in natural waters were determined by Nemtseva et al. [74]. A 0.5-1 liter sample of water was passed through an ion-exchange column. The fatty acids were extracted from the resin with 100 cm³ of 2N $(NH_4)_2CO_3$. After the addition of 2 cm³ of 1N NaOH, the extract was evaporated to dryness. The residue was dissolved in 0.5 cm³ of water, acidified with HCl, and extracted with 2 cm³ of diethyl ether. Alkaline alcohol (0.2 cm³) was added to the ether layer and the solvent was evaporated to dryness. Ethyl alcohol (0.3 cm³) and 6N HCl (0.1 cm³) were added and an aliquot was taken for gas chromatographic analysis. The columns used were 1.5 m × 6 mm stainless steel packed with hydrochloric acid-treated "zikeevsk" earth coated with 20% polyethylene glycol adipate plus 1% H_3PO_4. The column temperature was 146°C, the injection block was 200° C, and the carrier gas was nitrogen at 1.5 liters/hr. Acetic, propionic, butyric, caproic, and enanthic acids were found.

Kawahara [13, 75] prepared pentafluorobenzyl thioesters and esters of organic acids for detection by an electron capture detector. This made detection of subnanogram quantities possible. An 8 ft × 1/8 in. stainless steel column packed with 10% FFAP on Chromosorb T and Chromosorb W was used for the separation. The column was maintained at 195°C. The carrier gas was He at 40 cm^3/min.

Low molecular weight (C_2-C_{10}) straight-chain carboxylic acids were determined in aqueous solutions by gas chromatography of their p-bromophenacyl and p-phenylphenacyl esters by Umeh [76]. Six-foot glass columns of 60/80 mesh Chromosorb G coated with 2.5% sodium dodecylbenzenesulfonate or 1% Apiezon L, or a 1:1 mixture of the two, were used depending on the acids present.

G.   Phenols and Mercaptans

Phenols have also been analyzed by direct injection of aqueous solutions. Baker and Malo [77] investigated a variety of columns for separating phenol; o-, m-, and p-cresol; o- and m-chlorophenol; 2,3-, 2,4-, 4,5-, 2,6-, and 3,4-dichlorophenol; o-nitrophenol; tymol; and guaiacol. A column of 5-10% FFAP on Teflon 6 was preferred. Addition of steam to the carrier gas reduced but did not eliminate ghosting.

Hermann and Post [78] analyzed indole and o-phenylphenol by direct aqueous injection gas chromatography. Glass columns, 1 m × 4 mm i.d., packed with 5% butendiol succinate on 100/120 mesh Gas Chrom Q were used. The column temperature was 170°C, that of the injection block was 185°C, and that of the flame ionization detector was 225°C. Nitrogen was used as the carrier gas at 120 cm^3/min. No ghosting was observed.

Phenols and mercaptans have been recovered from aqueous solutions by driving them off in vapor form. Kaplin et al. [79] added 1-1.5% $CuSO_4$ to the water sample, acidified it, and then drove off the volatile phenols. The distillate was saturated with NaCl and acidified with 14 cm^3 of HCl per 100 cm^3 of solution. The sample was extracted twice with 30 cm^3 of diethyl ether. The ether extracts were combined and extracted three times with 30 cm^3 of a solution of 5% NaOH and 20% NaCl. The solution was cooled and saturated with CO_2 until $NaHCO_3$ began to precipitate. This solution was then extracted with diethyl ether. A brass column, 1.5 m × 4 mm, packed with 20-30% dinonyl phthalate coated on diatomaceous firebrick particles was used for analysis. The column was maintained between 150°C and 160°C. Phenol, o-cresol, m- and p-cresol; 2,4- and 2,5-xylenol, 2,3-xylenol, 3,5-xylenol, and 3,4-xylenol were separated. A column of 10% tricresyl phosphate separated the m- and p-cresols and the 2,4- and 2,5-xylenols.

Le Rosen [80] neutralized caustic solutions of mercaptans under reduced pressure. The acidic gases were collected and analyzed on a 2-m column packed with 30% Silicone 200 coated on 60/100 mesh Celite.

Phenols can be collected on activated carbon filters. Eichelberger et al. [33] used a Florisil column to clean up the chloroform extracts from carbon filters for phenol analysis. Aluminum columns 10 ft × 1/8 in. o.d. packed with 10% Carbowax 20M on HMDS-treated Chromosorb W were used for separation of the phenols.

Goren-Strul [81] extracted phenols from a carbon filter with one volume of methanol and 25-50 volumes of chloroform. The solvent was evaporated and the residue was redissolved in diethyl ether and separated into acid, phenol, neutral, basic, and amphoteric substances by back extraction. The phenol phase was chromatographed on columns 1.5 m × 4 mm i.d. packed with 5% tris (2,4-xylenyl) phosphate on Chromosorb W.

The detection limits of phenol and mercaptan analyses have been increased by preparation of halogenated derivatives that can be detected by an electron capture detector. Kawahara [12-13] prepared pentafluorobenzyl derivatives of phenols and mercaptans. Argauer [82] prepared chloroacetyl derivatives of microgram quantities of phenols for detection by an electron capture detector.

H. Sterols

Coprostanol (5 β-cholestane-3 β-ol) has been suggested as a tracer for fecal water pollution. Tan et al. [9] extracted water samples with hexane, evaporated the solvent to dryness, and dissolved the residue in dioxane. The extracts were analyzed either directly or after a TLC cleanup step. A U-shaped glass column 180 cm × 0.3 cm packed with 3% OV-1 and 3% QF-1 on 100/120 mesh Gas Chrom Q was used for separation of the coprostanol. The column temperature was 230°C and the injector temperature was 265°C.

Matthews and Smith [10] analyzed hexane extracts of water samples for cholesterol, stigmasterol, and β-sitosterol. The gas chromatographic columns used were 1.83 m × 6 mm o.d. silanized glass packed with either 3% SE-30 on 80/100 mesh Gas Chrom Q or 3% QF-1 on 100/120 mesh Gas Chrom Q. The column was maintained at 245°C and the injection port at 250°C. Nitrogen was used as the carrier gas at 21 cm^3/min.

I. Amines

The gas chromatographic analysis of amines in aqueous solutions is complicated by the fact that most column packing materials tend to adsorb amines and cause tailing of the chromatographic peaks [83]. Coating the solid support with KOH in addition to the liquid phase has solved this problem for many [83-85]. The use of Teflon solid support also seems to prevent the tailing phenomenon [86].

Smith and Radford [83] tested firebrick, Chromosorb, Chromosorb W, and Celite with and without KOH coating in addition to the Carbowax 20M coating for the gas chromatography of diamines. All columns without the KOH produced tailing peaks, while those with the KOH showed greatly reduced tailing. Rinsing the support with a KOH solution and then rinsing it with water before packing did not provide the desired effect. The KOH had to be incorporated in the column packing. They found that a 2-m column packed with 10% Carbowax 20M and 5% KOH on Chromosorb W at 160°C was effective for separating hexamethylenediamine, tetramethylenediamine, pentamethyl-enediamine, and 1,2-diaminocyclohexane. An 8-m column packed with 10% DC 710 and 5% KOH on Chromosorb W at 120°C was effective in separating the cis and trans isomers of 1,2-diaminocyclohexane.

Umbreit et al. [84] analyzed the C_1-C_5 alkylamine hydrochlorides in aqueous solutions. The hydrochloride salts were converted to free amines in the injection port and then separated on a 6 ft × 3 mm i.d. glass column packed with 10% Amine 2204 and 10% KOH on 80/100 mesh acid-washed Chromosorb W. Helium was used as the carrier gas at 60 cm^3/min. Retention times were studied for column temperatures of 60°C, 90°C, and 110°C. The flame ionization detector was held at 250°C and the injection port at 200°C. Concentrations of 17-250 ppm were detected.

Allylamine, propylamine, butylamine, di-n-butylamine, and benzylamine were determined in aqueous solution by Arad et al. [85]. A 2.5 m × 5 mm i.d. copper column packed with 20% Carbowax 20M and 5% KOH on 30/60 mesh Chromosorb W was used. The column was held at 70°C until after the elution of the allyl-, propyl-, and butylamines and then raised at 10°C/min to 180°C. A thermal conductivity detector was used.

Landault and Guiochon [86] used Teflon as a solid support in the analysis of aqueous solutions of amines. A column of 4.4% Apiezon L and one of 20% polyglycol 1500 were both used for the separation of water, ammonia, primary alcohols, and primary, secondary, and tertiary amines.

J. Petroleum Products

Jeltes and Veldink [87] analyzed water containing milligrams per liter levels of petroleum products. The hydrocarbons were extracted into $C_6H_5NO_2$ and then separated on a 3-m copper column packed with 10% Polyethylene Glycol 1500 on 60/80 mesh silanized Chromosorb W. The carrier gas was a combination of nitrogen and hydrogen at 25 cm^3/min. A flame ionization detector was used.

Boylan and Tripp [88] prepared samples in the lab to simulate solutions of dissolved hydrocarbons. These were extracted into pentane, concentrated by evaporation, and separated on an 8 ft × 1/16 in. column packed with 80/100 mesh texturized glass beads coated with 0.2% Apiezon L. The column temperature was programmed from 60° to 220°C at 4°C/min. A mass spectrometer was used for detection and identification of the peaks.

Two columns were employed by Adlard et al. [89] in the identification of hydrocarbon pollution on seas and beaches. The first was a 1 m × 3 mm i.d. stainless steel column packed with 3% OV-1 on 85/100 mesh AW-DMCS Chromosorb G. Helium was the carrier gas at 35 cm^3/min. The effluent was split to be analyzed by both flame ionization and flame photometric detectors. The other column used was a 20 m × 0.25 mm stainless steel capillary column coated with OV-101. The temperature was programmed from 60° to 300°C at 5°/min.

K. Miscellaneous Organic Compounds

Many papers [90-97] have discussed the use of pyrolysis-gas chromatography for the determination of the total organic content of water. Most of these can be only loosely classified as gas chromatography since no chromatographic separation is achieved or even attempted. The sample is injected into a CuO tube, or a tube packed with CuO wire, where it is oxidized at 850°-900°C to CO_2. The CO_2 is reduced to CH_4 at 300°-400°C in the presence of hydrogen gas with nickel-coated firebrick as a catalyst. The methane is then passed through a short tube to a flame ionization chamber where the total carbon content is measured as one peak. Parts per million levels of carbon can be measured in this manner.

Lysij and Nelson [97] have attempted to provide some degree of separation of pyrolyzed samples. Samples of 0.1-0.25 cm^3 containing 100 ppm starch, 100 ppm gelatin, or 10 ppm heptanoic acid were pyrolyzed at 700°C in a tube filled with granular Ni and then chromatographed on a column of 20% Carbowax 20M on Chromosorb W-(AW-DMCS). Steam was used as the carrier gas. The method could be used for the identification of classes of compounds but not for specific compounds.

One of the advantages of gas chromatography in the analysis of organic impurities in water is its sensitivity. Novitskaya [98] has analyzed aqueous solutions containing 0.009-0.091% ethanol, 0.017-0.17% crotonaldehyde, and 0.5-3.2% acetaldehyde. A column 85 cm × 6 mm i.d. packed with 28% of a 3:1 mixture of β, β'-thiodipropionitrile and polyethylene glycol on firebrick was found to be most effective for the separation of these compounds.

Levels of 40 mg/liter of pyridine bases and 100 mg/liter of various ketones and aromatic hydrocarbons in waste water have been determined by Novotny [99]. Columns used were 86 cm × 0.6 cm packed with Rysorb BKL V on glass balls etched with HF. For separation of pyridine bases, triethanolamine was used as the liquid phase. For the separation of ketones and aromatic hydrocarbons, a mixture of triethanolamine and Tween 60 was used.

Wasik and Tsang [100] used aqueous solutions of complexing metal ions as the liquid phase. An example of one such column is a 50% aqueous solution of 5M $AgNO_3$ and 0.05M $Hg(NO_3)_2$ coated on 60/80 mesh acid-washed Chromosorb P. The metal ions complex the hydrocarbons with varying efficiencies and retard their progress through the column. Alkanes and aromatics are separated; olefins are retained and do not elute.

Several researchers have developed methods of removing the water from aqueous samples prior to the gas chromatographic column. Jacobs [101] used a 16 × 1/4 in. copper tube filled with 3.7 g of P_2O_5 and 0.4 g of 60/80 mesh firebrick coated with Desicote and Siliclad to adsorb the water from up to 50 successive 10-μl samples.

Kung et al. [102], analyzed samples containing up to 90% water by first converting the water to acetylene on a column at 220°C packed with glass beads and CaC_2. Alcohols and mixtures of aldehydes, esters, and alcohols were completely recovered. Methanol was not completely recovered, but recovery was sufficiently reproducible to allow quantitation.

Blum and Koehler [103] prepared trimethylsilyl derivatives of glycols and mono- and diglycerides prior to chromatographic analysis. Cholesteryl acetate was used as an internal standard. A 1 ft × 4 mm i.d. glass column packed with 3% OV-1 on 80/100 mesh Chromosorb W was used for the separation. The column was programmed from 50° to 300°C at 12°C/min. The flame ionization detector was maintained at 300°C.

Nitrilotriacetic acid (NTA) has been proposed as a substitute for phosphates in detergents. Chau and Fox [104] collected NTA by concentration on Dowex 1 ion-exchange resin. It was eluted from the resin with formic acid, subjected to propyl esterification, and analyzed by gas chromatography. The column used was 6 ft × 1/4 in. stainless steel packed with 3% OV-1 on 80/100 mesh Chromosorb WHP. The carrier gas was nitrogen at 65 cm^3/min. The column temperature was programmed from 180° to 225°C at 3°/min. A flame ionization detector at 250°C was used to detect levels in the 10-ng range. A glass, U-shaped, 1.9 m × 2 mm i.d. column packed with 0.65% ethylene glycol adipate on 80/100 mesh Chromosorb W has also been used in conjunction with a flame ionization detector for NTA determinations [105].

REFERENCES

1. P. Cukor and H. Madlin, Water and Water Pollution Handbook, Vol. 4 (L. L. Ciaccio, Ed.), Marcel Dekker, New York, 1973, pp. 1557-1615.

2. S. D. Faust and I. H. Suffet, Water and Water Pollution Handbook, Vol. 3 (L. L. Ciaccio, Ed.), Marcel Dekker, New York, 1972, pp. 1263-1264.

3. A. Bevenue, T. W. Kelley, and J. W. Hylin, J. Chromatog., 54, 71 (1971).

4. F. L. Beezhold and V. F. Stout, Bull. Env. Contam. Toxicol., 10, 10 (1973).

5. S. Gorbach, R. Haarring, W. Knauf, and H-J. Werner, Bull. Env. Contam. Toxicol., 6, 40 (1971).

6. M. L. Schafer, J. T. Peeler, W. S. Gardner, and J. E. Campbell, Env. Sci. Tech., 3, 1261 (1969).

7. L. Rudling, Water Res., 4, 533 (1970).

8. J. M. Devine and G. Zweig, J. Assoc. Off. Anal. Chem., 52, 187 (1969).

9. L. Tan, M. Clemence, and J. Gass, J. Chromatog., 53, 209 (1970).

10. W. S. Matthews and L. L. Smith, Lipids, 3, 239 (1968).

11. A. Bevenue and C. Y. Yeo, Bull. Env. Contam. Toxicol., 4, 68 (1969).

12. F. K. Kawahara, Anal. Chem., 40, 1009 (1968).

13. F. K. Kawahara, Env. Sci. Tech., 5, 235 (1971).

14. B. B. Chakraborty and R. Long, Env. Sci. Tech., 1, 828 (1967).

15. F. C. Wright, B. N. Gilbert, and J. C. Riner, J. Agr. Food Chem., 15, 1038 (1967).

16. P. A. Frank and R. J. Demint, Env. Sci. Tech., 3, 69 (1969).

17. J. G. Konrad, H. B. Poinke, and G. Chesters, Analyst, 94, 490 (1969).

18. J. W. Eichelberg and J. J. Lichtenberg, Env. Sci. Tech., 5, 541 (1971).

19. T. T. Schmidt, R. W. Risebough, and F. Gress, Bull. Env. Contam. Toxicol., 6, 235 (1971).

20. P. M. Williams, Nature (London), 189, 219 (1961).

21. R. G. Ackman and S. N. Hooper, Lipids, 5, 417 (1970).

22. C. E. McKone and R. J. Hance, Bull. Env. Contam. Toxicol., 4, 31 (1969).

23. A. A. Rosen, L. R. Musgrave, and J. J. Lichtenberg, Calif. State Water Pollution Control Board Publ., No. 21, 1959, pp. 47-61.

24. B. Versino, M. Th. Van der Venne, and H. Vissers, J. Assoc. Off. Anal. Chem., 54, 147 (1971).

25. A. M. Kadoum, Bull. Env. Contam. Toxicol., 3, 354 (1968).

26. A. M. Kadoum, Bull. Env. Contam. Toxicol., 3, 65 (1968).

27. L. M. Law and D. F. Goerlitz, J. Assoc. Off. Anal. Chem., 53, 1276 (1970).

28. J. R. W. Miles, J. Assoc. Off. Anal. Chem., 55, 1039 (1972).

29. O. W. Grussendorf, A. J. McGinnis, and J. Solomon, J. Assoc. Off. Anal. Chem., 53, 1048 (1970).

30. A. A. Rosen and F. M. Middleton, Anal. Chem., 31, 1729 (1959).

31. J. A. Armour and J. A. Burke, J. Assoc. Off. Anal. Chem., 53, 761 (1970).

32. J. A. Armour and J. A. Burke, J. Assoc. Off. Anal. Chem., 54, 175 (1971).

33. J. W. Eichelberger, R. C. Dressman, and J. E. Longbottom, Env. Sci. Tech., 4, 576 (1970).

34. F. K. Kawahara, R. L. Moore, and R. W. Gorman, J. Gas Chromatog., 6, 24 (1968).

35. A. Sodergren, Bull. Env. Contam. Toxicol., 10, 116 (1973).

36. R. A. Baker, Amer. Soc. Test. Mater. Spec. Tech. Publ., No. 448, 1969, pp. 65-67; through Chem. Abstr., 71, 73887v (1969).

37. R. A. Baker, Water Resources, 3, 717 (1969); through Chem. Abstr., 71, 94642b (1969).

38. S. Nishi and Y. Horimoto, Bunseki Kagaku, 20, 16 (1971) (in Japanese); through Chem. Abstr., 75, 58434j (1971).

39. M. Shafik, D. Bradway, and H. F. Enos, Bull. Env. Contam. Toxicol., 6, 55 (1971).

40. H. B. Poinke, Diss. Abstr., 28, 4383B (1968).

41. P. A. Greve and S. L. Wit, J. Agr. Food Chem., 19, 372 (1971).

42. S. C. Lau and R. L. Marxmiller, J. Agr. Food Chem., 18, 413 (1970).

43. J. J. Lichtenberg, J. W. Eichelberger, R. C. Dressman, and J. E. Longbottom, Pest. Monit. J., 4, 71 (1970).

44. A. W. Breidenbach and J. J. Lichtenberg, Science, 141, 899 (1963).

45. J. W. Dolan, R. C. Hall, and T. M. Todd, J. Assoc. Off. Anal. Chem., 55, 537 (1972).

46. E. S. Windham, J. Assoc. Off. Anal. Chem., 52, 1237 (1969).

47. H. B. Poinke, J. G. Konrad, G. Chesters, and D. E. Armstrong, Analyst, 93, 363 (1968).

48. A. S. Y. Chau and W. P. Cochrane, J. Assoc. Off. Anal. Chem., 52, 1220 (1969).

49. L. Kahn and C. H. Wayman, Anal. Chem., 36, 1340 (1964).

50. K. H. Deubert, Bull. Env. Contam. Toxicol., 5, 379 (1970).

51. J. W. Biggar, G. R. Dutt, and R. L. Riggs, Bull. Env. Contam. Toxicol., 2, 90 (1967).

52. A. Bevenue and C. Y. Yeo, J. Chromatog., 42, 45 (1969).

53. E. Hindin, D. S. May, and G. H. Dunstan, Residue Rev., 7, 130 (1964).

54. S. C. Lau and D. R. Schultz, J. Chromatog. Sci., 8, 681 (1970).

55. M. Ragab and H. Tawfik, Bull. Env. Contam. Toxicol., 3, 155 (1968).

56. A. C. Tas and R. H. de Vos, Env. Sci. Tech., 5, 1216 (1971).

57. E. T. Wilder, J. Amer. Water Works Assoc., 60, 827 (1968); through Chem. Abstr., 69, 42916h (1968).

58. A. A. Kilner and G. A. Ratcliff, Anal. Chem., 36, 1615 (1964).

59. R. Navone and W. D. Fenninger, J. Amer. Water Work Assoc., 59, 757 (1967); through Chem. Abstr., 67, 93855d (1967).

60. J. W. Swinnerton, V. J. Linnenbom, and C. H. Cheek, Anal. Chem., 34, 483 (1962).

61. J. W. Swinnerton, V. J. Linnenbom, and C. H. Cheek, Anal. Chem., 34, 1509 (1962).

62. J. W. Swinnerton and V. J. Linnenbom, Science, 156, 1119 (1967).

63. J. W. Swinnerton and V. J. Linnenbom, J. Gas Chromatog., 5, 570 (1967).

64. S. Nishi, Y. Horimoto, and Y. Umezawa, Bunseki Kaguka, 19, 1646 (1970) (in Japanese); through Chem. Abstr., 74, 134799d (1974).

65. J. G. Bergmann and R. L. Martin, Anal. Chem., 34, 911 (1962).

66. R. Bock and H. J. Semmler, Fresenius' Z. Anal. Chim., 230, 161 (1967) (in German); through Chem. Abstr., 67, 96585b (1967).

67. R. F. Addison and R. G. Ackman, J. Chromatog., 47, 421 (1970).

68. D. A. M. Geddes and M. N. Gilmour, J. Chromatog. Sci., 8, 394 (1970).

69. R. G. Ackman and R. D. Burgher, Anal. Chem., 35, 647 (1963).

70. T. A. Scher, Amer. Lab, July, 1970, p. 24.

71. T. J. Lemoine, R. H. Benson, and C. R. Herbeck, J. Gas Chromatog., 3, 189 (1965).

72. E. M. Emery and W. E. Koerner, Anal. Chem., 33, 146 (1961).

73. J. F. Slowley, L. M. Jeffery, and D. W. Hood, Geochim. Cosmochim Acta, 26, 607 (1962) (in German); through Chem. Abstr., 57, 10944d (1962).

74. L. I. Nemtseva, T. S. Kishkinova, and A. D. Semenov, Gidrokhim. Mater., 41, 129 (1966) (in Russian); through Chem. Abstr., 67, 93851z (1967).

75. F. K. Kawahara, Anal. Chem., 40, 2073 (1968).

76. E. O. Umeh, J. Chromatog., 56, 29 (1971).

77. R. A. Baker and B. A. Malo, Env. Sci. Tech., 1, 997 (1967).

78. T. S. Hermann and A. A. Post, Anal. Chem., 40, 1573 (1968).

79. V. T. Kaplin, L. V. Semenchenko, and N. G. Fesenko, Gidrokhim. Mater., 41, 142 (1966) (in Russian); through Chem. Abstr., 68, 16019c (1968).

80. H. D. Le Rosen, Anal. Chem., 33, 973 (1961).

81. S. Goren-Strul, H. F. W. Kleijn, and A. E. Mostaert, Anal. Chim. Acta, 34, 322 (1966); through Chem. Abstr., 64, 17250f (1966).

82. R. J. Argauer, Anal. Chem., 40, 122 (1968).

83. E. D. Smith and R. D. Radford, Anal. Chem., 33, 1160 (1961).

84. G. R. Umbreit, R. E. Nygren, and A. J. Testa, J. Chromatog., 43, 25 (1969).

85. Y. Arad, M. Levy, and D. Vofsi, J. Chromatog., 13, 565 (1964).

86. C. Landault and G. Guiochon, J. Chromatog., 13, 327 (1964) (in French); through Chem. Abstr., 60, 13855f (1964).

87. R. Jeltes and R. Veldink, J. Chromatog., 27, 242 (1967).

88. D. B. Boylan and B. W. Tripp, Nature (London), 230, 44 (1971).

89. E. R. Adlard, L. F. Creaser, and P. H. D. Matthews, Anal. Chem., 44, 64 (1972).

90. I. Lysyj, K. H. Nelson, and P. R. Newton, Water Res., 2, 233 (1968); through Chem. Abstr., 69, 5108a (1968).

91. I. Lysyj, K. Nelson, and P. R. Newton, J. Water Poll. Contr. Fed., 40, 181 (1968); through Chem. Abstr., 69, 54143s (1968).

92. I. Lysyj and K. H. Nelson, J. Gas Chromatog., 6, 106 (1968).

93. F. R. Cropper, D. M. Heinekey, and A. Westwell, Analyst, 92, 433 (1967).

94. F. R. Cropper, D. M. Heinekey, and A. Westwell, Analyst, 92, 436 (1967).

95. R. A. Dobbs, R. H. Wise, and R. B. Dean, Anal. Chem., 39, 1255 (1967).

96. D. L. West, Anal. Chem. , 36, 2194 (1964).

97. I. Lysyj and K. H. Nelson, Anal. Chem. , 40, 1365 (1968).

98. R. N. Novitskaya, Izv. Sibirsk. Otd. Akad. Nauk SSSR Ser. Khim.
     Nauk, 3, 154 (1965) (in Russian); through Chem. Abstr. , 64, 18378g
     (1966).

99. J. Novotny, Chem. Prum. , 20, 575 (1970) (in Czech.); through
     Chem. Abstr. , 74, 79301h (1971).

100. S. P. Wasik and W. Tsang, Anal. Chem. , 42, 1648 (1970).

101. E. S. Jacobs, Anal. Chem. , 35, 2035 (1963).

102. J. T. Kung, J. E. Whitney, and J. C. Cavagnol, Anal. Chem. , 33,
     1505 (1961).

103. J. Blum and W. R. Koehler, Lipids, 5, 601 (1970).

104. Y. K. Chau and M. E. Fox, J. Chromatog. Sci. , 9, 271 (1971).

105. C. B. Warren and E. J. Malec, J. Chromatog. , 64, 219 (1972).

106. D. P. Paris, D. L. Lewis, and N. L. Wolfe, Env. Sci. Tech. , 9,
     135 (1975).

Chapter 12

LIQUID CHROMATOGRAPHIC ANALYSIS
IN WATER POLLUTION

Irwin H. Suffet

Department of Chemistry, Environmental
Engineering and Science
Drexel University
Philadelphia, Pennsylvania

and

Edward J. Sowinski
Western Electric Company, Inc
Allentown, Pennsylvania

I. INTRODUCTION

The development of chromatographic methods of analysis has provided
aquatic scientists with new qualitative and quantitative tools for the determination of organics in water. It has become possible to identify trace
amounts of specific compounds (a) in highly complex organic mixtures, such
as domestic waste water, industrial wastes, and sludges; (b) in dilute
inorganic solutions, such as drinking waters and natural waters; and (c) in
concentrated salt solutions, such as oceanic waters. Knowledge of the
specific organic compounds involved in water quality can be invaluable in
leading to more effective control of water quality problems.

In its inception, chromatography was oriented strictly to liquid chromatography (LC) [1]. With the introduction of gas chromatography, the liquid
chromatographic applications for nonpolar, volatile, and low molecular
weight compounds became inadequate as the selectivity, resolution, sensitivity, and analysis speed of gas chromatography and its associated detectors
were far superior. However, liquid chromatography maintained its usefulness in the areas of preparative methodology, preliminary functional group
separation, and cleanup before other methods of final separation. Fertile
ground for the use of liquid chromatography remains for very polar molecules, macromolecules, nonvolatiles, and temperature sensitive molecules.
The intent of this chapter is to present a framework for the application of
existing liquid chromatographic techniques to water (pollution) and outline
potential applications for new high-speed LC that have been developed.

Generally, liquid chromatography is used for the isolation and the systematic identification of unknown organic compounds and for the isolation and
quantitation of a known compound or a group of compounds. Analytical
separation methods concerning liquid chromatography in aqueous analysis
can be classified as (a) sample collection; (b) preliminary sample separation
of interferences, called cleanup; (c) preliminary group classification; and
(d) final separation and quantitation. These uses of liquid chromatography
are explored in depth in Section II of this chapter.

Section III reviews details of liquid chromatographic methods by compound type. In this section, the new area of high speed is highlighted. The liquid chromatographic techniques presently developed but not utilized for water systems are drawn from as examples where their applicability exists. Competitive procedures are noted.

Liquid chromatography is now in a renaissance because of the recent development of gel chromatography and high-speed liquid chromatographic systems [2, 3]. Gel chromatography has mainly been applied to separate macromolecules. High-speed LC is developing into a collaterate technique to gas chromatography. Therefore, an important dimension to be considered within this review is, where applicable, how can liquid chromatographic procedures developed at atmospheric pressure be used with high-pressure technology?

A. Definition of Liquid Chromatography

Liquid chromatography is a column process. All liquid chromatographic systems include a mobile solvent phase, a means of producing solvent flow, a means of sample introduction, a liquid or solid stationary phase, a fractionating collector and a detector.

"Liquid-liquid chromatography," "liquid-solid chromatography," "gel chromatography," and "ion-exchange chromatography" are terms usually used to differentiate the forms of liquid chromatography when the nature of the stationary phase or mechanism in the stationary phase is not known [4]. Many authors have used the terms "adsorption" and "participation" are pertaining to the description of liquid chromatography of the liquid-solid and liquid-liquid types, respectively [5-7]. However, these mechanisms do not appear to be exclusive for any type of liquid chromatographic process [4]. For example, Determann [8] and Altgelt [9] describe the gel chromatographic mechanism as a liquid-liquid partition of solute molecules distributed between liquid phases in pores and outside the gel structure as well as a steric exclusion of molecules and a restricted diffusion of molecules. Even ion-exchange resins have been shown to have adsorptive character [10]. Adsorption and partition characteristics have been amply described [11]. The use of ion-exchange chromatography in water pollution is reviewed in a separate chapter.

B. Theoretical Considerations Related to Qualitative and Quantitative Analysis by Liquid Chromatography

To present a framework for liquid chromatographic applications, it is necessary to describe the "general elution problem." In the rapidly changing field of liquid chromatography any meaningful discussion of the applications of liquid chromatography to water pollution must involve the theory and future capabilities of the method. The theoretical elements on which the capabilities of the method are based will be defined briefly on a practical basis.

Resolution of the components in a liquid chromatographic elution system of two gaussian distributed peaks is a function of the separation of band centers and the width of each band. This is defined to be theoretically based on Eq. (1), where $1/4[1 - (K_1/K_2)]$ is the selectivity of the chromatographic system to separate band centers (the distribution coefficients K_1 and K_2), $\sqrt{N}$ is the efficiency of the column and therefore band width (Van Deemter equation) and $K_2/[V_0/(W + K_2)]$ is the column capacity of the stationary phase (V_0/W) [6]. V_0 and W are the column void space and weight of packing, respectively, or the net retention time relative to the nonsorbed time of the second compound. The resolution of early eluted sample components is a problem when K_2 is small [12-15]. N is the number of theoretical plates.

$$ R = \frac{1}{4}\left[1 - \frac{K_1}{K_2} \right] \sqrt{N} \left[\frac{K_2}{V_0/W + K_2} \right] \tag{1} $$

The control of K_1 and K_2 in LC is accomplished for a particular sample type by the choice of elution solvent and stationary phase. The choice of elution solvent in liquid chromatography determines K_1 and K_2 in a manner analogous to temperature change in gas chromatography. The control of K's in LC can be accomplished by the new technique of automatic gradient elution in a manner similar to temperature programming in GC, which smooths early and late peaks [16]. An order of magnitude increase in peak capacity is possible with solvent programming. N and the capacity factor are effected by the column technology: stationary phases (type, activity, velocity, viscosity), column type (length, bore diameter, column packing procedure, pressure, temperature), sample size, type, and separation time. Generally, the greater the difference in polarity of two compounds the more need for solvent programming.

The mobile and stationary phases are generally selected on the basis of selectivity and capacity factor and, happily, these are near optimum for N_{max}. Recycling chromatography can be utilized with or without sequential columns of different stationary phases when a separation is difficult to accomplish.

Thin-layer chromatography and paper chromatography have been suggested as screening methods for the selection of mobile and stationary phases for "classical" LC [17]. Silica gel, alumina, etc., cannot be used for high-speed LC as a low surface area (thin film) is necessary for quick phase equilibria at high speed [18]. Gel chromatography with rigid stationary phases (e.g., Poragel) can be used as high-speed LC supports.

Maximum resolution per unit time is critical for all chromatographic analyses. The retention time (tR) has been described as inversely proportional to pressure drop in a column (ΔP)

$$tR = \frac{1}{\Delta P} \times 1000\frac{V_0}{W}(1 + c)_{\eta}\left(\frac{L}{d_P}\right)^2 \tag{2}$$

where η is viscosity of the mobile phase, V_0/W is the porosity of the column packing, c is the capacity factor, L is the column length, and d_P is the particle size diameter of the column packing. It has been observed that the larger the change of pressure and permeability of the column, the less time is necessary for a chromatographic separation [19]. A pressure increase decreases the time of analysis but a limit is reached under high-pressure operation. Many separations may be done in 1-20 min with a good separation factor at high pressure [20]. Classical liquid chromatography is not carried out optimally. Column pressures are less than 1 atm and short columns of large diameters (125 mm) are packed with 100-150 μ size particles. This limits the number of theoretical plates to about 50 plates in a few hours. Thin-layer chromatography, by comparison, is capable of developing more plates in less time. However, at pressures greater than 10 atm (flows of 1-5 ml/min) with small diameter columns (1-5 mm) packed with less than 50 μ size particles liquid chromatography can develop up to 2500 theoretical plates in 1-4 hr [18, 21]. Gas chromatography can develop 10,000 theoretical plates.

Liquid chromatography is a complementary technique to gas chromatography. Current developments involving automated liquid chromatographs with thin columns and sensitive detectors indicate that rapid quantitative analysis comparable to gas-liquid chromatography are becoming routine.

II. THE USEFULNESS OF LIQUID CHROMATOGRAPHY IN WATER STUDIES

Table 1 shows a general flow diagram for analytical routes utilized for qualitative and quantitative analysis of organic compounds in aquatic environments. The versatility of classical LC allows it to be considered as a backbone of this scheme, although this is not usually recognized.

First, liquid chromatography has been used as a general collection method for trace analysis where direct analysis of organic components in water is not feasible because of sensitivity and specificity requirements. Collection methods are exemplified by the carbon adsorption method (CAM) [22], and reversed-phase partition method [23]. For example, the CAM has isolated many types of organics from various rivers [24]. Large concentrations of an organic compound (1 mg/liter) in water do not require these techniques; e.g., phenols of concentrations greater than 1 mg/liter can be analyzed by direct aqueous injection into a gas chromatographic column [25].

Second, liquid chromatography is the prime method of sample cleanup. "Cleanup" is defined as a technique that separates the sample of interest

TABLE 1

A General Analysis Flow Diagram for Trace Organics
in Aquatic Environments

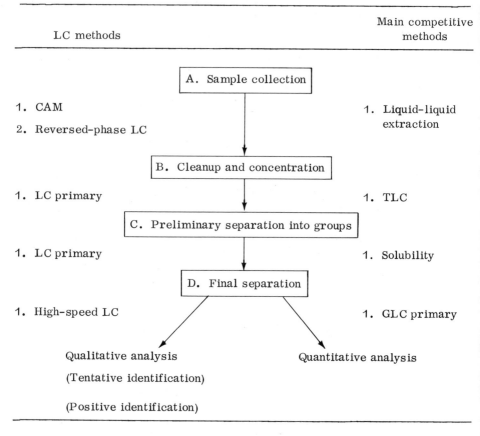

LC methods		Main competitive methods
	A. Sample collection	
1. CAM		1. Liquid-liquid extraction
2. Reversed-phase LC		
	B. Cleanup and concentration	
1. LC primary		1. TLC
	C. Preliminary separation into groups	
1. LC primary		1. Solubility
	D. Final separation	
1. High-speed LC		1. GLC primary
	Qualitative analysis (Tentative identification)	Quantitative analysis
	(Positive identification)	

from coextracted organic interferences that may interfere with a specific analysis, usually in the microgram to nanogram per liter concentration range. There are extensive applications of LC cleanup procedures of organic extracts from the CAM and liquid-liquid extraction collection methods.

Third, preliminary separation by LC of water extracts has also been extensively utilized, e.g., the separation of natural color components of surface water [26].

Fourth, classical LC has primarily been used for the final separation of large amounts of polar, heat sensitive, and/or nonvolatile compounds, e.g., organic acids [27], and organophosphate pesticides [28]. High-speed LC is now being utilized for these and other compounds at microgram to nanogram per liter concentration levels [29].

A. Liquid Chromatography as a Collection Method

Direct measurement and identification of trace organic chemicals in water is not usually feasible because of the sensitivity and specificity requirements of detection systems. Sample collection (extraction) and subsequent concentration of the sample usually must be employed prior to a quantitative analysis to attain the necessary concentration of the organics for the detection system. The primary methods of sample collection are a frontal liquid chromatographic method, the carbon adsorption method (CAM), and its main competitive method, the liquid-liquid extraction (LLE) method (Table 1).

1. Carbon Adsorption Technique (CAM)

Carbon adsorption has been employed as a field device for collection of organics from water. The continuous technique was developed and introduced by the U.S. Public Health Service [30] for isolation of synthetic organic contaminants affecting taste and odor qualities of surface waters. This field device is known as the carbon adsorption method. The CAM provided one of the first insights into the composition of trace organic contaminants since it yielded the necessary quantities of trace compounds for infrared identification (Table 2).

The CAM utilizes a prefiltering device that consists of a borosilicate glass pipe 3 in. in diameter and 18 in. in length packed with 4/10 mesh and 30 mesh carbon. Sample volumes of 1000 and 5000 gallons are pumped through the filter at rates of 0.25 and 0.56 gal/min (5.1 and 10.2 $gal/min/ft^2$), after which the carbon is removed and oven dried at $40°C$ for approximately 2 days prior to solvent extraction with chloroform [31]. An important point of technique is the proper carbon pretreatment, e.g., acidification, to eliminate interfering inorganics and organics adsorbed on the carbon surface.

The CAM has many well-documented serious limitations for quantitative recovery of organic compounds from natural waters. These limitations were demonstrated in the original U.S. Public Health Service studies [22-24]. A higher recovery of "total" organics from river water was accomplished by modifications of the CAM: (a) complete removal of particulate matter by diatomaceous earth prefiltration; (b) acidification of the river water to pH 2.8; (c) operation of several carbon filters in series; (d) utilization of acetone, ethanol, and benzene in addition to chloroform for carbon extraction; and (e) slower sampling rate [32-34].

Lee et al. [34] found that the carbon chloroform extract (CCE) and carbon alcohol extract (CAE) from the CAM actually recovered less than 5% of the dissolved organics in a eutrophic lake. An unusual variable developed during the summer months, when a slime layer of microorganisms developed on the carbon surface. Microbiological transformation of the isolated organic matter may occur with release of organics different than those in the original water. Lee suggested that the drying step ($40°C$ for 2 days) would lose

TABLE 2

Specific Classes of Organic Materials Collected from
Natural Waters by the CAM

	References
Phenols	35
Chlorinated hydrocarbon pesticides	22, 24, 36
Organophosphate pesticides	36
Numerous organic compounds, e.g., o-nitrochlorobenzene, phenyl ether, phenols, ketones, aldehydes	24
Numerous organic compounds, including taste and odor formers: naphthalene, tetralin, styrene, acetophenone, ethyl benzene, bis (2-chloroisopropyl) ether, 2-ethyl hexanol, bis(2-chloroethyl) ether, diisobutyl carbinol, phenyl methyl carbinol, 2-methyl-5-ethyl pyridine	39
Polynuclear hydrocarbon carcinogens, carbon filter, 2 years: 3,4-benzopyrene, 3,4-benzofluoranthene, 1,12-benzofluoranthene	40, 41
Noncarcinogen hydrocarbons: fluoranthene, 10,11-benzofluoranthene, 10,12-benzofluoranthene	
Polynuclear hydrocarbons: Raw water supply carbon ($CHCl_3$ eluants)	42

low-boiling compounds and that the extraction step (with chloroform and ethyl
alcohol) furnished an opportunity for volatilization and polymerization. Lee
speculated at length on errors in recovery of specific organic compounds by
the CAM that might have occurred from (a) removal by the sand prefilter; (b)
preferential adsorption for certain types of organic compounds as influenced
by type of carbon, pH, temperature, and mineral constituents in the water;
and (c) change of selectivity during the course of a carbon column run.
(Some organics with a high affinity for carbon can displace compounds with
lower affinities.) It was reasoned that the CAM might perform well for those
organics that were sorbed quickly and had high affinities. All of these errors
and limitations lead to caution in interpreting data collected by the CAM.

The primary concern for use of the CAM is the adsorption efficiency by
carbon and the desorption efficiencies from carbon by solvents. Hoak [35]
found 94% adsorption but only 22% desorption of phenol from activated carbon.

Recently Eichelberger and Lichtenberg found that the CAM (CCE) is not suitable for recovery of heptachlor and aldrin chlorinated hydrocarbon pesticides and six organophosphate pesticides of 21 pesticides tested for adsorption-desorption efficiency [36].

In spite of the aforementioned and well-documented limitations, the CAM may have qualitative usefulness for continuous surveillance of ground and surface water pollution. Also, the CAM might be utilized in the continuous monitoring of treated municipal waters [37]. The CAM does carry two major advantages as an extraction technique: (a) it is a continuous process and (b) large volumes of water are sampled. Table 2 shows the compounds isolated by the CAM and subsequently identified by qualitative analysis of the extract   The CAM, in the form of the carbon chloroform extraction (CCE), is an accepted drinking water standard for the measurement of total organics; less than 0.2 mg/liter CCE is acceptable in a drinking water supply [38].

2. Reversed-Phase Liquid-Liquid Partition

Ahling and Jensen [23] developed a reversed-phase liquid-liquid chromatographic collection method for collecting trace chlorinated hydrocarbon pesticides and polychlorinated biphenyls in tap water, river water, and waste water at the part per quadrillion level. A column of 3 g of 30% n-undecane and 10% Carbowax 4000 monostearate on Chromosorb W (HMDS) was used. Samples of 176 liters of tap water, 166 liters of river water, and 10 liters of waste water, with and without sediments, were analyzed. The column was eluted with 10 ml of petroleum ether and then prepared for quantitative GC analysis. Recoveries on 6 liters of distilled water varied from 40 to 100% for different pesticides depending on the experimental conditions. Interferences were noted in distilled water blanks. Cleanup problems were observed with natural water samples.

Burnham and co-workers [43] reported a macroreticular resin could also collect neutral trace organic compounds. Alkylnaphthalenes, benzenes, and benzothiophenes were collected from a chemically contaminated well water. In a recent report Vinson [44] evaluated quantitative collection capabilities of the absorbent. The methodologies can be extended to collecting other pollutants and specificity can be developed by selecting different liquid chromatographic phases or resins. Qualitative analysis of collected samples appears possible for specific organics at the nanogram and picogram levels.

B. Cleanup

Coextracted organic interferences may arise from domestic and industrial pollution. Also, naturally occurring color substances and microbiological metabolites in surface or ground waters can act as analytical interferences.

For example, the CCE of the CAM contains oxygenated organics that interfere with the analysis of chlorinated hydrocarbon pesticides [31].

A cleanup step becomes imperative when grossly polluted waters are
extracted. Two groups of workers, Hindin et al. [45] and Epps et al. [46],
while analyzing for pesticides in water, found that surface and ground waters
contained a considerable amount of organic matter that interfered with their
analyses. Epps et al. [46] set a detection limit of 0.001 ppm for chlorinated
hydrocarbon pesticides for water samples and 0.01 ppm for bottom muds;
however, "in some cases, interferences were so great that amounts less
than 0.1 ppm could not be reported with confidence."

Liquid chromatography is a prevalent cleanup technique for water extracts.
Table 3 summarizes recent examples of trace aqueous analysis. Column
chromatographic supports have been investigated in order to standardize
cleanup procedures [47, 48]. Reproducible chromatographic conditions must
be carefully controlled and the activation procedure followed in detail. Solvent purity, purity of LC support, and reproducibility of the activation of the
LC support should be carefully checked as trace impurities can change
separations; e.g., trace amounts of benzene in hexane of pesticide quality
have caused incomplete separation of polychlorinated biphenyl during cleanup
of DDT [49].

The criterion for successful cleanup is considered to be satisfied by many
authors if the detection system exhibits minimum interferences when samples
are compared to reference standards. In many cases heavy reliance is laid
upon the success of the final separation step, i.e., GLC, TLC. The current
philosophy is to utilize some type of liquid chromatographic procedure for
gross cleanup. Cleanup data should be reported to indicate effectiveness and
percent recovery of a given cleanup system. A good approach are reported
by Eichelberger et al. [50] who presented cleanup recoveries of phenolic com
pounds obtained from a CCE. Previous cleanup techniques utilizing solubility
separations had resulted in loss of volatile phenols, poor recoveries, and
generally insufficient cleanup of phenols from water for subsequent TLC and
GLC.

The exact chemical species or organic interference that may be separated
from the organic material of interest is usually not considered in an analysis.
The interferences can vary from one water extract to another in any flowing
water system or in any microbially active one. Therefore, cleanup procedures are never universal and may have to be modified in particular situations.
Also, some of the total cleanup responsibility is shifted to the qualitative or
quantitative technique. This is not always acceptable since some interferences may have similar characteristics. A case in point is the polychlorinated biphenyls (PCB's) which have been identified as an impurity and not
separated but which have tentatively been identified by GC as a chlorinated
hydrocarbon pesticide. Subsequent spectroscopic techniques proved the GC
peaks were PCB's and not chlorinated hydrocarbon pesticides [52].

TABLE 3

Exemplary Column Chromatography Cleanup Systems Applied to
the Analysis of Organics from Aqueous Environment

Compounds	Cleanup sample	Support	Elution	Quantitative analysis	Reference
Phenols (alkyl, chloro, amino, miscellaneous)	(Water) CAM	Florisil	Ether	GLC	50
Organophosphate pesticides	(Water) LLE	Charcoal	CHCl₃	GLC	28
Chlorinated hydrocarbon and organophosphate pesticides	(Water) LLE	Silica gel	1. Hexane and 20, 40, 60, and 70% benzene in hexane; 2. benzene in ethyl acetate	GLC	51
Pesticides	(Water) LLE or CAM	Silica gel		GLC	52
Polychlorinated biphenyls	Water, sediments, and biota blend; soxhlet-solids, Water-LLE	Florisil	Cleanup method of Mills et al. [54] pet. ether	GLC	53

TABLE 4

Preliminary Separation of Organic Compounds into Groups by Liquid Chromatographic Techniques

Compounds	Sample source	Support	Elution solvent	Groups	References
Mainly hydrocarbons and lipids	Sea water and estuary (LLE)	Silicic acid	Varied (Table 5)	Eight groups	55,58
Organophosphate pesticides	River water sewage effluent (LLE)	Sephadex	Ethanol	Three groups	28
Humic acid and lignin sulfonates	Natural waters	Sephadex	Distilled H_2O	Ten groups	59
Proteins and amino acids	Synthetic mixture	Sephadex	Water	Eleven individual compounds	60
Surface[a] water organics	Surface water (CAM)	Silica gel	Isooctane–aliphatic benzene–aromatic 1:1 $CHCl_3$– CH_3OH–oxygenated	Three groups	57
High molecular weight organics	Secondary sewage effluent after: 1. Freeze conc. 2. Ether LLE, which is 3. Nondialyzable and 4. Water soluble	Sephadex		Compounds MW 156,000 100,000 60–70,000 3,000	61

[a]Cronin, John T., Techniques of Solvent Extraction of Organic Materials from Natural Waters, Oregon State University, 1967. Ph.D. Thesis–Dis. Abstr. 67–17, 573 (55).

C. Preliminary Separation of Organic Compounds into Groups

The collection of organic compounds from water is usually nonspecific by LLE and CAM. Liquid chromatography can be used for preliminary separation of chemical groups. Table 4 shows a series of preliminary class separations used during water pollution analysis. Alternate schemes of group separation by differential solubility are more time consuming and require more manipulation for similar reproducibility [52].

Table 5 represents the specific elution scheme for a separation of lipids into different groups. The residue lipid concentration of sea and estaurine water was studied by this method after organic compounds were collected by liquid-liquid extraction [55]. Further separation is necessary for analysis of one particular compound.

A general micromethod for the isolation and characterization of natural products that can be applied to aqueous extracts should be highlighted [56]. The method uses LLC on an impregnated inert phase for subfractionation of fatty acids, amines, thiols, alcohols, carbonyls, etc., from complex natural products. The method features an inert support impregnated with a colorimetric reagent dissolved in an aqueous phase. An eluant containing the unknown in aqueous immiscible solvent is the second liquid phase. The eluant is passed through the column, reaction occurs, and tentative group separation and qualitative identification is completed.

The preliminary group separation of organics from a CAM collection for chlorinated hydrocarbon pesticides demonstrates a combination of the solubility and liquid chromatographic approaches to preliminary class separation [52]. In very general terms, the dry carbon from the CAM is placed in a large scale soxhlet for continuous extraction (35 hr) with redistilled chloroform. This chloroform extract (CCE) is subsequently evaporated to dryness, whereupon a residue weight is obtained. At this point the dry residue (presumably all organic matter) can be taken up by ether for subsequent separation by classical solubility techniques into bases, strong acids, neutrals, and weak acids [52]. Most chlorinated hydrocarbons are found in the neutrals. The neutrals are dissolved in ether and placed on a silica gel column for separation into aliphatic, aromatic, and oxygenated compounds by selective elution with osooctane, benzene, and a 1:1 mixture of chloroform-methanol, respectively. The aromatics are then cleaned up by TLC for GLC analysis.

Alternatively, column chromatography can be used to obtain the neutrals directly. The CCE is redissolved in chloroform and placed on a silica column and eluted into groups as described above. The methanol-chloroform fraction contains many other compounds besides oxygenated compounds. The separation procedure of CCE neutrals into aliphatic, aromatic, and oxygenated groups was first used to characterize petroleum refinery wastes by IR spectra [57]. In this case, the aliphatic and aromatic hydrocarbons were desired for IR analysis and the oxygenated groups and other compounds were not desired.

TABLE 5

Elution on Silica Column[a]

Fraction	Solvent	Compounds eluted
I	50 ml of 1% diethyl ether in petroleum ether	Paraffinic hydrocarbons
II	75 ml of 1% diethyl ether in petroleum ether	Squalene, β-carotene, unsaturated hydrocarbons
III	225 ml of 1% diethyl ether in petroleum ether and 60 ml petroleum ether	Sterol esters, α-tocopherol
IV	240 ml of 4% diethyl ether in petroleum ether and 200 ml of 8% ethyl ether in petroleum ether	Triglycerides, free fatty acids, fatty alcohols
V	450 ml of 8% diethyl ether in petroleum ether and 50 ml of 25% diethyl ether in petroleum ether	Unesterified steroids, vitamin D_3, vitamin A alcohol
VI	200 ml of 25% diethyl ether in petroleum ether	Diglycerides, tributyrin
VII	300 ml of diethyl ether	Monoglycerides, vitamin A acetate, lithocholic acid, chimyl alcohol
VIII	400 ml of absolute methanol	Phospholipids, monoacetin

[a]Scheme of Hirsch and Ahrens [58].

D. Final Separation

Liquid chromatography has been used as the primary method of choice for final separation of large amounts of polar, heat sensitive, and/or nonvolatile compounds, e.g., organic acids [27] and organophosphate pesticides [28]. Whereas gas chromatography has been the primary method of choice for final separation of volatile and/or nonpolar compounds, the advent of high-speed liquid chromatographic systems provides competition for all types of GC systems [29]. The tentative qualitative identification by LC is accomplished by relative retention volumes. Quantitation is usually performed with a detection system placed directly at the end of the chromatographic column.

New high-speed commercial LC systems that are presently available incorporate the concept of tentative qualitative analysis and quantitative analysis similar to the operation of a gas chromatographic system. Byrne [18] has tabulated the state of the art for high-speed LC detectors, e.g., refractive index as a bulk detector and UV as a solute property detector. Each has a sensitivity to specific compounds in the microgram to nanogram per liter concentration range. A universal detector is not yet available.

III. SYSTEMATIC CLASSIFICATION OF ORGANICS IN WATER POLLUTION

The need exists for isolation, identification, and quantification of organic compounds present in aquatic systems (including natural water and waste water treatment processes) for an understanding of how these systems operate. For example, a better understanding of the nature of organic-metal complexation in waters and its effect on the growth of natural biota is desirable. Also, the mechanics of change in natural food cycles and improved methods for classifying organic compounds present in pollution control processes are of significance. Methods for determining the kind of organics present at different stages of water and waste water treatment are of particular significance.

The CAM [30-37] and other isolation methods for organic compounds indicate that only a small amount of the total organics collected are known. Vallentyne [62] has shown the diverse nature of compounds present in natural waters. Many thousands of compounds have been shown to be present.

Every water type (from pure lake water to industrial effluents) has its distinctive physiochemical and biological characteristics and offers its own share of analytical problems, especially for trace analysis. The chromatographic scheme (Table 1) is open to modifications based on water quality characteristics and the organics studied.

Most organic compounds can be present in water in a dissolved, adsorbed, or particulate form. The organic compounds that have been most studied are those which have particularly undesirable effects on the environment. Table 6 outlines an arbitrary frame of reference for classification of organic compounds by their environmental effect and source. Table 7 outlines organic compounds of high molecular weight by origin. These lists are made arbitrarily and are by no means complete. The intent is to enable a logical presentation of the applications of LC to water pollution.

A detailed discussion of topics directly related to water pollution is given below. Typical methods from other fields also are included where we feel they have applicability to water pollution systems. The reader is referred to review articles for general schemes of specific LC analyses [63-65]. The annual reviews Anal. Chem., Journal of Water Pollution Federation, Water-Pollution Abstracts, and Chemical Abstracts [66-69] were primarily utilized for obtaining references to LC as applied to water pollution.

TABLE 6

Classification of Organics in Aqueous Environments

Classification based on environmental effects

		Metabolic processes at all trophic levels	
Toxicological effects	Aesthetic effects	Biodegradable	Nonbiodegradable
Pesticides	Color (humic substances)	Carbohydrates	Detergents
Phenols	Foam (detergents)	Proteins	Petroleum products
Polycyclics	Oil spills (petroleum products)	Lipids	Pesticides
Polychlorinated biphenyls		Nucleic acids	Polychlorinated biphenyls
and related compounds		Vitamins	and related compounds
Petroleum products			

Classification Based Upon Environmental Source

Domestic waste water	Industrial wastewater	"Natural" origin
Lipids	Coal-tar products	Carotenoids
Carbohydrates	Cellulose	Chlorophyll
Proteins	Carbohydrates	Lipids
Detergents		Proteins
		Carbohydrates
		Nucleic acids
		Vitamins
		"Humic substances"

TABLE 7

High Molecular Weight Compounds of Interest
in Water Studies

Natural origin[a]
  Proteins (enzymes)
  Cellulose
  Lipids
  Carbohydrates
  Nucleic acids
  Toxins produced by the biota
  Virus
  Carotenoids
  Chlorophyll
  Plant, wood, and soil extracts (humic substances)

Waste water origin
  Coal-tar products — polycyclics and petroleum products
  Paper mill wastes — lignin sulfonates
  Sewage treatment plant effluents
  Oil effluents
  Others

---

[a]Based on a literature review by Vallentyne [62].

A. Macromolecules

The characterization of naturally occurring high molecular weight organics
found in aquatic environments involves the separation and identification of a
large number of diverse compounds at trace concentrations from an extra-
ordinarily complex matrix. This is of course only a reflection of the
diversity and complexity of the life systems that natural aquatic systems
must support. Only a minor fraction of the typical dissolved organic fraction has been experimentally identified as specific macromolecules. Those
compounds that have been consistently separated and identified have been
reviewed in the works of Vallentyne [62] and others. It is obvious that most
macromolecules are delineated to no greater extent than classes or groups.
Natural organic macromolecules are important from the point of view of
their source, their use, and their effect on man or on the metabolic
processes at all trophic levels of natural waters. Table 7 outlines some
groups of important macromolecular compounds. Any macromolecular compound, known or unknown to man, could be found in water pollution; for
example, Ciaccio [61] separated five groups of high molecular weight
organics from the water-soluble, hydrolyzable fraction of a freeze concentrate (20:1) of secondary sewage effluent.

Gel chromatography has become the primary tool for the separation of high molecular weight compounds into groups [70]. Multiple column chromatography utilizing up to three grades of Sephadex has been used for the separation of compounds with a wide range of molecular weights [71]. The recent application of efficient, high-speed techniques on Poragel 100 is indicative of escalating interest as well as improved data from the separation of naturally occurring, high molecular weight organics [72].

B. Organic Acids

Aerobic and anaerobic biological decomposition causes a mixture of low molecular weight metabolites (Kreb cycle intermediates) to form from macromolecules (e.g., carbohydrates, proteins, and lipids). The metabolites are primarily fatty acids, simple alcohols, amines, and mercaptans. Further aerobic metabolism forms CO_2 and water as products; anaerobically, CO_2, CH_4, H_2S, and hydrogen are formed. The amount of each specific component depends on the environment and nature of the original material.

The importance of the metabolic process is exemplified in the anaerobic sludge digestion of solids in sewage treatment and their aerobic and anaerobic degradation in polluted streams and sediments. The extent of fatty acids found in sewage treatment plant effluents and streams has been suggested as a gage of pollution [73]. The possible interrelationship between low molecular weight organics and general stream water quality therefore is recognized. Vallentyne [62] had classified the organic acids found in the water environment (seston, i.e., decayed vegetable matter; dissolved matter; sewage; and sediments) before 1957.

The "standard method" of analysis of total volatile acids in sludges is performed using butanol-chloroform elution from an acidified silicic acid column. Recovery of 95% volatile fatty acids above 200 mg/liter as HOAc is possible [74]. Hattingh and Hayward [75] completed a comprehensive evaluation of the analysis. Scherfig et al. [76] modified the extraction step of the standard method to obtain a sensitivity of 30 mg/liter as HOAc.

Table 8 shows an analysis of organic acids done in water pollution; the separations reported are: Group (1) butyric, crotonic; (2) propionic, chloracetic acid; (3) acetic; (4) adipic, pyruvic, phthalic; (5) formic, fumeric, ABS (alkylbenzene sulfonate); (6) lactic, succinic; (7) malonic, gallic, aconitic, oxalic; (8) L-malic; (9) citric, isocitric; and (10) tartaric. This represents the state of the art reported by Mueller et al. in 1960 [77] and expanded by Freeman in 1967 [78]. Waters Associates has automated the process for a 5-hr separation [79].

Polar fatty acids can be analyzed directly by GLC in the milligram per liter range with a direct water injection as reviewed by Geddes and Gelmour [80]. Ellerker et al. studied volatile fatty acids in sludge gas by GLC [81]. Below 1 mg/liter concentration, extraction, and derivatization to more volatile and sensitive compounds for GLC electron capture detection is necessary [82].

TABLE 8

Fatty Acids

Sample	Eluant	Support	Results	Reference
(1) Sewage sludge	Butanol in CHCl$_3$ (gradient elution)	0.5M H$_2$SO$_4$ on silicic acid	0.5–5 mg of each acid as HOAc; butyric, propionic, acetic, pyruvic, formic acid (in order elution)	27
(2) River water, sewage effluent, anaerobic digestor liquor	Same	Same	Butyric, propionic, acetic, pyruvic, and formic acid, in order of elution; detergents which are eluted with formic acid can be separated with a subsequent elution of this fraction with 15% butanol in CHCl$_3$	83
(3) River water	Same	Same	Here ppb conc. for the first time; more compounds added to above	73, 77
(4) Mineral water	Same	Same	Same as (1)	84
(5) Natural waters	Same	Same	Hydrolyze and extract interfering esters; phenol elutes before the acids with chloroform; naphthenic acids also elute	85
(6) Natural waters	Same	Same	Nonvolatile fatty acids; four fractions, then PC for identification; this is really a preliminary separation for PC analysis	86

TABLE 8. — Continued

Sample	Eluant	Support	Results	Reference
(7) Ground water (near petroleum deposits)	Same	Same	LC followed by GLC; amines also are eluted; this is really a preliminary identification	87
(8) Plant organic acids	Same	Same	39 Mono-, di-, and tricarboxylic acids; summation of above	78

C. Lipids

Glycerides and fatty acid salts constitute the major fraction of relatively insoluble organics in waste water. The principal constituents are the long-chain saturated fatty acids — lauric, myristic, palmitic, and stearic acids — and the long-chain unsaturated fatty acids — oleic and lineoleic acids [88, 89]. Lipid-related species also constitute considerable fractions of insoluble waste organics. These include compound lipids, sterols, and hydrocarbons. Lipids and lipid-related species are insoluble in water and are less easily degraded in waste treatment facilities than carbohydrates and proteins. Therefore, significant quantities of lipids pass through waste treatment plants and contribute significantly to the organic pollutants in surface waters.

Knowledge of the fate of relatively insoluble organic matter (as lipids and lipid-related species or "grease") that is put into surface waters as a part of domestic and industrial waste effluents is meager. To obtain a better understanding of lipid degradation and removal patterns in waste water treatment facilities and natural waters it is necessary to have at hand analytical methods for the separation of lipid classes and identification of specific components in waste water. This is in contrast to the conventional approach in which lipids are considered a single broad class including "fats, waxes, oils, and any other nonvolatile material extracted by hexane from an acidified sample of sewage or industrial waste" [74].

Liquid chromatography, with silicic acid, has been successfully applied to the separation of lipid classes and to the isolation of specific lipid materials. For example, Loehr and Kukar [90] isolated eight major lipid classes from samples of sewage and sludge with LC. These included saturated hydrocarbons, unsaturated hydrocarbons, sterol esters, methyl esters, triglycerides, diglycerides, monoglycerides, and compound lipids. Gel chromatography recently has been utilized to separate lipid classes. Table 9 outlines some examples. Examples of specific materials isolated by liquid chromatography include the nonvolatile fatty acids. Fatty acids having a chain length of eight carbons or more are only slightly soluble in water. However, in waste water the solubility of a fatty acid salt will vary with the cation. In most waste waters calcium and magnesium cations predominate and relatively insoluble fatty acid salts are produced. Consequently, one may desire to pass a sample directly through a cation-exchange resin to remove Ca(II) and Mg(II) and eliminate precipitation of insoluble salts during subsequent concentration steps [64].

In 1969, Marinetti edited a comprehensive book, "Lipid Chromatography Analysis," that includes comprehensive liquid chromatographic analysis of lipid group separations and specific analyses of plant, animal, and microorganism extracts [64]. Morris and Nichols (in Heftmann [63]) also reviewed the analytical methods for lipid classification.

TABLE 9
Lipids

Sample	Eluant	Support	Result	Reference
1. Water and effluent samples	Sample passed directly through cation-exchange resin to remove Ca and Mg.	Amberlite IRC-50	Synthetic samples with known acid content showed no detectable loss of organics after passage through ion-exchange resin to eliminate precipitation of insoluble salts during subsequent concentration steps	73
2. Treated sewage, activated sludge, digested sludge, and raw sewage	Solvents with increasing polarity	Silicic acid	Eight lipid classes Saturated hydrocarbons Unsaturated hydrocarbons Sterol esters Methyl esters Triglycerides Diglycerides Monoglycerides Compound lipids	90
3. Polar and non-polar fats from sewage		Alumina columns		91

4. Complex fatty acid mixture	Petroleum ether	Silicic acid	General classification scheme — eight major classes separated: Saturated hydrocarbons Unsaturated hydrocarbons Cholesterol esters Triglycerides Nonesterified cholesterol Diglycerides Monoglycerides Phospho species	58
5. a) Lipid mixture	Chloroform–methanol mixture	Methylated G-25 Sephadex	Separation of: Phospholipids Triglycerides Cholesterol esters Bile acids Fatty acids	92
b) Lipid-soluble compounds	Several solvent mixtures	Methylated G-25 Sephadex		

TABLE 10

Amines and Amino Acids

Sample	Eluant	Support	Result	Reference
1. a) Mixture of amino acids, peptides, and proteins	$Na_2B_4O_7$ solution	Sephadex G–25	Amino acids were separated as copper chelates from peptides and proteins	99
b) Amino acids and proteins	$Na_2B_2O_7/H_2O$ and 2N HCl	Sephadex G–25	Amino acids were isolated from proteins	99
2. Asparagine and glutamine mixed with other amino acids, biological fluids	Lithium citrate buffers, pH 2.80 and 3.82	Beckman PA–28	Asparagine and glutamine were isolated from other amino acids	100
3. Simple amino acids, peptides, and amines	Various HOAc, NaCl mixtures with added. Pyridine citrate, NaOH, tris, H_2S, H_2SO_4	Sephadex G–10	Any pair of small water-soluble molecules that differ in charge or size should be separable	101
4. Mixture of aromatic amines	Methanol–water mixtures	Teflon 6 (polyfluorocarbon polymer)	Binary mixtures of aniline derivatives were separated	102

D. Amines and Amino Acids

Dissolved free amines and amino acids have been found in parts per billion concentrations in lake, estuary, and ocean waters [62, 93-95]. Amino acids are involved in the biota-energy flux at all trophic levels. In a review up to 1957 Vallentyne has noted that very little attention has been given to protein-like material in water [62]. Recent reports still indicate this lack of attention.

Table 10 outlines liquid chromatographic systems for the separation of amines and amino acids. The use of Sephadex enables separation from protein materials and separation of one amino acid from another by differences in charge and size. In view of this, the application of gel chromatography to water pollution studies should be easily accomplished. Ion-exchange chromatography is extensively used for automatic analyzers [96] and it has been used for water pollution studies [93, 94]. Automated silica gel chromatography has also been used for amino acid analysis [97], but ion exchange is generally preferred [60].

E. Peptides, Proteins, and Enzymes

The classification of peptides and proteins into subgroups is accomplished as shown for the lipid groups. Table 11 shows Sephadex to be the most commonly utilized column. The application of this type of separation to separate enzymes, toxins, and plant virus is also demonstrated.

Enzyme purification and isolation has been extensively accomplished by gel chromatography [98] and ion-exchange chromatography on modified cellulose columns. These are of interest in water pollution and natural processes. Recently affinity chromatography has also been used for enzyme purification. In this method an inhibitor is covalently bonded to Sepharose 4B (activated with cyanogen bromide). The inhibitor selectively reacts with the enzyme and separates it from other proteins, which pass directly through the column. Staphylococcus aureus extracellulose enzymes have been separated in a one-step procedure by this method [98].

F. Carbohydrates

Carbohydrates are of prime importance because of metabolic processes in natural water systems. Understanding the nature of isolated carbohydrates is critical for understanding food cycle relationships. Final separation of simple sugars has been accomplished by ion-exchange chromatography [107, 108]. Reducing sugars present in aqueous environments have been studied by this method [109]. TLC and ion exchange are primarily used with colorimetric detection. Future use of high-speed LC for the separation of saccharides seems to be inevitable.

TABLE 11

Peptides and Protein

Sample	Eluant	Support	Result	References
1. Protein	pH = 6.0 0.1M acetate buffer 0.4M NaCl	Sephadex G-100	Automated system: Globulin Bovine serum Albumin Pepsin Cytochrome c	103
2. Peptides	0.1M NaCl 0.1M Na_3PO_4, pH = 7	Sephadex G-15	Peptides utilization by Escherichia coli (a typical water-borne bacterium)	104
3. Enzymes	0.01M phosphate, pH = 6.8	Sephadex gel	Adsorption of enzyme then elution	105
4. Plant virus	Dextrane polyethylene glycol, NaCl, $MgCl_2$, glucose, phosphate pH = 7 buffer	Cellulose	Virus purification procedure by adsorption and elution	106

TABLE 12

Carbohydrates: Group Classification

Sample	Support	Eluant	Results	Reference
Carbohydrate mixtures (plant material)	1:1 Charcoal-Celite	a. Water b. 5% Ethanol c. 15% Ethanol d. Gradient elution	Monosaccharides, disaccharides, trisaccharides, etc., are eluted	110
Carbohydrate mixtures	Sephadex (G-75 to G-200) or Biogel	1% n-Butanol	Opposite elution order from above; high MW components first	111

Preliminary sample classification of carbohydrates in water pollution studies should be considered. For example, oliogosaccharides have been separated into groups by chromatography on 1:1 carbon-Celite columns with gradient elution of alcohol in water [110] (Table 12).

G. Miscellaneous Compounds

The analysis of such miscellaneous as nucleic acids, carotenoids, chlorophylls, biological toxins, antibiotics, and vitamins isolated from natural waters includes the necessary steps of trace analysis (Table 1). GLC and TLC primarily have been used for final identification and quantitation. Classical chromatography has been used for cleanup and preliminary group separations.

Vallentyne [62] reviewed the importance of miscellaneous biochemical materials, which have been shown to exist free and within the biota of natural waters. Carotenoids have been shown to be a stable part of the sediments of natural water. Vitamins and chlorophyll have been shown by biological assay to be dissolved in water and in sediments. Phytoplankton have been assayed by their chlorophyll content. Algae blooms have been shown to be the source of toxins that can cause death to domestic animals and of antibiotics that can kill other microorganisms. Some recent examples of isolation are: Holm-Hansen's isolation of DNA [112] and Garside's and Riley's [113] separation of chlorophyll and carotenoid pigments from sea water. Chlorophyll, the green tetrapyrrolic pigment of autotrophic plants, is a type of porphyrin found in the chloroplasts of plants. Carotenoids are another type of plant pigment.

Chlorophyll and carotenoids are weakly polar molecules that decompose on vaporization. Therefore, TLC and LC are the primary methods of analysis. Chlorophyll also can decompose on extraction from sample sources unless precautions are taken (such as the use of cold solvents). Only inactive sorbents as polysaccharides (e.g., sucrose and cellulose) have been extensively used with alcohol elution [114]. Table 13 indicates the kind of separation of these pigments that is possible on powdered sugar. A complete review of LC of chlorophyll can be found in Heftmann [63].

Toxins are protein material and their isolation usually involves gel chromatography on modified cellulose. For example, a toxin from Staphylococcus aureus (a common water-borne bacterium) has been purified on carboxymethyl cellulose and Sephadex (Table 13).

LC is used for cleanup of vitamin extracts. Procedures have been described for different vitamins by Katsui in Heftmann [63]. LC is also used for isolation of large quantities of antibiotics. Nucleic acids are primarily separated by ion exchange. An example of classical LC separation of t-RNA for Escherichia coli is shown in Table 13.

TABLE 13

Miscellaneous Group

Sample	Eluant	Support	Result	Reference
Free and bound vitamin B_{12} in sea water	Sea water	Sephadex G-25	An estimation of free and bound vitamin B_{12} in water samples and sea water	115
Carotenoids in complex mixtures	Ethyl acetate–methanol (5:1) in benzene; gradient elution	1:1 Silica gel-celite	Automated analysis with a spectrophotometer	116
Chlorophyll	Petroleum ether–acetone–1–propanol (9:1:45 by volume)	Polyethylene powder, cellulose powder	Purification by LC of chlorophyll a and b preparations. Chlorophyll a and b were demonstrated to be altered during TLC with certain adsorbents	117
Chlorophyll and carotenoids in diatoms	0.5% 1-propanol in petroleum ether	Powdered sugar	Time of analysis is less than 30 min	118
Chlorophyll and carotenoid mixtures	(7:3) Petroleum ether-acetone or (3:1:1) iso-octane-acetone-ether some with 1% N,N-dimethyl-aniline	Various silicas or Florisil, Celite, Microcell	Mild silica supports give the same sequence as sugar or cellulose; others alter the compounds. Elution order as 3-carotenes (B and α), chlorophyll a, xeaxanthin and lutein, chlorophyll b, violaxanthin, neoxanthin	63

TABLE 13. — Continued

Sample	Eluant	Support	Result	Reference
Antibiotic activity of Asterionella japonica (diatom) vs S. aureus Strep. feelis sarc. intia		Sephadex	Separation of 2 antibiotic active zones from inactive zones (similar expts. for Chaetoceros teres)	119, 120
Biological toxin Enterotoxin C, from Staphylococcus (a possible water-borne organism from domestic waste water)	0.01M pH = 5.5, phosphate buffer	Ion exchange-carboxymethyl cellulose, Sephadex G-50 or G-75	Gross separation from other materials; purification of a preparation by both methods	121
t-RNA from E. coli (a water-borne organism)	Step or gradient elution-phosphate buffers	Hydroxyapatite	Several t-RNA species detected by amino acid specificity	122

H. Humic Substances

A specific area of water pollution concern that has been extensively studied is the high molecular weight organics which cause the color of natural water, the so-called "humic substances." This area of research exemplifies the usefulness of gel chromatography for water pollution studies.

The coloration of natural water results from the aqueous extraction of the soluble fraction of wood tissues, dissolution of decomposition products of decaying wood, and/or leaching of soluble soil components. Early attempts to separate the humic substances by classical liquid chromatography met with many problems. Elution solvents appeared to alter the character of the compounds and in some cases they eluted only with difficulty. Temperature, drying, base addition, and neutralization affect the stability of humic substances.

Oden classified all humic substances into three groups as shown in Table 14. Fulvic acids have the lowest molecular weight, while humic acids have the highest. This of course gives only the broadest classification into molecular weight ranges.

Several recent studies have attempted to elucidate the molecular size characteristics of natural color using gel filtration. Gel LC, in theory, can distribute the groups more or less continuously as a function of molecular size and shape. When conducted with ionic eluants gel filtration essentially works by sieving the compounds randomly through a lattice of various size holes. Therefore the smallest compounds travel the greatest pathlength and the largest are eluted first. Size varies with molecular weight differently depending on the compounds. Christman and Minear [124] have obtained the elution profiles for a variety of colored waters on Sephadex G-25. Exclusion and inclusion limits are approximated by the elution positions of Blue Dextran 2000 (MW 2×10^6) and glucose. The tailing beyond glucose is due to reversible adsorption.

Other authors have attempted to compare Oden's classification scheme with elution patterns from Sephadex gels. The pretreatment necessary to fractionate according to Oden's scheme causes an obvious shift in molecular

TABLE 14

Humic Substances by Oden Solubility Classification[a]

Steps	Strong acid	Alcohol
1. Fulvic acid	Soluble	—
2. Hymatomelconic acids	Insoluble	Soluble
3. Humic acids	Insoluble	Insoluble

[a]From Ref. 123.

TABLE 15

Classification of Humic Cubstances by Gel Chromatography

Sample	Eluant		Result	Reference
Extract of S.Bohemium turf	0.1N NaOH	G-25 G-50	Three major peaks	125
Water from the Norweigen moorland	Distilled water	G-25, -50, 2 peaks G-75, 2 peaks G-100, 2 peaks G-200, 1 peak	Retention on column (1) $10^5 - 2 \times 10^4$ MW (2) $<10^4$ MW "possible" two types of humic acids are present	127
Lake water	0.5M NH$_3$		Classify the substances causing color in lake water	128
1. Marshland creek water 2. Lake water 3. Aquatic extract of lake mud 4. Filtered algae extract	Distilled water	G-25 G-75 G-100 G-200	Ten peaks multistage Gel Chrom; each peak has a different color. Many $>5 \times 10^4$ MW; pH has a great effect on elution patterns	59
1. Few years old peat extract 2. Few months old peat extract 3. Natural water (HA, HMA, FA-groups)	0.1M NaCl 0.1M NaCl 0.5M NH$_3$ 0.1M NaHCO$_3$ 0.01M NaOH	G-50 G-75	0.01M NaCl G-75 is the best combination; several peaks	26

Sample	Eluent	Gel	Observations	Ref.
Moorland water	NaCl	G-25	Two peaks	129
Red-brown earth distilled	Distilled water	G-25, -50	Two peaks	130
a. Pasture since 1949		G-75		
b. Fallow/wheat rotation since 1925 (NH_4Cl added to sample)		G-100 G-200	Cultivated land showed a larger fraction of low MW substances	
Fulvic acids	a. Distilled water	G-25	G-25 best; diverse structural constituents found; adsorption effects on the gel were observed	131
	b. 0.01M $KMnO_4$	G-50		
Soil extract three types	0.1N NaOH	G-25, -75, -200	Successive fractionation by different gels; similar elution patterns were observed	132
1. Six colored waters	Clark and Lub buffer pH = 6	G-25	Observed differing elution profiles	124
2. Three colored waters	pH = 6	G-10	Color excluded	124
3. Colored water	0.01N NaCl at pH = 3.5, 5.5, and 8.0	G-75	Different elution patterns at different pH's resulting from alteration of effective size	124
4. Colored water	pH = 7 phosphate buffer pH = 7 borate buffer	G-75	Shift to higher molecular weights for borate buffer	124

size distributions. A time functional relationship of the Oden extract and its
Sephadex pattern was observed when an eluant was isolated, fractionated,
redissolved in NaOH, and neutralized. The elution pattern differed tremen-
dously after 1, 12, and 43 days. Apparently, the extract partially ionizes on
dissolution in alkali and then slowly resumes its colloidal consistency.

Humic substances are being actively investigated by gel chromatography
in order to determine the structure and size of organic constituents. Separations into fractions by solubility and gel chromatography are still being made.
Table 15 outlines the gel chromatographic work.

Many investigators have concluded that Oden's solubility classifications of
organic matter match only to a limited extent the gel chromatographic concept
of size distribution by apparent molecular weight. The breakdown of constituents to functional groups and further fractionation for actual identification
of material is also being carried out. Generally, organic matter from
different waters has different gel chromatographic elution patterns [62, 126].
Thus it appears that organics in water can be classified by the Sephadex
elution patterns as well as by the classical Oden solubility scheme. Gjessing
and Lee found the more highly colored water was, the greater the amount of
high molecular weight fraction eluted [59].

An effect of pH and complexation with the elution buffer on the elution
pattern is possible as shown by Christman and Minear [124]. Different
patterns were obtained at pH 7 for phosphate versus borate buffered eluants.
The effect of pH is primarily on size and shape characteristics. This is
consistent with the known functional groups, as pointed out by Christman and
Minear [124]; as pH increases, the ionization of carboxyl and acid phenolic
groups increases, causing polymeric extension. A larger apparent molecular
weight is thus possible. A pH effect on light absorption characteristics must
also be recognized in order to validly interpret the data.

Sephadex columns have been used to fractionate the organics in natural
waters and characteristics other than color have been studied for each fraction. Early studies by Gjessing and Lee [59] in 1967 correlated the
distribution of carbon as COD, color, and organic nitrogen in fractions eluted
from Sephadex columns.

In closing, a few salient points are worthy of note with regard to Sephadex
data. Many possible effects have not been delineated to any significant degree.
These include the interaction of color acids and other sample types with the
gel. Significant interaction with the column could greatly alter the molecular
size distribution. Even if no column interactions are present, the effect of
sample concentration on the organics is unknown. Many data suggest that
such effects are important. For example, some researchers have reported a
color decrease on concentrating their sample. It is entirely possible that the
natural size distribution is entirely different than that of the sample presented
to the column. Until Sephadex data are confirmed by other methods of molecular size analysis, these data can be considered useful and indicative but not
conclusive.

TABLE 16

Separation of Lignin Sulfonic Matter by Gel Chromatography

Sample	Eluant	Sephadex	Results	Reference
Lignin sulfonic matter (commercial Meadol NWS)	0.1M NaCl 0.01M NaCl 0.5M NH$_3$	G-50 G-75	Several peaks; coincidence of peaks in humic acids of natural origin are observed	26
Lignin sulfonic matter (paper mill waste)	0.1M NaHCO$_3$ 0.01M NaOH			
Spent sulfite liquor group I from an ion exclusion fractionation of six groups	Distilled water	1. G-75	G-75 can separate the medium high and high MW Ligno-sulfonic acid	135
1. Lignosulfonic acids fraction only 2. Aromatic fraction only		2. G-75, G-25	G-25 can separate low MW compounds	
Kraft lignin	1:7.0.1M NaOH-0.1M NaHCO$_3$	G-25 G-50 G-75 G-100	Extracts are of increasing MW as Kraft process proceeds; the general elution pattern is similar to the sulfite process	136
Sodium lignin sulfonates isolated chemically from a spent sulfite liquor	10^{-1}-10^{-4} M NaCl	G-50	MW 70,000 initially to final eluates of a few thousand was maintained; reproducibility by a constant eluant ionic strength	134

I. Lignosulfonic Acids from Paper Mill Wastes

The manufacture of paper products is one of the largest industries in the
United States. The soluble organic effluent from the production of these
products, e.g., sulfite liquor, includes saccharides and a large quantity of
lignin (a phenylpropane-type polymer) in the form of sulfonic acids. Lignin
sulfonic acids fractionate as humic acids according to the Oden system.

Spent sulfite liquor has been fractionated by ion-exclusion chromatography
into five groups of saccharides and a sixth group that contains the aromatic
fraction assumed to contain strong electrolytes and high molecular weight
substances [133]. At low flow rates the aromatic fraction could be fractionated into six groups, of which the first two contained the lignosulfonic acids.
Lignosulfonic acids have also been isolated by other means such as by a
combination of steam stripping, sugar fermentation, ion exchange, and
precipitation [134]. Sephadex columns have been used for further fractionation of lignosulfonic acids and sulfonic acids. Table 16 reviews the work.

J. Pesticides

Pesticide analysis of aqueous environments has been actively investigated in
the last few years. Three different uses of liquid chromatography have been
developed: (a) cleanup of aqueous extracts, (b) preliminary separation of
pesticides into groups for subsequent TLC and GLC analysis, and (c) high-
speed liquid chromatographic analysis since 1968. Pesticide analysis
exemplifies trace analysis at the microgram and nanogram per liter
concentration levels and illustrates the general analytical flow diagram
(Table 1).

Cleanup of aqueous environmental pesticide samples has been thoroughly
reviewed through 1971 by Suffet and Faust [65]. LC is the method of choice.
Recent LC is illustrated in Table 17, in which all methods appear to show
good recovery.

Table 17 also shows preliminary group separations, which are used to
share the burden of complete separation with GLC and to help aid identification.

High-speed liquid chromatographic procedures for pesticides are in their
infancy. Table 17 also presents some separations utilized by manufacturers
to show the utility of their chromatographs, chromatographic support, and
detection systems. Detailed analytical methods for the best high-speed
chromatographic systems are yet to be perfected. Presently the sensitivity
of high-speed liquid chromatography for pesticides is at the microgram to
nanogram level. High-speed liquid chromatography for pesticides will have
its greatest impact with polar pesticides and herbicides as the carbamates,
and phenolic metabolites of pesticides and herbicides. These compounds are
usually made into derivatives to enable electron capture or flame photometric
GLC detection at sensitivities in the nanogram range. The elimination of the
derivative step would enhance recovery and simplify analyses.

TABLE 17

Pesticides Analysis

Sample	Eluant	Support	Result	Reference
Preliminary separation to aid identification				
Organophosphates (procedure for river waters and effluents)	Ethanol	Sephadex LH-20	Three groups are developed as a subclassification; the column gives repeatable results and can be reused; this method aids identification	28
Mixtures of organophosphates and chlorinated hydrocarbons applied to soils and crops	a. Petroleum ether b. 5:1 Benzene in petroleum ether c. Chloroform	Florisil	Free elution into groups as these compounds are hard to resolve by GLC directly	137
Pesticides in water	a. 10% Benzene in hexane b. 60% Benzene in hexane	Silica gel 60/200	Separated chlorinated hydrocarbons into two groups for better quantitative GLC analysis; no recovery data reported	138
Neutral fraction of 1. CAM aqueous sample-chlorinated hydrocarbon pesticides 2. As (1) no solubility separation	a. Isooctane b. Benzene c. 1:1 Methanol-CHCl$_3$	Silica gel	CCE separated a. Aliphatic b. Aromatic c. Oxygenated after solubility separation	52

TABLE 17. — Continued

Sample	Eluant	Support	Result	Reference
Fish extracts, Missouri river	Cyclohexane	Bio-Bead-S-X-2	Gel chromatography separation from lipids; greater than 95% recovery of chlorinated hydrocarbon pesticide; for some samples, the sole cleanup	139
		Cleanup		
Aqueous extracts organophosphates	CHCl₃	a. Most Nuchar C b. Specific supports, e.g., alumina, MgO	Use specific P detector; usually no problem for water extracts but this technique is used anyway	28
Cleanup of 65 organophosphates of crops	a. 1:2 Ether-benzene b. Acetone	Florisil	Compounds are cleaned up and divided into six groups based on chemical groups; recovery data reported	140
Cleanup of crop extracts — all pesticides	25% Ethyl acetate in benzene	1:2:4 Charcoal-Sea Sorb-Celite 545	Recovery data reported	141

High-speed liquid chromatography

Salt marsh water abate and impurity (used for mosquito control)	Heptane or CHCl$_3$	1% BOP on Zipax	High-speed LLC; sensitivity 1 ng (UV detector)	141
a. Pure methoxychlor and analog b. Pure substituted urea herbicides	a. Isooctane b. Dibutyl ether	a. 4% β,β-oxy-dipropionitile (BOP) on 160/170 mesh Anakrom A b. 4% BOP on 230/270 mesh Gas Chrom P	a. High-speed LLC; these are examples of capability (UV detector) b. 1-μg concentrations were detected (UV detector)	29
Pure pesticides — chlorinated hydrocarbon	n-Hexane	37.5 μm Corasil	High-speed LSC; a typical analysis in 60 sec at μg levels (refracto-meter detector)	142
Parathion, methyl parathion, p-nitrophenol (pure compounds)	Water, 60.1%; ethanol, 38.8%; acetic acid, 0.80%; potassium chloride, 0.09%; sodium hydroxide, 0.21%	10% isooctane on silanized diatomaceous earth 28/40 μm	High-speed LLC; 10 min; sensitivity 10^{-8} moles/liter (polarographic detector)	143
Organophosphate pesticide (EPN)	n-Hexane	4% BOP on a controlled surface porosity support	High-speed LLC; (UV detector)	144
Pesticides (fortified fish and spinach extracts)	Isooctane	10% BOP on 37 μm Porasil 60	Refractive index and UV detectors; 45-min analysis; sensitivity 0.4 μg/RI 0.15	145

TABLE 17. — Continued

Sample	Eluant	Support	Result	Reference
Pesticides (cont.)	n-Hexane	Corasil II	High-speed LSC; difficult separation pattern for similar pesticides as compared to above	146
Carbaryl and 1-naphthol (hydrolysis product) (plant extract)	n-Hexane saturated TMG	Trimethylene glycol (TMG) on Zipax	High-speed LLC (with GLC, a derivative must be made; with TLC no quantitation possible	147
2,4-D and esters	60% Methanol in H_2O	Permaphase ODS	High-speed LLC	147
Pure substituted herbicides (linuron, neburon, diuron, monuron, fenuron)	a. 1% Dioxane in hexane b. 35% Methanol in water	ETH permaphase	a. High-speed LLC (UV detector) b. High-speed reversed-phase LLC (UV detector)	148

K. Polychlorinated Biphenyls (PCB's) and Related Compounds

Polychlorinated biphenyls (PCB's) are used in many industrial applications [149]. PCB's have been found in all parts of the environment since their presence was noted as an interference during the analysis of chlorinated hydrocarbon pesticides in 1966. PCB's are toxic to aquatic organisms and they are cleanup problems for the analysis of chlorinated hydrocarbon pesticides. This highlights two points in the analysis of trace materials in the environment; (a) positive identification of a chromatographic peak by spectroscopic means is highly desirable and (b) cleanup should be studied systematically to determine what the common interferences are and not to rely on the chromatographic retention times. The appearance of peaks for chlorinated hydrocarbon pesticide in samples where it was known that chlorinated hydrocarbon pesticides were not present led to the discovery of PCB's by spectroscopic means. The cleanup of a sample of PCB's is reviewed in Table 18. The method of Zitko [49] appears best.

Chlorinated naphthalenes and polychlorinated terphenyls and halogenated biphenyls have uses similar to those of PCB's and may be present in environmental samples. Problems of toxicity associated with these compounds have not been delineated. However, it is recommended that these compounds be considered in analytical studies of the water environment. Table 19 gives available cleanup data for these compounds.

L. Petroleum Products Including Polycyclic Hydrocarbons

Any of the various classes of petroleum products can become a water pollutant. Oil pollutants can cause taste and odor problems at low concentration and ecological disasters by giant oil slicks at high concentrations.

At present, preliminary group separation of petroleum hydrocarbons is by classical LC schemes (Table 1). A 1969 in-depth review by Oro is noteworthy [64]. Specific examples of water pollution applications of LC to petroleum products at low concentration are described by Lively et al. [156] and Rosen et al. [57] (Table 20). The identification and characterization of oil spill types have recently been studied in detail. GLC [157], infrared [158], fluorescence [159], and high-speed liquid chromatography (see Table 22) have been considered for characterization. Preliminary LC is not utilized for general characterization of an oil type. However, the need for specific component analysis may necessitate classical group classification

GLC at present is the primary method of final identification of specific volatile petroleum products. Desty and Goldup [63] have reviewed this methodology. New high-speed LC techniques have recently been used to supplement GLC (Table 22).

Polycyclics from oil refinery wastes have been shown to be carcinogenic to rats. Borneff has isolated polycyclics from many areas of the water environment, including their natural production by water plants [164]. Polycyclics

TABLE 18

Polychlorinated Biphenyls

Sample	Eluant	Support	Results	Reference

Pure compounds	a. Hexane
	b. 1. 10 ml hexane
	   2. 20 ml hexane
	   3. 10% ether : hexane
		a. Alumina
		b. Silica
			Cleanup and separation of PCB from chlorinated hydrocarbon pesticides except p,p-DDE; p,p-DDE interference is usually small	150

Pure compounds	a. Hexane b. Ether; hexane	Florisil	Separation of PCB's from chlorinated hydrocarbons; DDT is not eluted with hexane	151
Pure compounds	a. Hexane b. Ether : hexane (various ratios)	Florisil	Unable to repeat Reynolds' [151] work; DDT partially eluted with hexane; recommends checking Florisil before use (Zitko [49] suggests benzene impurities caused problems)	152
Pure compounds	a. Petroleum ether b. CH_2Cl_2 : CH_3CN : hexane	Silicic acid-Celite	When carefully controlled, PCB's and p,p-DDE can be separated; need large volumes of solvent	153

Sample	Solvent	Adsorbent	Comments	Ref.
Fish samples	a. Hexane b. 1. 10 ml hexane 2. 20 ml hexane 3. 10% ether : hexane	a. Alumina b. Silica	Evaluated PCB's found in the batches of silica gel; the amount of benzene in the hexane is the cause of poor separation	49
Water, sediments, biota sample	Petroleum ether	Florisil	Used Mills and Olney, cleanup, JAOAC, 46, 982 (1962).	53
Fish sample, Missouri River	Cyclohexane	Bio Beads S-X2	Separation from lipids used with [153] for quantitation	139
Natural waters	Hexane	Alumina	First aliquot contains PCB's aldrin and chloronaphthalenes	154

TABLE 19

Chlorinated Naphthalenes

Sample	Eluant	Support	Result	Reference
Fortified fish extracts	a. Petroleum ether b. $CH_2Cl_2 : CH_3CN :$ hexane	Silicic acid Celite	Separates chlorinated naphthalenes from chlorinated hydrocarbon pesticides which are not not separated by GLC; this procedure is the same as that used for the fourth sample in Table 18; PCB's are not separated from chlorinated naphthalenes	153
Fish samples	a. Hexane b. . 10 ml hexane 2. 20 ml hexane 3. 10% ether : hexane	a. Alumina b. Silica	Separates as PCB in the hexane; the presence of chlorinated naphthalenes with PCB can be detected by UV at 306 nm in very clean samples	49
Extracts of fish and food products	6% Ether in petroleum ether	Florisil	Uses general multipesticide method; results showed that the compounds were eluted with chlorinated hydrocarbon pesticides and could interfere with analysis	155

TABLE 20

Petroleum Products: Preliminary Group Classification

Sample	Eluant	Support	Results	Reference
General samples	Heptane soluble CCl_4 soluble Benzene soluble Methanol soluble Each elutes with the solvent of preferred solubility	Silica gel (alumina similar)	Many applications of original work	160
			Group classification: elution order — primary groups, saturates, olefins, aromatic, polar nonhydrocarbons; there is an overlap between subgroups of each group	161
Subgroup petroleum hydrocarbon mixtures	Acetone is best; tetrahydrofuran is next best	Sephadex LH-20 (200-2000 MW)	Example of a separation of subgroups, paraffins from cycloparaffins and alkyl benzenes from cyclobenzenes	162
Oil refinery effluent of surface water (CAM extract) CCE used	a. Isooctane or 2-methylpentane b. Benzene c. 1:1 Methanol-water	Silica gel	a. Aliphatic b. Aromatic c. Oxygenated and other compounds IR of fractions a and b for identification of oil	57
Oil refinery effluent in surface water (LLE)	a. Isooctane or 2-methylpentane b. Benzene c. 1:1 methanol-water	Silica gel	a. Aliphatic b. Aromatic c. Oxygenated and other compounds IR of fractions a and b for identification of oil	156
Mineral oil in waste water	Light petroleum ether or pentane	Silica gel or alumina	Same as above but GLC of fractions a and b for oil identification	163

TABLE 21

Polycyclic Aromatic Hydrocarbons

Sample	Eluant	Support	Results	Reference
Polycyclics in oysters from polluted waters (LLE)	a. Initial gradient elution of 2% acetone in cyclohexane then b. different solvent mixtures	a. Partially deactivated (3% H$_2$O) alumina; b. a complex scheme of rechromatographing with silica gel and alumina	Nonsaponifiable matter was refractionated to purify six components for identification by UV detection; a major cleanup problem; this is not recommended for present use but indicates LC use; benzoahthracene, benzo(α)pyrene, benzo(ϵ)pyrene, benzo($\varkappa$)fluoranthene, benzo(g,h,i)perylene	169
Polycyclics in gas works and sewage effluents (liquid and sludges) (LLE)	Cyclohexane (trace benzene)	Alumina	Separation of many types of polycyclics as above after an acid and base wash; a cleanup problem	170
Polycyclics in air samples	Ethyl ether	Deactivated alumina	Recovery data; fluorescence is used — more sensitive than UV	166
Polycyclics general	Isopropanol	Sephadex LH-20	Benzene Napthalene Anthracene Fluoranthene 3,4–Benzpyrene 1,2–Benzpyalene This is a reverse of normal order of gel chromatography as adsorption controls; direct fluorescence	171

TABLE 22

High-Speed Liquid Chromatography of Petroleum Products — Especially Polycyclics

Sample	Eluant	Support	Results	Reference
Aromatic hydro-carbons	1:1 H$_2$O-isopropanol	Zipax (hydrocarbon polymer-HCP)	1 μg each of: benzene, naphthalene, phenanthrene, pyrene, benz[α]pyrene (UV detector)	172
Fused-ring aromatic analysis	40% methanol in H$_2$O	Zipax (hydrocarbon polymer-HCP)	Benzene Naphthalene Anthracene Pyrene Chrysene Benz[e]pyrene Benz[α]pyrene (UV detector)	173
Fused-ring aromatic analysis	25, 50, and 75% methanol in H$_2$O	Zipax (hydrocarbon polymer-HCP)	Naphthalene Anthracene Pyrene Benz[α]anthracene 75% methanol is best (UV detector)	173
Motor oil	Tetrahydrofuran	Poragel	2.5-mg sample; definite patterns were observed for three types of motor oil (RI and UV detectors)	72
Polynuclear aza heterocyclics			Separate six compounds	174
Aromatic hydro-carbons (Charles River basin)	Heptane	Porasil T	Preliminary report; LC quantitation of compounds identified by GLC-MS	175

have been identified in natural waters from CCE (Table 2) and LLE (Table 22).

Analysis of polycyclic hydrocarbons can be accomplished by classical LC on alumina, silica gel, and Sephadex with subsequent UV or fluorescent detection (Tables 21 and 22). Before 1965, classical column chromatography was described by Sawicki [165] as "usually the first step in the analysis of a solution or an extract of a complex mixture for polycyclic aromatic hydrocarbons." Zdrojewski et al. [166] reviewed and investigated classical column chromatographic procedures for trace polycyclics (Table 21 and 22).

Recently, direct GLC and TLC with UV and fluorescence detectors have been used for final separation and quantitative analysis of water extracts from coke oven effluents [167] and soot samples [168]. No cleanup was used. High-speed LC techniques have been suggested that include direct spectrophotometric detection (Tables 21 and 22).

M. Phenols

Phenols are introduced into surface waters from industrial effluents, domestic waste waters, agricultural runoff, and chemical spills. Phenols can cause off tastes and odors at parts per billion concentrations in drinking water. They exhibit toxic effects in the environment at higher concentrations.

Preliminary separation of phenols from other components in natural water can be accomplished by gel chromatography. Sephadex gels strongly adsorb phenols from aqueous solutions [176], apparently through the hydroxyl group [177]. Table 23 outlines those separations of interest for water pollution analysis. Free phenol can be separated from high molecular weight protein and phenol polymers by Sephadex methods [178]. The application of these methods to cleanup of phenols could be an advantage for water analysis of phenols in industrial effluents, as from oil refineries.

TLC and GLC (after a derivative sensitive to electron capture detection has been made) are the primary qualitative and quantitative tools for final analysis of phenol [179]. New high-speed LC systems may offer a superior method by eliminating the derivatization step (Table 23).

N. Detergents

Detergents in natural waters are usually analyzed directly by spectrophotometric methods as total detergent [74]. Specific separation of LAS (linear alkyl sulfonates) from the formic acid fraction of a fatty acid separation is illustrated in Table 24. Identification of specific detergent species has been obtained by LC (Table 24). These may have application to water pollution studies.

TABLE 23

Phenols

Sample	Eluant	Support	Results	Reference
21 phenols (including nitro, halide, and amino phenols)	1. pH 4 acetate buffer 2. 0.1M NaOH	Sephadex G-10	A suggested use of the method: cleanup from high M.W. compounds as phenols adsorb to the gel in water solutions	177
17 phenols and phenolic acids	1. H_2O 2. 0.5M NaCl 3. 0.1M NH_3 4. 0.1M HOAc 5. 0.01M Na_2MoO_4	Sephadex G-25	As above; an electrolyte is necessary for phenolic acid separation	178
Monohydroxy-phenols and high MW phenols	1. 0.5M NaCl 2. Water–CH_3OH mixtures with gradient elution	a. Few % 1-hexanol in cyclohexane or b. cyclohexane alone; both on 80/100 mesh microporous polyethylene or 70/80 mesh Teflon 6	Reversed-phase chromatography	180
Trace dihydric phenols in the presence of high concentrations of monohydric phenols (industrial effluent)	Cyclohexane	60/80 mesh silica gel	Aqueous phenolic effluent from coal-tar distillate; a classification procedure for phenolic effluents	181

Table 23. — Continued

Sample	Eluant	Support	Results	Reference
River water CCE	Ethyl ether	Florisil	Cleanup procedure; excellent recovery of 31 of 37 phenols; competitive solubility procedures give poor cleanup for TLC and GLC	50
Industrial effluents phenol, from paper mills	—	—	—	182
Seven simple methyl-substituted phenols	2.5% methanol in cyclopentane	0.88% ether bonded phase <37 μm	High-speed LC	143

TABLE 24

Detergents

Sample	Eluant	Support	Results	Reference
Detergent raw materials, including impurities	Chloroform/propanol and ethanol/water	Sephadex G-10	The active detergent in a detergent composition was isolated; Sephadex G-10 elutes the detergents in reverse order to that of ion exchange	183
Anionic detergents in sewage effluent	a. Butanol in $CHCl_3$ (gradient elution)	Silicic acid	The detergent LAS eluted in formic acid fraction of a fatty acid procedure; titrates total anionic detergent	77
	b. Butanol in $CHCl_3$ (gradient elution)	Silicic acid	Separation of formic acid and lauryl hydrogen sulfate from eluant formic acid fraction; elimination of colored interferences	77
Nonionic detergents	Successive elution; ethyl ether, acetone, and methanol-chloroform mixtures	Silica gel	Separate after ion exchange removes ionic materials	184
Surfactants Span and Tween	Tetrahydrofuran	Poragel	Observed that each mixture has a characteristic elution pattern; the elution order was monolaurate, monopalmitic, tri- and monostearate, and tri- and monooleate	185

REFERENCES

1. A. J. P. Martin and R. L. M. Synge, Biochem. J., 35, 91 (1941).

2. J. Cazes, J. Chem. Ed., 47, A461 (1970).

3. Gas Chrom. News Letter, Applied Science Labs, Sept., Oct. (1970).

4. E. Bayer and Committee, Chromatographia, 2, 153 (1969).

5. L. R. Snyder, Principles of Adsorption Chromatography, Marcel Dekker, New York, 1968.

6. L. R. Snyder, Anal. Chem., 39, 698 (1967).

7. H. Veening, J. Chem. Ed., 47, A549 (1970).

8. H. Determann, Advan. Chromatog., 8, 3 (1969).

9. K. H. Altgelt, Advan. Chromatog., 7, 3 (1968).

10. J. S. Fritz and A. Tateda, Anal. Chem., 40, 2115 (1968).

11. G. M. Gaucher, J. Chem. Ed., 46, 729 (1969).

12. B. L. Karger, J. Chem. Ed., 43, 47 (1966).

13. J. C. Giddings, in Dynamics of Chromatography, Part I, (Giddings and Keller, eds.), Marcel Dekker Inc., New York, p. 195, 1965.

14. L. R. Snyder, Anal. Chem., 39, 705 (1967).

15. S. Dal Nogare and R. S. Juvet, Gas Liquid Chromatography, Interscience, New York, 1962.

16. K. J. Bombaugh, Amer. Lab., July, 43 (1969).

17. E. Soczewinski, Advan. Chromatog., 8, 91 (1969).

18. J. J. Kirkland, Ed., Modern Practices of Liquid Chromatography, Interscience, New York, 1971.

19. L. R. Snyder, J. Chromatog. Sci., 7, 352 (1969).

20. J. L. Waters, J. N. Little, and D. F. Horgan, J. Chromatog. Sci., 7, 293 (1969).

21. Varian Corp, Res. Notes, July (1969).

22. A. A. Rosen and F. M. Middleton, Anal. Chem., 31, 1729 (1959).

23. B. Ahling and S. Jensen, Anal. Chem., 42, 1483 (1970).

24. F. M. Middleton and J. J. Lichtenberg, Ind. Eng. Chem., 52, 99A (1960).

25. R. A. Baker, J. Amer. Water Works Assoc., 58, 751 (1966).

26. A. Brodsky, J. Prochazka, and H. Vydrova, J. Amer. Water Works Assoc., 62, 386 (1970).

27. H. F. Mueller, A. M. Buswell, and T. L. Larson, Sewage Ind. Waste, 28, 255 (1956).

28. J. Askew, J. H. Ruzicka, and B. B. Wheals, Analyst, 94, 275 (1969).

29. J. J. Kirkland, Anal. Chem., 40, 391 (1968).

30. H. Braus, F. M. Middleton, and G. Walton, Anal. Chem., 23, 1160 (1951).

31. A. W. Breidenbach, J. J. Lichtenberg, C. F. Henke, D. J. Smith, J. W. Eichelberger, Jr., and H. Stierli, PHS No. 1241, U.S. Dept. of Health, Education and Welfare, Washington, D.C., 1964.

32. R. L. Booth, J. N. English, and J. N. McDermott, J. Amer. Water Works Assoc., 57, 215 (1965).

33. R. T. Skrinde and H. D. Tomlinson, J. Water Poll. Cont. Fed., 35, 1292 (1963).

34. G. F. Lee, G. W. Kumke, and S. L. Becker, Int. J. Air Water Poll., 9, 69 (1965).

35. R. D. Hoak, Intern. J. Air Water Poll., 6, 521 (1962).

36. J. W. Eichelberger, Jr. and J. J. Lichtenberg, J. Amer. Water Works Assoc., 63, 25 (1971).

37. H. P. Nicholson, Limnol. Oceanog., 9, 310 (1964).

38. U.S. Public Health Service, Drinking Water Standards, 1962, Publ. No. 956, U.S. Government Printing Office, Washington, D.C., 1962.

39. A. A. Rosen, R. T. Skeel, and M. B. Ettinger, J. Water Poll. Contr. Fed., 35, 777 (1963).

40. J. Borneff and R. Fischer, Arch. Hyg. Bakt., 145, 334 (1961); Water Poll. Abstr., No. 2099 (1962).

41. J. Borneff and R. Fischer, Arch. Hyg. Bakt., 146, 1 (1962); Water Poll. Abstr., No. 2304 (1962).

42. W. C. Hueper and W. W. Payne, Amer. J. Clin. Pathol., 39, 475 (1963).

43. A. K. Burnham, G. V. Calder, J. S. Fritz, G. A. Junk, H. J. Svec, and R. Willis, Anal. Chem., 44, 139 (1972).

44. J. A. Vinson, 164th American Chemical Society Meeting, New York, N.Y., August 1972.

45. E. Hindin, D. S. May, and G. H. Dunstan, Residue Rev., 7, 130 (1964).

46. E. A. Epps, F. L. Bonner, L. D. Newsom, R. Carlton, and R. O. Smitherman, Bull. Env. Contam. Toxicol., 2, 33 (1967).

47. J. A. Burke and B. Malone, J. Assoc. Off. Agr. Chem., 49, 1004 (1966).

48. P. A. Mills, J. Assoc. Off. Agr. Chem., 51, 29 (1968).

49. V. Zitko, 54th Canadian Institute of Chemistry Conference, Halifax, Nova Scotia, May, 1971.

50. J. W. Eichelberger, Jr., R. C. Dressman, and J. E. Longbottom, Env. Sci. Tech., 4, 576 (1970).

51. A. M. Kadoum, Bull. Env. Contam. Toxicol., 2, 264 (1967).

52. A. W. Breidenbach, C. F. Henke, D. J. Smith, J. W. Eichelberger, Jr., and H. Stierli, U.S. Dept. of Interior W.P.-22, 1966.

53. T. W. Duke, J. I. Lowe, and A. J. Wilson, Jr., Env. Contam. Toxicol., 5, 171 (1970).

54. P. A. Mills, J. H. Onley, and R. A. Gaither, J. Assoc. Off. Agr. Chem., 45, 186 (1963).

55. J. T. Cronin, Techniques of Solvent Extraction of Organic Material from Natural Waters, Oregon State Univ., 1967; Diss. Abstr. Order No. 67-17, 573, Ph.D., Diss. Abstr. B28,2975 (1968).

56. D. P. Schwartz, Agr. Serv. Rev., 8, 41 (1970); and ACS-MARM, Philadelphia, Pa., 1971.

57. A. A. Rosen and F. M. Middleton, Anal. Chem., 27, 790 (1955).

58. J. Hirsch and E. H. Ahren, Jr., J. Biol. Chem., 233, 311 (1958).

59. E. T. Gjessing and G. F. Lee, Env. Sci. Technol., 1, 631 (1967).

60. K. Hasegawa, T. Kusano, and H. Mitsuda, Agr. Biol. Chem. (Japan), 27, 878 (1963).

61. L. L. Ciaccio and L. Kirschner, Amer. Lab., Dec. 21 (1971).

62. J. R. Vallentyne, J. Fish Res. Board Can., 14, 33 (1957).

63. E. Heftmann, Chromatography, 2nd ed., Rheinhold, New York, 1966.

64. L. Marinetti, Lipids Chromatography Analysis, Marcel Dekker, New York, 1967.

65. S. D. Faust and I. H. Suffet, Water and Water Pollution Handbook (L. L. Ciaccio, ed.), Marcel Dekker, New York, (1971), p. 1250.

66. Ann. Rev., Anal. Chem., through 1970.

67. Ann. Rev., Anal. Methods, J. Water Poll. Contr. Fed, through 1970.

68. Water Poll. Abstr. through 1970.

69. Chem. Abstr. through 1970.

70. J. C. Moore, J. Polymer Sci., A2, 835 (1964).

71. K. J. Bombaugh, W. A. Dark, and R. N. King, Rev. Devel., Sept. 28 (1968).

72. K. J. Bombaugh, Amer. Lab., May, 61 (1971).

73. J. J. Murtaugh and R. L. Bunch, J. Water Poll. Contr. Fed., 37, 410 (1965).

74. American Public Health Association, Standard Methods for the Examination of Water and Wastewater, 12th ed., 1971.

75. W. H. J. Hatlingh and F. V. Hayward, Int. Air Water Poll. J., 8, 411 (1964).

76. J. Scherfig, C. F. Reid, and E. A. Pearson, Water Sewage Works, 115, 316 (1968).

77. H. F. Mueller, T. E. Larson, and M. Ferretti, Anal. Chem., 32, 687 (1960).

78. G. G. Freeman, J. Chromatog., 28, 338 (1967).

79. Waters Associates, Publication PL-1001, 1969.

80. D. A. M. Geddes and M. N. Gelmour, J. Chromatog. Sci., 8, 394 (1970).

81. R. Ellerker, H. J. Dee, F. G. I. Lax, and D. A. Sargent, Water Poll. Contr. (Brit.), 67, 542 (1968).

82. F. K. Kawahara, Anal. Chem., 40, 2073 (1968).

83. H. F. Mueller, T. E. Larson, and W. J. Lennarz, Anal. Chem., 30, 41 (1958).

84. A. G. Stradomskaya and I. A. Goncharova, Gidrokhim Materialy, 43, 62 (1967); Chem. Abstr., 68, 72107s (1968).

85. A. G. Stradomskaya and I. A. Goncharova, Gidrokhim Materialy, 41, 78 (1966); Chem. Abstr., 66, 58750y (1967).

86. I. A. Goncharova, A. Nikhomenko, and A. D. Semenov, Gidrokhim Materialy, 41, 116 (1966); Chem. Abstr., 66, 68772w (1967).

87. A. D. Semenov, Org. Vesh. Podyemn. Vod. Ego. Zrachenic Neft. Geol., 235, 67 (1969).

88. G. Zweig, J. H. Nair, and B. Compton, A Study of Fat and Oil Pollution of New York State Waters, Research Report No. 16, New York State Dept. of Health, September, 1967.

89. R. C. Loehr and J. C. Roth, J. Water Poll. Contr. Fed., 40, R385 (1968).

90. R. C. Loehr and T. J. Kukar, Int. J. Air Water Poll., 9, 479 (1965).

91. L. M. G. Rios, Saneamento, 30, 45 (1966).

92. E. Nystrom and J. Sjovall, Anal. Biochem., 12, 235 (1965).

93. J. E. Hobbie and C. C. Crawford, Science, 159, 1463 (1968).

94. Y. C. Chair and J. P. Riley, Deep Sea Res. Oceanog. Abstr., 13, 1115 (1966); Water Poll. Abstr., 40, 204 (1967).

95. A. Siegel and E. T. Degens, Science, 151, 1098 (1966).

96. K. Dus, S. Lindroth, R. Pabst, and R. M. Smith, Anal. Biochem., 14, 111 (1966).

97. L. Kesner, E. Muntwyler, G. E. Griffin, and J. Abrams, Anal. Chem., 35, 83 (1963).

98. P. H. Cuatrecasas, M. Wilchekayd, and C. B. Anfinsen, Proc. Nat. Acad. Sci. U.S., 61, 636 (1968).

99. S. Fazakerley and D. R. Best, Anal. Biochem., 12, 290 (1965).

100. J. H. Peters, B. J. Berridge, Jr., J. G. Cummings, and S. C. Lin, Anal. Biochem., 23, 459 (1968).

101. D. Eaker and J. Porath, Separation Sci., 2, 507 (1967).

102. C. E. Hedrick, Anal. Chem., 37, 1044 (1965).

103. W. T. Roubal and A. L. Tappel, Anal. Biochem., 9, 211 (1964).

104. J. W. Payne and C. Gilvarg, J. Biol. Chem., 23, 6291 (1968).

105. I. T. Oliver, Nature (London), 190, 810 (1961).

106. J. H. Venekamp and W. H. M. Mosch, Virology, 23, 394 (1964).

107. J. I. Ohms, J. Zec, J. V. Benson, Jr., and J. A. Patterson, Anal. Biochem., 20, 51 (1967).

108. J. G. Green, Nat. Cancer Inst. Monographs, 21, 447 (1966).

109. A. D. Semenov, Trudy Kom. Analit. Khim., 3, 66 (1963); Water Poll. Abstr., 39, 374 (1966).

110. J. Davy, Arch. Pharmacol. Franc., 24, 703 (1966).

111. K. Granth and P. Flodin, Makromol. Chem., 48, 160 (1961).

112. V. Holme-Hansen, Limnol. Oceanog., 13, 507 (1968).

113. C. Garside and J. P. Riley, Anal. Chem. Acta, 46, 179 (1969).

114. H. H. Strain, J. Sherma, and M. Grandolfo, Anal. Biochem., 24, 54 (1968).

115. K. W. Daisley, Nature (London), 191, 868 (1961).

116. B. H. Davies, Biochem. J., 103, 51P (1967).

117. M. F. Bacon, Biochem. J., 101, 34C (1966).

118. H. H. Strain, J. Sherma, and M. Grandolfo, Anal. Chem., 39, 926 (1967).

119. M. Aubert and M. Gauthier, Rev. Int. Oceanog. Med. (France), 2, 53-63 (1963); Water Poll. Abstr., 41(1376), 7 (1968).

120. M. Aubert and M. Gauthier, Rev. Int. Oceanog. Med. (France), 4, 23-33 (1966); Water Poll. Abstr., 41(1378), 9 (1968).

121. C. R. Borja and M. S. Begdoll, Biochemistry, 6 1467), 71 (1967).

122. U. Harding, H. Schauer, and G. Hartmann, Biochemistry, 346, 212 (1966).

123. S. Oden, Kolloidchem. Biochem. Zeitschrift Beihefte, 11, 75 (1919).

124. R. F. Christman and R. A. Minear, in Organic Compounds in Aquatic Environments (S. D. Faust and J. V. Hunter, eds.), Marcel Dekker, New York, 1971, p. 119.

125. M. Soukup, Coll. Czech. Chem. Comm., 29, 3182 (1964).

126. D. Povoledo, Mem. 1st Ital. Idrobiol., 17, 21 (1964).

127. E. T. Gjessing, Nature (London), 208, 1091 (1965).

128. J. Shapiro, in A. Brodsky (Ref. 28), Symp. Hungarian Hydrologic Soc., Budapest, Tirany, September, 1966.

129. R. Obenaus and H. J. Neumann, Naturwissenschaften, 52, 131 (1965).

130. A. M. Posner, Nature (London), 198, 1161 (1963).

131. W. Wildenhain and G. Henseke, Z. Anal. Chem., 229, 271 (1966).

132. J. R. Bailly and H. Margulis, Plant Soil, 29, 343 (1968).

133. K. Fross, The Composition of a Spent Spruce Sulfite Liquor, Dist. Abo Akademi (1961); referred to by W. Jensen, K. E. Fremer and K. Forss, Tappi, 45, 122 (1962).

134. P. R. Gupta and J. L. McCarthy, Macromolecules, 1, 236 (1968).

135. W. Jensen, K. E. Fremer, and K. Forss, Tappi, 45, 122 (1962).

136. J. G. McNaughton, W. Q. Yean, and D. A. I. Goring, Tappi, 50, 548 (1967).

137. W. W. Sans, J. Agr. Food Chem. , 15, 192 (1967).

138. L. G. Johnson, Bull. Env. Contam. Toxicol. , 5, 542 (1970).

139. D. L. Stallings, R. C. Tindle, and J. L. Johnson, Paper 79, Pesticide Chemistry Division, 161st American Chemical Society Meeting, Los Angeles, Calif. , March 1971.

140. H. Beckman and D. Garber, J. Assoc. Off. Agr. Chem. , 52, 286 (1969).

141. R. R. Watts, R. W. Storherr, and J. R. Pardue, J. Assoc. Off. Agr. Chem. , 52, 523 (1969).

142. J. L. Waters, Amer. Lab. , May, 61 (1967).

143. J. G. Koen, J. F. K. Huber, H. Poppe, and G. der Boef, J. Chromatog. Sci. , 8, 192 (1970).

144. J. J. Kirkland, J. Chromatog. Sci. , 7, 7 (1969).

145. D. F. Horgan, Jr. , W. A. Dark, and K. J. Bombaugh, Waters Associates Tech. Rept. Framington, Mass. , 19531, 1969.

146. K. J. Bombaugh, R. F. Levangie, R. N. King, and L. Abrahams, J. Chromatog. Sci. , 8, 657 (1970).

147. R. A. Henry, J. A. Schmit, J. F. Dieckman, and J. F. Murphy, Anal. Chem. , 43, 1053 (1971).

148. J. J. Kirkland, J. Chromatog. , 43, 36A (1971).

149. G. D. Veith and G. F. Lee, Water Res. , 4, 265 (1970).

150. A. V. Holden and K. J. Marsden, J. Chromatog. , 44, 481 (1969).

151. L. M. Reynolds, Bull. Env. Contam. Toxicol. , 4, 128 (1969).

152. A. Bevenue and J. N. Ogata, J. Chromatog. , 50, 142 (1970).

153. J. A. Armour and J. A. Burke, J. Assoc. Off. Agr. Chem. , 53, 761 (1970).

154. D. F. Goerlitz and E. Brown, Methods for Analysis of Organic Substances in Water, Chap. A-3, TWI 5-A3, 1972; U.S. Geological Survey Report, 1972.

155. Food and Drug Administration, Pesticide Analysis Manual, U.S. Govt. Printing Office, Washington, D.C. , 1969.

156. L. Lively, A. A. Rosen, and C. I. Mashni, Purdue University Eng. Bull. Ext. Ser. 118, Industrial Waste Conference 657 (1965).

157. F. K. Kawahara, American Chemical Society, 161st Annual Meeting, Los Angeles, California (March, 1971).

158. F. K. Kawahara, Env. Sci. Tech., 3, 150 (1969).

159. A. D. Thurston, Jr., and R. W. Knight, Env. Sci. Tech., 5, 64 (1971).

160. W. G. Meinschein and G. S. Kenny, Anal. Chem., 29, 1153 (1957).

161. E. D. Evans, G. S. Kenny, W. G. Meinschein, and E. E. Bray, Anal. Chem., 29, 1858 (1957).

162. B. J. Mair, P. T. R. Hwang, and R. G. Ruberto, Anal. Chem., 39, 838 (1967).

163. H. E. Burmeister, Erdol Kohle, 19, 129 (1966); Water Poll. Abstr., 41, 205 (1968).

164. J. Borneff, F. Selenka, H. Kunte, and A. Maximos, Env. Res., 2, 22 (1968).

165. E. Sawicki, Chemist-Analyst, 53, 24 (1964).

166. A. Zdrojewski, L. Dubois, G. E. Moore, R. S. Thomas, and J. L. Monkman, J. Chromatog., 28, 317 (1967).

167. T. D. Searl, F. J. Cassidy, N. H. King, and R. A. Brown, Anal. Chem., 42, 954 (1970).

168. B. B. Chakraborty and R. Long, Env. Sci. Tech., 1, 828 (1967).

169. H. J. Cahnmann, Anal. Chem., 29, 1307 (1957).

170. P. Wedgewood and R. L. Cooper, Analyst, 79, 163 (1954).

171. M. Wilk, J. Rochlitz, and H. Bende, J. Chromatog., 24, 414 (1966).

172. Dupont Data Sheets No. 820 M-2, E. I. du Pont de Nemours, Wilmington, Del. (1969).

173. Dupont Data Sheets No. 820 M-4, E. I. du Pont de Nemours, Wilmington, Del. (1970).

174. R. Vivilecchia and R. W. Frei, 54th Annual Canadian Institute of Chemistry Conference, Halifax, Nova Scotia, May 1971.

175. R. A. Hites and K. Bieman, Paper presented at the 164th ACS Meeting, New York, N.Y., 1972.

176. B. Gelotte, J. Chromatog., 3, 330 (1960).

177. A. J. W. Brook and S. Housley, J. Chromatog., 41, 200 (1969).

178. J. B. Woof and J. S. Pierce, J. Chromatog., 28, 94 (1967).

179. S. Goren-Strul, H. F. W. Kleijn, and A. E. Mostaert, Anal. Chem. Acta, 34, 322 (1966).

180. J. S. Fritz and C. E. Hendrick, Anal. Chem., 37, 1015 (1965).

181. J. H. Young, Analyst, 86, 520 (1961); based upon the work of W. H. Blackburn et al., Gas Council Res. Comm., 24 (1955).

182. C. de Chouden, Tech. Res. Paper, 4, 127 (1966); Anal. Chem. Abstr.,
     344 (1967).

183. M. Mutter, Chromatographia, 2, 208 (1969).

184. M. J. Rosen, Anal. Chem., 35, 2074 (1963).

185. K. J. Bombaugh, W. A. Dark, and R. F. Lavangie, Z. Anal. Chem.,
     236, 443 (1968).

Chapter 13

PAPER CHROMATOGRAPHIC ANALYSIS IN WATER POLLUTION

Osman M. Aly

Department of Environmental Engineering
Campbell Soup Company
Camden, New Jersey

and

Samuel D. Faust

Department of Environmental Sciences
Rutgers University
New Brunswick, New Jersey

I. INTRODUCTION

Paper chromatography has found some application for the separation and identification of various pollutants in fresh and saline waters. Although this chromatographic procedure chronologically preceded thin-layer techniques, it has not been extensively developed. One reason may be that paper chromatography does not have the capacity to resolve interferences from the substances being sought. This somewhat eliminates its use as a cleanup technique. Also, paper chromatography is not quite as sensitive as thin-layer chromatography. Despite these limitations, paper chromatography can be an extensively valuable analytical tool as a technique of identification and confirmation. As the R_f values are specific for a given pollutant under specific analytic conditions, paper chromatography provides supplementary data to other procedures.

This chapter provides the procedural detail for a few organic pollutants in aquatic environments: hydrocarbons, organic acids, phenols, and pesticides and herbicides. Most of the procedures were developed for polluted surface waters.

II. ACCURACY, PRECISION, AND DATA EVALUATION

As applied to polluted waters, one cannot speak of accuracy and precision of paper chromatography in the true analytical sense. The procedures reported in this chapter are qualitative or semiquantitative at the best. Sensitivity, precision, and accuracy data were often not reported. The R_f value was reported with its attendant standard deviation, which may or may not be an indication of the procedure's precision and accuracy. The investigators were concerned with the R_f value as the confirmatory feature of the procedure. Another consideration, which influences precision and accuracy, is the resolution of the pollutant from interferences. The analyst must use a cleanup procedure prior to the paper chromatographic step if reproducible R_f values are to be obtained. The analyst must be aware also that the use of a cleanup procedure results in the partial loss of the pollutant. Therefore, recovery values will be less than 100%.

III. SAMPLING

No special sampling techniques or equipment were employed in the analytical procedures reported in this chapter. For the most part, the "grab" sample may vary from as little as 1 liter to several gallons. Size is dependent, of course, on the constituent and the desired sensitivity. A most important consideration is the time interval between collection and analysis. Most organic pollutants are subject to either biological or chemical degradation of some sort. Thus, the analyst should process the sample as soon as possible after collection. The problems of sampling and storage of samples are discussed by Faust and Suffet [1].

IV. SEPARATION OF INTERFERENCES AND CONCENTRATION TECHNIQUES

Coextracted interferences may arise from domestic and industrial waste waters, naturally occurring substances in surface waters and bottom muds, and particulate matter. Hindin et al. [2] found it impossible to interpret a paper chromatogram of chlorinated hydrocarbons in lake water and lake sediment samples due to coextracted interferences. These analysts suggested the use of an activated aluminum silicate for sample cleanup. The problems of sample cleanup, recovery, and concentration techniques are discussed by Faust and Suffet [1].

V. METHODS

A. Hydrocarbons

Maehler and Greenberg [3] utilized paper chromatographic techniques for the identification of petroleum products in estuarine waters. First the oil is separated from the water by decantation, or by extraction with $CHCl_3$ solution with anhydrous sodium sulfate. This solution is filtered through a Whatman No. 40 paper and is evaporated on a steam bath just to dryness. Reference samples are treated in a similar manner. Two techniques were used. However, in each technique, the sample is applied to the paper in the same manner. A small circle, 1 cm in diameter, is drawn on Whatman 3MM chromatographic grade filter paper. A droplet of the sample is placed at the center. Oil is added until the circle is just filled. Viscous oil tends to enter the paper slowly at room temperature but can be spread easily by warming the paper under an infrared heat lamp. In one of the techniques, called "radial chromatography," three concentric cicles 1 cm, 3 cm, and 7 cm in diameter are drawn on a 10-cm square of Whatman 3MM paper. The sample is placed on the small circle. Next, a solvent, such as benzene, is added at a rate of about one drop every 7 sec until the solvent reaches the 3 cm circle. The benzene is allowed to evaporate and a second solvent, methanol, is added at one drop every 10 sec until the solvent reaches the outside circle. The rate of solvent addition is selected so that solvent flows evenly through the paper rather than across the surface in order to avoid distortion of the chromatogram. After the solvent evaporates, the developed patterns are noted for the color of the spots and the manner of the spreading. Chloroform followed by acetonitrile is another solvent system that may be used. Figure 1 shows a typical radial chromatogram of some oil samples under visible light. The chromatograms were

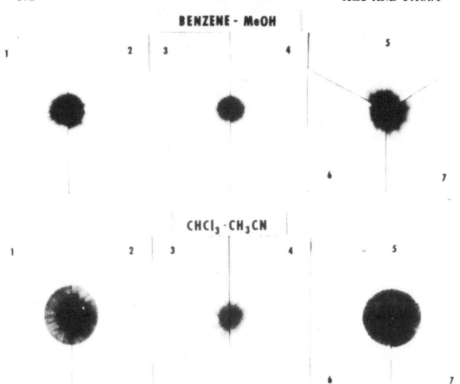

Figure 1. Radial chromatogram viewed under visible light. Reprinted from Ref. 3 with the permission of the American Society of Civil Engineers, New York, New York.

sectioned and placed together for easy comparison. This shows the identificative features of these types of paper chromatograms.

In the second technique, that of ascending chromatography, the circles are drawn about 1 in. apart and 3/4 in. from one edge of an 8 × 8 in. sheet of paper. After it is spotted, the sheet is suspended sample side down in a chromatographic tank with the lower 3/8 in. immersed in n-heptane. The solvent slowly rises up the paper, which dissolves and separates the sample, leaving characteristic patterns. The paper is left in place about 1-2 hr, until the chromatograms are well developed. Then the paper is removed and the solvent is allowed to evaporate. The chromatograms are inspected under visible and ultraviolet light. Figure 2 shows the ascending chromatograms from seven different samples (No. 1 is duplicated to show reproducibility). The characteristics of these streaks suggest that the samples are of different origin.

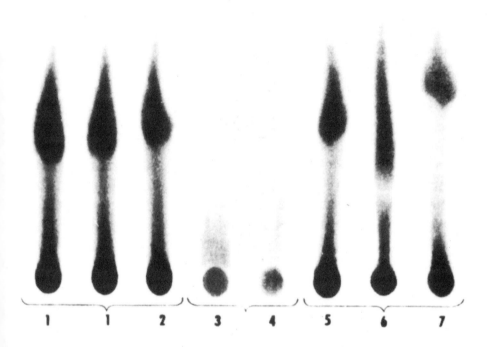

Figure 2.   Ascending chromatogram, n-heptane solvent system, viewed under visible light. Reprinted from Ref. 3 with permission of the American Society of Civil Engineers, New York, New York.

B. Organic Acids

Mueller et al. [4] developed a paper chromatographic technique for the separation and identification of organic acids in river waters. The water samples are collected in a container to which a few pellets of NaOH have been added. After filtration, 3–4 liters are heat concentrated to approximately 30 ml, which is acidified with a mineral acid. The organic acids are then extracted with ether for 15 hr. After extraction, the acids are neutralized with dilute NaOH. The resulting solution is warmed on a steam bath to volatilize the ether layer and the aqueous phase is concentrated to a

TABLE 1

Comparison of R_f Values of Separated Acids with Reference Acids Showing Similar Elution Patterns [a]

| | | R_f values | | | | | |
| | | Sodium salt[b] | | Hydroxamate[c] | | Acid[d] | |
Peak	Reference acid	Reference	Unknown[e]	Reference	Unknown[e]	Reference	Unknown[e]
A	Butyric	0.37	0.14 0.16	0.78	—	—	—
	Crotonic	0.31	0.07	0.78	—	—	—
B	Propionic	0.28	0.15	0.66	0.40 0.47 0.50	—	0.55
C	Acetic	0.16	0.16	0.58	0.58	—	—
D	Pyruvic		0.18 0.13 0.11 0.07	0.55	0.56	0.06 0.62	0.04
E	Formic	0.14	0.14	0.48	0.48 0.26	—	0.72 0.04
	ABS (alkylbenzene-sulfonic acid)	0.77		—		0.72 0.04	
F	Lactic	0.14	0.14	0.49	0.48 0.39 0.54	0.55	0.55
	Succinic			0.55 0.52 0.47		0.60	

G	Gallic	0.09	0.18	—	0.23	0.48	0.39
	Malonic	0.02	0.15	—	0.47	0.50	0.58
	Oxalic	—	0.07	0.59	0.44		0.68
	Aconitic			0.55 0.53 0.66	0.54 0.59 0.38	0.67	
H	1-Malic		0.18	0.57 0.52 0.44 0.49	0.53 0.44	0.32	—
J	Citric	0.23	0.52 0.59	—	0.23	—	
	Isocitric	—	0.45	—	—	—	
K				0.54 0.44			
L	Tartaric		0.52	0.54 0.61	0.15		

[a] Reproduced from ref. [4], p. 688, by courtesy of the American Chemical Society, Washington, D.C.

[b] Solvent system: 5% ethanol–45% 1-butanol–50% NH_4OH (conc.).

[c] Solvent system: 1-butanol–acetic acid–water (4:1:5).

[d] Solvent system: 1-pentanol–5M formic acid (1:1, v/v).

[e] R_f values obtained on different samples.

volume of 1 ml. This sample extract is acidified and adsorbed on silicic acid for quantitative separation and grouping of the acids by column chromatography. The fractions associated with each of the identifying peaks are combined and the sodium salts are extracted into water from the eluting solvent. After concentration to dryness over steam, the acids may be converted to the acid salts or hydroxamate derivatives or left as the free acids for subsequent chromatographic separation on paper. The dry organic acid salts are solubilized by 0.4N alcoholic HCl. The acids are esterified to hydroxomates at 0°C by the addition of the etheral diazomethane to the appearance of a yellow coloration, which should persist for a minimum of 10 min. Excess diazomethane is removed by the addition of several drops of the 0.4N alcoholic HCl. For 20 μmoles of the organic acids (this quantity was found to be limiting for the successful preparation of the acid hydroxamates), 8 ml of ether are added to the ester, followed by 0.5 ml of hydroxylamine reagent (prepared at 5°C just prior to use from equal volumes of 5% $NH_2OH \cdot HCl$ and 12.5% NaOH in absolute methanol). After 30 min at 25°C, 0.02 ml of glacial acetic acid is added and the solution is filtered again. After evaporation to dryness over steam, the hydroxamates are dissolved in 0.1–0.2 ml of absolute methanol and aliquot containing a minimum of 0.05 μmole of the unknown acid is chromatographed on Whatman No. 1 filter paper by the ascending technique. The chromatogram is developed with a mixture of n-butanol-acetic acid-water (4:1:5). After 15 hr of development, the chromatograms are removed, dried for 1–2 hr, and sprayed with 1% alcoholic ferric chloride containing 0.1% HCl.

Hydroxamates appear as red-purple spots on a yellow background; a specific test for this type compound. The sensitivity of this technique falls in the 0.03–0.05 μmole range for various organic acids the R_f values of which are seen in Table 1. In contrast, identification of the dicarboxylic acids as their hydroxamate derivatives is limited to a few acids as R_f values often overlap. In this case, the dicarboxylic acids may be identified by direct chromatography of the acid fraction. A developing solvent of 1-pentanol and 5M formic acid (1:1, v/v) is utilized. The aqueous phase of the solvent system is retained in a beaker in the chromatographic jar. After development overnight, the chromatograms are removed and dried for 1–2 hr in a flowing air current to remove the formic acid. The acids are located by spraying with a 0.04% solution of Bromophenol Blue in 95% alcohol adjusted to pH 6.7. The acids appear as yellow spots against a blue background. The method is sensitive to 0.1 μmole of acid. R_f values are seen in Table 1.

Goncharova et al. [5] were able to separate several nonvolatile carboxylic acids in natural waters by paper chromatography. A 500-ml sample of water is adjusted to a pH of 8.0–9.0, evaporated to dryness, and dried at 60°C. The residue is dissolved in 1:1 HCl and extracted twice with 50-ml volumes of ether. This ether extract is neutralized with 0.01N NaOH and the ether is removed. The aqueous phase is then evaporated to a 1-ml volume, acidified with H_2SO_4, and passed through a silica column. The acids are separated into four fractions (a, b, c, d) by subsequent elution with 5, 10, 20 and 30% butanol in $CHCl_3$. Each fraction is separated by paper chromatography with these development solvents: 7:3 Phenol-1% formic acid for fractions a and b and 18:2:9 butanol-1% formic acid-water for fractions c and d. The spots are located by spraying with 0.04% Bromocresol Green. The fractions are (a) pyruvic, fumaric, glutaric, and adipic; (b) succinic and lactic; (c) oxalic and gallic; and (d) citric acids. No sensitivity data were given.

C. Phenols

Hoak [6] was concerned with the recovery and separation of phenolic compounds from river waters. Several organic solvents were tried for the extractive step, of which methylisobutylketone was chosen (by what criterion was not stated). In field trials, a Scheibel 18-stage countercurrent continuous extractor was employed. A typical extractive run was 27.5 hr duration, with 825 gallons of water processed and a extract volume of 25 gallons. A 75-ml aliquot of the extract was distilled to remove the solvent. An extract concentrate (2 ml) was made acid with H_2SO_4 and steam distilled. Enough NaCl was added to make a 20% solution, which was extracted with diethyl ether. This extract was separated into strong- and weak-acid fractions by differential solubility. The latter fraction was analyzed by paper chromatography. The phenols in this fraction were coupled with diazotized p-nitroaniline and spotted on Whatman No. 1 paper that had been impregnated with Na_2CO_3 and dried. The chromatogram was developed by the descending technique with a mixture of diethyl ether, petroleum ether, and 50% methanol (10:10:4). Phenol, m-cresol, and quaiacol were identified by comparison of R_f values to standards. 3,5-Xylenol was found from a chromatogram of the strong acid fraction. Another portion of the weak-acid fraction was analyzed by a second chromatographic procedure in which the phenols were converted to phenylazobenzenesulfonic acid dyes. The chromatogram was developed with a 1:1 mixture of s-butanol-2% aqueous Na_2CO_3. This procedure confirmed the presence of quaiacol, phenol, m-cresol, and 3,5-xylenol. No sensitivity data were given.

D. Pesticides and Herbicides

Sumiki and Matsuyama [7], in an early application of paper chromatography, attempted to resolve parathion and related compounds. The parathion, in a 200-liter water sample, was extracted with 2-4 liters of benzene and hexane (1:1, v/v) by means of a continuous process. The water was passed through and extractor at the rate of 100-200 liters/hr. The solvent was subsequently removed by vacuum distillation until a volume of 200 ml was obtained.

TABLE 2

R_f Values of Parathion and Related Compounds [a]

Compound	Mobile solvent: n–Hexane / Stationary solvent: Ethylenechlorohydrin-methanol (1:1 v/v)	Ethanol / Soybean oil	Ethanol–nitromethane (4:1 v/v) / Soybean
Parathion	0.23	0.56	0.53
Methylparathion	0.10	0.65	0.63
4124 (Isochlorothion)	0.17	0.54	0.63
Ethyl 4124	0.36	0.45	0.41
Chlorothion	0.13	0.56	0.50
Ethylchlorothion	0.30	0.54	0.43
O–(2–Nitrophenyl)–O, O–diethylthiophosphate	(0.02), (0.06), 0.18 0.32	0.60, 0.81	0.61, 0.76
EPN	0.22	0.40	0.37
Dipterex	0.02	0.79	0.81
Malathion	0.12	0.69	0.75
TEPP, technical	0, 0.03	0.71, 0.75	0.70, 0.75
Diazinon	0.03	0.63	0.60
Systox, technical	0.02, 0.10	(0.04), (0.64), 0.78	(0.04), (0.64), 0.79
Metasystox, technical	0.02, 0.04	0.74	(0.02), 0.74

Thimet, technical	(0.02), (0.11), (0.19)	0.56, (0.82)	0.46, (0.80)
Trithion, technical	0.75		
Trithion, technical	(0.0), 0.64	(0.06), 0.28	(0.08), 0.38, (0.86)
DDVP, technical (Dichlorvos)	0, 0.02, (0.05), 0, 10	0.05, 0.82	0.05, 0.85
p,p′-DDT	0.74	0.26	0.28
γ-BHC (benzene hexachloride)	0.32	0.41	0.42
Orthophosphate	0	0	0
Pyrophosphate	0	0.01	0.02

[a] Development of spot: by UV lamp irradiation, Hanes–Isherwood reagent, and UV radiation, and indicator for chlorinated pesticides. Parentheses indicate the light spot. Reproduced from Ref. 7, p. 330, by courtesy of the Nippon Nogei Kagaku Kaishi, Tokyo, Japan.

Evaporation to oils is continued with an agitating evaporator. A 97% recovery of parathion through these steps of extraction and concentration was claimed. At this point, the parathion in the extracts was isolated from other insecticides by two separate paper chromatographic systems employed in conjunction by the descending technique. The first system uses soybean oil as a stationary solvent and either ethanol–nitromethane (4:1 v/v) or ethanol only as the mobile solvent. The second system uses an ethylenechlorohydrin: methanol mixture (1:1, v/v) as the stationary solvent and n-hexane as the mobile solvent. The spot of parathion can be detected by an ultraviolet lamp. The first solvent system resolves parathion from EPN (O-ethyl-O-p-nitrophenyl phenylphosphonothioate), whereas the second system does not. R_f values of parathion and related compounds are given in Table 2. No sensitivity data were given.

Erne [8] utilized paper chromatography for the detection and determination of chlorophenoxyacetic acid derivatives (4-D, MCPA, and 2,4,5-T) in water. The pH of a 300-ml sample of water was adjusted to a value greater than 12 with 5M NaOH and the sample was evaporated on a steam bath to about 25 ml. It was then transfered to a separating funnel, acidified with 5M H_2SO_4 to a pH value below 2.0 and extracted with three 10-ml portions of $CHCl_3$. The combined $CHCl_3$ portions were extracted with a phosphate buffer, pH 6.2 (3 × 10 ml). The combined buffer extracts were washed with 3 ml of $CHCl_3$, adjusted to a pH value below 2.0, and extracted with $CHCl_3$ (3 × 10 ml). These latter extracts were filtered through anhydrous Na_2SO_4 (rinse with $CHCl_3$). The extracts were concentrated to about 5 ml on a steam bath and transfered to a graduated concentration tube, and the evaporation was continued at $30°-40°C$ in a weak current of air to 0.1-0.2 ml. The samples were rinsed with a little methanol and aliquots of this solution were used for the chromatograms. On circular filter paper (Whatman No. 1, 16-18 cm in diameter), a concentric starting circle 40-45 mm in diameter was marked; the paper was washed with 5M NH_4OH and distilled water and air dried. The sample and relevant standards were applied on the starting circle. The paper was equilibrated in a 15-cm petri dish for 5-10 min in the presence of the developing solvent, s-butanol-5M NH_4OH-water (25:5:20 by volume). A wick of ether-washed and dried cotton was attached to the center of the paper and the chromatogram developed with the clear upper phase of the solvent mixture kept in a smaller dish inserted into the larger dish. It takes about 1 hr for the solvent to approach the rim of the larger dish. The paper was dried for about 30 min in a hood. This paper was sprayed evenly with the silver-phenoxyethanol reagent (1.70 g of $AgNO_3$ in 5 ml of water, 20 ml of 2-phenoxyethanol, diluted with acetone to 200 ml) and immediately exposed to strong ultraviolet light until the dark spots were developed fully. The spots were marked and washed to remove excess silver ions once in very dilute NH_4OH, then repeatedly with distilled water, and finally once with methanol. The paper was air dried. Dark spots in the R_f region 0.6-0.7 usually appear after about 1 min of exposure and indicate the probable presence of chlorophenoxyacetic acids. If spots appear, a duplicate chromatogram is prepared for quantitative

determination (10-50 μg of the phenoxy derivative should be applied here).
The chromatogram is equilibrated and developed as described above, dried
in air, and sprayed lightly with the 4-methylumbelliferone reagent (0.02%
w/v in 30% ethanol adjusted to a pH of 8 with ammonia). The phenoxy acids
are located by observation of the chromatogram in filtered UV light. The
relevant areas and a blank are cut out and the phenoxy acids are eluted by
refluxing for 10 min at 50°C with 3 ml of methanol containing three drops
of concentrated ammonia. The eluate is filtered through a small pledget
of glass wool into a test tube and the flask and the filter are rinsed with two
1-ml portions of methanol. The solvent is removed at 30°-40°C in a slow
current of air. Exactly 5.00 ml of chromotropic acid reagent (0.40 g of
disodium salt in 100 ml of concentrated H_2SO_4) are measured into each of
the tubes containing eluted samples, standards, paper blank, and reagent
blank. All the tubes are heated in a thermostatically controlled oven at
150°C for 30 min and then cooled in running water. The absorbance, in
1-cm cells, at the wavelengths 505, 580, and 655 nm is measured against
the reagent blank. For each solution a baseline absorbance is calculated
by this formula:

$$A_{corr} = A_{580} - 0.5(A_{505} + A_{655})$$

This procedure was tested in recovery experiments with the compounds
added to well waters in concentrations of 0.15-1.00 mg/liter. For example,
the recovery of 2,4-D (0.15 mg/liter, 6 trials) ranged from 84 to 101%
with a mean of 93 ± 5.1%. The "practical" sensitivity of this procedure is
0.5 μg per 100 ml or 0.005 mg/liter. The separating efficiency of the paper
chromatographic step has been tested with a number of compounds that may
or may not be in natural waters. These R_f values and sensitivities are
seen in Table 3.

Goodenkauf and Erdie [9] utilized paper chromatography for the identification of chlorinated hydrocarbon pesticides in river water. The carbon
adsorption method (see Chapter 12 of this volume for details) was employed
for separation and concentration of the pesticides from the water phase.
Standards of aldrin, dieldrin, DDT, chlordane, heptachlor, endrin, lindane,
rhothane, kelthane, chlorobenzilate, atrazine, and propazine were added,
2 μg/liter, to $CHCl_3$ extracts of the carbon filtered river water. In turn,
the extract was separated into (a) water- and ether-soluble, (b) weak-acid,
(c) strong-acid, (d) basic, and (e) neutral fractions. The neutral fraction
was eluted from a silica gel column with successive aliquots of trimethylpentane, benzene, and a 1:1 $CHCl_3$-CH_3OH mixture. After the solvents were
evaporated and the samples weighed, each fraction was redissolved in ethyl
acetate and diluted to a volume of 10 ml. Suitable aliquots (5-35 μg) were
spotted on 8 × 8 in. paper (Whatman No. 1). The paper was washed several
times with distilled water and dried at 100°F. Suitable aliquots (generally
10 μl of standard and 10-60 μl of sample) were pipetted into 15-ml conical

TABLE 3

Approximate R_f Values (in the System s–Butanol : 1M Aqueous Ammonia) and Behavior Toward the Silver Reagent of Selected Acidic Compounds [a]

Compounds	R_f	Silver reaction	
		Response	Sensitivity (μg)
Oxalic acid	0.08	(+)[b]	5
Tartaric acid	0.08	(+)	2
Succinic acid	0.10	(+)	5
Dialuric acid	0.16	(+)	2
Phthalic acid	0.18	(+)	1-2
Uric acid	0.19	—	
Glycolic acid	0.22	(+)	5
Lactic acid	0.26	(+)	2
Acetic acid	0.28	—	
Thioglycolic acid	0.29 (0.14)	(+)	2
Fluoroacetic acid	0.29	(+)	1-2
Propionic acid	0.30	—	
Pantothenic acid	0.33	—	
Chloroacetic acid	0.38	+	0.3
Bromoacetic acid	0.39	+	0.3
Nicotinic acid	0.39	(+)[c]	2
Pyruvic acid	0.40 (0.54)	(+)[c]	3
Butyric acid	0.41	—	
Iosoacetic acid	0.44	+	0.2
Mandelic acid	0.44	(+)	1-2
Hippuric acid	0.45	(+)	5
3-Indoleacetic acid	0.46	(+)[c]	1
Dichloroacetic acid	0.47	+	0.2
Phenylacetic acid	0.48	(+)	3
Benzoic acid	0.50	(+)	3
Phenooxyacetic acid	0.51	(+)	3

TABLE 3. — Continued

Compounds	R_f	Silver reaction	
		Response	Sensitivity (μg)
2,2-Dichloropropionic acid	0.52	+	0.2
Sorbic acid	0.53	—	
Trichloroacetic acid	0.56	+	0.1
Salicyclic acid	0.57	(+)	1-2
2,2-Dichlorobutyric acid	0.58	+	0.3
4-Nitrophenol	0.58	— (yellow)	0.2
MCPA (2-methyl-4-chlorophen-oxyacetic acid)	0.62	+	0.3
2,4-D (2,4-dichlorophenoxyacetic acetic acid)	0.63	+	0.2
2,3,6-Trichlorobenzoic acid	0.64	+	0.5
2,4,5-T (2,4,5-trichlorophen-oxyacetic acid)	0.69	+	0.2
Warfarin	0.72	—[c]	
4-Methylumbelliferone	0.74	—	
4,6-Dinitro-2-methylphenol	0.78	— (yellow)	0.1
2,4,5-Trichlorophenol	0.81	+	0.2
Pentachlorophenol	0.82	+	0.5
Coumachlor	0.87	—[c]	
4,6-Dinitro-2-secbutylphenol	0.91	— (yellow)	0.1
2,4-Dichlorophenol	0.97	+	0.3
4-Chlorophenol	0.98	+	0.3
2-Methylphenol (o-cresol)	0.99	(+)	5

[a]Reproduced from Ref. 8, p. 1672, by courtesy of Acta Chemica Scandinavica, Stockholm, Sweden.

[b]A plus within parenthesis indicates that spots appeared, although not of the characteristic, dark colored type. Compounds not detectable by the silver reagent were located by means of the methylumbelliferone reagent.

[c]Fluorescent after silver reaction.

centrifuge tubes. These were evaporated to dryness and finally put into solution with about 0.1 ml of ether. These ethereal solutions were spotted on appropriately marked paper. The paper was dipped into the immobile solvent (350 ml of dimethylformamide diluted to 1 liter with ethyl ether) to the origin line (7/8 in. from edge of paper). The paper was removed, a glass rod was dipped to the top, and the paper was transferred to the mobile solvent (2,2,4-trimethyl pentane) tank with the origin end allowed to dip about 1/4 in. into the solvent. When the solvent front has climbed to within 1.5 in. from the top of the paper, it is removed and hung to dry in a hood. When dry, the paper is sprayed lightly, but uniformly, with the chromogenic agent (1.7 g of $AgNO_3$ in water, 10 ml of 2-phenoxyethanol, diluted to 200 ml with acetone). The paper is exposed to UV light for 10 min. Table 4 gives the R_f values for the 12 chlorinated hydrocarbons and the solubility fraction from which they have been chromatographed.

TABLE 4

R_f Values Using Solvent System of Trimethylpentane:
Dimethylformamide Diethyl Ether [a]

Pesticide	R_f	Solubility fraction chromatographed
Aldrin	0.90-0.97	Aromatic
Atrazine	0.08-0.13	Oxygenated and basic
DDT	0.52-0.60	Aromatic
Kelthane	0.19-0.28	Aromatic
Rhotane	0.23-0.30	Aromatic
Endrin	0.63-0.74	Aromatic
Chlorobenzilate	0.13-0.18	Aromatic and oxygenated
Heptachlor	0.84-0.91	Aromatic
Dieldrin	0.52-0.68	Aromatic
Propazine	0.14-0.18	Oxygenated and basic
Lindane	0.24-0.28	Aromatic
Chlordane	0.52-0.81 (streak)	Aromatic

[a]Reproduced from Ref. 9, p. 604, by courtesy of the American Water Works Association, New York, New York.

Abbott et al. [10] describe a paper chromatographic technique for the detection and determination of six organic herbicides in water. These herbicides are 2,4-dichlorophenoxy acetic acid (2,4-D), 2,4,5-trichlorophenoxyacetic acid (2,4,5-T), 2-methyl-4-chlorophenoxyacetic acid (MCPA), 4-(2-methyl-4-chlorophenoxy)butyric acid (2,4-DB), and 2,2-dichloropropionic acid (Dalapon). A 100-ml water sample is acidified with 5 ml of 6N H_2SO_4 and extracted with two 100-ml portions of ether. These ethereal extracts are combined and are extracted with two portions consisting of a mixture of 150 ml of a Na_2SO_4 solution (8%, w/v) with 5 ml of a 2.5N NaOH solution. These alkaline aqueous extracts are acidified with 10 ml of 6N H_2SO_4 and then extracted with successive portions of 150 and 100 ml of ether. These ethereal extracts are filtered through a column containing 50 g of anhydrous Na_2SO_4 into a Kuderna-Danish evaporator [Anal. Chem., 23, 1835 (1951)] fitted with a 5-ml flask. This volume is reduced on a warm water bath and the residual herbicides are dissolved in 40 μl of ethyl acetate immediately before the chromatographic spotting. The paper chromatographic procedure utilizes Whatman No. 3 chromatographic paper cut into 19 × 30 cm sheets. An origin line, 20 mm from and parallel to one short edge, is drawn. One-microliter portions of the ethyl acetate solutions (standards and samples) are transferred to previously marked spots on the paper. Next, 200 ml of mobile solvent A (2.0 ml NH_3 solution, sp. gr. 0.88, and 18 ml of t-butanol to 40 ml of water, diluted to 200 ml with ethylmethylketone) are placed in the stainless steel tray, the lid is replaced on the chromatographic tank, and the tank is allowed to equilibrate for 1 hr. The frame of papers is carefully lowered into the tank so that the bottom edges just enter the solvent. The lid is replaced and the chromatograms are allowed to develop for 100 min. The papers are removed from the tank and dried in a current of air at 80°C for 10 min. The paper is drawn through the chromogenic reagent (1.7 g of Ag NO_3 in 5 ml of water, 20 ml of 2-phenoxyethanol, diluted to 200 ml with acetone). One drop of 30 vol % H_2O_2 is added. The solvent is evaporated in a current of air and each side of the paper is irradiated for 10 min with an UV light. The chromatogram is washed three or four times with successive volumes of water. The paper is immersed in 0.5% sodium thiosulfate for 10 sec and then immediately washed with several successive portions of water. The paper is dried at 80°C. The chromatograms are evaluated either by area measurement or by a reflectance densitometer. The spots are colored orange to brown, set on an off-white background. Standard curves (linear) of micrograms of herbicide vs reflectance may be constructed. A sensitivity of 0.5 μg is claimed for each herbicide.

An alternate paper chromatographic separation uses Whatman No. 1 paper as described above, spotted with ethyl acetate solutions of samples and standards, and developed with mobile solvent B (10 ml of liquid paraffin, 30 ml of benzene, and 20 ml of glacial acetic acid added to 200 ml of cyclohexane) for 3 hr. The R_f values (MCPB, 0.81; 2,4-DB, 0.79; MCPA, 0.68; 2,4,5-T, 0.66; 2,4-D, 0.58; Dalapon, 0.45) of the unknown spots are compared to those of standards. Figure 3 shows the chromatograms of several mixtures of the herbicides.

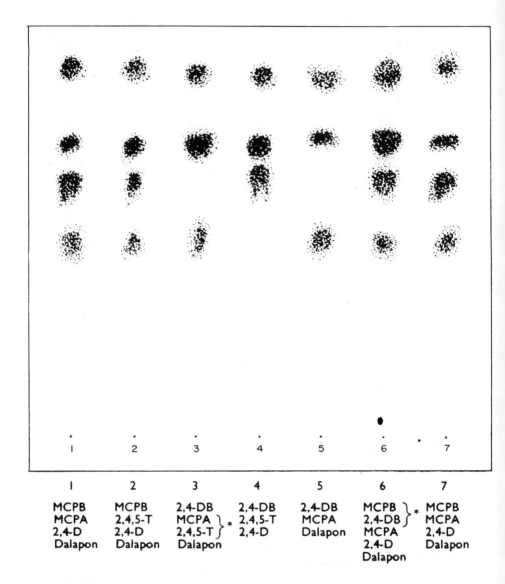

Figure 3.   Appearance of a chromatogram on Whatman No. 1 chromatographic paper developed in mobile solvent B.   Reprinted from Ref. 10 with permission of the Society for Analytical Chemistry, London, England.

Hindin et al. [2] examined surface and ground waters, sediments, and bottom muds for organic pesticides by paper and gas-liquid chromatographic techniques. The pesticides in water samples were collected on activated carbon as previously described. Usually, a volume of 18,925 liters of water was passed through the carbon column, after which the carbon was air dried and placed in a modified soxhlet extractor for extraction by petroleum naphtha or petroleum ether. After the extraction process, the solution was concentrated to about a 10-ml volume by evaporation of solvent. The pesticides from the stream sediments and muds were air dried, crushed by a mechanical mortar grinder, and thoroughly mixed with petroleum naptha for extraction. Three extractions were made, combined, and filtered initially through cheesecloth and then through Whatman No. 1 filter paper. All but 10 ml of the solvent was removed by evaporation. It was necessary to clean up the water and soil concentrates on a column of activated aluminum silicate prior to paper or gas-liquid chromatographic analysis.

Hindin et al. [2] used the paper chromatographic technique of Mitchell [11] for separating and identifying the chlorinated hydrocarbons in the above aqueous and soil extracts. Four sets of duplicate 8 x 8 in. papers (Whatman No. 1) are spotted with 0.001-ml portions of the test solutions along the baseline. Three sets of papers are sprayed with immobile solvent (1 ml of refined soybean oil diluted to 200 ml with ether). One set of papers is transferred to the first tank and developed for 2.5 hr with aqueous acetone (30 ml of water added to 100 ml of acetone). The second set of papers is transferred to a second tank and developed (4.75 hr) with aqueous 2-methoxyethanol (30 ml of water diluted to 100 ml with 2-methoxyethanol). The third set of papers is transferred to the third tank and developed (4.5 hr) with aqueous pyridine (45 ml of water diluted to 100 ml with pyridine). The fourth set of papers is sprayed with immobile solvent (20 ml of practical grade dimethylcanamide diluted to 100 ml with ether), transferred to the fourth tank, and developed (2.5 hr) with 2,2,4-trimethylpentane. The papers are removed from the tanks, the solvent front is marked, and the papers hung until they are dry. All the papers are sprayed with the chromogenic agent (1.7 g of Ag NO_3 in H_2O, 10 ml of 2-phenoxyethanol, and 50 ml of ethanol, diluted to 200 ml with H_2O) and allowed to air dry. The papers are exposed to strong ultraviolet light until the spots are fully defined.

Table 5 gives the R_f values obtained by Hindin et al. [2] for several chlorinated hydrocarbons at two concentrations in water. They claimed a sensitivity of 0.5 μg for DDT but the minimum detectable limit for most other chlorinated compounds was 2.0 μg. There were, however, several limitations to this procedure: (a) "some chlorinated pesticides appeared as streaks rather than spots, (b) several pesticides had nearly identical R_f values, and (c) a few chlorinated compounds did not respond to the method."

Hindin et al. [2] used the paper chromatographic procedure of Mitchell [12] for the separation and identification of organophosphorus pesticides in the above aqueous and soil extracts. A one-dimensional procedure was used

TABLE 5

R_f Values of Samples of Pesticides for an Aqueous Solvent with
Silver Nitrate-2-Phenoxyethanol in Acetone as Chromogenic Agent;
Temperature, $20^\circ C$ [a]

| Sample | Quantity of substance in micrograms (solids) or microliters (liquids) | |
	2	50
Insecticides:		
Aldrin	0.26 ± 0.06	0.26 ± 0.06
Aramite	F	
BHC, high gamma	0.70 ± 0.1	0.50-0.91
BHC, low gamma	0.70 ± 0.1	0.54-0.91
Captan	—	0.51-0.96
Chlordane	—	0.26-0.54
Chlorobenzilate	—	0.77-0.99
DDD(1,1-dichloro-2,2-bis-(p-chlorophenol)ethane	0.69 ± 0.06	0.68 ± 0.06
DDT, technical	0.40 ± 0.06	0.40 ± 0.06
DDT, wettable	—	0.43 ± 0.05
Dibrom	1.0	1.0
Dieldrin	0.51 ± 0.09	0.50 ± 0.09
Dyrene	F	F
Endrin	0.47 ± 0.03	0.49 ± 0.04
Fenson	F	F
GC-2466	F	F
Genite	F	F
Heptachlor	0.40 ± 0.09	0.39 ± 0.01
Kelthane	0.70 ± 0.08	0.72 ± 0.06
Kepone	F	F
Korlan	0.69 ± 0.1	0.69 ± 0.12
Lindane	0.70 ± 0.09	0.72 ± 0.06

TABLE 5. — Continued

Sample	Quantity of substance in micrograms (solids) or microliters (liquids)	
	2	50
[Insecticides]		
Methoxychlor	0.77 ± 0.15	0.77 ± 0.15
Ovex	0.83-0.99	0.81-1.00
Phygon	0.94 ± 0.06	0.49-0.98
Spergon	0.74-0.98	0.10-0.98
Tedion	0.72 ± 0.12	0.72 ± 0.17
Terraclor	—	0.38 ± 0.11
Thiodan	0.69 ± 0.06	0.83 ± 0.10
Toxaphene	0.30-0.54	0.22-0.61
Other:		
Atrazine	F	F
Diuron	1.0	1.0
GC-6499	1.0	1.0
GC-6690	—	1.0
GC-6691	1.0	1.0
Neburon	1.0	1.0
NIA-5996	—	0.85-1.00
Telvar	1.0	1.0
2,4-D acid	0.92 ± 0.04	0.92 ± 0.05
2,4-D butyl ester	1.0	1.0
2,4-D amine	0.94 ± 0.05	0.92 ± 0.05
2,4-D isooctyl ester	0.63 ± 0.1	0.61 ± 0.10
2,4-D isooctyl ester	—	0.36 ± 0.10
2,4-D isopropyl ester	0.87-1.0	0.82-1.0
2,4,5-T acid	0.95 ± 0.05	0.78-1.0
2,4,5-T isooctyl ester	0.50 ± 0.06	0.49 ± 0.08
2,4,5-T isopropyl ester	0.97 ± 0.03	0.68-1.00

TABLE 5. — Continued

Sample	Quantity of substance in micrograms (solids) or microliters (liquids)	
	2	50
Urab	0.98 ± 0.01	0.96 ± 0.04
Urox	0.98-1.00	0.90-1.00
Zobar	0.94 ± 0.08	0.91 ± 0.07

[a]Reproduced from Ref. 2, p. 132, by courtesy of Springer Verlag, New York, New York. F indicates fluorescences at the solvent front.

wherein the 8 × 8 in. Whatman No. 1 paper is serrated with a hard pencil and a template. The papers are spotted along the numbered serrations with standard solutions (ethyl acetate) of the pesticides or of the extracted samples. A dipping tank is filled with the immobile solvent (20%, v/v, N,N-dimethylformamide in ether or 10%, v/v, formamide in acetone). The paper is inserted into the dipping tank until the solvent touches the starting line. The paper is immediately removed and transferred to the chromatographic tank (containing the mobile solvent, 2,2,4-trimethylpentane) for development in about 75 min. The spots are developed with a spray of bromine and fluorescein (0.25% w/v, fluorescein in N,N-dimethylformamide, 2 ml of which are diluted to 200 ml with 95% ethanol). The spots were delineated by viewing under an ultraviolet light. Table 6 gives the R_f values obtained by Hindin et al. [2] for several organophosphorus compounds.

Gapotchenko and Nezdoimishapka [13] used radial partion paper chromatography to determine DDT and hexachlorocyclohexane in water. The sample is extracted twice with 30 ml of benzene that is washed with successive 5-ml portions of concentrated H_2SO_4 until it is colorless. Dry the benzene extracts with anhydrous Na_2SO_4 and evaporate to an appropriate volume for application to the paper. A 5% solution of castor oil in ether is used to wet the paper. After the ether is evaporated, the chromatogram is developed with a 7:3 acetone-water mixture. The spots are made visible by spraying with 0.05N alcoholic $AgNO_3$, drying for 30 min at 37°C, spraying with 40% formaldehyde, drying for 15-20 min, spraying with 2N KOH in methanol, drying at 120-130°C for 30 min, and finally spraying on both sides with 1:1 perhydrol-concentrated HNO_3 mixture. The spots of DDT and hexachlorocyclohexane will be found at 1.6-1.8 and 3.6-3.8 cm., respectively, after 3 hr of development and at 2.2-2.5 and 4.4-4.8 cm. after 12 hr.

TABLE 6

R_f Values of Samples of Insecticides for an Aqueous Solvent
System with Fluorescein: N,N-Diethylformamide in Ethyl
Alcohol as Chromogenic Agent; Temperature $20^\circ C$ [a]

Sample	Quantity of substance in micrograms (solids) or microliters (liquids)	
	2	50
Delnav	0.30 ± 0.80	0.30 ± 0.10
Diazinon	0.33 ± 0.04	0.31 ± 0.10
Dibrom	0.90 ± 0.08	0.87 ± 0.07
Disyston	0.19 ± 0.05	0.18 ± 0.04
Disyston	—	0.41 ± 0.04
EPN	0.33 ± 0.04	0.29 ± 0.07
EPN	—	0.93 ± 0.03
Ethion	0.14 ± 0.03	0.12-0.06
Ethion	—	0.81-1.00
Guthion	0.87 ± 0.05	0.82 ± 0.09
Malathion	0.79 ± 0.06	0.77 ± 0.08
Methyl parathion	0.79 ± 0.05	0.75 ± 0.06
Methyl parathion	0.94 ± 0.03	0.94 ± 0.06
OMPA	—	0.91 ± 0.03
Parathion	0.47 ± 0.05	0.47 ± 0.10
Parathion	—	0.97 ± 0.05
Phosdrin	0.95 ± 0.02	0.94 ± 0.02
Phostex	0.18-1.00	0.05-1.00
Ronnel	0.18 ± 0.03	0.16 ± 0.03
Ronnel	—	0.27-0.61
Systox	0.90 ± 0.05	0.88 ± 0.05
Systox	0.29 ± 0.06	0.28 ± 0.03
Tepp	0.99 ± 0.01	0.93 ± 0.04

TABLE 6. — Continued

| | Quantity of substance in micrograms (solids) or microliters (liquids) | |
Sample	2	50
Thimet	0.23 ± 0.03	0.20 ± 0.03
Thimet	—	0.93-1.00

[a]Reproduced from Ref. 2, p. 136, by courtesy of Springer Verlag, New York, New York.

VI. DISCUSSION AND SUMMARY

Paper chromatography has been utilized in the identification and separation of various pollutants in water. This analytical technique does, however, have some limitations and restrictions. First, it is employed mainly as an identification and confirmation technique in conjunction with some other analytical procedure. Paper chromatography does not have the capacity to resolve gross interferences from the molecules being sought. This eliminates its use as a cleanup procedure. Second, many of the procedures reported in this chapter are cumbersome and time consuming. These aspects affect the sensitivity and recovery ability of paper chromatography. In fact, many authors did not include these data. In general, paper chromatography is less sensitive than thin-layer chromatography. Although some authors claim a sensitivity of 10^{-6} g for paper, a more realistic and practical level would be 10^{-4} g in systems where interferences are present. Hindin et al. [2] illustrate this point very nicely with their experiences with organic pesticides in water, sediments, and bottom muds. These authors found it necessary to clean up extracts vigorously before spotting them on paper. Severe streaking occurred on chromatograms from coextracted interferences in uncleaned samples. Another limitation is that several pesticides have nearly the same R_f values.

Despite these limitations, paper chromatography does have a useful place in the analysis for pollutants in aqueous systems. However, its application should be confined to a confirmatory role. It could be used very well with gas-liquid chromatography, for example, as aqueous samples should be very clean before injection.

ACKNOWLEDGMENTS

The authors acknowledge the New Jersey Agricultural Experiment Station and the Department of Environmental Sciences at Rutgers, the State University of New Jersey, New Brunswick, New Jersey.

REFERENCES

1. S. D. Faust and I. H. Suffet, in Water and Water Pollution Handbook (L. Ciaccio, ed.), Marcel Dekker, New York, 1971, p. 1249.

2. E. Hindin, D. S. May, and G. H. Dunstan, Residue Rev., 7, 130 (1964).

3. C. Z. Maehler and A. E. Greenberg, J. Sanit. Eng. Div. (ASCE), SA5, 969 (1968).

4. H. F. Mueller, T. E. Larson, and M. Ferretti, Anal. Chem., 32, 687 (1960).

5. I. A. Goncharova, A. N. Khomenko, and A. D. Semenov, Gidrolchim. Water, 41, 116 (1966); through Chem. Abstr., 66, 68772W (1967).

6. R. D. Hoak, Int. J. Air Water Poll., 6, 521 (1962).

7. Y. Sumiki and A. Matsuyama, Bull. Agr. Chem. Soc. Japan, 21, 329 (1957).

8. K. Erne, Acta Chem. Scand., 17, 1663 (1963).

9. A. Goodenkauf and J. Erdei, J. Amer. Water Works Assoc., 56, 600 (1964).

10. D. C. Abbott, H. Egan, E. W. Hammond, and J. Thomson, Analyst, 89, 480 (1964).

11. L. C. Mitchell, J. Assoc. Off. Agr. Chem., 39, 980 (1956).

12. L. C. Mitchell, J. Assoc. Off. Agr. Chem., 43, 810 (1960).

13. P. A. Gapotchenko and N. A. Nezdoimishapka, Gig. Sanit., 34, 56 (1969); through Chem. Abstr., 71, 128527k (1969).

Chapter 14

THIN-LAYER CHROMATOGRAPHIC ANALYSIS
IN WATER POLLUTION

Osman M. Aly

Department of Environmental Engineering
Campbell Soup Company
Camden, New Jersey

and

Samuel D. Faust

Department of Environmental Sciences
Rutgers University
New Brunswick, New Jersey

I. INTRODUCTION

For fresh and saline waters that may be polluted, thin-layer chromatography offers considerable promise as a cleanup technique prior to other methods of separation and identification of a pollutant, for separation of several species of a specific class of pollutants, and as a supplement in the positive identification of a given pollutant. This analytical procedure has a high capacity to resolve pollutants from extraneous matter. One of the more enthusiastic analysts in Kovacs [1] states: "thin-layer chromatography is superior to paper chromatography because spots are more compact and resolution is much sharper." In connection with the organic pesticide pollution problem, Smith and Eichelberger [2] report: "the most important advantage of the method is that it provides corroborative evidence of the presence or absence of a particular pesticide. The first part of the evidence is that the pesticide occurs in a specific TLC-separated section. The second part is that the peaks of the gas chromatogram obtained from a particular section possess the required retention times of pesticides known to migrate to this section." Thus, thin-layer chromatography can play a definite analytical role in the detection, separation, and identification of water pollutants.

This chapter provides the procedural detail for only a few of the many pollutants that may be found in water. These are all of organic origin; namely, amino acids, detergents, hydrocarbons, plant pigments, phenols, pesticides, and sterols. Most of the procedures were developed from polluted surface waters and waste water effluents. Some procedures, however, are included from the marine environment because they may find application to fresh water situations.

II. ACCURACY, PRECISION, AND DATA EVALUATION

It is difficult to speak of accuracy and precision of thin-layer chromatographic procedures because most of the procedures are qualitative or semiquantitative at best. In fact, many authors fail to report sensitivity, precision, and accuracy data. Most frequently, the R_f value is reported with its attendant standard deviation, which may or may not be an indication of precision and accuracy. Most investigators are concerned with the R_f value because of its role in the identification of a pollutant. Another consideration is the emphasis placed on the resolution of the pollutant from interferences. This often requires a cleanup prior to TLC, which results in some loss of the pollutant. Therefore, any precision and accuracy data given in this chapter are confined to R_f values where available and applicable.

The water pollution analyst is concerned with the significant level of detection of any analytical technique. This is a question of sensitivity. The true analytical procedure is that which gives reproducible data for the minimum detectable concentration of a substance that may be determined. In clean systems TLC can detect and resolve pollutants in the order of 10^{-8} g.

More important, however, is the sensitivity in the presence of interferences. Examples are provided by Kovacs [1] and Askew et al. [3]. Kovacs developed a thin-layer technique for 14 organothiophosphate pesticides in "pure" systems and reported sensitivities of 0.05 μg for 11 compounds and 0.1 μg for three compounds. In contrast, Askew et al. [3] were concerned with the determination of organophosphorus pesticides in river waters and sewage treatment plant effluents. In these aqueous systems a sensitivity of 1 μg was reported. Thus, the sensitivity for a polluted system is one or two orders of magnitude above those reported for clean systems.

III. SAMPLING

In the analytical procedures reported in this chapter, no special sampling techniques or equipment were employed. In almost all cases, the "grab" sample was utilized. The size or volume of the "grab" sample may vary from as little as 1 liter to several gallons. Size is dependent, of course, on the constituent being sought and on the desired sensitivity. Perhaps the most important consideration is the time interval between collection and analysis. Most of the organic pollutants reported in this chapter are subject to either biological or chemical degradation. Consequently, the analyst should process the sample as quickly as possible after collection.

The problems of sampling and storage of samples have been discussed by Faust and Suffet [4] for the organic pesticide. The precautions recommended in that situation also apply to the organic pollutants cited in this chapter. The major concern is that the samples be analyzed as quickly as possible after withdrawal from the source.

IV. SEPARATION OF INTERFERENCES

Coextracted organic interferences may arise from several sources: domestic and industrial waste water, naturally occurring substances in surface waters and bottom muds, and particulate matter. For example, the carbon chloroform extracts (CCE's) from the concentration technique of carbon adsorption contain oxygenated organics that interfere with the identification of chlorinated hydrocarbon pesticides. A cleanup step is absolutely necessary when grossly polluted waters are extracted.

Thin-layer chromatography is, on occasion, used as a cleanup technique. However, if it is used as the quantitative or confirmative step, then some form of partition chromatography, with columns of the silica gel type of support should be used for the cleanup. Other supports that have been used for cleanup area alumina, MgO-Celite, and Florosil.

The inclusion of a cleanup step raises the question of recovery because of the desirability of achieving a 100% recovery of the constituent under analysis. However, losses occur in each analytical step in a given procedure. It is therefore general practice to report efficiencies of recovery by techniques of fortification, i.e., the addition of a known quantity of a constituent to water in a laboratory test prior to processing the sample through each step of the analytical procedure. Fortification provides data only on the theoretical efficiency of recovery of the total analytical procedure. Some qualitative indication of losses may be obtained from the work of Askew et al. [3]: chlorfenvinphos, 1 μg/liter, was added to river water, extracted with CHCl$_3$, separated on a Sephadex column, and quantitatively determined by gas chromatography. Recoveries of 82 and 87% were recorded in two separate trials. In another trial, 10 μg/liter of chlorfenvinphos was added to a water containing 4200 mg/liter of suspended solids (a mixed liquor from an activated sludge tank). In this case, the recoveries were 33 and 36%. This illustrates very nicely the effect of natural aquatic environments on the recovery of any pollutant. It also raises the question of whether the percentages of recovery from the laboratory fortification experiments represent the actual efficiencies from field samples. The answer would be yes if the constituent did not interact with any portion of the environment. However, as all pollutants interact with its surroundings in some physical, chemical, or biological manner, a 100% of recovery is somewhat difficult to achieve.

V. CONCENTRATION TECHNIQUES

There are many innovations available for concentrating the organic pollutant for subsequent thin-layer chromatography. By "concentrating" is meant that the constituent is placed in a volume smaller than when sampled. This is done primarily to increase the sensitivity of the analytic method and to separate the constituent from interferences. There are, of course, coextractives that make the concentration step analytically fallible.

In the extraction and concentration of organic pollutants from aqueous systems, several techniques are available: carbon adsorption, liquid-liquid extraction, and evaporation. These were discussed thoroughly by Faust and Suffet [4] for the recovery of organic pesticides from water. As the principles are the same for other organics, this discussion is limited to a brief summary. Also, procedural details are given below under specific pollutants.

Carbon has been employed for extraction of organics from water in two different techniques: as a continuous device in the field and as a batchwise concentration and preliminary cleanup in the laboratory. The continuous system is the most widely used system. It was introduced by Braus et al. [5] as the carbon adsorption method (CAM) for isolating synthetic organic contaminants affecting taste and odor qualities of surface waters. Despite many serious limitations, the CAM has qualitative usefulness for extraction and concentration of organics from surface and ground waters.

Liquid-liquid extraction is, perhaps, the predominant technique for the extraction and concentration of organics from aqueous systems. The state of the art has been discussed thoroughly by Faust and Suffet [4]. The major consideration is selection of the proper solvent. Often this selection is arbitrary, with the result that inefficient recoveries are obtained. The p-value concept of Bowman and Beroza [6, 7] offers a quantitative way for proper selection of the solvent and of such environmental conditions as pH, temperature, and ionic strength. The p-value concept was extended to the extraction of organophosphorus pesticides from water by Suffet and Faust [8]. The reader is referred to the references in this paragraph for a detailed discussion of solvent extraction of organics from water.

VI. METHODS

A. Amino Acids

Chau and Riley [9] used a thin-layer chromatographic technique for the separation of amino acids from sea water samples. First, the sea water is filtered through a 0.5 μm membrane filter immediately after collection. Then the water is heated to 60°C for 1 hr, cooled, and adjusted to a pH value of 4.0 by the addition of HCl. One milliliter of $CHCl_3$ per liter is added to inhibit microbiological activity. A 2.5-liter sample of the sea water is then concentrated to 400-800 ml by passage through a climbing film evaporator. A rotary film evaporator (liquid temperature >40°C) is used to continue the concentration. The salts are removed periodically by filtration with a wash by a mixture of ethanol, water, and concentrated HCl (80:20:1). The concentrate and washings are returned to the evaporator where the evaporation is continued to dryness. This residue is dissolved in 500 ml of water and then passed through a column of the acid form of Amberlite CG120 cation exchanger (100/200 mesh).

A 0.1M piperidine solution (8.5 g restilled piperidine/liter) is passed through the column. The first 600 ml of eluate are discarded. The elution is continued until the eluate becomes alkaline to litmus and the piperidine front has passed through the column. An additional 200 ml of eluate is collected. The combined eluate (ca. 3 liters) is evaporated to a small volume in a rotary evaporator, transferred then to a 10-ml pear-shaped flask, and evaporated almost to dryness. The resulting brown residue is dissolved in 10 ml of water and passed through the ion-exchange column again with subsequent elution by the 0.1M piperidine. This eluate is evaporated almost to dryness. These residues must be prepared for TLC as follows: by means of a dropping tube the almost colorless concentrate (ca. 0.5 ml) is transferred to a glass tube (7.5 cm long and 6 cm i.d.) with small volumes of water used for rinsing the flask. The tube is heated in a bath of boiling water and the evaporation of the water is assisted by passing a stream of nitrogen into the tube. When all water has been evaporated 50 μl of 0.1N hydrochloric acid containing 10% by volume of isopropanol are added. The tube is sealed 2.5 cm from its bottom; it is immersed in a beaker of hot

water and the solvent is allowed to reflux up and down for 5-6 hr. After it
is cool, the tube is centrifuged and cut open with a file. If desired, the
sealed tube may be stored at 20°C. The thin-layer chromatography separation is conducted as follows: best results are obtained with 0.5-1.0 μg of
most amino acids. The plates, which measure 20 × 24 cm, are coated with
0.25-mm layers of Merck Silica Gel G and are dried at room temperature.
In order to keep the size of the amino acid spots as small as possible, the
appropriate volume of the concentrate is applied to the silica gel layer 1 μl
at a time using a micropipet and each application is allowed to dry before the
next aliquot is added. The concentrate should be applied in the right-hand
corner with the plate lying with the shorter edge as base. The spot is placed
1.5 cm from the adjacent edges. After the spot has dried, the plate is stood
with this shorter edge in a mixture of phenol and water (75:25 v/v) in a
chromatographic tank. When the solvent front has traveled 20 cm, the plate
is dried in a current of air for 15 min and then placed in an oven at 60°C for
20 min. When it has cooled, the chromatogram is developed at 90°C using a
solvent consisting of n-butyl alcohol, acetic acid, and water (60:20:20, v/v)
containing 0.2%, v/v, of ninhydrin. When the solvent front has moved 18 cm,
the plate is removed from the chromatographic tank and dried in an oven at
60°C for 30 min to develop the color spots. The plates are individually
removed from the thin-layer plate using a micro suction device consisting of
a 3-mm glass tube with a small plug of glass fiber filter wedged against a
constriction ~3 cm from the lower end; its other end is connected to a filter
pump. The colored compound is eluted by forcing 60% ethyl alcohol through
the tube. The extract is diluted to 2 ml with 60% alcohol and centrifuged to
remove glass fibers. The absorbance of the solution is measured in a 40 cm
microcell at 570 nm (at 440 nm for the spots arising from proline and hydroxyproline). A blank solution is prepared by scraping off a white area of the plate
roughly the same size as the spots and eluting it as described above. The
absorbance of the blank should always be very low.

A calibration run is carried out using 1 μl of a standard solution of amino
acids (containing 30 mg of each of 18 amino acids dissolved in 3 ml of 0.1N
hydrochloric acid containing 10%, v/v, of isopropanol; 1 μl = 1 μg of each
amino acid). The spots are eluted and their absorbances measured as described above. The weight of each amino acid in the sample is measured using
the corresponding calibration value and the result is multiplied by 1.1 to
correct for losses occurring in the separation process.

B. Detergents

The determination of commonly used nonionic detergents in sewage effluents
and river waters by TLC was effected by Patterson et al. [10]. First, a
250-ml sample of unfiltered sewage effluent is added to a 250-ml volume of
a $MgSO_4$ solution (150 g of $MgSO_4$ in 250 ml of water, $CHCl_3$ washed) which

is extracted four times with 50 ml of $CHCl_3$. If an emulsion persists, the $CHCl_3$ layer is run through a pad of 0.8 g of white absorbent cotton wool in a filter funnel. The combined $CHCl_3$ extracts are "acid washed" by shaking with a solution of 50 g of NaCl in 200 ml of 1N HCl (previously washed with $CHCl_3$. The $CHCl_3$ layer is run into an "alkaline-wash" solution (50 g of NaCl in 200 ml of 2N NaOH, $CHCl_3$ washed) and shaken. The alkaline-washed $CHCl_3$ extracts are then run through a small, dry filter paper and into a suitable container for subsequent evaporation to approximately 2 ml. This solution is transferred to a small test tube or conical centrifuge tube and evaporated carefully to dryness over boiling water. After it is cool, the residue is dissolved with 0.5 ml of $CHCl_3$ and spotted on a thin-layer plate. Twenty grams of Kieselgel G are shaken with 60 ml of water and this is spread on 20 × 20 cm plates to a thickness of 0.25 mm. The plates are allowed to dry overnight and then activated in an $110^\circ C$ oven for 1 hr. Two chromatographic tanks are prepared one-half hour before insertion of the plates. The walls of the first tank are lined with filter paper. This tank receives 100 ml of solvent mixture A (40 ml of ethyl acetate, 30 ml of water, and 30 ml of glacial acetic acid). The second tank (walls unlined) receives 100 ml of solvent mixture B (70 ml of ethyl acetate, 15 ml of water, and 16 ml of glacial acetic acid). Spots of 10 and 20 μl of the $CHCl_3$ extract are made along with spots of standard solutions of the detergents (1-5 μg). The plates are developed in the two tanks until the solvents have risen to a 10-cm line. The plates are air dried and then oven dried at approximately $100^\circ C$ for 10 min. The plates are cooled and then sprayed with the color reagent (10 ml of the Burger reagent, which is 1.7 g of bismuth oxynitrate in 220 ml of glacial acetic acid and 40 g of KI in 100 ml water, 1 ml of H_3PO_4, 10 ml of ethanol, and 5 ml of a 20%, w/w, aqueous solution of $BaCl_2 \cdot 2H_2O$). The chromatogram is covered with a clean glass plate immediately after it has been sprayed. A pink color should develop if nonionic detergents are present. This procedure "enables an estimate of the content over a range of 0.1 to 1.0 mg/liter." There are no major interferences from sewage effluents and river waters. Approximately 20-30% of the anionic detergents present is extracted from effluents. These detergents, however, run ahead of the non-ionics on the plates and do not give the pink color.

C. Hydrocarbons

Lambert [11] was able to separate and to identify some oil and greases in water by thin-layer chromatography. The sample of water is adjusted first to a pH value in the range 4.0-4.5 and is extracted with a mixture of $CHCl_3$ and methanol. When more water and $CHCl_3$ are added, a two-phase system results. This $CHCl_3$ layer is evaporated to dryness. The dry residue is redissolved in $CHCl_3$ so that 1 ml contains approximately 1 mg of sample. A silica gel plate (13% $CaSO_4$) is activated for 2 hr at $110^\circ C$. The controls and unknowns are spotted and the plate is heated for 10 min more. After equilibration for several hours in a chromatographic tank, 100 ml of a mixed solvent are added: (a) hexane, ethyl acetate, acetic acid (90 : 10 : 2); (b) hexane, pentane, and ethyl acetate (50 : 40 : 10); or (c) hexane, chloroform, and acetic acid (90 : 10 : 1). The best color developers are (a) H_2SO_4 (sp. gr. 1.84) + 0.5% HCHO and (b) 80 vol. CCl_4 + 20 vol. $SbCl_5$.

Borneff and Kunte [12] used a TLC procedure for the separation of several carcinogenic substances in water and sewage. Ten liters of the water to be examined are extracted with 600 ml of highly purified benzene by stirring for 15 min with a high-speed stirring apparatus. After separation of the two phases, the benzene layer is collected and concentrated to a small volume (0.05-0.2 ml). For five plates (200 × 200 mm) 14 g of Silica Gel G, 14 g of Aluminum Hydroxide G, and 12 of acetylated cellulose powder (40% acetylated) are mixed thoroughly with 70 ml of ethanol and the plates prepared in the usual manner. The air-dried plates are activated at 110°C for 1 hr and stored in a desiccator. The extract, or an aliquot (according to the degree of pollution of the water), is applied as a spot in one corner of the plate (about 15 mm from the edge). The plate is then developed 40 min with isooctane and dried an additional 40 min with n-hexane-benzene (9 + 1 vol.). The plate is again dried, turned by 90° and developed in the second direction with methanol-diethyl ether-water (4 + 4 + 1 vol.) twice for 1 hr with drying in between. During the whole procedure the plates must be protected from light and they must be viewed under an UV lamp (365 nm) only after they have been developed. They are then sprayed with Neatan (produced by Merck, Darmstadt), after which the whole layer can be detached from the glass plate and fixed on a sheet of black paper (fluorescence shows up better on black background).

A mixture of pure substances, in different concentrations from 0.01 to 1.0 μg, is chromatographed in the same way, and the chromatograms used for the quantitative determination by comparing size and intensity of the fluorescing spots under UV light. The reference chromatograms should be kept in the dark and may be used for about 2 months or less if used very frequently, because fluorescence diminishes in the course of time owing to destruction of the substances by UV light, sublimation, and loss of material from the thin layer.

D. Plant Pigments

A thin-layer chromatographic method for the determination of plant pigments in sea waters and cultures was reported by Garside and Riley [13]. A known volume of the sample (normally 0.5-5 liters of sea water or 10-100 ml of culture), sufficient to provide approximately 12 μg of chlorophyll a, is filtered through a 7-cm diameter Whatman GF/C glass fiber filter. All work on the pigment and its extracts must be conducted as rapidly as possible in subdued light. A magnesium carbonate suspension (3 g of $MgCO_3$ to 100 ml of water) first is placed on the filter to give a layer 1-2 mm in thickness. The filter and $MgCO_3$ mat are transferred to a test tube, where 3-5 ml of acetone are added and glass rod is used to break up the suspension. The test tube is placed in an ultrasonic bath for 4-5 min, 10 ml of methanol are added and mixed, and the agitation is continued for an additional 10 min. The supernatant is passed through a 2-cm layer of anhydrous sodium sulfate. The residue and sodium sulfate are washed with a few milliliters of methanol and ether, which is combined with the extract. The combined extracts are evaporated with a rotary evaporator, the bath temperature of which should not exceed $50°C$. The residue is dissolved immediately in 1-2 ml of diethyl ether containing 1% (v/v) of diethylamine. The entire solution is applied as a small spot about 2 cm from the end of a 6 × 20 cm plate coated with silica gel (Merck PF254, 0.25 mm thick, dried in a $100°C$ oven for several hours, and allowed to cool in a relative humidity of 50-60% for 30 min before use). While the rotary evaporation stage is proceeding, a chromatographic tank is allowed to come to equilibrium with the first developing solvent [light petroleum $(60°-80°C)$, ethyl acetate, and diethylamine (55:32:13). The plate is developed until the center of the chlorophyll a spot reaches a 10-cm mark. The plate is removed from the tank and the solvent is allowed to evaporate for 2 min. The plate is scanned with a Chromoscan in the reflectance mode, using light through an Ilford 601 filter (maximum transmission at 430 nm). The integrator reading for each peak is noted and the R_p values are relative to chlorophyll a. Further identification may be made by removing the spots and extracting with methanol. This extract is evaporated in a rotary unit, the pigment is dissolved in 1-2 ml of hexane, and the absorption spectrum is obtained. Chlorophyll c remains at the origin of this thin-layer system but can be developed by a two-dimensional technique with the second developing solvent [light petroleum $(60°-80°C)$, ethyl acetate, and formdimethylamide (1:1:2)]. R_p values for the various pigments are seen in Table 1. These values were calculated by measuring the distance between the peak of a given pigment and that of the origin and dividing this by the distance moved by the chlorophyll a peak. These ratios are somewhat specific for a given pigment.

E. Phenols

p-Nitrophenol, a hydrolysis product of parathion, was determined in human urine by a thin-layer chromatographic technique developed by Moye and Winefordner [14]. This technique can be applied, presumably, to polluted waters that may contain p-nitrophenol and/or parathion. Ninety milliliters of urine is refluxed with 10 ml of concentrated HCl for 1 hr and then stored in a refrigerator. At least 6 ml of this urine solution are put into a 10-ml centrifuge tube and spun down at high speed (6000 rpm). Exactly 5 ml (the minimum volume) is carefully drawn off and is pipetted into a snap-on

TABLE 1

Averages and Standard Deviations for R_p Values
and Calibration Values for Pigments[a]

Pigment	R_p	ng of pigment/unit integrator reading[b]
Chlorophyll a	1	47.4 ± 0.4
Chlorophyll b	0.82 ± 0.02	14.5 ± 0.7
Chlorophyll c	0.00	22.4 ± 0.5
Phacophytin a	1.07 ± 0.04	42.8 ± 0.4
β-Carotene	1.17 ± 0.04	11.8 ± 0.4
Fucoxanthin	0.64 ± 0.03	29.1 ± 1.0
Lutein	0.55 ± 0.02	33.6 ± 0.9
Violaxanthin	0.41 ± 0.01	23.8 ± 1.2
Diadinoxanthin + dinoxanthin	0.44 ± 0.02	19.2 ± 0.9
Neoxanthin	0.21 ± 0.02	26.7 ± 0.7

[a]Reproduced from ref. 13, p. 185, by courtesy of Elsevier Publishing Company, Amsterdam, The Netherlands.

[b]On the most sensitive setting of the instrument.

polyethylene capped vial to which 6 ml of ether is added. After the sample has been vigorously shaken for 5 min, the aqueous phase is removed with a hypodermic syringe. After the ether solution has been evaporated to about 2 ml in vacuo, an appropriate aliquot is applied to the activated thin-layer plate. Merck Silica Gel G was heated at $700°C$ for at least 12 hr to remove a phosphorescent background. To a graduated Erlenmeyer flask with an aluminum-wrapped cork, enough 0.1M oxalic acid is added to 150 ml of the cleaned silica gel to make a 150-ml volume slurry. This is immediately poured onto plates (2 mm thickness) and room dried. The plates are activated at $115°C$ for at least 1 hr. After the ether extracts have been spotted, the thin layers are developed in ether by the ascending technique. Development is continued until the ether has moved exactly 5.5 cm above the origin. All thin-layer material is removed except an area that is 3.4-4.2 cm above the origin. This should be directly above a strong blue fluorescent band, which shows up clearly on the thin-layer plate under an ultraviolet light.

The retained portion of silica gel is scraped into a capped vial and prepared for subsequent determination by a phosphorimetric measurement. For 0.071-7.1 μg of p-nitrophenol per 5 ml of urine, recoveries ranged from 80 to 97% and averaged 88%.

Zigler and Phillips [15] presented a rapid, unsophisticated thin-layer technique for the estimation of several chlorophenols in water. These compounds are 3-chlorophenol, 2,4-dichlorophenol, 2,4,5-trichlorophenol, 2,4,6-trichlorophenol, and pentachlorophenol. A 1-liter sample of water is acidified with phosphoric acid to a pH value of approximately 1.5. This acidified water sample is extracted with four separate 100-ml volumes of petroleum ether. These extracts are combined in a 600-ml beaker and are evaporated to about 5-10 ml on a steam bath with the aid of a gentle current of air. This concentrated extract is dried over sodium sulfate and then quantitatively transferred to a 15-ml graduated centrifuge tube for evaporation to 0.1 ml. This volume is spotted on a thin-layer plate (8 in. $\times$ 8 in.) covered by Aluminum Oxide G with gypsum binder (30 g of Al_2O_3 in 60 ml of distilled water, dried 15 min at room temperature, and at $75°C$ for 30 min) approximately $\frac{3}{4}$ in. from the lower right-hand corner. The thin-layer plate sandwich is assembled and mobile solvent I (benzene, ACS grade) is allowed to ascend 3-4 in. up the plate. The plate is removed and allowed to dry. It is washed repeatedly with benzene (three or four times) until the sample spot is sufficiently clean. Chlorophenol standards of 0.05, 0.1, and 0.5 μg are spotted ahead of the benzene front. The sandwich is reassembled and mobile solvent II (6 ml of 1N NaOH, 94 ml of acetone) is run perpendicular to the direction of benzene. It is allowed to ascend 0.5 in. from the top of the plate, the sandwich is disassembled, and the plate is dried and developed at room temperature for 30-40 min. The chromatogram is heavily sprayed with the silver nitrate reagent (to 0.5 g of analytical reagent grade $AgNO_3$ in 5 ml of distilled water, 100 ml of 2-phenoxyethanol are added; the solution is made up to 1 liter with acetone and three drops of 30% H_2O_2 are added) and dried in a forced draft oven at $80°C$ for 1-2 min. The plate is exposed to ultraviolet light for 15 min. Spots will appear light to dark brown. Confirmation of the presence of chlorophenols may be made on a duplicate sample by the 4-aminoantipyrine reaction. The water sample is extracted and the plate is spotted as described above. The 4-aminoantipyrine reagent (2% in acetone freshly prepared) is sprayed moderately heavily and the plate is dried with the aid of a warm air flow. Then the plate is lightly sprayed with 1N NaOH, dried with warm air, and sprayed moderately with potassium ferricyanide solution (8% in distilled water, stored in refrigerator). The chlorophenols, pentachlorophenol excepted, will appear as pink to red spots against a yellow background. The R_f values and sensitivities obtained for the five chlorophenols are seen in Table 2. Recoveries were evaluated from several fortified waters: raw river water, carbon filtered water, well water, surface waters before and after chlorine dioxide treatment, and distilled water. Below 1 μg/liter, the recoveries ranged from 75 to 95%, whereas above this level and up to 10.0 μg/liter they ranged from 80 to 100%.

TABLE 2

Values and Sensitivities for Five Chlorophenols[a]

Compound	R_f (solvent front)	Sensitivity (μg/liter)	
		AgNO$_3$	Antipyrine
3-Chlorophenol	0.94	0.1	0.5
2,4-Dichlorophenol	0.71	0.1	0.5
2,4,5-Trichlorophenol	0.62	0.1	0.5
2,4,6-Trichlorophenol	0.42	0.1	0.5
Pentachlorophenol	0.09	0.5	—

[a]Reproduced from ref. 15, p. 66, by courtesy of the American Chemical Society, Washington, D. C.

Gebott [16] used the Gelman-Camag coarse silica gel with 13% gypsum binder and 2% ultraviolet indicator for the thin-layer chromatographic separation of p-chlorophenol, 2,4-dichlorophenol, and 2,4,5-trichlorophenol in water samples. Gelman-Camag alumina DSF-5 with 5% gypsum binder and 2% ultraviolet indicator was also used. These adsorbents were spread to a thickness of 0.3 mm onto 20 x 20 cm plates, air dried for 24 hr, and activated at 150°F for 60-90 min. For one directional chromatography, samples and standards containing 1 μg of chlorophenol per millileter were dissolved in acetone, from which 2 μl were spotted on silica gel-coated plates, dried, and equilibrated in a chamber. The chromatograms were developed in benzene to a height of 15 cm, air dried, and viewed under UV at 254 nm. The same technique was used on the Al$_2$O$_3$ layer with a petroleum ether-benzene (50 : 50) solvent. Two-dimensional chromatography was performed on both the silica gel and alumina plates. After equilibration, the chromatograms were developed to a height of 8 cm in benzene for silica gel and petroleum ether-benzene (1 : 1) for alumina. For development in the second direction, chlorophenol standards were applied ahead of the benzene solvent front and in line with the origin of sample. A solvent of 6 ml of 1.0N NaOH in 94 ml of acetone was used. This direction was run for 15 cm. The fluorescent detection was semiquantitative down to 0.5 μg of the chlorophenol.

Aly [17] presented a rather comprehensive study in the separation and identification of several phenols in waters by thin-layer chromatography. These compounds were separated as their 4-aminoantipyrine dyes on neutral silica gel plates or as their p-nitrophenylazo dyes on basic silica gel plates. These dyes were prepared as follows:

1. 4-Aminoantipyrine Dyes

To a suitable volume of water sample (100 ml) containing about 100 μg of phenol, 5 ml of phosphate buffer were added, followed by 5 ml of 0.5N ammonium hydroxide solution. The pH of this solution should be about 8.0 ± 0.2. Two milliliters of 4-aminoantipyrine reagent were added, followed by 1 ml of potassium ferricyanide solution. The reaction mixture was shaken after each addition of the reagents. The red color produced was allowed to develop for 15 min. The dye was then extracted with two 5-ml portions of chloroform and dried over anhydrous magnesium sulfate, and the solvent was evaporated to about 0.5 ml. The antipyrine dyes were chromatographed on neutral silica gel plates. The neutral plates are prepared by slurrying 25 g of Silica Gel G with 65 ml of water and grinding for 60 sec in a mortar. The 20 x 20 cm plates are coated to a thickness of 250 μm, air dried for 15 min, and then activated at 110°C for 30 min.

2. p-Nitrophenylazo Dyes

To a suitable volume of water sample (100 ml) containing about 100 μg of phenol 5 ml of phosphate buffer are added, followed by 1 ml of p-nitro-benzenediazonium fluoroborate reagent. The solution is shaken, allowed to stand for 2 min, and then made alkaline by the addition of 4 ml of 1.0N sodium hydroxide solution. Turbidity may develop at this step due to the precipitation of calcium and magnesium salts, but this does not interfere with the reaction. In the presence of phenols, orange-purple p-nitrophenylazo dyes are formed. The solution is then acidified by the addition of dilute hydrochloric acid. The azo dyes are extracted with two 10-ml portions of ether and dried over anhydrous magnesium sulfate, and the solvent to evaporated to dryness. The azo dyes are then redissolved in 0.5 ml of chloroform. A blank is always run. The azo dyes are chromatographed on basic silica gel plates. The basic plates are prepared by slurrying the silica gel with 0.1N NaOH.

Chloroform solutions of the dyes (5-10 μl containing 1-2 μg) are applied at positions 1 cm apart along the starting line, which is 1.5 cm from the lower edge of the plate. Equilibrium conditions inside the chromatographic chamber are insured by lining the walls with strips of filter paper dipped in the solvent system. The plates are developed by the ascending technique at room temperature (20°-25°C). Usually 15-20 min were required for the solvent to reach a distance of 10 cm from the starting line. The resulting chromatograms are then air-dried.

The solvent systems for the 4-aminoantipyrine dyes were (1) methylene chloride-acetylacetone (70:30), (2) chloroform-ethyl acetate (80:20), and (3) ethyl acetate-hexane-acetic acid (70:20:10); and for the p-nitrophenylazo dyes; chloroform-acetone (90:10). Table 3 shows the R_f values of the

TABLE 3

$R_f \times 100$ Values for 4-Aminoantipyrine Derivatives of Some
Phenols on Neutral Silica Gel Plates [a]

Compound[b]	Solvent system		
	1	2	3
Phenol	40	15	35
2-Chlorophenol (OCP)	64	29	45
3-Chlorophenol (MCP)	52	32	52
4-Chlorophenol (PCP)	40	15	35
2,4-Dichlorophenol (2,4-DCP)	64	28	45
2,6-Dichlorophenol (2,6-DCP)	90	44	53
2,4,6-Trichlorophenol (2,4,6-TCP)	90	44	53
2-Nitrophenol	42	3	49
3-Nitrophenol	25	5	40
4-Nitrophenol	—	—	—
1-Naphthol	53	36	56
2-Naphthol	60	66	80
Resorcinol	—	—	—
Catechol	—	—	—
Hydroquinone	—	—	—
2-Cresol	50	18	43
3-Cresol	45	22	45
4-Cresol	—	—	—
2,5-Xylenol	53	30	54
2,6-Xylenol	47	21	44

[a]Reproduced from Ref. 17, p. 590, by courtesy of Pergamon
Press, Ltd., London, England.

[b]Quantities applied: 1-2 μg of phenol as derivative.

TABLE 4

$R_f \times 100$ Values of p-Nitrophenylazo Dyes of Some Phenols
on Basic Silica Gel Plates[a]

Compound	$R_f \times 100$[b]	Color of azo dye	
		Before exposure to ammonia	After exposure to ammonia
Phenol	45	Orange	Reddish orange
2-Chlorophenol (OCP)	21	Orange	Reddish orange
3-Chlorophenol (MCP)	30	Orange	Reddish orange
4-Chlorophenol (PCP)	82	Yellow	Purple
2,4-Dichlorophenol (2,4-DCP)	79	Pink	Pink
2,6-Dichlorophenol (2,6-DCP)	5	Orange	Red
2,4,6-Trichlorophenol (2,4,6-TCP)			
2-Nitrophenol	10	Orange	Orange
3-Nitrophenol	20	Orange	Orange
4-Nitrophenol	45	Orange	Orange
1-Naphthol	86	Pink	Blue
	55	Yellow	Blue
2-Naphthol	83	Orange	Reddish orange
Catechol	34	Dark green	Dark green
	4	Pink	Pink
Resorcinol	85	Yellow	Yellow
	8	Orange	Orange
Hydroquinone	30	Dark green	Dark green
2-Cresol	54[c]	Yellow	Mauve
3-Cresol	43[c]	Yellow	Magenta
4-Cresol	94[c]	Yellow	Purple
2,5-Xylenol	62[c]	Yellow	Lilac
2,6-Xylenol	74[c]	Yellow	Pale lilac

TABLE 4. — Continued

[a]Reproduced from Ref. 17, p. 591, by courtesy of Pergamon Press, Ltd.,
London, England.

[b]Solvent system, chloroform-acetone (9:1). Quantities applied: 1-2 μg of
the phenol as derivative.

[c]After Crump [18], using Silica Gel G (Merck) treated with 0.5N sodium
hydroxide.

4-aminoantipyrine derivatives of 20 phenols on neutral silica gel plates with
the three solvent systems. The best solvent for these derivatives appears to
be the methylene chloride-acetylacetone system. Table 4 shows the R_f values
of the p-nitrophenylazo dyes. A significant feature of this system is the resolution of the chlorophenols, which was not observed with the 4-aminoanti-
pyrine dyes. The change in color of the various azo dyes after exposure of
the chromatograms to ammonia vapor enables the identification of overlapping
spots. The detection limits of 4-aminoantipyrine dyes are 0.05-0.1 μg of
phenol. The detection limits for phenol and ortho-substituted phenols, and for
para-substituted phenols as their p-nitrophenylazo dyes after exposure to
ammonia vapor, are 0.1 μg and 0.5 μg respectively.

Smith and Lichtenberg [19] also presented a rather comprehensive study
in the thin-layer chromatographic separation of phenols extracted from surface waters either by the carbon adsorption method (CAM) or by solvent
extraction from 1-liter grab samples. The CAM samples are collected by
passing 1000-20,000 liters of water through an activated carbon column (3 in.
in diameter and 18 in. in length). Subsequently, the carbon is air dried in an
oven at 40°C for about 2 days. Then the carbon is extracted with $CHCl_3$ in a
large soxhlet extractor for 35 hr. Sequential evaporation yields 20 ml of the
carbon chloroform extract (CCE), which is transferred to a tared 50-ml
beaker, evaporated to dryness at room temperature, and weighed. The 1-
liter grab samples are acidified to a pH value of 2.0 with concentrated HCl
and extracted three times with 100-ml aliquots of ethyl ether or $CHCl_3$.
These aliquots are combined in a 300-ml beaker and carefully evaporated to
about 10 ml on a steam bath. This concentrated extract is transferred to a
15-ml graduated centrifuge tube and carefully evaporated to 0.1 ml in a
warm-water bath. Layers of Silica Gel G (30 g in 60 ml of water) 0.25 mm
thick are prepared on 20 × 20 cm glass plates, which are allowed to stand for
5 min and then dried in an oven at 100°C for 1 hr. The CCE is subjected to
the classical organic solubility class separation in order to obtain the weak-
acid group. This fraction is dissolved in $CHCl_3$, transferred to a 15-ml
graduated centrifuge tube, and evaporated to an appropriate volume (0.5-1.0
ml) for spotting. Two silica gel layers are spotted with one developed in
solvent system I (benzene-cyclohexane-diethylamine, 5:4:1) and the other in

solvent system II (chloroform). The developing tanks are lined with filter paper wicks and allowed to equilibrate. Both layers are developed to the 10-cm line, which occurs in about 18 and 40 min, respectively. The layer developed in system I is removed from the tank and is allowed to dry thoroughly at room temperature. Nitrophenols (natural yellow color) are determined at this time. Then the layer is sprayed with a light, even coat of p-NBDF (1% nitrobenzenediazonium fluoroborate in acetone) and over-sprayed with 1N KOH. The layer from solvent II is removed from the tank and is allowed to dry for about 10 min. It is developed a second time and again allowed to dry. This layer is sprayed with a light, even coat of Gibb's reagent (1% 2,6-dibromoquinone chlorimide in $CHCl_3$) and heated in an oven at 110° C for 10 min. Quantitative estimation is made by visually comparing the intensity and area of the sample spots to those of standards. The entire extract of the water grab sample is spotted whenever possible. Duplicate samples are spotted on separate layers, developed in systems I and II, and sprayed as described above. R_f values for the phenols developed in solvent system I are given in Table 5, whereas those for solvent system II are given in Table 6. Solvent system I is selective for the alkylphenols since the chloro-, nitro-, and aminophenols generally do not migrate or migrate only slightly in this system. Chloroform, however, serves as a more general developing solvent as the alkyl-, chloro-, and nitrophenols migrate to various degrees and are resolved. The color reactions from the two spray reagents are given in Table 7 and sensitivities are seen in Table 8.

Wallwork et al. [20] utilized thin-layer chromatography for the identification of phenols extracted from river waters. These waters are acidified with $0.1N\ H_2SO_4$ and extracted with 4 × 25 ml portions of $CHCl_3$. The combined $CHCl_3$ extracts are extracted with 10% NaOH (2 × 10 ml), which portions are combined, acidified, and extracted with 2 × 10 ml portions of ether. These ethereal solutions are evaporated in vacuo to dryness, leaving a residue of a pale yellow oil. The thin-layer plates, 200 × 200 mm, are precoated with silica gel (Merck) to a thickness of 0.25 mm. These plates are saturated in 0.5N NaOH and dried in air for 15 min and at 105° C for 1 hr. They are stored in a desiccator over silica gel and activated at 105° C for 1 hr immediately before use. Appropriate aliquots of acetone solutions of the oily residue (~10 μg) and of the phenol and o-, m-, and p-cresol (2 μg) standards are spotted. Methylethylketone is the developing solvent. These spots are air dried and then overspotted with 2 μl of an acid diazotizing solution [solution (1) is 1.9 g of p-nitroaniline dissolved in 45 ml of concentrated HCl and diluted to 1 liter; solution (2) is 5% $NaNO_2$, stored at 0° C; 1 ml of (2) is mixed with 50 ml of (1) immediately before use]. These diazotized spots are air dried before development (170 mm for 65-70 min at 20°-22° C. R_f values (phenol, 0.18; o-cresol, 0.31; m-cresol, 0.26; and p-cresol, 0.83) were measured immediately after removal of the plate from the developing tank since the color and intensity of the spots diminished as the solvent evaporated. No sensitivity data were given.

TABLE 5

R_f Values for Phenols Developed in Benzene-
Cyclohexane-Diethylamine $(5:4:1)$ [a]

Compound	R_f	Compound	R_f
Phenol	0.23	o-Chlorophenol	0.05
o-Cresol	0.32	m-Chlorophenol	0.07
m-Cresol	0.21	p-Chlorophenol	0.09
p-Cresol	0.22	2,3-Dichlorophenol	0.00
2,3-Dimethylphenol	0.48	2,4-Dichlorophenol	0.05
2,4-Dimethylphenol	0.38	2,5-Dichlorophenol	0.00
2,5-Dimethylphenol	0.42	2,6-Dichlorophenol	0.00
2,6-Dimethylphenol	0.40	3,4-Dichlorophenol	0.00
3,4-Dimethylphenol	0.21	2,4,5-Trichlorophenol	0.00
3,5-Dimethylphenol	0.22	2,4,6-Trichlorophenol	0.00
2,4,5-Trimethylphenol	0.36		
2,3,5-Trimethylphenol	0.42	4-Chloro-2-nitrophenol	0.00
2,3,5,6-Tetramethylphenol	0.53	o-Nitrophenol	0.00
		m-Nitrophenol	0.00
Catechol	0.02	p-Nitrophenol	0.00
Resorcinol	0.06	o-Aminophenol	0.05

[a]Reproduced from Ref. 19, p. 82, by courtesy of the American Society
for Testing and Materials, Philadelphia, Pennsylvania.

Aly [21] extended his previous work [17] to the separation of various
substituted phenols on neutral and basic silica gel plates. The procedure,
similar to the one cited above, involves conversion of the phenols in the
aqueous phase into their antipyryl or p-nitrophenylazo derivatives. These
dyes are extracted then with the appropriate solvent ($CHCl_3$ for the antipyryl
dyes and ether for the p-nitrophenylazo dyes), concentrated, and then applied
to the thin-layer plates. The chromatographic behavior of 53 substituted
phenols was investigated. Separation of the antipyrine dyes of the o-, m-, and
p-isomers of halogenated phenols was accomplished on neutral plates. The

TABLE 6

R_f Values for Phenols Developed in Chloroform[a]

Compound	R_f	Compound	R_f
Phenol	0.19	o-Chlorophenol	0.56
o-Cresol	0.33	m-Chlorophenol	0.20
m-Cresol	0.22	p-Chlorophenol	0.20
p-Cresol	0.21	2,3-Dichlorophenol	0.55
2,3-Dimethylphenol	0.35	2,4-Dichlorophenol	0.55
2,4-Dimethylphenol	0.34	2,5-Dichlorophenol	0.58
2,5-Dimethylphenol	0.35	2,6-Dichlorophenol	0.69
2,6-Dimethylphenol	0.57	3,4-Dichlorophenol	0.21
3,4-Dimethylphenol	0.23	2,4,5-Trichlorophenol	0.50
3,5-Dimethylphenol	0.23	2,4,6-Trichlorophenol	0.67
2,4,5-Trimethylphenol	0.34		
2,3,5-Trimethylphenol	0.37	4-Chloro-2-nitrophenol	0.73
2,3,5,6-Tetramethylphenol	0.63	o-Nitrophenol	0.80
		m-Nitrophenol	0.08
Catechol	0.03	p-Nitrophenol	0.07
Resorcinol	0.01		
		o-Aminophenol	0.02

[a]Reproduced from Ref. 19, p. 84, by courtesy of the American Society for Testing and Materials, Philadelphia, Pennsylvania.

R_f values (Table 9) were wide enough apart to permit resolution. A mixture of chloroform-acetonitrile (8 : 2) was the developing solvent. Separation of halogenated cresols or xylenols as their antipyryl dyes was achieved also (Table 10). Since some para-substituted phenols do not react with 4-amino-antipyrine, the p-nitrophenylazo derivatives were prepared and their chromatographic behavior on basic silica gel plates was investigated. These R_f values are seen in Table 11 for the solvent system chloroform-acetone (9:1). The combination of these two procedures can be used for the identification and separation of a wide variety of phenols in water.

TABLE 7

Color Reactions Produced by Phenols[a]

Compound	Reagent[b] Gibbs'	Reagent[b] p-NBDF	Compound	Reagent[b] Gibbs'	Reagent[b] p-NBDF
Phenol	Pur	Or-R	o-Chlorophenol	Bl-Gr	R-Or
o-Cresol	Pur	R-Pur	m-Chlorophenol	Bl	Or-P
m-Cresol	Pur	R	p-Chlorophenol	Bl-Pur	Or
p-Cresol	Or	Or	2,3-Dichlorophenol	Bl	R
2,3-Dimethylphenol	Pur	R-Pur	2,4-Dichlorophenol	Bl	Or-R
2,4-Dimethylphenol	Or	Or	2,5-Dichlorophenol	Bl	R
2,5-Dimethylphenol	R-Pur	R	2,6-Dichlorophenol	Bl	R
2,6-Dimethylphenol	Pur	R-Pur	3,4-Dichlorophenol	Bl	Y
3,4-Dimethylphenol	Or	Or-R	2,4,5-Trichlorophenol	Bl	Or
3,5-Dimethylphenol	Pur	Or-R	2,4,6-Trichlorophenol	Bl	—
2,4,5-Trimethylphenol	Y-Or	Or	4-Chloro-2-nitrophenol	Y-Or	Y
2,3,5-Trimethylphenol	Pur	R-Pur	o-Nitrophenol	Y	Y
2,3,5,6-Tetramethylphenol	R-Pur	Pur	m-Nitrophenol	Y	Y
			p-Nitrophenol	Y	Y
Catechol	Gy	Br	o-Aminophenol	Y-Br	Pur
Resorcinol	Br	Br			

[a]Reproduced from Ref. 19, p. 88, by courtesy of the American Society for Testing and Materials, Philadelphia, Pennsylvania.

[b]Pur, purple; Or, orange; R, red; Y, yellow; Gy, gray; Br, brown; Bl, blue; Gr, green

Limits of Detection on the Thin Layer[a]

Compound	μg Detected		Compound	μg Detected	
	Gibbs'	p-NBDF		Gibbs'	p-NBDF
Phenol	0.1	0.5	o-Chlorophenol	0.5	1.0
o-Cresol	0.1	0.5	m-Chlorophenol	0.1	5.0
m-Cresol	0.1	0.5	p-Chlorophenol	0.5	25.0
p-Cresol	2.0	10.0	2,3-Dichlorophenol	0.2	2.0
2,3-Dimethylphenol	0.1	0.5	2,4-Dichlorophenol	2.0	20.0
2,4-Dimethylphenol	0.5	1.0	2,5-Dichlorophenol	1.0	20.0
2,5-Dimethylphenol	0.1	0.5	2,6-Dichlorophenol	0.5	2.0
2,6-Dimethylphenol	0.1	0.2	3,4-Dichlorophenol	1.0	5.0
3,4-Dimethylphenol	0.5	1.0	2,4,5-Trichlorophenol	2.0	10.0
3,5-Dimethylphenol	0.2	0.5	2,4,6-Trichlorophenol	10.0	—
2,4,5-Trimethylphenol	1.0	20.0	4-Chloro-2-nitrophenol[b]	1.0	1.0
2,3,5-Trimethylphenol	0.2	1.0	o-Nitrophenol	1.0	1.0
2,3,5,6-Tetramethylphenol	0.2	0.5	m-Nitrophenol	10.0	10.0
			p-Nitrophenol	2.0	2.0
Catechol	0.5	1.0	o-Aminophenol	0.5	1.0
Resorcinol	0.5	1.0			

[a] Reproduced from Ref. 19, p. 89, by courtesy of the American Society for Testing and Materials, Philadelphia, Pennsylvania.

[b] Nitrophenols are determined prior to spraying.

TABLE 9

$R_f \times 100$ Values for 4-Aminoantipyrine Dyes of Halogen
Derivatives of Phenol on Neutral Silica Gel Plates[a]

Position of substituent	Substituent		
	Cl	Br	I
2-	42	48	54
3-	37	45	47
4-	20	20	20
2,3-Di-	52	—	—
2,4-Di-	42	48	—
2,5-Di-	58	—	—
2,6-Di-	61	64	—
3,4-Di-	37	—	—
3,5-Di-	25	—	—
2,4,5-Tri-	58	—	—
2,4,6-Tri-	61	64	—
2,3,4,6-Tetra-	66	—	—
2,3,4,5,6-Penta-	70	73	—

[a]Solvent systems: chloroform-acetonitrile (8:2).

Table 12 summarizes the conditions used in the thin-layer chromatographic
separation of phenolic compounds extracted from water. Relatively few procedures have been reported for these common pollutants, especially in
surface waters. Silica gel appears to be the adsorbent of choice. There is a
choice of developing solvent. Two techniques are available for the detection
of the separated phenols: (a) application of a spray reagent to the thin-layer
plate after resolution of the compounds and (b) preparation of the phenol-dye
complex in the aqueous phase, extraction, and subsequent resolution on the
thin-layer plates. No special advantage is realized by either technique, although the latter may permit the detection and separation of a greater number
of phenols. Sensitivities appear to be in the order of 10^{-7} to 10^{-8} g. All of
the techniques are suggested as confirmatory procedures in conjunction with
some other analytical procedure.

TABLE 10

R_f Values for 4-Aminoantipyrine Derivatives of
Some Halogenated Cresols and Xylenols on
Neutral Silica Gel Plates

	A[a]	B[b]
o-Cresol	30	37
4-Chloro-o-cresol	30	37
6-Chloro-o-cresol	55	67
4,6-Dichloro-o-cresol	55	67
m-Cresol	25	19
4-Chloro-	25	19
6-Chloro-	53	53
4,6-Dichloro-	53	53
3,5-Xylenol	20	17
4-Chloro-	20	17
2,4-Dichloro-	43	38
2,4,6-Trichloro-	60	65
2,3-Xylenol	35	37
4-Chloro-	35	37
2,5-Xylenol	36	35
4-Chloro-	36	35
2,6-Xylenol	28	28
4-Chloro-	28	28

[a]Chloroform-acetonitrile (8:2).

[b]Methylene chloride-acetylacetone-pentane
(7:3:2).

TABLE 11

$R_f \times 100$ Values of p-Nitrophenylazo Dyes of Some
Halogenated Phenols on Basic Silica Gel Plates[a]

Position of $R_f \times$ 100 substituent		Substituent		
		Cl	Br	I
Phenol	55			
2-		35	44	35
3-		42	48	40
4-		80	87(79)	75
2,3-Di-		13	—	—
2,4-Di-		77(53)	83(60)	—
2,5-Di-		16	—	—
2,6-Di-		9	13	—
3,4-Di-		83(83)	—	—
3,5-Di-		39	—	—
2,4,5-Tri-		81(74)	—	
2,4,6-Tri-		NR	NR	
o-Cresol	28			
4-Cl-		90		
6-Cl-		16		
4,6-Di-		NR		

[a]Solvent system: chloroform-acetone (9:1). NR, no reaction. Parentheses indicate value from another trial.

F. Pesticides and Herbicides

One of the earliest applications of thin-layer chromatography to aqueous extracts was made by Abbott et al. [22]. Six organic acid herbicides were separated: 2,4-dichlorophenoxy acetic acid (2,4-D), 2,4,5-trichlorophenoxy-acetic acid (2,4,5-T), 4-chloro-2-methylphenoxyacetic acid (MCPA), 4-(4-chloro-2-methylphenoxy)butyric acid (MCPB), 4-(2,4-dichlorophenoxy)-butyric acid (2,4-DB), and 2,2-dichloropropionic acid (dalapon). Initially, a

TABLE 12

Summary of Conditions Required for Thin-Layer Chromatographic Separation of Phenolic Compounds

Species	Adsorbents	Developing solvents	Detectors	Sensitivities	References
p-Nitrophenol	Silica Gel G	Ether	UV	—	14
Chlorophenols	Alumina G and gypsum	Benzene (I); NaOH-acetone (II)	AgNO$_3$, UV; 4-AAP	0.1, 0.5 μg/liter	15
Chlorophenols	Silica gel and gypsum	Benzene	UV	0.5 μg	16
Chlorophenols	Alumina and gypsum	Petroleum ether–benzene; NaOH-acetone	UV	0.5 μg	
Various phenols	Silica Gel G basic and neutral	a. Methylene chloride, acetylacetone; b. Chloroform, ethyl acetate; c. Ethyl acetate, hexane, acetic acid	4-AAP[a]	0.05 μg	17, 21
Various phenols	Silica Gel G	a′. Chloroform, acetone	p-NPA[b]	0.1 μg	21
Various phenols	Silica Gel G	Benzene, cyclohexane, diethylamine (I); Chloroform (II)	p-NBDF[c], KOH; Gibb's reagent	0.2 μg; 0.1 μg	19
Phenol; o-, m-, p-cresol	Silica Gel G	Methylethylketone	p-NA,[d] NaNO$_2$	—	20

[a] 4-AAP, 4-Aminoantipyrine dye.
[b] p-NPA, p-Nitrophenyl azo dye.
[c] p-NBDF, p-Nitrobenzenediazoniumfluoroborate.
[d] p-NA, p-Nitroaniline.

100-ml sample was acidified with 5 ml of 6N sulfuric acid that was extracted
subsequently with two 100-ml portions of ether. These ethereal extracts were
combined and were extracted with two portions, each, that consisted of a
mixture of 150 ml of a sodium sulfate solution (8%, w/v) with 5 ml of a 2.5N
sodium hydroxide solution. These alkaline aqueous extracts were acidified
with 10 ml of 6N sulfuric acid and then were extracted with successive portions of 150 ml and 100 ml of ether. These ethereal extracts were filtered
through a column containing 50 g of anhydrous Na_2SO_4 into a Kuderna-Danish
evaporator fitted with a 5-ml flask. This volume was reduced subsequently on
a warm-water bath. The residual herbicides were dissolved in 40 μl of ethyl
acetate immediately before chromatographic spotting. The thin-layer plates
were a mixture of 30 g of Kieselguhr G-Silica Gel G (60 + 40) and 60 ml of
water. After the usual plate preparation, 1-μl portions of the ethyl acetate
solutions of the unknown sample and of the standard solutions were spotted
along one edge. A 200-ml volume of the mobile solvent (10 ml of liquid paraffin, 30 ml of benzene, and 20 ml of glacial acetic acid to 200 ml of cyclohexane) was placed in the tray of the chromatographic tank. After the
appropriate equilibration time, the plate was removed from the tank and dried
in an oven at 120°C for 10 min. Then the plates were sprayed with a 0.5%
alcoholic $AgNO_3$ solution (1.7 g of $AgNO_3$ in 5 ml of water, 20 ml of 2-
phenoxyethanol diluted to 200 ml with acetone; then one drop of 30% H_2O_2 is
added), and again dried at 120°C for 10 min. The plates were irradiated with
a germicidal UV lamp for 10 min whereupon the herbicides were observed as
black spots on an off-white background (Fig. 1). Some research subsequent
to the above extended the technique to include 2-(1-methyl-n-propyl)4,6-
dinitrophenol (dinoseb) and 2-methyl-4,6-dinitrophenol (DNOC). Unfortunately,
no sensitivity data were given but some recovery values were. For example, a
95% recovery of 2,4-DB was recorded from 0.1 mg per 100 gram solution.

Smith and Eichelberger [2] were also early applicators of thin-layer
chromatography to the separation of organic pesticides. Their technique was
unique in the sense that TLC was used as a cleanup procedure prior to gas
chromatographic analysis. The pesticides of concern were dieldrin, lindane,
heptachlorepoxide, aldrin, heptachlor, and DDT. These pesticides were
separated from water by the carbon adsorption method (CAM) and subsequently eluted from the carbon by extraction with $CHCl_3$; this was labeled the
"CCE fraction." This latter fraction contained, in addition to the chlorinated
hydrocarbon, organic interferences of unknown identity. The thin-layer plates
were composed of Silica Gel G (30 g in 60 ml of water). The entire aromatic
fraction of the CCE (generally 5-15 mg) is dissolved, first, in acetone in a
15-ml centrifuge tube and made up to 2 ml. A 100-μl aliquot of this acetone
solution is spotted and the plates are placed in the appropriate chamber with
CCl_4 as the solvent. After separation and equilibration, the plates are
sprayed rather heavily with Rhodamine B dye (0.10 mg/ml methanol) and then
allowed to dry for 5 min. In natural light, the chlorinated hydrocarbons are
seen as purple spots on a pink background, whereas under a UV lamp they

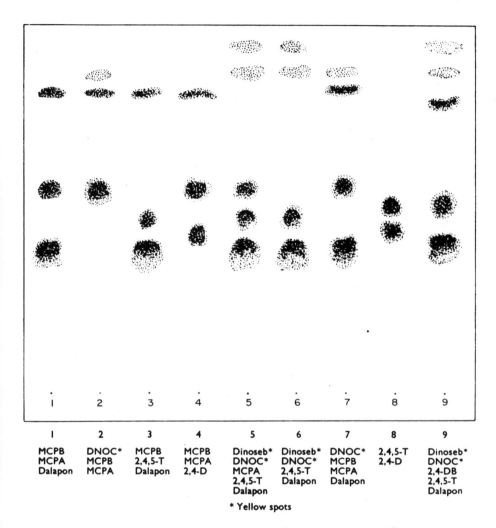

Figure 1. Appearance of chromatogram of 60% Kieselguhr G–40% Silica Gel G developed in mobile solvent B (see text). Reprinted from Ref. 22, p. 480, with permission of the Society for Analytical Chemistry, London, England

appear as quenched areas on the fluorescent background. The spots are marked, sectioned (as seen in Fig. 2), scraped, and eluted with ether–petroleum ether (1 : 1). This elution solvent selectively removes the pesticide from the spot whereupon it is ready for gas chromatographic analysis. A sensitivity of 10 μg of the pesticide was reported. Recoveries of the six chlorinated hydrocarbons, after separation by TLC, averaged 85–97.7%.

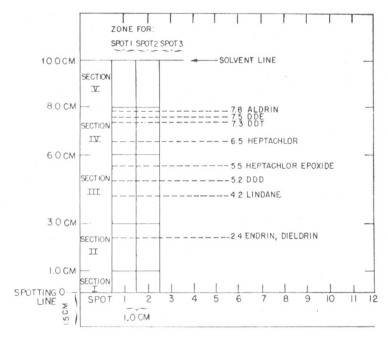

Figure 2. Diagram of designation of sections in the cleanup and separation of CCE-aromatics on silica layers. Reprinted from Ref. 2, p. 77, with permission of the Water Pollution Federation, Washington, D.C.

An important aspect of this work was the evaluation of several solvents used in their developing system. The results of repetitive testing of a number of multicomponent systems are seen in Table 13 and single-component systems in Table 14.

Abbott et al. [23] described several TLC systems for the cleanup of eight triazine herbicides extracted from water. These herbicides were atraton, atrazine, desmetryne, prometron, prometryne, propazine, simazine, and simetryne. A 200-ml sample is adjusted to a pH value of 9.0 with an ammonia solution (sp. g., 0.88) and is extracted successively with two 25-ml portions of dichloromethane. This solvent is dried by passage down a short column of anhydrous sodium sulfate and subsequently is reduced to dryness in a Kuderna-Danish evaporator. The residue is dissolved in 50 μl of hexane, from which a suitable aliquot is applied to a 250 μm thick Silica Gel G plate. Development is with a 9:1 chloroform-acetone mixture for 35 min. After the chromato-plate is allowed to dry, it is sprayed with a 0.5% solution of Brilliant Green, in acetone, which is quickly exposed to an atmosphere of bromine vapor. After the plate is removed from the oxidative atmosphere, the spots are

TABLE 13

TLC Separation of Pesticides by Multicomponent Solvent Systems[a]

Solvent	Adsorbent	Results
Benzene-acetone, 9:1	Silica gel	Poor separation of chlorinated pesticides; good separation of organophosphate pesticides
Petroleum ether-ethyl ether, 9:1	Alumina	Poor separation of pesticides
Petroleum ether-ethyl ether, 19:1	Silica gel	Fair separation, erratic results; no separation of organophosphate pesticides
	Alumina	No improvement in separation of chlorinated pesticides; organophosphate pesticides separate; very little tailing
Chloroform-petroleum ether, 1:1	Silica gel	Small range of R_f values for chlorinated pesticides
Chloroform-petroleum ether, 1:2	Silica gel	Results no better than 1:1 mixture; more erratic
	Alumina	Poor separation of chlorinated pesticides but no tailing; organophosphate pesticides separate
Chloroform-petroleum ether 1:3	Silica gel	Separation is fair; increased tailing of chlordane, toxaphene
Chloroform-petroleum ether 1:4	Silica gel	Good separation; dieldrin and endrin separate; very bad tailing of chlordane and toxaphene; results quite variable from plate to plate
Carbon tetrachloride-isooctane, 1:2	Silica gel	Fair separation; all R_f values below 0.50
Carbon tetrachloride-isooctane, 1:3	Silica gel	Results essentially same as 1:2 mixture
Chloroform-benzene-petroleum ether, 1:1:1	Silica gel	No separation

[a]Reproduced from Ref. 2, p. 79, by courtesy of the Water Pollution Control Federation, Washington, D.C.

TABLE 14

Separation of Pesticides [a]
(Single-Component Solvent Systems)

Solvent system	Adsorbent	Results
Chloroform	Silica gel	No separation; R_f values, 0.64 → 0.80
Petroleum ether	Silica gel	No separation; R_f values, 0.0 → 0.21
Hexane	Silica gel	No separation; R_f values, 0.02 → 0.43
Benzene	Silica gel	
Carbon tetra-chloride	Silica gel	Good separation; R_f values, 0.10 → 0.75
	Alumina	Fair separation; dieldrin and endrin separate; no tailing of chlordane, toxaphene

[a]Reproduced from Ref. 2, p. 80, by courtesy of the Water Pollution Control Federation, Washington, D.C.

outlined and their areas are subsequently determined. A typical plate is seen in Fig. 3. This thin-layer procedure was quantitated supposedly through a standard curve of the square root of the area plotted against the logarithm of the weight of material applied to the plate. The method was applied to several fortified samples of London tap water. Recoveries, at concentrations of 20 μg per 200 ml, ranged from 74 to 100% for the eight triazine herbicides.

Katz [24] was able to qualitatively determine linuron, monuron, diuron, neburon, and fenuron (substituted urea herbicides) in surface waters by a thin-layer chromatographic technique. A 1-liter sample of water is extracted successively with five 100-ml portions of chloroform. These extracts are collected in a beaker, evaporated to near dryness, and transferred to a small test tube. The beaker is washed several times with 5-ml portions of $CHCl_3$ with the evaporation and transfer repeated. The remaining $CHCl_3$ is removed very carefully with a stream of air. Then 0.2 ml of acetone is added to the test tube and 10 μl of this acetone extract are spotted onto an Eastman Chromatogram Sheet (No. K301R2). Standard solutions, 10 μl, of the herbicides are also spotted. The chromatogram is developed by an ascending technique in a solvent system of 5 ml of methanol, 35 ml of methyl chloroform, and 60 ml of 2,2,4-trimethylpentane. The plate is removed from the chamber and the solvent is evaporated with the aid of a hair dryer. Ninhydrin spray reagent (0.5 g of ninhydrin dissolved in 95 ml of n-butanol and diluted to 100 ml volume with 10% aqueous acetic acid) is applied, whereupon the plate is

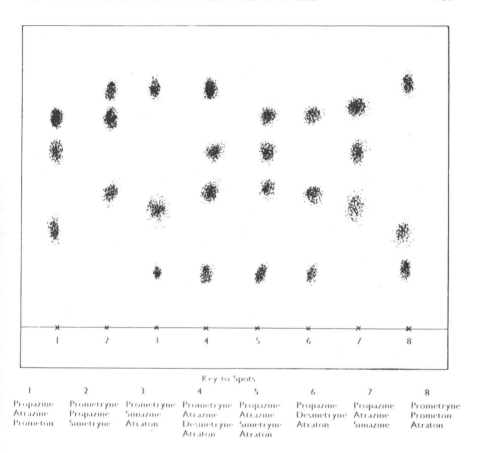

Key to Spots

1	2	3	4	5	6	7	8
Propazine	Prometryne	Prometryne	Prometryne	Propazine	Propazine	Propazine	Prometryne
Atrazine	Propazine	Simazine	Atrazine	Atrazine	Desmetryne	Atrazine	Prometon
Prometon	Simetryne	Atraton	Desmetryne	Simetryne	Atraton	Simazine	Atraton
			Atraton	Atraton			

Figure 3. Chromatogram showing the separation of the herbicides on a silica gel chromatoplate with carbon tetrachloride-nitromethane (1 + 1) mixture as the solvent system. Reprinted from Ref. 23, p. 365, with permission of the Society for Analytical Chemistry, London, England.

placed in an oven at $140°$ C for 10 min. The sample plate and standards plates are then compared. Levels as low as 0.2 μg of herbicide may be detected.

Miller et al. [25] utilized thin-layer chromatography to trace [^{14}C] diazinon and [^{35}S]parathion translocation in a model cranberry bog. These organophosphorus insecticides were extracted successively from 1000-ml volumes of water with 1:1 ether-petroleum ether or $CHCl_3$ (100 ml then 4×50 ml). The thin-layer plates were coated with Silica Gel G and were developed in 0.7% ethanol in $CHCl_3$. Also, the plates were exposed to X-ray film and autoradiographs were made. The spots were isolated and scraped from the plates; the diazinon and parathion were eluted with acetone. The concentration of each insecticide was determined from the radioactive counts per minute. In an effort to determine the presence of metabolites, the plates were sprayed with palladium chloride (0.5 g of $PdCl_2$ and 2 ml of concentrated HCl in 98 ml of distilled water) for parathion and p-aminoparathion. Counterspraying with 5N NaOH resolved paraoxon, p-aminophenol, and p-nitrophenol. No sensitivity values were given but a recovery of 84% for both compounds was reported.

Kawahara et al. [26] investigated thin-layer chromatography as a means to resolve parathion and methyl parathion in the presence of dieldrin, endrin, lindane, DDD, chlordane, heptachlor, DDT, DDE, and aldrin. This system was developed for Missouri River water when a chemical warehouse burned in Omaha, Nebraska. The fire runoff water was suspected of containing these pesticides. Three solvent systems were studied for the extraction from water namely, ether-hexane (1:1), benzene-hexane (1:1), and hexane. One-liter samples were extracted successively with 100, 50, 50, and 50 ml of the solvent mixture. The final extraction was completed with 50 ml of hexane to remove the ether from the water layer. These extracts were combined in a 300-ml Erlenmeyer flask and dried by pouring through a 5.1-cm column of anhydrous sodium sulfate. The extract was concentrated to an "appropriate volume" (0.1-1.0 ml) in the Kuderna-Danish apparatus for thin-layer spotting Aliquots of the extracts were spotted on a thin-layer plate (Silica Gel G, 250 μm thick) and developed with CCl_4 to the 10-cm line. The R_f values for 11 pesticides are given in Table 15. The TLC zones were scraped and eluted with 5-ml ether-petroleum ether (1:1), collected in a 15-ml graduated centrifuge tube, and concentrated to an "appropriate volume" (0.1-15 ml) for gas chromatographic analysis. Recoveries of the pesticides (10 μg) from the thin-layer plates were methyl parathion, 94.2 ± 10.4%; parathion, 97.2 ± 6.82%; and aldrin, 97.4 ± 15.3%.

Kawahara et al. [27] extended their aforementioned thin-layer technique to the microanalysis of 14 chlorohydrocarbons in waste water. These chlorohydrocarbons were 1-hydroxy chlordene, an endrin rearrangement product, endrin, heptachlorepoxide, γ-chlordane, nonachlor, heptachlor, isodrin, hexachlorobicycloheptadiene, chlordene, aldrin, heptachloronorbornene, and hexachlorocyclopentadiene. This technique was used to resolve coemerging gas chromatographic pairs of the chlorohydrocarbons. Solvent extraction (cited above) was used for separation from the aqueous phase. The chromatographic procedure is as follows: 30 g of Aluminum Oxide G powder are added to 45 ml of distilled water in a 300-ml Erlenmeyer flask. After the appropriate mixing, the slurry is spread to a thickness of 0.24 mm on 8 × 8 in. glass plates. Before use, the plates are dried in a clean air atmosphere over a period of 16 hr at room temperature. Prior to spotting, each plate is dried at 75°C for 10 min. The spots are developed with a mixture of 20% CCl_4 and 80% hexane and made visible by spraying with a Rhodamine B solution (0.10

TABLE 15

R_f Values of Pesticides Developed with CCl_4 on Silica Gel G Thin-Layer Plate[a]

Pesticide	R_f value	Zones
Methyl parathion	0.05	II
Parathion	0.07	II
Dieldrin	0.17	II
Endrin	0.20	II
Lindane	0.37	III
DDD	0.54	III
γ Chlordane	0.55	III
Heptachlor	0.67	IV
DDT	0.68	IV
DDE	0.72	IV
Aldrin	0.73	IV

[a]Reproduced from Ref. 26, p. 449, by courtesy of the Water Pollution Control Federation, Washington, D.C.

mg/ml in ethanol). One column of spots, however, is left unsprayed since these developed zones are used for gas chromatographic analyses. Each zone of the TLC plate is scraped and the collected adsorbent material is eluted with 4 ml of a mixture of 50% ether, 40% hexane, and 10% acetone for subsequent gas chromatographic analysis. R_f values for the chlorohydro-carbons are seen in Table 16 and recovery data are seen in Table 17.

Howe and Petty [28] describe a procedure for the quantitative measurement of Abate $(0,0,0',0'$-tetramethyl-$0,0'$-thiodi-p-phenylene phosphorothioate) in surface waters. First, the desired quantities of the standard Abate solution were diluted to 1500 ml with distilled water, surface water, or a 4:1 mixture of surface water and settled sewage. The samples were acidified with 5 ml of 6M sulfuric acid, filtered through glass wool, and shaken for 1 min with 500 ml of $CHCl_3$. The $CHCl_3$ layer was filtered through 50 g of anhydrous Na_2SO_4 and evaporated to dryness with a stream of dry nitrogen on a hot water bath at $60°$-$70°$C. These residues were dissolved in acetone and 15 μl were spotted on the thin-layer plates (a slurry of 5.0 g of Silica Gel G and 19.0 ml of water spread on 20 × 20 cm plates to a depth of 250 μm, activated for 1 hr at $90°$C, and then cooled to room temperature). Each plate contained six standards and duplicate aliquots of the sample.

TABLE 16

R_f Values of Chlorohydrocarbons Developed with 20% CCl_4
in Hexane on Aluminum Oxide G[a]

Compound	R_f value	Zone
1-Hydroxychlordene	0.000	I
Endrin rearrangement product	0.045	I
Endrin	0.200	II
Heptachlor epoxide	0.200	II
γ-chlordane	0.400	III
Nonachlor	0.490	III
Heptachlor	0.620	III
Isodrin	0.640	III
Hexachlorobicycloheptadiene	0.645	III
Chlordene	0.660	III
Aldrin	0.685	III
Heptachloronorbornene	0.690	III
Hexachlorocyclopentadiene	0.850	III

[a]Reproduced from Ref. 27, p. 25, by courtesy of Preston
Technical Abstracts Co., Evanston, Illinois.

The plates were developed for 1 hr in a tank presaturated with a 10 : 1
mixture of hexane and acetone, air dried for 1 min, exposed to Br_2 vapors
for 1 min, reaerated for 3 min, and sprayed with 1% N,N-dimethyl-p-
phenylazoaniline in 95% 2-propanol. This procedure yielded sharp red spots
with R_f values of 0.11 ± 0.01 on a bright yellow background. The quantitation
was accomplished by outlining the spots, transferring them to tracing paper,
and measuring them by superposition on millimeter graph paper. Percentage
recovery was determined by an average of the spot areas from duplicate aliquots and comparison to a standard curve. A typical spot area for 9 μg of
Abate in 15 μl of acetone was 51 mm^2. A typical recovery of 68% was
recorded from a 4 : 1 mixture of surface water and settled sewage spiked with
53 μl/liter of Abate. Any interfering substances in the sewage were apparently chromatographed away.

   Vylegzhanina and Kalmykova [29] describe a TLC method for the separation of DDVP (0,0-dimethyl-2,2-dichlorovinyl phosphate) from soil, plant,
and water extracts. The generalized procedure is: (a) $CHCl_3$ extraction from
the aqueous phase is followed by cleanup on a column of carbon; (b) the eluate
from the carbon column is spotted on silica gel grade KSK plates; (c) the

TABLE 17

Recovery Data of Chlorohydrocarbons[a,b]

Compound	Percent recovery
Hexachlorocyclopentadiene	10
Hexachlorobicycloheptadiene	45
Heptachloronorbornene	21
Chlordene	66
Heptachlor	62
Aldrin	62
Isodrin	61
1-Hydroxychlordene	91
Heptachlorepoxide	95
γ-Chlordane	103
Nonachlor	91
Dieldrin	73
Endrin	82
Endrin rearrangement product	98

[a]Overall process yield includes concentration from 100 ml to 0.5 ml, TLC spotting, TLC developing, and TLC eluting, concentration, and gas chromatographic procedures. These recovery values are averages obtained after three persons have utilized the thin-layer chromatographic procedure. The concentration levels of chlorohydrocarbons varied from 1.5 μg/liter.

[b]Reproduced from Ref. 27, p. 27, by courtesy of Preston Technical Abstracts Co., Evanston, Illinois.

developing solvent is a 2:1 mixture of n-hexane and acetone; (d) after development 2N NaOH is applied and the plates are dried at 110°C–120°C for 20 min, which is followed by a spray of 0.05N AgNO$_3$ in dilute nitric acid; and (e) exposure to ultraviolet light yields the DDVP spot. No recovery or sensitivity figures were given.

Askew et al. [3] developed a general method for the determination of organophosphorus pesticide residues in river waters and sewage effluents by gas, thin-layer, and gel chromatography. A 1-liter sample of the water or sewage effluent is extracted with three portions each of 50 ml of $CHCl_3$. These extracts are combined and dried by passage through a column of anhydrous sodium sulfate. The eluate is collected in a Kuderna-Danish evaporator and is evaporated to a small volume. Occasionally, a cleanup of the $CHCl_3$ extracts is necessary. This may be done on an activated carbon column or on alumina and magnesium oxide when some of the pesticides are retained by the carbon. Silica Gel G plates (250 μm thick) are prepared on 20 x 20 cm glass plates and are activated by heating at 120°C for at least 2 hr. The concentrated extracts are spotted and are developed in any one of three solvent systems: (1) hexane-acetone, 5:1; (2) $CHCl_3$-acetone, 9:1; and (3) $CHCl_3$-acetic acid, 9:1. The first of these systems is generally useful with the R_f values of the 40 pesticides spreading over the range 0 to 0.9, whereas solvent (2) is intended for the resolution of those pesticides with lower R_f values in solvent (1). The third solvent is applicable to the same group of pesticides that show a separation in solvent (2). The observed R_f values are given in Table 18. After the solvent has traveled about 10 cm from the origin, the plates are removed from the chromatographic tank, dried, and sprayed uniformly with the hydriodic acid solution (25 ml of hydriodic acid, sp. g. 1.7; 25 ml of glacial acetic acid; and 50 ml of water). A glass plate is clipped over the sprayed surface and it is heated in an oven at 180°C for 30 min with the plates standing in a vertical plane. When the plate is cool, they are sprayed with the ammonium molybdate solution [2 g of ammonium molybdate in 20 ml of water and concentrated HCl (1:1) with gentle heating, volume adjusted to 100 ml with water] and replaced in oven for 5 min. They are removed and, when cool, sprayed with the stannous chloride solution (1 g of $SnCl_2 \cdot 2H_2O$ in 10 ml concentrated HCl, heat, 40 ml of water, and 50 ml of acetone). The background is bleached by placing the plate in a tank containing an atmosphere of ammonia vapor. Organophosphorus pesticides appear as blue spots on a buff background. Compounds containing no phosphorus do not give this reaction but large amounts of some coextractives can appear as light brown charred areas. This procedure can be applied to waters containing 0.001 mg/liter of the pesticide.

El-Dib [30] presents a thin-layer technique for the detection, separation, and identification of 16 N-phenyl-N-methylcarbamates and related ureas in natural waters (Table 19). The aqueous samples (500 ml) are acidified with HCl and are extracted with $CHCl_3$ or dichloromethane. These extracts are spotted on silica gel (25 g in 50 ml of 0.5% H_2SO_4, 250 μm thickness). The plates are developed at room temperature (20°-30°C) in any one of six solvent systems, as seen in Table 20. The plates are air dried and then heated for 15 min at 150°C. In order to detect carbamates and phenylureas, the plates were sprayed with a p-dimethylaminobenzaldehyde (0.5% in methanol) reagent and were reheated at 150°C for 10 min. In order to detect phenyl

TABLE 18

Usage and Chromatographic Results of
Organophosphorus Pesticides [a]

| Compound | Thin-layer chromatographic [b] results | | |
	Solvent (1)	Solvent (2)	Solvent (3)
Azinphos-ethyl	0.33	0.90	—
Azinphos-methyl	0.19	0.88	—
Bromophos	0.85	0.93	—
Carbophenothion	0.83	0.96	—
Chlorfenvinphos	0.24	0.79	—
Coumaphos	0.33	0.90	—
Crufomate	0.06	0.43	0.86
Demeton-S	0.33	0.93	—
Demeton-S-methyl	0.17	0.73	—
Diazinon	0.61	0.95	—
Dibrom	0 to 0.22[c]	0 to 0.89[c]	—
Dichlofenthion	0.77	0.96	—
Dichlorvos	0.22 to 0.27	0.73	—
Dimefox	0.08	0.44	0.66
Dimethoate	0.05	0.37	0.59
Disulfoton	0.82	0.97	—
Ethion	0.77	0.97	—
Ethoate-methyl	0.07	0.61	0.75
Fenchlorphos	0.84	0.93	—
Fenitrothion	0.49	0.91	—
Formothion	0.15	0.75	—
Haloxon	0.04	0.71	0.86
Malathion	0.37	0.95	—
Mecarbam	0.42	0.95	—

TABLE 18. — Continued

Compound	Thin-layer chromatographic[b] results		
	Solvent (1)	Solvent (2)	Solvent (3)
Menazon	0	0.02	0.38
Mevinphos	0.10	0.64	0.69, 0.82
Morphothion	0.06	0.49	0.77
Oxydemeton-methyl	0	0.05	0.20
Parathion	0.57	0.91	—
Phenkapton	0.74	0.97	—
Phorate	0.80	0.97	—
Phosalone	0.39	0.97	—
Phosphamidon	0.04	0.34	0.60
Pyrimithate	0.62	0.96	—
Schradan	0	0.02	0.16
Sulfotep	0.75	0.92	—
Tepp	0	0 to 0.50[c]	0.03, 0.62
Thionazin	0.45	0.92	—
Trichlorphon	0.03	0.18	0.61
Vamidothion	0.01	0.16	0.30

[a] Reproduced from Ref. 3 by courtesy of the Society for Analytical Chemistry, London, England.

[b] Thin-layer chromatographic results: the values shown are the R_f values in the solvent (1) hexane-acetone, 5:1; (2) chloroform-acetone, 9:1; and (3) chloroform-acetic acid, 9:1.

[c] Denotes streaking.

TABLE 19

TLC Characteristics of Carbamates and Phenylureas[a]

Compound[b]	Limits of TLC detection (μg)	Color of derivative	
		Spray 1[c]	Spray 2[d]
Ethyl N-phenylcarbamate	4	Yellow	Blue-violet
Ethyl N-(4-nitrophenyl)carbamate	4	Yellow	Blue-violet
Ethyl N-(2-chloro-4-nitrophenyl)carbamate	4	Yellow	Blue
n-Propyl N-phenylcarbamate	5	Yellow	Blue-violet
Isopropyl N-phenylcarbamate (IPC)	5	Yellow	Blue-violet
Isopropyl-N-(3-chlorophenyl)carbamate (CIPC)	4	Yellow	Blue
3-(3,4-Dichlorophenyl)-1-methoxy-1-methylurea (linuron)	5	Yellow	Blue-violet
3-(Phenyl)-1,1-dimethylurea (fenuron)	5	Yellow	Blue
1-Naphthyl N-methyl carbamate (carbaryl)	4	Grey-yellow	None
2-Isopropoxyphenyl N-methyl-carbamate (baygon)	4	Grey	None
1-Phenyl-3-methyl-5-pyrazolyl-N,N-dimethyl carbamate (Pyrolan)	2	Rose red	None
2-Dimethylcarbamoyl-3-methyl-5-pyrazolyl-N,N-dimethylcarbamate (Dimetolan)	4	Red	None
Isopropyl-3-methyl-5-pyrazolyl-N,N-dimethyl carbamate (Isolan)	4	Orange	None
4-Dimethyl-m-tolylmethylcarbamate (Metacil)	5	Orange	None
4-(Methylthio)-3,5-xylylmethylcarbamate (Mesurol)	5	Orange	None
1-Naphthol	4	Grey	None

[a] Reproduced from Ref. 30 by courtesy of Association of Official Agricultural Chemists, Washington, D.C.

[b] The first four compounds are not pesticides; they are included for comparison.

[c] 0.5% p-Dimethylaminobenzaldehyde.

[d] 5% Sodium nitrite in 0.2N HCl and 5% naphthol in methanol.

553

TABLE 20

R_f Values of Carbamates and Phenylureas on Silica Gel Plates[a]

Compound[c]	$R_f \times 100$ in solvent systems[b]					
	1	2	3	4	5	6
1	30	60	75	35	55	65
2	15	50	63	20	45	52
3	43	72	70	62	63	65
4	40	65	70	50	60	72
5	35	65	75	45.	60	75
6	46	68	75	45	60	75
7	12	40	60	25	40	55
8	2	7	28	5	15	20
9	15	33	55	32	35	45
10	15	35	60	32	42	52
11	2	5	30	27	30	32
12	8	15	42	20	22	25
13	2	5	8	12	15	15
14	0	3	5	5	5	10
15	10	42	58	38	50	55
16	35	52	65	40	45	58

[a]Reproduced from Ref. 30, p. 760, by courtesy of Association of Official Agricultural Chemists, Washington, D. C.

[b]Solvent systems: 1, benzene; 2, benzene-acetone (95:5); 3, benzene-acetone (85:15); 4, cyclohexane-ethanol (85:15); 5, hexane-toluene-acetone (60:20:20); 6, hexane-acetone (70:30).

[c]Numbers refer to corresponding compounds in Table 19.

carbamates and related ureas, the plates were sprayed with a sodium nitrite (5% in 0.2N HCl) solution and 1-naphthol (5% in methanol) solution in succession. The sensitivities and derivative colors are seen in Table 19 and the R_f values are seen in Table 20 for the six solvent systems. It was reported that DDT, lindane, endrin, dieldrin, 2,4-D, kelthane, methyl parathion, and

azinphosmethyl did not interfere. An ultraviolet spectrophotometric determination may follow the separation by TLC. If this option is used, then the
spots are scraped selectively from the plates and extracted with 2 x 5 ml
portions of $CHCl_3$. Maximum absorbance is recorded at the appropriate
ultraviolet wavelength. A 98% recovery of Baygon from tap water was
reported when the TLC and UV techniques were used.

Koppe and Rautenberg [31] outlined a method of determining the less
volatile lipophilic chlorinated (insecticides) compounds in water. First, a
1-liter water sample is extracted with 20 ml of isooctane. After separation
of the solvent phases, the extract is poured through a dry black band filter
into a weighed flask. The extract is then evaporated to one-tenth of its
original weight. A series of comparison solutions is prepared by dissolving
hexachlorobenzene in isooctane so that 0.01 ml of the solutions contains 0.1,
0.2, 0.3, and 0.5 μg of organically bound chloride in the solutions. Exactly
0.01 ml of each of these solutions is slowly transferred to spots on the thin-
layer plate to give circles of 6-8 mm diameter. The solution is evaporated
rapidly. The same volume of concentrated extract is transferred to a series
of spots on the test plate and evaporated; then enough ethyl alcohol is added
to yield circles of the same diameter as in the calibration series. After
drying, a nitrogen stream is used to spray the plate with $AgNO_3$ solution (34
ml of distilled water, 10 ml of 0.2N $AgNO_3$, 2 ml of 10% nonylphenol-8-
ethoxalate, and 4 ml of concentrated ammonia solution). The plate is exposed
to ultraviolet light for 4 hr at a distance of 50 cm. The intensity of the gray
outer ring of the spot is compared with the standards. This method is applicable to 1-1000 μg Cl^- per liter. The thin-layer plates (0.15 mm thick) are
prepared from a mixture of 3.5 g of Kieselguhr G (size <10 μm) and 15 g of
Al_2O_3 (size <40 μm) in 40 g of water. The plates are dried for 30 min at
110° C.

Fenthion [0,0-dimethyl-0-4-(methylthio)-m-tolyl phosphorothioate] and
seven of its degradation products were separated successfully in a thin-layer
chromatographic technique by Suffet et al. [32]. This procedure was developed for application to aqueous systems where chemical hydrolysis and
microbial degradation may yield any one of the products. Either Eastman
chromatogram sheets, Silica Gel 100 μm, type K201R2, with fluorescent
indicator, or Mallinkrodt's Silica Gel Ar-TLC-FGF plates with fluorescent
indicator may be used. The pesticide standard solutions were made to 1.0
μg/μl in acetonitrile and were spotted on the sheets or plates by conventional
techniques. The glass plates were developed in a tank and the sheets in the
conventional thin-layer sandwich with a mixture of acetone and benzene, 1:9
(v/v). When the solvent reached the marked front, the plates or sheets were
removed and air dried for at least 10 min. The spots were marked on exposure by an ultraviolet lamp and subsequently sprayed with NBP reagent [2%
4-(p-nitrobenzyl)pyridine in acetone]. Table 21 presents the R_f values and
limits of detection of fenthion and the seven degradation compounds. Each of
the eight compounds is detectable by fluorescence. All of the phosphorus-
containing compounds show distinctive blue spots with the NBP spray reagent.

TABLE 21

R_f Values for Fenthion and Seven of its Degradation Products[a]
(Solvent System, Acetone–Benzene, 1:9, v/v)

No.	Compound[b]		R_f value	Sensitivity (μg)	
				NBP spray reagent	Fluor-escence
I	Venthion	P=S, S	0.94	0.3	0.5
II	Fenthion oxygen analog (oxon)	P=O, S	0.54	0.1	0.5
III	Fenthion sulfoxide	P=S, SO	0.35	0.1	0.5
IV	Fenthion oxygen analog sulfoxide	P=O, SO	0.06	0.1	0.5
V	Fenthion sulfone	P=S, SO₂	0.73	0.1	0.5
VI	Fenthion oxygen analog sulfone	P=O, SO₂	0.30	0.1	0.5
VII	Fenthion, S-methyl isomer	S-CH₃, P−S, S	0.65	0.1	0.5
VIII	Hydrolysis product MMTP	MMTP	0.67	—	0.5

[a]Reproduced from Ref. 32, p. 473, by courtesy of Pergamon Press, Ltd., London, England.

[b]General structure:

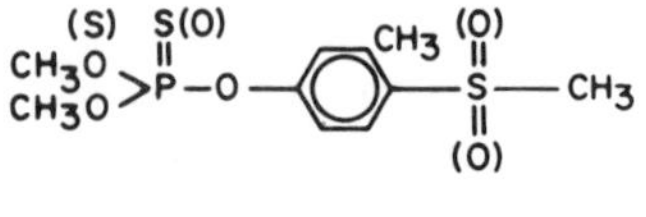

The blue color is not stable but may be regenerated with a respray of NaOH. Seven of the eight compounds are separated in the solvent system. Compounds VII and VIII have similar R_f values. If only MMTP (VIII) is present, fluorescence occurs at the R_f value and a blue spot does not appear due to overspraying with NBP reagent. If compound VII is present, however, fluorescence does occur and a blue spot appears with the NBP reagent. Differentiation may be made between compounds VII and VIII by the spray reagent (ferric sulfate-potassium ferricyanide in sulfuric acid). An intense blue color is produced immediately with MMTP (VIII), whereas the color appears slowly, more than 3 min, for the S-methyl isomer (VII).

Table 22 summarizes the various conditions required for the thin-layer chromatographic separation of organic pesticides. There is no uniformity or standardization of procedures for a given class of pesticides. Each procedure is unique for a given investigator. Silica gel appears as the major adsorbent with an occasional use of alumina and chromatogram sheets. The selection of developing solvents is related, more or less, to the polarity of the pesticide molecule. There is a tendency to select nonpolar solvents for such relatively nonpolar molecules as the chlorinated hydrocarbons. However, one cannot generalize on this point and the current state of the art is whatever "works best for me." There is an obvious reluctance by the investigators to cite sensitivities for the various procedures. From what has been published, the sensitivity order of magnitude is 10^{-5} to 10^{-7} g. Perhaps the reluctance to give sensitivity data may lie with the qualitative application of most of the procedures. Most of the investigators used R_f values as a means of confirming the pesticidal molecule. Herein lies the major use of thin-layer chromatography in pesticide residue in water analysis, namely, the confirmatory aspects.

G. Sterols

Murtaugh and Bunch [33] used thin-layer chromatography as a cleanup technique prior to gas-liquid chromatography for cholesterol and 5β-coprostan-3β-ol in domestic waste waters. These sterols are extracted first from a 2-liter sample of the water (to which 5 ml of concentrated HCl and 10 ml of 20% NaCl have been added) with 100 ml of hexane. This solvent layer is washed with 50 ml of 70% ethanol and evaporated to dryness. The sterol esters are refluxed for 3 hr with 7.5% KOH in 70% ethanol and diluted with an equal volume of water. The free sterols are extracted with two volumes of hexane, which is in turn washed with 5 ml of 50% ethanol and finally evaporated to dryness in a 5-ml test tube. The thin-layer plates (20 x 20 cm) have been coated with a 0.25-mm layer of methanol-washed Silica Gel G and activated for 1 hr at $110°C$. The hydrolyzed sterols are dissolved in 20-50 μl of acetone and then spotted on the plates along with standards. Development occurs with a mixture of $CHCl_3$ and ether (9:1). The sterols are located by spraying with a 10% solution of phosphomolybdic acid in 95% ethanol. After the plates were heated at $100°C$ for 5 min, the sterols appeared as dark spots on a yellow background. R_f values for coprostanol and cholesterol were 0.65 and 0.50, respectively. The areas for unknown samples corresponding to locations of reference standards are removed and are prepared for gas-liquid chromatography. Coprostanol can be detected in water at 20 ng/liter and quantitative measurements are possible above 100 ng/liter.

The detection and tentative identification of three sterols, cholesterol, stigmasterol, and β-sitosterol, has been achieved by Matthews and Smith [34]

TABLE 22

Summary of Conditions Required for Thin-Layer Chromatographic Separation of Organic Pesticides

Pesticides	Adsorbents	Developing solvents	Detectors	Sensitivities	Reference
Phenoxy acids	Kieselguhr G + Silica Gel G	Liquid paraffin, benzene, + glacial acetic acid	$AgNO_3$: 2-Phenoxy-ethanol	—	22
Chlorinated hydrocarbons	Silica Gel G	CCl_4	Rhodamine B, UV	10 μg	2
Triazines	Silica Gel G	$CHCl_3$-acetone	Brilliant Green: Br_2	—	23
Substituted ureas	Chromatogram sheets	CH_3OH-methylchloroform-2,2,4-trimethylpentane	Ninhydrin	20 μg/liter (linuron)	24
[^{14}C]-Diazinon [^{35}S]parathion	Silica Gel G	0.7% EtOH in acetone	Radioactive counts	—	25
OTP, CH[a]	Silica Gel G	CCl_4	Gas-liquid chromatography	—	26
CH	Alumina G	CCl_4-hexane	Rhodamine B	—	27
Abate	Silica Gel G	Hexane-acetone	Br_2, N,N-dimethyl-p-phenylazoaniline	—	28
DDVP	Silica Gel KSK	Hexane-acetone	NaOH, $AgNO_3$ in HNO_3, UV	—	29

OTP	Silica Gel G	Hexane-acetone $CHCl_3$-acetone $CHCl_3$-acetic acid	HI, $NH_4Mo_7O_{24}$· $4H_2O$, $SnCl_2$, NH_3 vapor	1 μg/liter	3
Carbamates	Silica Gel	Benzene Benzene-acetone Cyclohexane-ethanol Hexane-toluene-acetone Hexane-acetone	p-Dimethylamino benzaldehyde, $NaNO_2$, 1-naphthol	2-5 μg	30
CH	Kieselguhr G + Al_2O_3	Isooctane	$AgNO_3$, UV	—	31
Fenthion and metabolites	Chromatogram sheets or silica gel	Acetone-benzene	UV, NBP	0.1-0.5 μg	32

[a] CH = chlorinated hydrocarbons.

in marine waters with thin-layer and gas chromatographic procedures. Water
samples (17-45 liters) were taken in lipid-free glass bottles and in Negal
plastic bottles. Each water sample was filtered through a layer of Celite 545
diatomaceous earth, after which the pH value was adjusted to 2.5-3.5 with 6N
HCl. Serial extraction of ten volumes of sea water with one volume of hexane
was repeated until no sterol components could be detected by TLC. These
hexane extracts were dried by filtering them through a layer of anhydrous
sodium sulfate and concentrating them to near dryness in an all-glass rotary
evaporator. The total-extractables fraction was analyzed qualitatively by
one-dimensional TLC with ethyl acetate-heptane (1:1) to show sterol-like
components. Two-dimensional TLC analysis with ethyl acetate-heptane (1:1)
and acetone-heptane (1:1) was employed to determine the extent of autoxidation, where such was suspected, and to characterize more satisfactorily the
more mobile components of certain samples. Gas chromatography (1.0-μl
samples) was conducted on suitable dilutions of the initially analyzed preparation, which was evaporated under nitrogen and redissolved in methylene
chloride. Analytical TLC was conducted with 250-μm thicknesses on 20 × 20
cm chromatoplates of Silica Gel HF$_{254}$ irrigated with ethyl acetate-heptane
(1:1) or acetone-heptane (1:1). The resolved components were routinely
examined under UV light (254 and 366 nm) prior to visualization by spraying
with 50% aqueous H$_2$SO$_4$ and heating. The chromatoplates were heated on an
electric hotplate to full color display and then charred. Preparative TLC was
conducted on 20 × 20 and 20 × 40 cm chromatoplates with 1 and 2 mm thick
layers of Silica Gel HF$_{254}$. The sterol samples were applied as solutions in
CHCl$_3$-CH$_3$OH (7:1). Sterol zones were detected under UV light and by
spraying a 1-cm end portion of the chromatogram with 50% H$_2$SO$_4$ and inter-
polating the zones. The located zones were scraped from the chromatoplate
and individually extracted with methanol, which was evaporated under a
vacuum and the residue was dissolved in methylene chloride. This concentrated sterol preparation was considered to be pure enough for gas chromatographic analysis.

Smith and Gouron [35] applied their thin-layer and gas-liquid chromatographic techniques (described above) for the detection of the human fecal
sterol 5β-cholestan-3β-ol (coprosterol) in natural waters. The solvent
extracts were made by stirring vigorously each water sample (8 liters) with
one-half of its volume of redistilled hexane (4 liters) for 15 min. After phase
separation, a second and third extraction were performed. Each extract was
evaporated under vacuum and examined by thin-layer and gas chromatography
prior to combining as one preparation for further analysis. Lipid extract
residues from these solvent extractions were combined as CHCl$_3$-CH$_3$OH
(7:1) solutions. Further concentrations and evaporations of these solutions
were conducted under nitrogen. Subsequent analytical and preparative TLC
was performed in exactly the same manner as by Matthews and Smith [34],
cited above.

VII. DISCUSSION

Thin-layer chromatography is not without limitations and analytical problems. It alone cannot suffice for the positive and absolute identification of organic pollutants extracted from aqueous systems. It should be utilized in conjunction with some other form of analytical identification. Also, it is not possible to develop a single, universally applicable system for the isolation and separation, for example, of all classes of organic pesticides. In fact, it is even difficult to separate the parent molecule and metabolites for a given pesticide. The same reasoning may be applied to phenols, detergents, etc. Because of the widely different polarities of individual compounds, the use of a single solvent system for TLC for resolution is very unlikely. The analyst should run several trials before satisfactory conditions are established for his or her particular system. R_f values must be determined for each system under consideration.

Often overlooked in residue analyses are the subtle interferences arising from poor manipulation in the laboratory. Bevenue et al. [36] present an excellent study of the many problems created during the manipulation of water samples in the analytical laboratory. These investigators have been concerned mainly with organic pesticides, but their observations may be applied to other pollutants as well. The first problem is the sample size. If small grab sample volumes (1 gallon or less) are extracted, then the small quantities of pollutants thus obtained may eliminate the use of the spray reagent technique in TLC for verificative purposes. This is because of the detectability limits of the stain reagent, which are more applicable in the microgram range. Second, there is the technique of sectioning the TLC spot and eluting it with a solvent for subsequent gas chromatographic analysis. Extraneous interferences are magnified on the recorder chart unless special precautions are taken with the organic solvents, glassware, and other equipment and the thin-layer adsorbent is completely free from organic contaminants. For example, organic solvents of the "reagent grade" quality cannot be used for pesticide residue analysis in the nanogram to picogram range because of contaminants. They must be redistilled in an all-glass system before use.

Additional precautionary measures suggested by Bevenue et al. [36] are as follow: the glass jars used for developing the TLC plates may not tolerate the stress of heat treatment ($300°$ and $400°$C, as recommended by Lamar et al. [37] and the FWQA [38]). Instead, the jars should be treated with the sodium dichromate-sulfuric acid solution, followed by rinses with water, acetone, and hexane. In another area, Whatman filter paper sheets are used as liners on occasion in the chromatographic task to saturate the interior with vapors of the solvent. Bevenue et al. [36] suggest that "this practice cannot be tolerated in water analyses confirmatory work because the paper may contaminate the developing solvent with organic materials which will be transferred to the TLC silica gel plate and finally to the concentrated and eluted extract."

Four glassware cleaning procedures were examined by Bevenue et al. [36], as seen in Table 23. Method 1 did not completely remove contaminants from

TABLE 23

Rinsing Solutions in Glassware Cleaning Procedures [a]

Method 1	Method 2	Method 3	Method 4
Ethanol [b]	Dichromate–H_2SO_4 [c]	Acetone	Dichromate–H_2SO_4 [d]
Acetone	Tap water	—	Tap water
Hexane	Distilled water	—	Distilled water
—	Acetone	—	Acetone
Air dry	Air dry	Air dry	Air dry
—	—	Heat	Heat

[a] Reproduced from Ref. 36, p. 73, by courtesy of the Elsevier Publishing Co., Amsterdam, The Netherlands.

[b] Glassware rinsed three times with each solvent in order of the listed sequence.

[c] Glass was soaked for 16 hr a solution of sodium dichromate–concentrated sulfuric acid.

[d] Glass was heated in an air oven for 16 hr at 200° C.

the glassware. Methods 2, 3, and 4 removed all contaminants. However, these investigators preferred method 3 because of its simplicity. In cases where organic solvents would not remove firmly bonded organic contaminants from glass, the more drastic oxidizing solution was recommended as a prerequisite treatment.

Bevenue et al. [36] cited the experiences of several analysts with interferences arising from contaminated silica gel. For example, Miller and Kirchner [39] noted that silica acid adsorbents contained as much as 100 mg of a yellow oily material in 100 g of adsorbent that would interfere with UV and fluorescein tests. Bevenue et al. [36] examined five different commercially available silica gels, some of which were shipped in plastic bottles or in aluminum bottles with plastic caps. Gas chromatographic curves (electron capture detector) of extracts of the silica gels revealed peaks that would interfere with the determination of chlorinated hydrocarbon pesticides. Heat treatment of the silica gels at 300° C for 16 hr effectively removed the contaminants. This treatment did not affect the TLC properties of the gels.

In summary, thin-layer chromatography can be a powerful analytical tool for separating organic pollutants from one another and from gross quantities of interferences. Furthermore, these separations occur where quantities of the pollutants are in the order of 10^{-5} to 10^{-7} g. However, there is neither a universal procedure for all pollutants nor is there a standardization of procedures for a given pollutant. It is also preferable to use TLC in conjunction with some other analytical technique where confirmation is the primary consideration.

ACKNOWLEDGMENTS

The authors acknowledge the New Jersey Agricultural Experiment Station and the Department of Environmental Sciences at Rutgers, the State University of New Jersey, New Brunswick, New Jersey.

REFERENCES

1. M. F. Kovacs, Jr., J. Assoc. Off. Agr. Chem., 47, 1097 (1964).

2. D. Smith and J. Eichelberger, J. Water Poll. Contr. Fed., 37, 77 (1965).

3. J. Askew, J. H. Ruzicka, and B. B. Wheals, Analyst, 94, 275 (1969).

4. S. D. Faust and I. H. Suffet, in Water and Water Pollution Handbook (L. Ciaccio, ed.), Marcel Dekker, New York, 1971, p. 1249.

5. H. Braus, F. M. Middleton, and G. Walton, Anal. Chem., 23, 1160 (1951).

6. M. Beroza and M. C. Bowman, J. Assoc. Off. Agr. Chem., 48, 358, 943 (1965).

7. M. Beroza and M. C. Bowman, Anal. Chem., 37, 291 (1965).

8. I. H. Suffet and S. D. Faust, J. Agr. Food Chem., 20, 52 (1972).

9. Y. K. Chau and J. P. Riley, Deep-Sea Res., 13, 1115 (1966).

10. S. J. Patterson, E. C. Hunt, and K. B. E. Tucker, J. Proc. Inst. Sewage Purif., Part 2, 3 (1966).

11. G. Lambert, Tribune CEBEDEAU, 20 (271-2), 279 (1966) (in French); Chem. Abstr., 66, 21993d (1967).

12. J. Borneff and H. Kunte, Arch. Hyg. Bakteriol., 153, 220 (1969).

13. C. Garside and J. P. Riley, Anal. Chim. Acta, 46, 179 (1969).

14. H. A. Moye and J. D. Winefordner, J. Agr. Food Chem., 13, 533 (1965).

15. M. G. Zigler and W. F. Phillips, Env. Sci. Tech., 1, 65 (1967).

16. M. D. Gebott, Solutions, 6(1), 8 (1967).

17. O. M. Aly, Water Res., 2, 587 (1968).

18. G. B. Crump, J. Chromatog., 10, 21 (1963).

19. D. Smith and J. J. Lichtenberg, A.S.T.M. Spec. Tech. Publ. No. 448, 78 (1969).

20. J. F. Wallwork, M. Bentley, and D. C. Symonds, Water Treat. Exam., 18, 203 (1969).

21. O. M. Aly, unpublished results, presented at the 158th National Meeting, American Chemical Society, New York, N.Y., September 1969.

22. D. C. Abbott, H. Egan, E. W. Hammond, and J. Thomson, Analyst, 89, 480 (1964).

23. D. C. Abbott, Mrs J. A. Bunting, and J. Thomson, Analyst, 90, 365 (1965).

24. S. E. Katz, J. Assoc. Off. Agr. Chem., 49, 452 (1966).

25. C. W. Miller, B. M. Zuckerman, and A. J. Chang, Trans. Amer. Fish. Soc., 95, 345 (1966).

26. F. K. Kawahara, J. J. Lichtenberg, and J. W. Eichelberger, J. Water Poll. Contr. Fed., 39, 446 (1967).

27. F. K. Kawahara, R. L. Moore, and R. W. Gorman, J. Gas Chromatog., 6, 24 (1968).

28. L. H. Howe III, and C. F. Petty, J. Agr. Food Chem., 17, 401 (1969).

29. G. F. Vylegzhanina, and R. G. Kalmykova, Gig. Sanit., 34, 75 (1969); Chem. Abstr., 71, 29522y (1969).

30. M. A. El-Dib, J. Assoc. Off. Anal. Chem., 53, 756 (1970).

31. P. Koppe and I. Rautenberg, Gas-Wasserfoch., Wasser-Abwasser, III, (1970); Chem. Abstr., 72, 124943r (1970).

32. I. H. Suffet, G. Dozsa, and S. D. Faust, Water Res., 5, 473 (1971).

33. J. J. Murtaugh and R. L. Bunch, J. Water Poll. Contr. Fed., 39, 404 (1967).

34. W. S. Matthews and L. L. Smith, Lipids, 3, 239 (1968).

35. L. L. Smith and R. E. Gowron, Water Res., 3, 141 (1969).

36. A. Bevenue, T. W. Kelley, and J. W. Hylin, J. Chromatog., 54, 71 (1971).

37. W. L. Lamar, D. F. Goerlitz, and L. M. Law, Geological Survey Water-Supply Paper 1817-B, U.S. Dept. Interior, Washington, D.C. (1965).

38. A. W. Breidenbach et al., WP-22, U.S. Dept. Interior, Washington, D.C., 1966; Chem. Abstr., 68, 98563 n (1968).

39. J. M. Miller and J. G. Kirchner, Anal. Chem., 24, 1480 (1952).

Chapter 15

ION-EXCHANGE ANALYSIS IN WATER POLLUTION

Harold F. Walton

Department of Chemistry
University of Colorado
Boulder, Colorado

I. INTRODUCTION

Water is an excellent solvent and in its natural state always contains dissolved impurities. If water is examined closely enough, most of the
chemical elements can be found in river and sea water, as well as a great
many organic compounds. "Pollutants," by definition, have found their way
into our water systems by the activities of man. The distinction between
"pollutants" and natural impurities is sometimes absolute, for example, as
with artificial radioisotopes and chlorinated pesticides, but is more often a
matter of degree. The choice of analytical methods to be described in this
chapter is likewise arbitrary. I have included as many as possible of the
references to ion-exchange chromatography in the past 10 years that have
the word "environment" in their titles or contents, but it must be recognized
that these are only illustrative.

Ion exchange is primarily a method of separation and concentration. In
water analysis it is used to concentrate trace quantities and to separate one
substance from other substances that may interfere in a determination. Ion-
exchange separations are discussed in the broad sense here and the discussion is not restricted to elution chromatography. Even the term "ion
exchange" is used rather loosely, for the nonionic properties of ion-
exchanging polymers are very important and are being consciously exploited,
particularly in the chromatography of organic compounds.

I shall review the methods according to the elements sought, finishing
with applications to organic contaminants.

II. ACCURACY AND PRECISION[1]

III. NONMETALS

A. Boron

Borate ions and boric acid occur in natural waters in concentrations ranging
from virtually zero up to 100 mg/liter and higher. They are not toxic, and
traces of borate are necessary for the healthy growth of apples and other
fruit trees. However, high borate concentrations could indicate pollution,
so borate is included in this account.

Boric acid is a weak acid and is therefore absorbed neither by weakly
basic anion-exchange resins, nor by strongly basic resins at low pH. It is,
however, absorbed by a special kind of anion-exchange resin that carries
mannitol groupings grafted on to a polystyrene matrix [1]. The functional
groups of this resin are:

_____

[1]See Part III, Chapter 10, Sec. II.

$-CH_2N(CH_3)C_6H_8(OH)_5$

This resin was used by Carlson and Paul [2] to remove boron from the aqueous extracts of soils and plant ash. The extract (if neutral or alkaline) was passed directly through a small column (6 mm in diameter by 30 mm) of this resin; if the extract was acidic it was first passed through a column of carboxylic resin (Bio-Rex 70) in the ammonium form, to neutralize the acid without absorbing borate. Borate was caught on the selective anion-exchange resin. The resin was then washed with dilute ammonia and water to remove anions other than borate. Then hydrofluoric acid was passed through the column, which stripped the boron from the resin as the fluoroborate ion, BF_4^-. Excess acid was neutralized and the fluoroborate concentration was measured by a selective ion electrode. This method measured concentrations of boron down to 0.1 ppm in the original aqueous solution.

A method that does not depend on a selective resin was described by Eristavi et al. [3]. They passed natural water through three resin columns: the first was a strong-acid cation exchanger in the hydrogen form to replace metallic cations by H^+; the second was a weakly basic anion exchanger to absorb anions other than borate; and the third was a strong-base anion exchanger in the hydroxyl form to absorb borate and concentrate it into a small volume. Borate was eluted from this last column by sodium hydroxide and determined by titration. An elegant isotope dilution method, which uses strong-acid and weak-base resins to remove all cations and anions other than borate, has also been described [4].

B. Nitrate and Nitrite

These ions occur naturally in water and may also be introduced by the excessive use of fertilizers. High concentrations are harmful to the health, especially for young children.

Nitrite has been determined in sea water by adding sulfanilic acid and N-(1-naphthol)ethylenediamine to a large volume of sea water, forming an azo dye. The dye was concentrated by passing the sample through a small column of strong-base anion-exchange resin in the chloride form. The dye was absorbed and then eluted by a small volume of 60% acetic acid and measured spectrophotometrically. Then 0.05 μg of nitrite per liter was measured [5].

Nitrate is strongly absorbed by anion-exchange resins, compared with chloride and sulfate, and can be concentrated from waters in this manner. Westland and Langford [6] used a small column of a weak-base resin in the chloride form to absorb nitrate ions from lake water and eluted the nitrate afterwards with 0.5M NaCl. Usually, ion exchange merely shows that nitrate may be present (from a quantitative electrolyte balance; cation exchange is routinely used to measure the total electrolyte content of natural water) or is used to exchange cations for hydrogen ions before nitrate is determined photometrically with strychnidine [6].

C. Fluoride

Fluorine is the trace element that has become a political issue. It is artificially added to many potable water supplies to reduce the incidence of dental caries. Concentrations greater than 1-2 ppm may be harmful. High concentrations may occur naturally or may result from industrial pollution.

There are two general methods for determining fluoride ions in the parts per million range. One is spectrophotometric, the other potentiometric, using the lanthanum fluoride membrane electrode. The potentiometric method is less subject to interferences and is generally used for potable water without any preliminary treatment. The spectrophotometric methods (there are several) are more sensitive and may be much more accurate but are more susceptible to interferences. A strong-acid cation-exchange resin serves to remove cations, exchanging them for hydrogen ions.

Traces of aluminum interfere badly by forming a complex ion with fluoride. The cations are best removed by making the solution alkaline to pH 9 and then passing the solution through a cation-exchange resin [7]. The best method for measuring low fluoride concentrations is the alizarin-complexon method [7, 8]. Chloride and sulfate ions interfere somewhat, reducing the intensity of the blue fluoride complex. These ions could be separated from fluoride by anion-exchange chromatography, because fluoride is very weakly held by anion-exchange resins. If the chloride and sulfate concentrations are known, however, a small correction may be applied to the fluoride readings.

Before ion exchange came into use, the distillation of SiF_4 was always used to separate fluoride from interfering substances. Distillation is tedious and difficult, but it is still the method of choice for extreme cases.

D. Phosphate

In spite of the great interest in phosphate pollution from detergents and fertilizers, little or no use has been made of ion exchange for determining phosphate in polluted waters. The common spectrophotometric methods for phosphate are not affected by normal water constituents, except for silicate, and silicate interference can be suppressed by control of acidity. However, these methods give the total phosphate and do not distinguish between orthophosphate, pyrophosphate, and the various condensed phosphates. If it is necessary to make this distinction, anion-exchange chromatography is an excellent tool [9]. A thin-layer technique has been described that uses polyethyleneimine with LiCl eluent; orthophosphate moves the fastest [10].

Phosphate and cyanide can be absorbed from salt brines and sea water by chitosan, an insoluble aminopolysaccharide obtained from crab shells [11]. Chitosan treated with molybdate absorbs nanogram quantities of phosphate·from 2-liter portions of sea water, while the silver form of chitosan absorbs cyanide from concentrations below 1 mg/liter.

E. Hydrogen Sulfide

A strong-base anion-exchange resin in its hydroxide form absorbs minute amounts of hydrogen sulfide from air or water [12]. The sulfide is eluted by 4M sodium hydroxide and determined colorimetrically by the product of Methylene Blue from N,N'-tetramethyl-1,4-diaminobenzene. It is claimed that one part of hydrogen sulfide in 10^{10} parts of water can be collected and detected.

F. Chlorate and Perchlorate

These are unusual constituents in natural water and nearly always arise from industrial pollution. Chlorate ions are fairly toxic, perchlorate less so. Both ions are very strongly held by anion-exchange resins. Loach [13] absorbed perchlorate ions selectively on a quaternary base resin loaded with trichloroacetate ions, eluting with ammonium trichloroacetate. Perchlorate in lake and sea water was undetectable, but vegetable extracts and urine had a few parts in 10^9. Chlorate is less strongly bound by a resin than perchlorate but is more strongly bound than bromate [14].

G. Bromide and Iodide

Both these ions occur naturally in sea water and some fresh waters. They are absorbed strongly by anion-exchange resins. Iodide ions are normally bound five or ten times more strongly than bromide, and some 50 times more strongly than chloride. The exact selectivities depend, of course, on the crosslinking and other factors of the column. Anion-exchange resin-impregnated paper was used to filter bromide ions from fresh water [15] for measurement by X-ray fluorescence. Radioactive iodide is selectively absorbed by strong-base resins [16].

IV. METALS

A. Aluminum

This is not a common constituent of natural waters because of the insolubility of its hydroxide at neutral pH, but industrial wastes may carry dissolved aluminum into rivers. Low concentrations of aluminum can be measured by extracting with 8-hydroxyquinoline in chloroform and measuring the absorbance or the fluorescence of the resulting complex; however, many metal ions interfere, especially Fe(III), chromate, and zinc, all commonly present in industrial wastes. Fluoride is added to form AlF_6^{3-} ; then the solution is shaken with a weak acid carboxylic resin. Other metal ions are absorbed by this resin, while aluminum is not. The AlF_6^{3-} complex is destroyed by raising the pH with an acetate buffer, and the aluminum, free from interfering ions, is extracted by 8-hydroxyquinoline [17].

B. "Heavy Metals": Cobalt, Nickel, Copper, Zinc, and Cadmium

These elements are found at trace levels in many natural waters. Copper and zinc are widespread and do no harm, provided their concentrations are well below 1 ppm. Higher concentrations may indicate pollution or, perhaps, the presence of an ore body. Cadmium is the most toxic of these elements, and its presence in easily detectable amounts definitely indicates pollution.

All the ions mentioned can be concentrated from sea water (and therefore from fresh water) by passing through a column of chelating resin [18]. They are then eluted by nitric or hydrochloric acid. A triaminophenol-glyoxal polymer has been described that has a very strong affinity for copper and can absorb it from waters containing less than 1 µg/liter [19]. Final determination is usually performed by atomic absorption.

A new approach to the concentration of trace metals in natural waters is due to Muzzarelli [20, 21], who uses chitosan, a modified natural polymer obtained from crab shells.

Common strong-acid cation-exchange resins serve to collect many trace metals from water; for example, Co, Ni, Cu, Zn, Sn, and Pb [22, 23]. Zinc was absorbed on a cation-exchange resin from waters containing five parts Zn in 10^9, after the waters were made 0.6M in acetic acid [24]. Cadmium, at one part or less in 10^9, was absorbed on an anion-exchange resin from water made 0.001M in cyanide [25], simply by stirring the resin with the water in a batch process.

Resin-impregnated papers and ion-exchanging membranes are often used as selective filters to collect traces of metals from dilute solutions, especially if the metals are to be measured by X-ray fluorescence. In the latter case they need not be removed from the filter; the paper or membrane, loaded with metals, is mounted in the instrument and the X-ray spectrum is excited. Thus cadmium was collected from industrial waste waters on an anion-exchange membrane [26]. An analysis, qualitative and quantitative, of minor elements in river water was made by suspending cation-exchange membranes in the river for several days, then removing them, irradiating with high-energy protons, and scanning the X-rays emitted [27].

C. Silver

Traces of silver appear in rain water and snow from clouds seeded with silver iodide. The silver can be absorbed by cation-exchange resins [28, 29] or by anion-exchange resins as negatively charged chloro complexes [30]. The silver is eluted by nitric acid or by ammonium thiocyanate dissolved in 90% acetone and then determined by atomic absorption spectroscopy or neutron activation. Concentrations can be measured down to 1.0 ng/liter of original sample.

Silver in sea water, at concentrations about 40 ng/liter, was collected on an anion-exchange resin in the thiocyanate form after the water was made 0.05M in thiocyanate [31]. It was eluted by thiourea and measured by neutron activation. Copper, gold, and other heavy metals were recovered at the same time.

D. Molybdenum and Vanadium

Traces of molybdenum are necessary for plant and animal growth, but higher concentrations are toxic. Molybdenum forms soluble compounds in an oxidizing environment and tends to spread through ground waters. It may come from mining operations or from buried ore deposits and does not necessarily come from molybdenum ores; molybdenum accompanies copper and other metals in their ores.

Molybdenum traces may be extracted from sea water or fresh water by a strong-base anion-exchange resin in the thiocyanate form [32, 33]. It is eluted with sodium hydroxide and determined spectrophotometrically with dithiol. Molybdenum, tungsten, and vanadium are taken up from sea water by a chelating resin [34]. Molybdenum is absorbed by chitosan or by an anion-exchanging cellulose [21] and vanadium is absorbed from sea water by a strong-base anion-exchange resin in the thiocyanate form, after the water is made 0.1M in ammonium thiocyanate and hydrochloric acid [35].

E. Mercury

Attention has been drawn in recent years to the presence of mercury in industrial effluents and the rivers and lakes into which they feed, and even in sea water. It was found that aquatic organisms convert mercuric ions into methyl mercury compounds, CH_3HgX, which are extremely toxic.

They are lipid soluble and enter the central nervous system, where they can cause permanent damage.

In the presence of chloride, mercuric ions form $HgCl_4^{2-}$, which is very strongly absorbed by anion-exchange resins. One approach is to pass the water (made 0.1N in HCl) several times through a disk of resin-impregnated paper. This absorbs the mercury, which is then determined by neutron activation, a very sensitive method for mercury [36]. X-Ray fluorescence is also used with resin-loaded paper [37]. Down to three parts Hg in 10^{11} parts of the original water can be measured in this way [36]. Organic mercury compounds must first be decomposed by treatment with chlorine for 30 min.

A special chelating resin, which contains the groups $-C(:NH)NH_3^+$ and $-CH_2-S-C(:NH)NH_3^+$, was prepared by Koster and Schmuckler [38] and has been available commercially. It is specially selective for the ions of Pd(II), Pt(II), and Au(III) and can absorb traces of these metals from very dilute solutions. It also has a high affinity for Hg^{2+} and CH_3Hg^+ [39]. It can be used in the form of resin-impregnated paper or in a short column. Inorganic Hg(II) may also be absorbed selectively on porous glass microbeads [40], leaving CH_3Hg^+ in solution.

Mercury is an easy element to detect. Activation by neutrons and the absorption of resonance lines by mercury vapor are extremely sensitive, and traces of mercury can be found almost anywhere.

V. RADIOACTIVE ELEMENTS

The fission products of uranium and the fallout from nuclear bombs include the long-lived species ^{137}Cs, ^{89}Sr, ^{90}Sr, and its daughter product ^{90}Y, and ^{137}Ba. A vast amount of literature deals with the separation of these elements from water, milk, and soil by ion exchange.

A. Cesium

Cesium is absorbed with amazing selectivity by a wide range of inorganic cation exchangers. Ammonium molybdophosphate (AMP) and zirconium phosphate are the most commonly used, but many other materials have been investigated, all for the purpose of separating ^{137}Cs from other fission products. Selectivities are so great that one gets good results by batch extraction, without using a column. Ammonium molybdophosphate is simply stirred with the water sample and allowed to settle, preferably with the help of a coagulant; cesium is absorbed and can be determined by radioactivity or by flame photometry [41, 42]. Ferrocyanide molybdate is used in the same way [43]. Boni [44] used a small column of potassium cobalt(II) ferrocyanide to absorb ^{137}Cs from milk, urine, sea water, and environmental samples. The cesium was not stripped from the column; instead, the column itself was placed in the scintillation counter. An auxiliary column of chelating resin removed other radioactive nuclides before they entered the ferrocyanide absorber.

A more elaborate system for concentrating and then separating radioactive cerium, yttrium, cesium, strontium, barium, and radium from rain water was devised by Senegacnik et al. [45]. The cations were absorbed on a sulfonated polystyrene resin and eluted successively, in the order named above, by ammonium glycolate followed by ammonium citrate.

B. Sodium

Radioactive ^{24}Na was collected from water by passing it through a column of KU-2 strong-acid cation-exchange resin, followed by elution with hydrochloric acid [46].

C. Strontium

The methods for determining ^{90}Sr all depend on absorbing the strontium ions on a cation-exchange resin, usually in the ammonium form. The strontium is then allowed to remain on the column for a week or two, as a rule, forming yttrium-90, half-life 64 hr, which has strong β activity. Yttrium is removed from the resin by lactate, citrate, or EDTA, which preferentially complex trivalent Y^{+3} over Sr^{2+} and elute it from the column first. The method has been applied to rain water [47, 48], sea water [49], food, milk, and agricultural material [50]. There are variations on the procedure; for example, milk is allowed to age for 2 weeks; then citrate and carriers are added and the sample is passed through a cation exchanger to hold back strontium and then through an anion exchanger to capture the yttrium-citrate complex [51]. Yttrium is stripped from this resin with hydrochloric acid, precipitated as the oxalate, and counted.

Another method is to add nonradioactive carriers and then to precipitate strontium, barium, and lead as their phosphates. These are dissolved in nitric acid and then passed into a cation-exchange resin column. The ions of interest are absorbed in the column and selectively eluted as follows: lead with ammonium acetate, strontium with ammonium lactate, and radium and barium with EDTA [52].

D. Barium and Radium

Barium and radium are absorbed very strongly by sulfonated polystyrene resins and can thus be recovered even from sea water [53-55]. Elution is done with hydrochloric acid, ammonium acetate, or EDTA.

E. Uranium

Uranium is present at microgram per liter levels in many natural waters, but it presents no health hazard and is not generally considered a pollutant. The element is measured by the fluorescence imparted to a sodium fluoride-sodium carbonate-potassium carbonate melt. Anion exchange in methyl glycol-aqueous HCl solutions is a very good way to recover microgram quantities of uranium from natural waters and sediments [56]. A simpler way, applicable to fresh water, is to make the water 0.01M in sulfuric acid and pass it through a strong-base anion-exchange resin (Dowex 1 × 8) in its sulfate form. Uranium is absorbed as its sulfate complex. The column, which need contain only 0.5 cm^3 of resin, is washed with 0.01M sulfuric acid to remove iron and manganese, which may otherwise later quench the fluorescence; then the uranium is eluted by a few milliliters 2M sulfuric acid. It is measured by the fluorescence of the solution itself. One part of uranium in 10^9 parts of water can be measured [57].

VI. ORGANIC COMPOUNDS

Since most organic and biological contaminants of water are not ionized, ion exchange has only limited use in their analysis. One contaminant that has become of interest recently is nitrilotriacetic acid, $N(CH_2COOH)_3$ or NTA, proposed as a replacement for phosphates in detergents. This could be absorbed on an anion-exchange resin. It would compete with the anions naturally present in water, of course. Interfering cations of metals that form complexes with NTA, particularly Fe, Cu, Zn, and Ni, have been successfully removed by passing the water sample through a column of sulfonated polystyrene resin in the sodium form. NTA is determined in the filtrate by AC polarography after a small excess of cadmium salt in an ammonia-ammonium chloride buffer has been added to give a final pH of 8.0. The Cd-NTA complex has its own characteristic peak at -0.8 V vs saturated calomel electrode compared with -0.6 V for the $Cd-NH_3$ complex [58].

High-performance ion-exchange chromatography has been used to determine NTA in water and sewage. One method uses a column of strong-base pellicular resin [59]. Anionic chelating agents are selectively absorbed on the column, then eluted successively with 0.02M borax at pH 9, and detected by absorption of ultraviolet light. Ethanol diglycine, EDTA, and other chelating agents elute before NTA. Traces of iron do not interfere.

A strong-base anion-exchange resin in the formate form retains NTA and citric acid from water or sewage made 0.1M in formic acid [60]. They are eluted (citric acid first) by concentrated formic acid and then esterified and determined by gas chromatography. Anion exchange was also used to collect NTA from raw sewage [61].

Amino acids occur naturally in river and sea water, and they also appear because of biological pollution. They may be concentrated and separated from water by absorption on a strong-acid cation-exchange resin at low pH. They are then separated from each other, if desired, by cation-exchange column chromatography, thin-layer chromatography, or paper chromatography, following standard techniques. In a typical procedure [62], 1-liter samples of water were passed through a small column of strong-acid cation-exchange resin in the hydrogen form. The amino acids were absorbed and then eluted by a small volume of concentrated aqueous ammonia (which converted all the amino acids into anions). The eluate was evaporated to dryness and the sum total of amino acids determined photometrically with a ninhydrin-cadmium chloride reagent. In another and somewhat more elaborate procedure [63], reservoir and lake waters were sampled by immersing a packet of strong-acid cation-exchange resin in them and then eluting the amino acids with ammonia, evaporating, and analyzing the residue by paper chromatography. Amino acid concentrations in unpolluted reservoirs were about 0.3-0.5 ppm; in polluted waters they were as high as 5 ppm. Aspartic acid accounted for half of the amino acids found.

Siegel and Degens [64] applied the principle of ligand-exchange chromatography to absorb and concentrate the amino acids from sea water. They passed large volumes of sea water through a small column of the iminodiacetate-polystyrene chelating resin Dowex A-1, loaded with $Cu(II)$. The amino acids formed complexes with the copper that remained absorbed in the resin. Then they were displaced from the copper by passing 5M ammonia. In such procedures as this it is common for a little copper to be displaced from the resin along with the amino acids. The copper can be retained or "filtered out" by passing the solution through a small column, say 1 cm in diameter by 2 cm high, of carboxylic, polymethacrylic resin (Amberlite IRC-50) in the sodium form. The amino acids pass and can be concentrated by evaporation and further analyzed if desired.

Finally, and with some hesitation, I mention the absorption of nonionic compounds by ion-exchanging polymers. It is becoming recognized that the polymer matrix of an ion-exchange resin is an effective and discriminating "solid solvent" for uncharged organic substances of all kinds. Carbohydrate mixtures are routinely analyzed by chromatography on cation- and anion-exchange resins, with 85-95% ethanol as eluent. The chief function of the ionic groups seems to be to make the resins swell (by solvation) and make them permeable. This principle has not been applied generally to environmental samples. However, in 1969 Riley and Taylor [65] used a nonionic, porous styrene-divinylbenzene copolymer, Amberlite XAD-1, to absorb organic compounds from sea water. Carbohydrates and amino acids were not absorbed, nor were inorganic ions, but long-chain fatty acids, cholesterol, surfactants, DDT, and other insecticides and pesticides were absorbed and could be eluted with 95% ethanol. A column 1 cm in diameter and 7 cm high was used, with 20/50 mesh resin beads.

Other macroporous resins, essentially nonionic, called XAD-2 and XAD-7 by their manufacturer (no other information is disclosed in the paper), were used in columns 1.5 cm in diameter by 7 cm high to absorb organic contaminants from well waters [66]. The columns were connected directly to the water supply. The model compounds, benzoic acid, phenol, cresol, phenylenediamine, and naphthalene, were eluted successively by sodium bicarbonate, which eluted benzoic acid; sodium hydroxide, which eluted phenol and cresol; and methanol, which eluted the remaining compounds. In actual polluted well waters some 17 compounds were separated, including hexanol, benzene, benzene sulfonic acid, aniline, and ketones. They were present in the micrograms per liter range.

This work is not really ion-exchange chromatography, but the absorbents are chemically related to ion-exchange polymers, and the boundaries of ion-exchange chromatography are becoming harder and harder to draw.

Phenols, such aromatic hydrocarbons as terphenyl, aniline, and many other compounds have been determined in industrial waste waters by high-efficiency liquid chromatography, using both ion-exchanging and non-ion-exchanging pellicular packings [67]. Eluents were aqueous alcohol or borate-nitrate buffers.

REFERENCES

1. F. Pinon, J. Deson, and R. Rosset, Bull. Soc. Chim. France, 3454, 4307 (1968).

2. R. M. Carlson and J. L. Paul, Soil Sci., 108, 266 (1969).

3. D. I. Eristavi, F. I. Brouchek, and L. I. Cheishuili, Ref. Zhur. Khim., 19GDE, Abstracts No. 5G181, 12G98, 12G99, 12G100 (1963); Anal. Abstr., 11, 1125 (1964); 12, 953 (1965).

4. B. Gorenc, J. Marsel, and G. Tramsek, Mikrochim. Acta, 24 (1970).

5. E. Wada and A. Hattori, Anal. Chim. Acta, 56, 233 (1971).

6. A. D. Westland and R. R. Langford, Anal. Chem., 28, 1996 (1956).

7. T. Kempf, Z. Anal. Chem., 244, 113 (1969).

8. O. S. Glaso, Anal. Chim. Acta, 28, 543 (1963).

9. F. H. Pollard, G. Nickless, D. E. Rogers, and M. T. Rothwell, J. Chromatog., 17, 157 (1965).

10. J. M. Tanzer, M. I. Koichevsky, and B. Chassy, J. Chromatog., 38, 526 (1968).

11. R. A. A. Muzzarelli and B. Spalla, J. Radioanal. Chem., 10, 27 (1972).

12. D. M. Paez and O. A. Guagnini, Mikrochim. Acta, 220 (1971).

13. K. W. Loach, Nature (London), 196, 754 (1962).

14. J. K. Sloss, H. A. Hudson, and C. J. Cummiskey, Anal. Chem., 37, 1240 (1965).

15. D. Radcliffe, Anal. Letters, 3, 573 (1970).

16. J. Konecky and D. Keltos, Radioisotopy, 11, 693 (1970).

17. C. A. Noll and L. J. Stefanelli, Anal. Chem., 35, 1914 (1963).

18. J. P. Riley and D. Taylor, Anal. Chim. Acta, 40, 479 (1968).

19. A. Zlatkis, W. Bruening, and E. Bayer, Anal. Chem., 42, 1201 (1970).

20. R. A. A. Muzzarelli and O. Tubertini, Talanta, 16, 1571 (1969); R. A. A. Muzzarelli and L. Sipos, Talanta, 18, 853 (1971).

21. R. A. A. Muzzarelli and R. Rocchetti, Anal. Chim. Acta, 64, 371 (1973).

22. Z. P. Miroshnikova, Trudy Voronezh. Gos. Univ., 82, 80 (1971); Anal. Abstr., 22, 4556 (1972).

23. H. Hellmann and A. Griffatong, Z. Anal. Chem., 257, 343 (1971).

24. H. Matsui, Anal. Chim. Acta, 66, 143 (1973).

25. T. Ashizawa and K. Hosoya, Bnnseki Kagaku, 20, 1416 (1971).

26. H. Tanaka and T. Yamamoto, Bunseki Kagaku, 20, 784 (1971).

27. C. H. Lochmueller, J. Galbraith, and J. Joyce, Analyt. Letters, 5, 943 (1972).

28. J. A. Warburton, J. Appl. Meteorol., 8, 464 (1969).

29. J. A. Warburton and L. G. Young, Anal. Chem., 44, 2043 (1972).

30. T. T. Chao, M. J. Fishman, and J. W. Ball, Anal. Chim. Acta, 47, 189 (1969).

31. K. Kawabuchi and J. P. Riley, Anal. Chim. Acta, 65, 271 (1973).

32. K. Kawabuchi and R. Kuroda, Anal. Chim. Acta, 46, 23 (1969).

33. M. J. Fishman and E. C. Mallory, J. Water Poll. Contr. Fed., 40, 67 (1968).

34. J. P. Riley and D. Taylor, Anal. Chim. Acta, 41, 175 (1968).

35. T. Kiriyama and R. Kuroda, Anal. Chim. Acta, 62, 464 (1972).

36. D. E. Becknell, R. H. Marsh, and W. Allie, Anal. Chem., 43, 1230 (1971).

37. W. B. Link, K. S. Heine, J. H. Jones, and P. Watlington, J. Assoc. Off. Agr. Chem., 47, 391 (1964).

38. G. Koster and G. Schmuckler, Anal. Chim. Acta, 38, 179 (1967).

39. S. L. Law, Science, 174, 285 (1971).

40. A. Gorgia and D. Monnier, Anal. Chim. Acta, 55, 247 (1971).

41. C. Feldman and T. C. Rains, Anal. Chem., 36, 405 (1964).

42. C. Sreekumaran, K. C. Pillai, and T. R. Folsom, Geochim. Cosmochim. Acta, 32, 1229 (1968).

43. S. Kawamura and K. Kwotaki, J. Chromatog., 45, 331 (1969).

44. A. L. Boni, Anal. Chem., 38, 89 (1966); see also W. E. Prout, E. R. Russell, and H. J. Groh, J. Inorg. Nucl. Chem., 27, 473 (1965).

45. M. Senegacnik and S. Paljk, Z. Anal. Chem., 232, 409 (1967); M. Senegacnik, S. Paljk, and K. Juznic, Z. Anal. Chem., 233, 81 (1968); M. Senegacnik, S. Paljk, and J. Kristan, Z. Anal. Chem., 249, 39 (1970).

46. R. Yu. Yasulenis, V. Yu. Luyanas, and V. P. Kekite, Radiokhimiya, 14, 650 (1972).

47. L. P. Gregory, Health Phys., 10, 483 (1964).

48. N. A. Talvitie and R. J. Demint, Anal. Chem., 37, 1605 (1965).

49. R. D. Ibbett, Analyst, 92, 417 (1967).

50. A. B. Strong, G. L. Rehnberg, and U. R. Moss, Talanta, 15, 73 (1968).

51. C. R. Porter and B. Kahn, Anal. Chem., 36, 676 (1964).

52. L. P. Gregory, Anal. Chem., 44, 2113 (1972).

53. N. R. Andersen and D. N. Hume, Anal. Chim. Acta, 40, 207 (1968).

54. B. J. Szabo and O. Joensuu, Env. Sci. Tech., 1, 499 (1967).

55. Y. Sugimura and H. Tsubota, J. Marine Res., 21, 74 (1963).

56. I. Hazan, J. Korisch, and G. Arrenius, Z. Anal. Chem., 213, 182 (1965).

57. A. Danielsson, B. Rönnholm, L. E. Kjellström, and F. Ingman, Talanta, 20, 185 (1973).

58. J. Wernet and K. Wahl, Z. Anal. Chem., 251, 373 (1970).

59. J. E. Longbottom, Anal. Chem., 44, 418 (1972).

60. W. A. Aue, C. R. Hastings, K. Gerhardt, J. O. Pierce, H. H. Hill, and R. F. Moseman, J. Chromatog., 72, 259 (1972).

61. L. Rudling, Water Res., 5, 831 (1971); Anal. Abstr., 23, 4253 (1972).

62. A. D. Semenov, I. N. Ivleva, and V. G. Datsko, Ref. Zhur. Khim., 19GDE, No. 3Gss7 (1964).

63. J. Paluch and S. Stangret, Biul. Zaki, Badan Nauk. Gornoslaskiegp Okregu Przem. Pol. Akad. Nauk, 12, 7-30 (1970); Chem. Abstr., 79, 108003y (1973).

64. A. Siegel and E. T. Degens, Science, 151, 1098 (1966).

65. J. P. Riley and D. Taylor, Anal. Chim. Acta, 46, 307 (1969).

66. A. K. Burnham, G. V. Calder, J. S. Fritz, G. A. Junk, H. J. Svec, and R. Willis, Anal. Chem., 44, 139 (1972).

67. K. Bhatia, Anal. Chem., 45, 1344 (1973).

PART V

WASTE CHEMISTRY SECTION

Chapter 16

GAS CHROMATOGRAPHIC ANALYSIS IN WASTE CHEMISTRY

Renato C. Dell'Acqua

Division of Laboratories and Research
New York State Health Department
Albany, New York

I. INTRODUCTION

Organic wastes are generated mainly by farms, industry, and urban con-
glomerates. These producers and the sewers that carry the wastes are the
sources of samples for analysis. These same compounds may then be
introduced into surface waters as pollutants from industrial effluents,
sewage disposal, agricultural runoff, and chemical spills. These compounds
are considered pollutants when they are present in sufficient concentration
to cause undesirable taste and odor problems. To help control this type of
pollution, it is important to have the ability to identify and measure individual
components. Specific identification is often required to determine the source
of pollution. Specific identifications also make it possible to carry out more
effective monitoring programs.

   Although quantitative analyses of organic wastes can probably be achieved
with gas chromatographic methods more accurately than with any other
method, quantitation is rarely required because wastes, per se, occur in
such great concentrations. This is fortunate in view of the infinite complexity of waste analyses arising from such variations as waste component
reactivity, degradability, and metabolization.

   The decision as to degree of analysis required rests with the definition
of the problem as one of waste discharge or one of environmental pollution.
For example, it might be necessary to establish the rate of dilution of a
contaminant in a body of water. Dissertations on pollution programs related
to the same contaminants that constitute waste discharges are offered in
other chapters of this monograph.

   Generally, all the wastes will be conveyed either to a treatment plant or
to a body of water. All organic components of wastes will undergo a biodegradation process and will be reduced to fixed, non-oxygen-demanding
compounds and thus will become harmless. This normal process of bacterial
bioequilibration can take place in an environment where the reducing agents
(bacterial) and oxidizing agent (oxygen) are at even capacity. The organic
components of wastes will eventually metabolize to inorganic salts. This
would be the ideal process of natural cleanliness. Unfortunately, the mani-
festation of such an ideal process is becoming more and more rare due to
our increase in industrialization, concentration of populated areas, lack of
adequate plants for waste treatment, and the need for laws to enforce
proper treatment. As a result the problem of eutrophication arises. The
wastes are depleting the surface waters of their oxygen content in an ever
increasing rate and menacing the existence of the natural reducing agents
so necessary to the purification of surface waters.

   The following individual wastes and their characteristics will be described
and optimal analytical methods suggested. The wastes carried in waters
dealt with are gasoline, kerosene, oils (fuel oils and lubricating oils),
benzene, toluene, styrene, xylene, fatty acids, phenols, glycerides, steroids,
pesticides, and organometallic salts. These compounds are not to be

considered as the total organic contaminants, but they do represent 90% or more of the total pollutants (based on my gross estimate according to work load at an analytical laboratory). Others, such as hydrogenated and nitrated compounds, complexed mono- and diasphalts, waxes, solid paraffins, carbonyl compounds, sulfur-containing compounds, chlorinated hydrocarbons, chlorinated biphenyls (these last two are industrial resins made in large amounts), and tri- and polyorganic salts are also contaminating the environment. This chapter does not include a detailed elucidation of their method of identification by gas chromatography, not because this is impossible (it is in fact being done) but because their nature and quantity make them of secondary importance to my purpose [1-5]. The foregoing list of wastes was made in direct ratio to the increasing difficulties encountered during the analyses. For example, it is quite easy to establish the quality of waste discharged by a factory that produces organic solvents; but trying to obtain the same result by analyzing the sewer contents a few miles downstream proves to be much more difficult because the original contaminant has been mixed with other compounds, has probably already reacted with appropriate agents, and may have been attacked by some others. The result of any analysis on organic waste must be interpreted as evidence of the components found at the moment of the analysis rather than as representative of the original contaminant or discharge.

The scope of this chapter prevents me from entering into the theoretical discussion of gas chromatography and its instrumentation (see Chapter 1); instead, my purpose is to aid the analyst in selecting the proper methodology. However, I must promptly qualify this statement by cautioning analysts that there is no foolproof formulary; the number of hydrocarbons of all kinds, of aromatic compounds, of phenols, of lipids, and of pesticides is more or less totally known, while the possible number of their derivatives is infinite and mostly unknown.

To point up one type of pollutant, consider sulfur. Each year an estimated 220 million tons of sulfur are discharged into the earth's atmosphere from various sources. This represents more than eight times the total annual consumption of sulfur. However, man and nature must share equally the responsibility for this problem. Nearly one-half of the total sulfur discharged, approximately 100 million tons, is in the form of H_2S from natural sources, such as swamps, oceans, and the like. Another 3 million tons enters the atmosphere as H_2S pollution, this being man made from pulp mills and related industries.

A second major natural contribution of sulfur to the atmosphere is an estimated 44 million tons as sulfate aerosols originating from sea spray. This means that the two principal sources of sulfur pollution account for 144 million tons a year, representing 65.6% of total sulfur pollution.

Looking at the remaining one-third, or about 73 million tons, this enters the atmosphere in the form of SO_2 air pollution. Of this, coal accounts for approximately 70%, or 51 million tons; petroleum 20% of SO_2 pollution, or 15 million tons; and finally metal smelting, 10% of SO_2 pollution, representing approximately 7 million tons [6].

While automated process hardware is of tremendous value to physical and computation scientists, analytical problems remain on such mundane subjects as the accurate and sensitive determination of H_2S, SO_2, halogen compounds, a variety of acid gases, eye irritants, and pesticides in the environment. With the exception of the measurement of car exhausts, monitoring of sources remains an undeveloped technology.

II. SAMPLING

The collection of samples for gas chromatographic analysis of wastes requires the use of grab rather than composite procedures. The reason is that waste analyses are performed on samples with generally high concentrations of pollutants and, consequently, the analyses are not quantitative but only qualitative. Quantitative analyses are required more for pollution problems. In such cases composite procedures of sample collection are necessary. Nevertheless, some precautions must be taken in order to assure that the samples are representative of the problem when it occurs or when it is noted.

Some guidelines may, however, be offered to illustrate sampling techniques. The organic compounds present in the samples collected for gas chromatographic analysis of waste are very susceptible to chemical and biological alterations and, for that reason, the samples should be analyzed as soon as possible. When required, the pH of the samples must be adjusted immediately after collection (see Sec. III, E). Only glass containers with ground glass stoppers should be used; as second best, a metallic one can be used; a rubber or plastic container should never be used. The sample container should never be filled completely; a 2- or 3-in. air pocket with all its oxidizing potential is better than the passage of the best contaminants through the stopper. The sample size is dependent upon the criteria of the collector with respect to the amount of the contaminants. On spills, at least three samples should be collected; one upstream, one at the site of contamination, and one downstream. In sewers of waste discharges the place of sampling must be representative of the problem and, if more than one sample is to be collected for consecutive analysis, the place of collection should always be the same.

Different sampling procedures have been used by various agencies for sampling automotive emissions. A short column containing Chromosorb W placed in the exhaust section was used to adsorb heavy hydrocarbons [7]. Mylar bags and gas washing bottles containing normal hexane have been evaluated also [8]. Gas chromatography with flame ionization detection was the method of choice in both cases.

In view of the complexity of the sample to be analyzed, a standard of the supposed contaminant obtained from the source will facilitate enormously the identification procedures. For example, the contaminant spilled by a barge carrying fuel oil No. 4 of a specific brand is surely identified if a standard of the same oil is available. A fuel oil No. 4 standard at the laboratory may be such that, in spite of being No. 4 with the same physical characteristics, it may well vary with respect to amount of components and thus never match the oil present in the collected sample. The analysis report in such a case would read "contaminant found similar to fuel oil No. 4" instead of "contaminant is fuel oil No. 4 and of the same type as for the standard supplied by the spiller."

III. ANALYSIS OF VARIOUS CONTAMINANTS

A. Identification of Gasoline

Gasoline is not found in chemical wastes very often because of its value as a fuel and because of its volatility; i.e., it is discharged only when strictly necessary and it vaporizes quickly. Occasionally gasoline becomes a by-product of the cleaning of reaction tanks and other vessels used in industry and generally it is mixed with low-grade kerosene. It could also become a waste during accidental spills. Gasoline is a mixture of saturated paraffins, olefins, naphthenes, and other aromatic compounds, with carbon number contents ranging from 5 to 12 at various concentrations according to type and brand. Its boiling range goes from 40° to $200^\circ C$. The saturation of its major components (normal, iso-, and cycloparaffins) makes gasoline a nonpolar chemical mixture for gas chromatographic purposes. If the unknown is to be identified as gasoline and it comes from a spill, it is easy to separate gasoline from its carrier by acidifying the sample to a pH of 1-2 with dilute sulfuric acid and extracting it with 5 ml/liter of pentane or hexane [9]. The size of the sample is suggested by its degree of contamination. The extraction is performed at room temperature by transferring the acidified sample to a separatory funnel, adding the solvent, and shaking for 4 min. The stopcock must be made of Teflon. The aqueous and organic phases will be clearly separated after 3-5 min. The bottom layer is discharged and the solvent is collected and dried in the following manner: The solvent and water remaining at the bottom of the separatory funnel are drained into a 50-ml beaker. The remaining water will be sharply visible. The supernatant solvent is decanted into a graduated centrifuge tube. The volume of the resulting solvent will be less than the original volume used. If quantitative results are desired additional extractions must be carried out on the original acidified sample. Once the required volume has been obtained, 1 g of granular anhydrous sodium sulfate, hexane washed, is poured into the tube. The extract is now ready for gas chromatographic analysis. The only required step will be to dilute the extract with hexane to a convenient volume before the chromatographic run.

If the unknown comes from a complex of wastes, a selective extraction must be applied and the use of a nonpolar column for its analysis will facilitate the identification [10]. The sample, 1 or 2 liters, which will probably be

very concentrated in its contamination, should be diluted with distilled
water to a 1:10 proportion. A 5-ml/liter quantity of saturated sodium
bicarbonate solution is added to the diluted sample, which is poured into a
separatory funnel and made acidic with dilute sulfuric acid to a pH of 1-2.
A 20-ml quantity of hexane per liter is added to the sample and the funnel
is shaken for 3-5 min. All the nonpolar components will be extracted into
the hexane phase, leaving the polar components such as phospholipids,
aromatic complexes, phenols, hydrocarbons substituted with chlorine,
sulfur, and nitrogen groups and others in the water phase. The two phases
are separated, the hexane is poured into a 500-ml separatory funnel, and
20 ml of acetonitrile is added. The two solvents are shaken for 4 min and
the gas formed is carefully released two or three times during the extraction.
The ionic characteristics of acetonitrile will permit the extraction of most
of the unsaturated compounds in the sample, leaving the normal hydrocarbons in the hexane phase. If a more sophisticated purification is
required, dilute sulfuric acid is added in the same proportions as above,
after the acetonitrile partition, to the hexane phase and the sample is
shaken for 4 min. The aqueous phase is discharged, the hexane dried and
the sample diluted if necessary in the same way as for the clean sample.
In both the cases the extracts can be separated on a 6 ft × 0.25 in. o.d. glass
column packed with 5% each of SE-30 and OV-1 liquid phases on either
Chromoport AB 80-90 or Gas Chrom Q 80-90.

Detection is accomplished with a flame ionization detector (FID). The
operating temperatures are: inlet, $200^{\circ}C$; detector, $275^{\circ}C$; and column
programmed from 35° to $210^{\circ}C$ at $7.5^{\circ}C$/min. At $114^{\circ}C$ the n-octane peak
will appear, at $143^{\circ}C$ the n-decane, at $169^{\circ}C$ the n-dodecane, and at $191^{\circ}C$
the n-tetradecane (see Fig. 1a). As can be seen, the baseline shift is negligible. This is due to the perfect conditioning of the column and to the
cleanliness of gases and detector. With the parameters described above as
little as 0.001 μl of gasoline diluted in hexane (1 ml per 10 ml) will give
identifiable peaks with a height of 8-10% on the chart (see Fig. 1b); for
qualitative analysis only, the chromatogram of the gasoline extract will
appear as represented in Fig. 2. More peaks may be seen by increasing
the sensitivity of the instrument. From this increased sensitivity the quality
of the gasoline may be established by observing the ratio of n-heptane to
isooctane given by the peak heights.

It is possible that during the selective extraction of the unsaturated from
the saturated compounds, traces of the former could be retained by the
saturated extract. This will not influence the results because the unsaturated
compounds will be eluted at the beginning of the gas chromatographic run
together with the solvent. By injecting the same amount of sample extract
and observing the resulting peaks (provided that the extract is not too
concentrated) it is very easy to identify the nature of the contaminant if it is
gasoline. Jelter and Veldink [11] used gas chromatography to determine

gasoline concentrations of 1 mg/liter in water. The sample was extracted in phenyl nitrate and then the extract was fed through a 3-m copper column packed with 10% Polyethylene Glycol 1500 on 60/80 mesh silanized Chromosorb W. Nitrogen plus hydrogen carrier gas at 25 ml/min and a flame ionization detector were used.

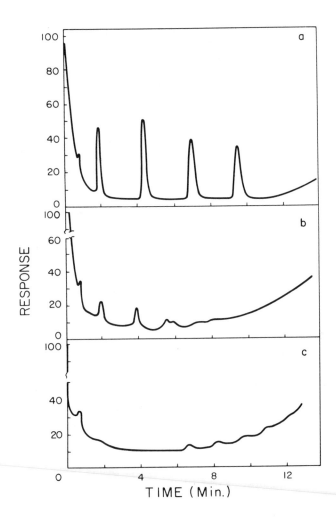

Figure 1. Paraffin-type compound responses to the FID. (a) n-C_6 to n-C_{14}; (b) Gasoline sample; (c) kerosene sample.

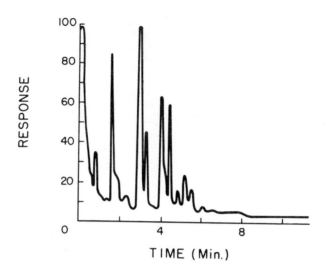

Figure 2. Chromatogram for the qualitative identification of gasoline extract.

B. Identification of Kerosene

Kerosene is mainly a saturated petroleum hydrocarbons, with carbon atom contents ranging from 12 to 20, and with a boiling range of 175° to 325°C. Either in a clean sample or in a complex waste, kerosene can be separated by the following procedures given for gasoline extraction and purification. By running a standard (0.001 ml per 20 ml of hexane) through the same column but extending, the temperature program to 250°C, it will be noticed that the first peak of kerosene will appear at 143°C as the n-decane peak. The rest of the peaks will elute regularly up to the tetradecane representative at 191°C. From there up to 250°C, at which temperature the elution of kerosene components will be finished, a series of small peaks with a bulking baseline will appear on the chromatogram (see Fig. 1c). This chromatogram shows an amount equal to a 0.001 μl of kerosene detected. For our qualitative analysis, the chromatogram represented by Fig. 3 identifies kerosene more appropriately. The appearance of those peaks is the consequence of the following two factors. First, the components of kerosene from pentadecane up to eicosane increase their isomers enormously (a C_{20} molecule can give up to more than 350,000 isomers) and a 6-ft column cannot even separate a hundredth of such components. Capillary columns from 300 to 1000 ft long are needed for the task. Second, unless the contaminant involved is kerosene for jet planes, the components of the ordinary kerosene after the pentadecane are quite impure and their ionic strength manifests a more polar coefficient affinity toward the solid support of the column, thus

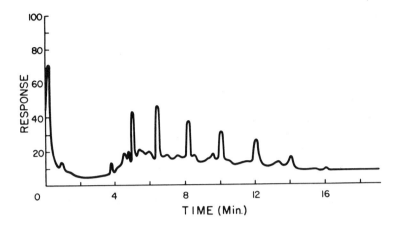

Figure 3.   Chromatogram for the qualitative identification of kerosene extract.

reducing the effectiveness of separation [12]. Nevertheless, the identification of kerosene has been proved by the fact that no carbon groups from 7 to 10 have shown their presence and oil is not observed even if the results are interpreted as comparing with the method that follows [10].

C. Identification of Mineral Oils

Mineral oils include both the burning and lubricating oils. These oils are represented by undistilled residuals of petroleum with a great variety of components. These components contain carbon groups 18-20 or higher and boiling points of 350°C and greater. The mixture is approximately a composite of 20-25% straight- and branched-chain paraffins, 40-50% alkylated naphthenes, 20% alkylated aromatic hydrocarbons, and 10% asphaltenes [13]. These values vary according to the type of oil. Mineral oils enter waste waters in many ways. Their complexity is made evident by the following facts: light oils, heavy oils (both as fuel oils), lubricating oils, dirty lubricating oils, and high-gravity oils, when mixed together in the presence of some reacting compound, can generate byproducts with structures impossible to individualize by the method given above (Sect. III,B). Crude oils have different characteristics; high or low nitrogen contents, low gravity or high gravity, few asphaltenes, and bottoms heavier than water. An infinite variety and ratio among aromatics, naphthas, and polycyclic compounds are present. On contact with organic solvents, oils oxidize, making the matching of the extracts with a standard almost impossible. If the sample to be analyzed is clear of other contaminants, such as greases, lipids, and glycerides, and if it has been recently generated, a quick qualitative way to identify it is to dilute the sample with hexane until the ratio of oil to hexane is approximately 1:100 and inject an adequate amount into the gas chromatograph [9]. These amounts should be such that, with the proper sensitivity of the instrument, a characteristic chromatogram for oil will be obtained. If the sample is complex, the task of the analyst is more difficult. No specific purification method is available because two samples with the same composition almost never occur. A few hints will help the analyst apply the most suitable procedure. With the following procedures some of the components of the oil are lost and some others are modified but the results are worth the loss. Diluting the sample with water and eluting it through activated charcoal washed with chloroform and rinsed with distilled water will allow the adsorption of all the organic compounds on the charcoal. These can be recovered by eluting the charcoal with chloroform and concentrating the eluate by evaporation.

The concentration is done "ad hoc." The extract is then fractioned according to solubility, pH, and polarity [10]. The fractions obtained should be base soluble, weak- or strong-acid soluble, and/or neutral. The neutral fraction is further separated by column chromatography into aliphatics, aromatics, and oxygenated and halogenated compounds. By mixing a suitable solution of sample with liquid sulfur dioxide or concentrated aqueous phenol solution, aromatic and unsaturated compounds can be regenerated. Normal paraffins can be separated by dissolving a portion of the dry residue from the chloroform extract into a mixture of methylethylketone-toluene (1:1). On cooling, the high molecular weight paraffins crystallize, or if a diluted hexane or chloroform extract of sample is centrifuged most of the interferences separate. The final solution should always be made with hexane, which is the most suitable solvent for gas chromatographic analyses. Once the purified extract has been obtained, the analysis can be accomplished by using the same nonselective column used for the gasoline and kerosene separation. The only alteration will be the range in temperature programming. As noted above, tetradecane elutes at $191^\circ C$ (see Sect. III, B). Some nonpolar naphthenes present in oil will elute at this temperature or slightly higher. For that reason, the programmed temperature should be started at approximately $140^\circ C$.

For a more precise analysis the extract to be run should be concentrated to approximately 20 μl for each milliliter of solvent [10]. Such a concentration will have two purposes: first it will eliminate the light fraction of the oil or contaminants, if present; and second it will permit using the instrument at low sensitivity. The low sensitivity will allow the analyst to obtain a steadier baseline and a marked deflection of the pen at the elution place of the oil. The concentration of the sample is easily achieved by heating the container of the sample (70°-80°) and passing a gentle stream of clean dry air or nitrogen over the surface of the sample. A 3.0-μl sample of the concentrated extract is injected onto the same column with the same settings used for the analysis of gasoline and kerosene (see Sect. III, A and B). The temperature program

will be 140°-235°C (at 7.5°C/min) and hold. The chromatogram will show
a steadily decreasing baseline from 140°C to approximately 195°C representing an absence of kerosene. At approximately this moment, depending on
the quality and characteristics of the oil being investigated, a group of
peaks will appear. The eluting peaks are masking each other and only small
perturbances will be observed. The bulge on the chromatogram represents
the totality of the oil components (Fig. 4). Some heavy asphaltenes will
remain in the column. The paraffins and naphthenes, which are the main
components of the oils, are high molecular weight, semisolid, and nonpolar.
If a short column, such as 6 ft (a nonpolar solvent on a nonselective phase),
is used the oil components do not elute separately. Alkylated naphthene-
aromatics with some degree of polarity will elute at the maximum temperature recommended but will require a longer time. That is the reason for
the holding of the temperature at 235°C until the baseline indicates the total
elution of the injected sample. Sometimes the asphaltenes are held by one
or both of the phases in the column and thus elute at varying times. To
minimize such an occurrence the column should be washed one or two times,
before the injection of the sample, with 5 μl of hexane and temperature programmed at 140°-235°C with a 5-min hold at the upper temperature. The
height and shape of the peak representing the oil will be in accordance with
the quality of the oil itself: light oils elute sharply and quickly; heavy oils
elute smoothly and slowly. Some characterization of oils can be approximated by such variations of elution. Elutions of some oil components take as
long as 35 min.

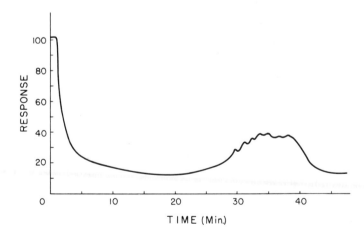

Figure 4. Typical chromatogram for the analysis of mineral oils.

D. Identification of Aromatic Compounds

Compounds such as benzene, toluene, xylene, and styrene are natural or
synthesized products derived from crude petroleum and from coal. Some
petroleums contain up to 35-37% of aromatic hydrocarbons and their alkyl
derivatives of benzene, such as isopropylbenzene, pseudocumene, and
diphenylmethane. For practical purposes, the identification of aromatic
hydrocarbons is carried out by their characteristic responses to ultraviolet
light absorption. The scientist involved in the analysis of aliphatic aromatic
hydrocarbons will deal with a long list of compounds either natural, synthetic,
or byproducts of chemical procedures in the preparation of some other
compounds. The complexity of the potentially infinite sources of hydrocarbons, lower or higher in their structural reactivity and formation of other
compounds, is illustrated by the fact that almost all aromatic hydrocarbons
react easily with acids, bases, oxygen, nitrogen and sulfur compounds,
hydrogen, chlorine, and bromine. These form even more complex compounds, ranging from the relatively simple phthalic acid with two carboxylic
acid groups attached to the benzene ring, up to the hexaphenylethane or
polystyrene, which is a long chain of $-CH-CH_2-$ groups with a benzene ring
attached to each CH group. The dipole moments of these compounds are
slightly positive, but dipole moment factors are not applicable as a whole
in the gas chromatographic technique. The affinity of substances for a
polar solvent depends greatly on their dipole moments as well as on their
molar volumes and steric factors. Benzene, for example, has not dipole
moment although it is accepted and used as a polar solvent in gas chromatography [14]. Giving proper consideration to the chemical complexity of
the aliphatic-aromatic hydrocarbons, it will be apparent that, in wastes,
the most probable aromatics to be found are benzene, toluene, xylene, and
styrene. Any nonpolar or low-polarity solvent will extract these compounds
from water. Unfortunately, all the other organic contaminants will also be
extracted. To separate the aromatics from these impurities, the extract is
diluted with an appropriate solvent and shaken with an equal mixture of 1:1
chloroethyl ether:acetonitrile. The partition will not be quantitative but will
serve the purpose of qualitative indentification. It is sufficient to extract
the sample with hexane in order to exclude the water from the hydrocarbons;
these are diluted at a ratio relative to their concentration with a final
proportion of approximately 99:1. The extract is shaken in a separatory
funnel with an equal volume of chloroethyl ether-acetonitrile and this last
portion is used for the chromatographic analysis. The use of standards in
this case is very important. These standards are prepared with the same
chloroethyl ether-acetonitrile solvent and run a few times on the selected
column with or without temperature programming in order to establish a
precise elution time for the aromatic to be identified [9]. These precautions
are necessary in order to avoid complications in the interpretation of
chromatograms resulting from injections of samples that may carry a great

number of side products. The side products so interfere with the analysis
that sometimes a liquid chromatographic separation is necessary. If this
is the case, use an activated alumina column prewetted with hexane. The
sample is extracted with a small portion of hexane and an aliquot is added
to the column. The column is eluted first with a small amount of hexane-
diethyl ether (85:15), second with hexane-diethyl ether (50:50), and finally
with a 1:1 portion of chloroethyl ether-acetonitrile. The three portions are
chromatographed separately and compared with proper standards prepared
with the same solvent as the eluates. The adsorption coefficients of the
aromatics, relative to the liquid phase behavior, are nonselective so a
great number of stationary phases can be used for their separation. With
a 6 ft × 0.25 in. o.d. nonpolar column loaded with 5% each of SE-30 and OV-1
on Gas Chrom Q, 80/100 mesh, a flame ionization (FID) detector with nitrogen carrier gas (85 cm^3/min), at 300°C, inlet at 220°C, and the oven at
210°C, the aromatics may be separated although their elution times are
very close to each other. By raising the injection port temperature to
240°C and temperature programming from 190° to 240°C at 5°C/min, a
much better separation is obtained. Even better separations are obtained by
using a column packed with a well-homogenized 1:1 mixture of OV-17
(slightly polar phase) on Gas Chrom Q, 100/200 mesh, and a permanently
bonded OPN (oxypropionitrile) phase on porous silica beads (Poracil C).
High temperatures are recommended for this column (150°C maximum).
The parameters are the following: detector, 290°C; inlet, 140°C; column,
120°C. The chromatogram shows benzene, toluene, and xylene (ortho
isomer) in that order; the total time of elution is approximately 8 min.
Before the reuse of the column or a subsequent run, it should be purged
twice with 20 μl of acetonitrile in order to flush the absorbed materials out
of the liquid phases. Temperature programming from 80° to 120°C at 5°C/
min and holding will separate the components of the mixture more efficiently
[9].

The elution of styrene requires a very long time. Chromatographic runs
with proper standards are required for the identification of such compounds.

E.   Identification of Phenols

These benzene derivatives have one or more hydroxyl groups attached
directly to the carbon atoms of the benzene ring. Some phenols are
products of natural origin, but since these compounds are very few and
small in amount, I shall consider only those products resulting from industrial processing of petroleum, paper and wood, dyes, drugs, photographic
developers, and resins. Phenols are all water soluble and slightly acidic.
The weakly acidic fraction obtained by chloroform elution of organics
adsorbed by activated charcoal will contain the phenols (as explained in
Section III, C). Phenols react easily to form salts, which undergo a quick

hydrolyzation to ethers and esters. These physical and chemical characteristics of phenols are very useful in gas chromatographic analyses. The
waste to be analyzed for phenolic content must first be diluted 1:1 and well
mixed with a solution of 1N NaOH followed by a 1:1 mixture of ether-
petroleum ether. The addition of NaOH solution converts the phenols into
their salts, which are soluble in water and insoluble in organic solvents.
The organic solvent will extract the other organic materials present from
the waste. The separation will be facilitated by swirling the two phases
gently. After separation, the lower aqueous phase is collected and acidified
with mineral acid in order to convert the salts to free phenols. The aqueous
solution can be used for direct gas chromatographic analysis or can be mixed
with a polar solvent that extracts the phenols if they are needed for esterification or for direct gas chromatographic analysis. The column to be used
for phenols separation varies according to the product and the solvent used.
If the water phase containing the phenolic acids is extracted with benzene,
the extracted phenols can be chromatographed on a hydrocarbon substrate
such as Apiezon L or N (5-15%) on Gas Chrom Q 80/100 mesh using a
6 ft × 0.25 in. o.d. glass column, hydrogen flame detector, column temperature $170^{\circ}C$, detector temperature $290^{\circ}C$, and nitrogen carrier gas (70
cm^{3}/min). The chromatogram obtained is shown in Fig. 5. This column
is nonselective, permitting the elution of highly polar phenol as well as
low-polarity esters of phenols, but the analyst must use good judgement in
planning the analysis of phenol wastes. Although the identification of such
pollutants is not difficult, the following consideration must be observed. A
phenolic waste may contain phenols or the converted ethers and/or esters.
On a straight analysis of such wastes with a highly selective polar column
(20% 20M-TPA, which is a mixture of Carbowax 20 and terephthalic acid),
only phenolic acids will be eluted. A hydrocarbon substrate, such as
the Apiezons, will elute both acids and ethers or esters but polyhydric
phenols are not eluted. If, for some reason the acid phenols must be

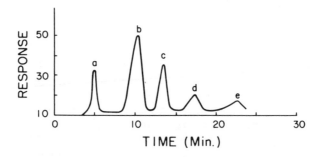

Figure 5. Chromatogram of phenol extract. (a) o-Chlorophenol, (b) phenol,
(c) m-cresol, (d) p-chlorophenol, (e) unidentified.

methylated, great care should be taken to prevent loss of the volatile esters; the polar polyester-type column, diethylene glycol succinate, is a good column for direct analysis of aqueous substrates of phenols. Methylation can produce a destruction of the sample by hydrolyzing it during the conversion of the acids to their salts. For some detailed quantitative identification of phenolic compounds given here as guidance only, excellent results have been achieved by many scientists in specific subjects, such as ether derivatives, ether and thioether derivatives of phenols, and mercaptans [15-17]. Again, the 6-ft column or even a 12-ft one will not be able to separate all the phenolic derivatives. Phenols and anisoles cannot be separated with 10% Carbowax 20M-TPA on Gas Chrom Q 80/100 mesh.

F. Lipids

The constituents of this family of contaminants are omnipresent. Their origins are widely spread from the isolated malodorous farm to the fragrance of an appetizing frying meal; from the ice cream factory to the discharge of a kitchen sink or a flushing toilet. With few exceptions, the attempted localization of lipid contamination sources is pointless. The exceptions are represented by such sources as slaughterhouses, canneries, dairies, tanneries, oil refineries (vegetable), and selected farms. If the sample is analyzed at the place of origin it is possible to establish the kind of lipid involved. The difficulty in localizing the source of contamination in a body of water is due to the complexity of the lipids, their reactivities (heterogeneous capabilities), oxidizability, and colloid-forming capacity. Bacterial metabolization, oxidation, and reduction of lipids play important roles in their fate depending on other factors that influence the bacterial life, such as available species or flow velocity of their carriers. The high volume of lipid contamination in a typical waste water from homes and municipal waste water is furnished by a study from associated researchers of the U.S. Water Pollution Control Federation, who reported 33-92 mg/liter and 16-200 mg/liter concentrations, respectively [18]. Where industrial wastes were present, larger amounts of lipids have been reported according to the size of the industry. The lipids are a family of substances that include glycerides, fats, oils, fatty acids, and sterols. All of these products are chemically and reactively interconnected. Glycerides include butter, seed oil, palm oil, olive oil, and lard and are esters of glycerol and related fatty acids such as butyric, linoleic, palmitic, oleic, and stearic. They are called fats when solids and oils when liquid. Hydrolysis or saponification of fats produces glycerol and fatty acids or glycerol and salts of fatty acids. The glycerides of saturated fatty acids are solids with low melting points: e.g., butter, palm oil, and lard. Those of unsaturated fatty acids are liquid, such as olive oil and seed oil. A single fat is usually made of one, two, three, or up to ten esters of different fatty acids.

Isolation of individual fatty acids from mixtures is extremely difficult. Chromatography of their methyl esters is possible [19, 20]. Palmitic and stearic acids are the main components of fatty acids commonly found; stearic acid makes up 10-30% of animal fats, whereas it is only present in traces in vegetables. Lauric acid constitutes up to 52% of palm oil. As one example of complexity, the composition of butter fat is [21]:

Volatile acids:	butyric	3.2%
	caproic	1.4
	caprylic	1.0
	capric	1.8
Nonvolatile acids:	lauric	1.9%
	myristic	22.6
	palmitic	19.3
	stearic	11.4
	oleic	27.4

Fats and oils, both vegetable and animal, consist of glycerol esters of the higher saturated and unsaturated acids and are distinguished from each other by the different acid radicals and different proportions in which they occur. Thus, fats are not pure chemicals but a mixture of several esters and this fact complicates the task of analysis. However, both the type of acid radicals present and the proportions found are definite and constant for any given fat being sought on analysis.

Fats undergo only two reactions: (a) hydrolysis that decomposes fats and oils into their constituents, which are an alcohol (always the same), glycerol, and acid or acids; and (b) hydrogenation, which is of no importance in this context. When analyzing fats and oils, the chemist must keep in mind the described complexity of such organic components and take advantage of their characteristics of differentiation as follows: fats and oils are water insoluble in a 99.9% ratio; they are all less dense than water; they are metabolized by microorganisms; they are emulsified and hydrolyzed by acids, alkalis, detergents, and soaps; organic solvents slowly but effectively oxidize them, thus changing their chemical and chromatographic characteristics. The solidification point of all saturated esters, with the exception of arachidin and the fats related to it, ranges from $49°$ to $27°C$; all the unsaturated esters form additional products with halogens.

The following procedures [10] should be observed. Considering the reactivity of the lipids, the sample to be analyzed must be collected and immediately preserved at $5°-10°C$. The pH must be checked at once and adjusted, as closely as possible, to the value of 7. If the sample must be adjusted for its pH, the result of the analysis must be interpreted not as the sample content at its origin, but as the content at the moment of the pH adjustment. The lower or higher pH will cause hydrolyzation of the lipids present in the sample. The sample to be analyzed will contain, with some exception and variation, mono-, di-, and triglycerides and/or free fatty acids. Given a "clean" sample, the triglycerides can be extracted and chromatographed without further treatment, but since they usually occur along with mono- and diglycerides the best way is to extract with chloroform. Three 20-ml portions of chloroform are used for the extraction. The portions are first combined and then separated into three equal parts: one for free fatty acids, one for glycerides, and one to be methylated. The first two parts are evaporated to dryness by gently blowing warm nitrogen over the chloroform. When just dry, the two residues are taken up with 1 ml of hexane each and chromatographed. For the free fatty acids a highly polar column must be used since this selective column does not respond to the many interferences present in the extract and is able to separate the acids, which otherwise are impossible to separate with a nonpolar column. The selection of the column and its conditioning are of paramount importance because the peaks of polar compounds tend to "tail" as well as give low responses due to the adsorption of the molecules by the solid support. The most practical solid phase is Teflon 6 (Chromosorb T). The conditioning must be done with a very slow programming temperature and the liquid phase load must be low. Careful preparation of such a column gives satisfactory results.

A great number of liquid phases with varied degrees of polarity can be used (DEGS, OV-17, OV-225, Carbowax 20M, and SP-1000). I prefer to use 8% Carbowax 20M with the modification applied by Metcalf [22], consisting of an additional 2% phosphoric acid added to the liquid phase prior to the conditioning. The phase is packed into a 1/8-in. column 5 feet long, run at 130°C with the injection port at 150°C and FID detector at 250°C, using helium as the carrier gas at 50 cm^3/min. The short-chain fatty acids will elute in the following manner: acetic acid, 0.8 min; propionic acid, 0.95 min; butyric acid, 1.3 min; and caproic acid, 3.3 min. By raising the temperature of the column to 195°C and the injection port to 210°C, washing the column twice with 10 μl of hexane, and allowing the column and temperature to equilibrate, it is possible to separate the long-chain fatty acids. The order and time of elution are as follows: myristic acid, 1.2 min; palmitic acid, 1.85 min; palmitoleic acid, 21.5 min; stearic acid, 3.1 min; oleic acid, 3.2 min; linoleic acid, 3.65 min; and arachidic acid, 5.1 min. The second hexane portion is chromatographed for the quantitation of the glycerides. A nonpolar column will not resolve the free acid, already analyzed by the previous method, but will resolve other nonpolar contaminants. To avoid such interferences, the analyst must judge the most practical applications for the problem at hand. A skillful set of runs with different sets of standard mono-, di-, and triglycerides and with different programmed temperatures on different polarity columns (one with 10% OV-1 and another 5% OV-1 plus 5% OV-225) will probably give good qualitative results. Another practical means of solving the problem is by partitioning the extract by two different solvents, shaking equal portions of extract with acetonitrile; and running the two solvents in the same manner as explained above (Sect. III, D). The ideal solution to such a complex critical analysis would be the use of a high-pressure capillary column long enough (300-400 ft) to permit the specific separation and individualization of all components. Nevertheless, given the optimum conditions on a well-conditioned 6-ft glass column with silanized Gas Chrom Q coated with 3% OV-1, the following triglycerides will be resolved in less than 10 min: caproin, caprylin, caprin, laurin, and myristin. The conditions are: column temperature, 190°C; injection port temperature, 210°C; and FID temperature at 250°C, with nitrogen as carrier gas at 90 cm^3/min [9]. The higher triglycerides, as well as the equivalent in mono- and diacid groups, will give responses analogous to comparable standards.

The third portion of the extract from the sample is used to separate total lipids (excluding the polar lipids) by means of esterification and trans-esterification. From the results of the three analyses and a little calculation, the correct identification of lipids present in the sample will be possible. The third extract is evaporated just to dryness and the residue is dissolved in 10 ml of methanol. Two milliliters of concentrated sulfuric acid are added and the mixture is boiled for 2 min in a glass vessel (preferably a closed one similar to a Kuderna-Danish evaporator). The free acids will be converted to their methyl esters and the glycerides also present in the extract will be converted to the same by transesterification. Extreme care must be taken during the boiling and cooling of the mixture. The short-chain acids and fatty acids are quite volatile. Thus, the reaction vessel should not be opened until room temperature has been reached. The chromatography of the esters is done by following the same procedures as for the glycerides, the only difference being the solvent involved, which is diethyl ether rather than hexane. The reacted mixture is extracted with three 5-ml portions of diethyl ether. The combined extracts are washed twice with 10 ml of pure water and dried by passing them through a small funnel filled with anhydrous sodium sulfate. They are washed repeatedly with diethyl ether and evaporated to a 1-ml volume using low temperature and nitrogen gently flowing over the extracts. This is the classical way to methylate lipids, the procedure is easy, and the results are good; yet many other methods are available to the analyst if more specifically detailed results are sought [23-25].

Very recently the introduction on the market of chemically bonded liquid phase has opened a new view on difficult gas chromatographic analyses, such as those related to lipids. The tool is new but very promising and a preview of the possibilities such phases offer is given by several excellent papers on the subject [26-29].

The last group of lipids to be considered is the polar lipids or phospholipids. These fats are nothing more than glyceride complexes of phosphoric acid groups and nitrogeneous bases, with sometimes a glucose or galactose group replacing the phosphoric acid group. Almost all the fatty acids found in common fats can be found in this kind of lipid. Their chemical complexity and minor roles (relative to the purpose of this chapter) suggest that the task be left to specialized researchers in gas chromatography.

G. Sterols

Sterols are lipids of crystalline alcohols, cholesterol being the most widely distributed. The reduction products of cholesterol are cholestanol and the two related homologs, cholestane and coprostane. They are water insoluble and can undergo the esterification process very slowly and under special circumstances; thus, they are not considered here. Qualitative separation of steroids from other contaminants is facilitated by the fact that they are unsaponifiable. Yet, this chemical property cannot be considered as an absolute analytical advantage, and the following facts must be evaluated when the analysis is reported. Sterols are insoluble in water but are sometimes colloidally suspended in it. They are soluble in water highly contaminated by glycerides and/or organic solvents. The analysis of the sample at the origin of contamination would ideally eliminate those serious inconveniences, since by saponifying the extract of the sample almost all the components would be separated from the unsaponifiable cholesterol, cholestane, and cholestanol. After the separation of all saponified matter, this portion of the extract must be repeatedly washed with diethyl ether in order to free the sterol particles occluded by the soap constituents. An even better way to achieve such separation is to pour the resultant mixture of the saponification into a suitable separatory funnel and extract it three times with diethyl ether. The final extract is evaporated with warm nitrogen to a volume dictated by the need of the specific sample and the degree of its contamination and chromatographed. The column and the parameters given for the free fatty acid analysis (see Section III, F) can be used for the separation of sterols, the retention times of the sterols being longer than those of the free fatty acids. The polar phase, 4% QF-1 on Gas Chrom Q solid support, will separate the sterols as follows: cholestane, 4.2 min; cholesterol, 11.0 min; and cholestanol, 12.0 min. Other polar phases can be used, and by adjusting temperature and carrier flows the retention times of the sterols can be varied considerably [9].

H. Pesticides

Wright et al. [30] described a gas chromatographic determination of Abate (0,0,0',0'-tetramethyl-0,0'-thiodi-p-phenylene phosphorothioate) residues in water. The sensitivity of the method is 0.05 ppm.

Pionke et al. [31] reported that extracted organochlorine insecticides could be determined on a 10% DC-200/Gas Chrom Q column at 195°C, with nitrogen carrier gas and an electron capture detector, and that organophosphate insecticides could be determined on a 10% DC-200/Chromosorb W column at 200°C, with helium carrier gas and a potassium chloride thermionic detector. In determinations on lake waters, recoveries of the organochlorine insecticides (at milligram per liter concentrations) were 94.4-102%, except for heptachlor, the recovery of which was ≤88.0%. Standard deviations ranged from 0.60% for the γ isomer of 1,2,3,4,5,6-hexachloro-cyclohexane to 1.36% for methoxychlor. Average recoveries of organo-phosphate insecticides were greater than 95%. Aly [32] outlined a quantitative determination of sevin in natural water. The determination involves chloroform extraction of sevin and hydrolysis of sevin and its hydrolysis product, 1-naphthol. The sensitivity of the method is 5 μg/liter.

Kadoun [33] described a simple technique for the removal of interferences before determination of chlorinated and organic phosphorus insecticides by gas chromatography. It involved passage of the sample through a micro-column of high-purity activated silica gel and selective elution of the insecticide with different solvent mixtures varying in polarity. Faust and Suffet [34] and Suffet et al. [35] described methods of recovery, separation, and identification of organic pesticides from natural waters. Diazinon, Baytex, parathion, and malathion were eluted on Reoplex 400 (polypropylene glycol adipate), a polyester liquid phase, which was coated on Chromosorb W. The pesticide systems were examined by gas chromatography using FID and electron capture detectors. Kawahara et al. [36] described a chromatographic procedure that permitted the rapid determination of 14 chlorinated hydrocarbons in a water grab sample, even though two pairs of compounds with identical or similar retention periods were involved. The co-emerging pairs were resolved by a preliminary thin-layer chromatographic procedure and the eluates then were concentrated and analyzed by gas chromatography, using electron capture or microcoulometric techniques.

Corcoran et al. [37] described a method for determining chlorinated hydro-carbons, such as chlordane, by analyzing head gas. The sample, together with anhydrous sodium sulfate, was placed in a bottle, sealed with a silicon cap, and heated in a water bath at 72°C. Some of the gas collected above the water was then extracted and analyzed in a gas chromatograph with a FID detector. Wright et al. [30] presented a detailed procedure for the determination by gas chromatography of the insecticide Abate, which is commonly used for the control of mosquito larvae. When the sample was extracted with chloroform, average recoveries of about 70% were obtained for water samples containing known amounts of Abate. The lower limit of detection was 0.05 mg/liter.

The fund of knowledge concerning pesticides, organomercuric salts, and their reactions is vast and continually changing. The reader is directed to the literature for some of the methods for the analysis of these contaminants [38-51].

I. Sulfur-containing Compounds

Sampling and direct gas chromatographic analysis of paper mill effluents for CH_3SH, CH_3SCH_3, and CH_3SSCH_3 have been reported [52]. Various organic solvents and mercuric cyanide solutions have been evaluated for collection of CH_3SH [53]. Methyl mercaptan, along with ethyl mercaptan and dimethyl sulfide, was separated on a 4-m tricresyl phosphate column prior to gas chromatographic analysis [54].

J. Carbonyl Compounds

Carbonyl compounds in automobile exhaust emissions have been collected in bubblers containing aqueous 2,4-dinitrophenylhydrazone reagent. The hydrazones were filtered, extracted, and dissolved in CS_2 and then analyzed by gas chromatography with a FID [55]. Sampling efficiencies, analytical errors, and potential interferences were evaluated.

Another investigation reported the analysis of aliphatic carbonyl compounds and benzaldehydes in exhaust gas by gas chromatographic separation of their 2,4-dinitrophenylhydrazones and by flame ionization detection [56].

K. Industrial Emissions and Related Compounds

1. Polynuclear Aromatic Hydrocarbons

Polynuclear aromatic hydrocarbons have been analyzed in emissions from several industrial processes, including asphalt air blowing, asphalt hot-road mix, and the manufacture of carbon black, steel, coke, and chemicals [57]. Effluents from the stack of a coal burning residence, from two industrial sources, and from air contaminated with coal-tar pitch fumes have been analyzed for 12 polynuclear aromatic hydrocarbons and 11 aza heterocyclics, including the animal carcinogens benzo[a]pyrene, benz[a]-anthracene, dibenz[a,h]acridine, and dibenz[a,j]acridine [58].

Several reports have been published describing the use of gas chromatography to separate and determine the widely publicized carcinogen, benzo-[a]pyrene (BaP); however, none of them describe a complete separation of BaP from its isomer benzo[e]pyrene (BeP) [59]. Only two of the reports [60, 61] describe a complete separation of benzpyrenes from benzo(K)-fluoranthene (B_KF) and perylene by using capillary columns with silicone gum rubber SE-52 as the stationary phase.

2. Nitrogen Compounds

Operating conditions have been evaluated in the use of pulsed voltage for electron capture gas chromatographic determination of parts per million quantities of NO_2 [62]. A temperature programmed molecular sieve 5A column has been used to obtain separations of a number of inorganic gases, including NO and N_2O [63]. NO_2 was not eluted as a peak but did not interfere. Limits of detection with the thermal conductivity detector used for NO and N_2O were 12 and 25 ppm, respectively.

The calibration conditions and analytical columns used for peroxyacyl nitrates and alkyl nitrate analyses in irradiation chamber experiments have

been described. In a different gas chromatographic analysis, lower
molecular weight compounds were separated from higher molecular weight
compounds by the use of different column lengths and flow rate conditions
but with the same substrate, consisting of 5% GE Versilube F-50 on 90/100
mesh Anakrom F-30, packed in a Teflon tube that was thermostatted at
$0°C$ [64].

L. Miscellaneous Gases and Vapors

Mono-, di-, and trichloroacetyl chloride at low levels in air were determined by a newly developed GC method. The sample was absorbed in 2-
propanol and the ester then formed was determined with an electron
capture detector. $COCl_2$ (phosgene) could be determined down to 10^{-3} ppm
in air [7].

The spacecraft industry is faced with removal of certain reactive gases
from the atmosphere of spacecraft simulators. The separation of mixtures
of reactive NO, NO_2, SO_2, HCl, H_2S, Cl_2, and NH_3, plus such other gases
as CO_2 and N_2O, in air has been investigated [65]. A 10% SE-96 methyl
silicone oil on Fluoropak 80 was used for the separation of NO_2 in air.
Morrison et al. [66] investigated five different columns for the separation
of N_2O, NO, NO_2, NH_3, O_2, and CO_2. Activated coconut charcoal was the
only column that provided satisfactory separations of CO_2 and N_2O.

Trowell [67] developed a three-column system for the separation of H_2,
O_2, N_2, NO, CO, N_2O, CO_2, H_2O, and NO_2. Another researcher used an
8-ft column packed with Davison No. 912 silica gel at $28°$-$31°C$ for the
separation of NO and N_2. A method for Cl_2, HCl, and HF required a novel
column packed with Kel-F No. 10 oil on Fluoropak at $50°C$. An argon
carrier was used [68]. Ellis and Forest [69] used an 11 ft × 6 in. column
packed with 50 wt% Kel-F No. 10 oil on PTFE powder for the quantitative
separation of ClF, HF, ClO_3F, Cl_2, ClO_2F, and ClF_3. Argon carrier
gas was used and the detection was by means of a Martin gas density
balance.

Chovin [70] has reviewed methods for the analysis of many of the common
pollutants of industrial atmospheres. Procedures for sulfur dioxide,
hydrogen sulfide dusts, aerosols, oxides of nitrogen, ammonia, ozone,
fluorine compounds, hydrocarbons, tars, carbon monoxide, nitrogen
peroxide, and aldehydes are covered. Bethea and Meador [65] have recommended a chromatographic system of three columns for the analysis of a
mixture of nitric and nitrous oxides, nitrogen dioxide, chlorine, hydrogen
chloride, fluoride, and sulfide, sulfur dioxide, and carbon dioxide in air.
Single columns and tandem combinations are suggested for less complex
mixtures.

Efforts to reduce atmospheric contamination have resulted in the enact-
ment of a solvent law in Los Angeles County, termed "Rule 66." It places
limitations on solvent emissions and on solvent formulations used within its

jurisdiction and as a consequence creates some analytical problems. MacPhee and Kuramoto [71] describe two procedures developed for use in conjunction with the rule. For emissions, a total combustion method is used. For solvent formulations, a column chromatographic screening is first performed to measure types of compounds, e.g., oxygenates, aromatics, olefins, and paraffins. This helps to select the conditions for a lengthier gas chromatographic method that provides the ultimate analysis of individual compounds.

Andrews [72] described two dual-column detector chromatographs that can be used for the analysis of nitrogen, carbon dioxide, oxygen, methane, hydrogen, and hydrogen sulfide in gas streams from anaerobic digesters, Warburg respirometers, and aerobic processes. With the addition of a zero suppressor the instruments also may be used to determine oxygen uptake rates and oxygen transfer efficiencies. Mayberry [73] discussed the use of several types of gas chromatographic columns capable of separating alcohols and free fatty acids of up to 18 carbons in length. Gas-liquid chromatographic studies were carried out to investigate the bacterial degradation of dodecyl sodium sulfate, aliphatic alcohols, members of the oligoethylene glycol series, benzoic acid, and phenylacetic acid.

Ellerker et al. [74] described methods for the use of gas chromatography in the analysis of sludge gas and the determination of individual lower fatty acids evolved in anaerobic digestion, higher fatty acids in waste waters containing grease, and petroleum fractions discharged to sewers.

A study was reported by Sugar and Conway [75] using gas-liquid chromatographic techniques for identifying and measuring organics occurring as complex petrochemical mixtures. Direct aqueous injections were practiced over a range of 1-100 mg/liter. For general utility in analyzing petrochemical waste organics, a 6.1 m × 3.2 mm carbon column was selected. Mass spectrographic and infrared identification of eluted components was used for confirmation.

Jaworski et al. [76] reported results of the determination of aldehydes, ketones, alcohols, esters, and ethers in waste water in concentrations of 10-100 mg/liter by gas-liquid chromatography, using a column containing particles of activated firebrick coated with polyoxyethylene glycol, nitrogen as carrier gas, and a flame ionization detector. The results were compared with suitable standards for quantitative analysis.

Jenkins et al. [77] reported gas chromatographic analyses of steam distillates of growing and decaying cultures of bluegreen algae, combined with spot tests to identify odorous sulfur compounds. Most of these were found to be associated with dimethyl sulfide and dimethyl disulfide. Actinomycete taste and odor compounds in water were investigated by Silvey et al. [78] using a gas chromatograph equipped with a hydrogen flame detector. Determinations were made of the liquid phases best suited for the study of actinomycete taste and odor compounds in aqueous samples.

Larson et al. [79] described apparatus and a procedure for determining low concentrations of organic matter in polluted streams or in sewage effluent after complete treatment. A sample is oxidized with the Van Slyke liquid reagent (containing chromium trioxide, potassium iodate, phosphoric acid, and fuming sulfuric acid); catalytic oxidation of any volatile compounds evolved follows. Carbon dioxide is trapped at $-195^{\circ}C$, purified by distillation under high vacuum, and measured manometrically. Except for a few volatile compounds that ordinarily do not occur in well-purified wastes, the method gives reasonably accurate results. One disadvantage of the method is the time required. Larson et al. are studying the use of gas chromatography for the final determination of the carbon dioxide.

Hermann and Post [80] reported that traces of indole and orthophenyl-phenol (10-200 ppm) in polluted water can be separated and determined by direct aqueous-injection gas chromatography on 5% butanediol succinate on Gas Chrom Q columns at $170^{\circ}C$.

Swinnerton and Linnenbom [81] used a gas chromatograph equipped with dual-hydrogen flame ionization detectors to determine small amounts of C_1-C_4 gaseous hydrocarbons. The hydrocarbons are concentrated in cold traps and then injected. Methane is detected on silica gel, whereas C_2-C_4 paraffins and olefins are separated on activated alumina containing 10% Nujol. Higher molecular weight hydrocarbons are separated on Chromo-sorb W (20% SF-96 or SF-30). As little as one part in 10^{13} by weight gaseous hydrocarbons can be detected in aqueous solutions.

Increasing public awareness of oil pollution has resulted in more attention to methods of detecting and measuring this contaminant. Alexander [82] discussed the problem of sampling oil-polluted waters and suggested that the best method of sampling the surface film of oil was by a cylinder fitted with filter paper, while a complete sample of the water was best obtained with a sliding cylinder. Crawford [83] described a method for detecting hydrocarbons in oil refinery waste waters by continuous monitoring in the air above. The use of gas-liquid chromatography for determining and identifying oil in waste and water and the preliminary treatment of the samples were discussed by Adams [84].

Sher [85] has successfully used gas chromatography to determine less than 5 ppm of hydrocarbons in cooling tower water. Samples were taken at atmospheric pressure in glass or low-porosity plastic containers, an internal standard was added, and 15-20 μl were charged to a chromatographic column of Apiezon L and detected with a FID.

Petrochemical waste water analysis has also received the attention of several investigators. Sugar and Conway [86] described a scheme for the gas chromatographic separation and identification of complex mixtures of organic compounds in plant effluents by direct injection onto various types of columns. By introducing an intermediate concentration stage in which nitrogen is used to strip the hydrocarbons from the water sample, Krichmar and Stepanenko [87] were able to effect a three to four order of magnitude improvement in the sensitivity of gas chromatographic methods for benzene, toluene, ethylbenzene, diethylbenzene, and isopropylbenzene.

Ramsdale and Wilkinson [88] report that gas chromatography alone can be used to distinguish the three major sources of beach pollution, i.e., materials resulting from the discharge of crude oils, fuel oils, or cargo oils at sea. Samples containing major amounts of sand and/or water can be analyzed rapidly without pretreatment. Brunnock et al. [89] also depend heavily on gas chromatography but in addition they have the use of measurements of the vanadium, nickel, and sulfur content.

IV. CONCLUSION

The prospect of analyzing an organic waste that contains all the pollutants mentioned in this chapter is staggering. Elucidation of methods for analyzing the various substances is offered to assist the chemist in selecting the gas chromatographic technique that is most likely to enable him to meet his challenge.

REFERENCES

1. J. A. Armour and J. A. Burke, J.A.O.A.C., 53(4), 761 (1970).

2. L. Sojak and A. Bucinska, Ropa Uhlie, 10, 572 (1968); B.A.A., 18, 1759 (1970).

3. K. Stransky, Feete Seinfen, Austrichm., 70, 543 (1968).

4. M. N. Chumachenko, Pakhomova I. Ye. Izv. Akad. Nank SSSR, Ser. Khim., 235 (1968).

5. W. C. Jones, Jr., Standards Methods of Chemical Analysis, Vol. IIIB, 6th ed. (F. J. Welcher, ed.), Van Nostrand, New York, 1966, pp. 1554-1578.

6. E. C. Sienwright, World Petrol., August (1971).

7. J. A. Dahlberg and I. B. Kelhman, Acta Chem. Scand., 24, 644 (1970).

8. A. L. Vander Kolb, Amer. Ind. Hyg. Assoc. J., 28, 588 (1967).

9. R. C. Dell'Acqua, unpublished data.

10. R. C. Dell'Acqua, unpublished work.

11. R. Jeltes and R. Veldink, J. Chromatog., 27, 242 (1967).

12. R. M. Gooding, Mechanical Engineer's Handbook (T. Baumeinster, ed.), McGraw-Hill, New York, 1958, pp. 7-21.

13. L. F. Fieser and M. Fieser, Organic Chemistry, 3rd ed., Reinhold, New York, 1956, p. 109.

14. Gunnel Westoo, Acta Chem. Scand., 20(8), 2131 (1966).

15. F. K. Kawahara, Anal. Chem., 40, 1009 (1968).

16. I. T. Clark, J. Gas Chromatog., 6, 53 (1968).

17. A. C. Bhattacharyya, Asit Bhattacharjee, O. K. Guna, and A. N. Basu, Anal. Chem., 40, 1873 (1968).

18. R. C. Loehr and C. T. Navarra, Jr., J. Water Poll. Contr. Fed., 41(5), R142 (1969).

19. M. E. Mason, M. E. Eager, and G. R. Waller, Anal. Chem., 36, 587 (1964).

20   S. J. Jankowski and P. Garner, Anal. Chem., 37, 1709 (1965).

21. J. S. Chamberlain, A Textbook of Organic Chemistry, P. Blakiston Son, Philadelphia, Pa., 1934.

22. L. D. Metcalf, Nature (London), 188, 142 (1960).

23. R. W. Kelly, J. Chromatog., 43(2), 229 (1969).

24. G. Rumpf, J. Chromatog., 43(2), 247 (1969).

25. J. A. Fioriti, M. J. Kanuk, and R. J. Sims, J. Chromatog. Sci., 7, 448 (1969).

26. V. Mahadevan and R. Decker, J. Chromatog. Sci., 8, 279 (1970).

27. G. E. Martin and J. S. Swinehart, J. Gas Chromatog., 6, 533 (1968).

28. D. B. Drucker, J. Chromatog. Sci., 8, 489 (1970).

29. A. Karleskine, G. Valmalle, and J. P. Wolff, J.A.O.A.C., 53, 1082 (1970).

30. F. G. Wright, B. N. Gilbert, and J. C. River, J. Agr. Food Chem., 15, 1038 (1967).

31. H. B. Poinke and J. G. Konrad, Analyst, 93, 363 (1968).

32. O. M. Aly, J. Amer. Water Works Assoc., 59, 906 (1967).

33. A. M. Kadoum, Bull. Env. Contam. Toxicol., 2, 264 (1967).

34. S. D. Faust and I. H. Suffet, Residue Rev., 15, 44 (1966).

35. I. H. Suffet, S. D. Faust, and W. F. Carey, Env. Sci. Tech., 1, 639 (1967).

36. F. K. Kawahara, R. L. Moore, and R. W. Gorman, J. Gas Chromatog., 6, 24 (1968).

37. E. E. Corcoran, J. F. Corwin, and D. B. Seba, J. Amer. Water Works Assoc., 59, 752 (1967).

38. H. Beckman and D. Garber, J.A.O.A.C., 52(2), 286 (1969).

39. J. A. Burke, J.A.O.A.C., 52(2), 270 (1969).

40. H. Egan, J.A.O.A.C., 52(2), 306 (1969).

41. R. R. Watts and R. W. Storherr, J.A.O.A.C., 52(3), 513 (1969).

42. A. M. Kadoum, Bull. Env. Contam. Toxicol., 3(6), 354 (1968).

43. C. C. Cassil, R. P. Stanovick, and R. F. Cook, Residue Rev., 26, 63 (1969).

44. D. M. Coulson, U.S. Agr. Res. Serv., (4), March 29, 1968.

45. H. B. Pionke and G. Chesters, Soil Sci. Soc. Amer. Proc., 32, 749 (1968).

46. V. Leoni and G. Puccetti, J. Chromatog., 43(3), 388 (1969).

47. W. E. Westlake and F. A. Gunther, Residue Rev., 18, 175 (1967).

48. J. O'G. Tatton and P. J. Wagstaffe, J. Chromatog., 44(2), 284 (1969).

49. J. R. Pardue, J.A.O.A.C., 54(2), 359 (1971).

50. L. M. Law and D. F. Goerlitz, J.A.O.A.C., 53(6), 1276 (1970).

51. Methods for Chlorinated Pesticides in Water and Wastewater, Fed. Water Poll. Contr. Adm., U.S. Dept. Interior, Washington, D.C., April 1969.

52. I. H. Williams and F. E. Murray, Pulp Paper Mag. Can. (Quebec), 67, T-347 (1966).

53. L. W. Niemeyer and R. A. McCormick, J. Air Poll. Contr. Assoc., 18, 403 (1968).

54. M. Oberg, Env. Sci. Tech., 2, 795 (1968).

55. P. K. Mueller, M. F. Fracchia, and F. J. Schette, Paper presented at the 152nd Nat. Meeting, Amer. Chem. Soc., New York, N.Y., September 11-16, 1966.

56. M. F. Fracchia, F. J. Schette, and P. K. Mueller, Env. Sci. Tech., 2, 464 (1968).

57. D. J. von Lehmden, R. P. Hangebrauck, and J. E. Meeker, J. Air Poll. Contr. Assoc., 15, 295 (1965).

58. E. Sawicki, J. E. Meeker, and M. J. Morgan, Int. J. Air Water Poll., 9, 291 (1965).

59. H. J. Dawson, Anal. Chem., 36, 1852 (1964).

60. V. Cantuiti and G. P. Cartoni, J. Chromatog., 15, 141 (1964).

61. N. Carugno and S. Rossi, J. Gas Chromatog., 5, 103 (1967).

62. M. E. Morrison and W. H. Corcoran, Anal. Chem., 39, 255 (1967).

63. R. N. Dietz, Anal. Chem., 40, 1576 (1968).

64. B. Dimitriades, J. Air Poll. Contr. Assoc., 17, 460 (1967).

65. R. M. Bethea and M. C. Meador, J. Chromatog. Sci., 7, 653 (1969).

66. M. E. Morrison, R. G. Rinkler, and W. H. Corcoran, Anal. Chem., 36, 2256 (1964).

67. J. M. Trowell, Paper presented at the 16th Annual Pitts. Conf. Anal. Chem. and App. Spectrosc., Pittsburgh, Pa., 1965.

68. J. F. Ellis and G. Iverson, Gas Chromatography 1958 (D. H. Desty, ed.), Butterworths, London, England, 1958, p. 300.

69. J. F. Ellis and C. W. Forrest, J. Inorg. Nucl. Chem., 16, 150 (1960).

70. P. Chovin, Bull. Soc. Chim. Fr., 2191 (1968).

71. R. D. MacPhee and M. Kuramoto, J. Air Poll. Contr. Assoc., 19, 443 (1969).

72. J. F. Andrews, Water Sewage Works, 115, 54 (1968).

73. W. R. Mayberry, Diss. Abstr., 27, B, 3388 (1967).

74. R. Elleker, H. J. Dee, F. G. I. Sax, and D. A. Sargent, Water Poll. Contr. (Brit.), 67, 542 (1968).

75. J. W. Sugar and R. A. Conway, J. Water Poll. Contr. Fed., 40, 9 (1968).

76. M. Jaworski, T. Stareczek, and J. Bobinski, Water Poll. Abstr. (Brit.), 41, 312 (1968).

77. D. Jenkins, L. L. Medsber, and J. F. Thomas, Env. Sci. Tech., 1, 731 (1967).

78. J. K. G. Silvey and W. H. Glaze, J. Amer. Water Works Assoc., 60, 440 (1968).

79. T. E. Larson, F. W. Sollo, and B. J. Gruner, Purdue Univ., Eng. Bull. Ext. Ser., 117, 761 (1964).

80. T. S. Hermann and A. A. Post, Anal. Chem., 40, 1573 (1968).

81. J. W. Swinnerton and V. J. Linnenbom, J. Gas Chromatog., 5, 570 (1967).

82. E. Alexander, Water Poll. Abstr. (Brit.), 41, 445 (1968).

83. H. M. Crawford, Hydrocarb. Pro. Petrol. Refin., 45, 130 (1966).

84. I. M. Adams, Process Biochem., 2, 33 (1967).

85. J. A. Sher, Oil Gas J. , 66(36), 93 (1968).

86. J. W. Sugar and R. A. Conway, J. Water Poll. Contr. Fed. , 40, 1622 (1968).

87. S. I. Krichmar and K. E. Stepanenko, Khim. Prom. , 235 (1969).

88. S. J. Ramsdale and R. E. Wilkinson, J. Inst. Petrol. , 54, 315 (1968).

89. J. V. Brunnock, D. F. Duckworth, and G. G. Stephens, J. Inst. Petrol. , 54, 299 (1968).

Chapter 17

LIQUID CHROMATOGRAPHIC ANALYSIS IN WASTE CHEMISTRY

Charles E. Hamilton

Waste Control, 628 BLDG
The Dow Chemical Company
Midland, Michigan

I. INTRODUCTION

The process of liquid chromatography is a traditional means of separation for many mixtures. Historically, the process has been limited in its application by the length of time required to perform the separation and by its ability to separate the constituents as eluted from the column. Until recently, the method had been retired to the role of a cleanup procedure of last resort.

Presently, most analyses, even in the most recent literature, are conducted using the more classical column methods wherein the eluent moves by gravity. In this capacity, liquid chromatography is not particularly popular for three reasons. First, reproducibility is poor. It is difficult, at best, to prepare two columns the same way. This can prove disasterous when several components must be separated as is the case with most "waste" analyses. The composition of the eluate must be determined by individual determinations on incremental cuts. The cut containing the parent compound is established and subsequent determinations are based on the parent compound always being present in the predetermined fraction of the eluate (cut). If column conditions change, so does the fraction of the eluate containing the parent compound. Second, recovery is generally incomplete. On an average, if 90% is recovered, it is regarded as good recovery. This can be acceptable if the recovery is reproducible, which is not the case in most instances. Third, the time for analysis is generally quite long, although this is a variable depending on the specific analysis.

The major contributing factor in all of these disadvantages is the inability to monitor the eluate during the course of the analysis. Interfering substances can easily go undetected depending on measuring techniques. As a result, most applications of liquid chromatography are those involving group classifications of compounds or cleanup procedures where interfering materials are adsorbed on an active packing essentially irreversibly. There are applications wherein one compound has been effectively separated by partitioning from an interfering matrix but this has been rare in the literature. The advent of high-speed, high-resolution columns and new detectors has sparked new interest in the technique for waste separations. The greater interest and concern for the environment has helped increase the need for better waste analysis. The demand for the separation of non-volatile or unstable compounds from complex waste mixtures (oils, water, and natural plant and animal materials) has increased significantly and gas chromatographic methods are unable in many cases to separate or detect these components.

II. SENSITIVITY AND SELECTIVITY

Sensitivity and selectivity (specificity) are distinct parameters although closely related in any form of chromatography. Both depend on the purity of the sample and reagents, technique, and the means of measurement or detection. Sensitivity may be regarded as the lowest concentration of a substance measurable in the presence of some sample and reagent blank. However, both sensitivity and selectivity must be determined according to the intended application of the analyst. In classical liquid chromatography the analysis of the eluate may limit or expand sensitivity or selectivity. Columns have been employed for gross isolation of a compound or class of compounds from a complex matrix that may have precluded analysis by a more sensitive or selective technique, such as GC. In most applications of this type, milligram quantities are usually the limit of the column, even though the means of detection may extend several orders of magnitude lower. There are other applications where columns are employed in concentration steps. That is, large volumes of eluant containing a complex matrix with a compound of interest at a low concentration are passed through a column that has an almost irreversible affinity for the compound of interest and very little else. Thousands of liters of eluant have been passed through columns, such as charcoal, to accumulate as little as a few milligrams of compound. These techniques are not popular as the adsorption is not completely irreversible and hence the analysis is not quantitative. Tandem columns have been employed to improve recovery but this is not practical in light of better techniques, such as thin-layer chromatography.

There are two classifications for the applications of liquid chromatography in waste analysis: (1) screening studies, and (2) cleanup procedures (the sample is purified by column techniques and then analyzed independently by another technique, such as GC).

A. Screening Studies

In screening studies, the analyst is concerned primarily with what is present rather than with how much. This is not the most common application because of more selective and faster techniques, as has been pointed out above. A column (ca. 0.5 in. i.d.) is packed with an active adsorbent, such as alumina, silica, or charcoal. An inert material, such as diatomaceous earth, may also be used although the former is more common. The usual particle size is about 150 μm. The mobile phase will consist of the sample matrix or an extract of the matrix. As the mobile phase passes through the column, the adsorbent selectively removes or retards almost irreversibly a compound or class of compounds contained in the eluent. After the eluent has passed through the column bed, the stationary phase is washed with the same solvent, ideally at the same ionic strength (if this does not interfere with the analysis). If the purpose of the column has been to remove contaminants then the eluant and the wash may be analyzed. If the purpose of the column has been to isolate the compound or compounds of interest, then the eluant and wash may be discarded but only after the analyst has determined what percentage of the compound(s) has been retained. Even though screening studies or qualitative analyses are being discussed here, when column techniques are used the sensitivity of the means of detection may be limited and a compound of interest may go undetected because of low recovery. Tandem columns are best employed for this purpose. A second eluant is now passed through the column(s) and the compounds are now removed (eluted) and subjected to further analysis. Flow rate is critical and must be determined for each analysis. The column bed must be penetrated thoroughly by fresh portions of the eluant. It is best to pass the eluant through in small volumes, allowing each volume to pass through the column so that the meniscus of the eluant is even with the top of the column bed before the next volume is added. This will prevent solute from dissolving in the eluant and diffusing away from the column bed, causing low recovery.

B. Cleanup Procedures

In cleanup procedures, the primary concern of the analyst is to separate one or more components from a matrix he or she knows can interfere with the quantitative analysis. Criteria of sensitivity and selectivity will depend on the degree of interference of the matrix. The behavior of a standard in the system must be determined first. The standard must represent the purest form of the compound of interest that is available. The portion of the eluant containing most of the compound must be determined along with the range of volume in which all the compound is found. The analyst must then determine whether this volume changes with sample concentration and with packing quantity. Sufficient determinations should be run to provide some statistical evaluation of the reproducibility, particularly if recovery is less than 100%. In this way, the analyst has established his or her accuracy and precision limitations under ideal conditions — in the absence of sample matrix. Next, the analyst must "spike" some sample with standard over some concentration range representative of his or her analysis. This accomplishes two things: the analyst determines whether recovery changes or whether the fraction containing the compound shifts; second, from a comparison of standard calibration curves, the analyst determines degree of matrix interference.

III. SAMPLING AND SAMPLE HANDLING CONSIDERATIONS

Sampling techniques are not particular for column techniques in waste analysis but are instead determined by the nature of the analysis. Waste systems analyses are naturally complicated, for each sample is based on a different matrix, however subtle the difference. Often overlooked, but extremely significant to the ultimate interpretation of the results, is the procedure for obtaining the sample. If the sample is obtained by someone else, it is even more important to know the why, where, when, and how of the sampling and handling procedure until it is in the analyst's hands.

Judicious consideration of the "why" significance of the ultimate results versus the sampling procedure, handling history, and the sample life of the desired constituents may determine that the analysis can yield no significant results. Sampling of industrial waste waters has been treated by Mancy and Weber [1]. Suffice it to say that the sampling in most waste analyses must be random and representative. The "grab" technique is most popular. In the grab technique, an isolated sample is obtained from a specific site and then analyzed independently or combined with other grab samples to form a regional composite sample. This may cause storage problems and the analyst must always be concerned with what may occur in the sample with time due to such things as chemical reaction of sample components, bacterial action, light effects, or oxidation. The number of samples will depend on the variation of component composition in the sample source.

A. Aqueous Waste Sample Systems

The first and foremost problem in aqueous systems is that of sampling the water representatively. Multiphase liquids and solids in water wastes are difficult to sample and sometimes result in very unrepresentative samples. Use of column chromatography as a sampling technique can greatly simplify and improve reliability of the results. The column acts as a filter for solids and insoluble liquids and tends to extract the soluble components into the separated phases as well as to separate them on the column with the aqueous eluant. Once the column is loaded with the nonaqueous materials, the problem becomes one of finding the solvent or solvents that can perform the desired separation or at least the partial splitting of the waste materials. Detection of organic loading on the column can be accomplished by use of nonspecific detection methods, such as total organic carbon or total oxygen demand of the aqueous effluent. Solvent gradient elution with several solvents has been made automatic, which increases the rate and ease of the inevitable trials for the desired separation solvent system. Available fraction collecting apparatus coupled with quick volatilizing techniques have made it possible to obtain nonvolatile residues on each fraction. The residue weight, although it does not indicate the degree of qualitative separation, does indicate the fractions for further chromatographic activity. Ultimately, qualitative or quantitative detection is required for some samples, but often solvent solubility groupings yield valuable information about the chemical characteristics of the residue. These chemical clues lead to choice of the better techniques available for the identification and quantification of the unknown. Scanning ultraviolet and infrared spectrophotometry can yield direct qualitative and quantitative results for many organic chemicals, particularly the phenolics, aromatics, and halogenated and oxygenated compounds [2]. Mass spectroscopy can be useful as a

detector if the compound is volatile under the conditions of the test [2].
Several spot tests or colorimetric reactions can be used to detect and
quantify many compounds [3].

B. Nonaqueous Waste Sample Systems

Distillations residues (tars) are one class of nonaqueous samples that lends
itself well to liquid chromatographic techniques. These wastes vary from
asphaltic solid tars, through semisolid gels, to almost liquids at room
temperatures.

Separation of the constituents can yield information regarding undesirable
side reactions, e.g., oxidations or condensations that have occurred in
the process of making the desired product. Identification and screening of
the compounds have yielded many new and useful "products."

Natural materials, such as human or animal excrement, vegetable
matter, biological solids, fish, and other aquatic organisms, are another
waste class of essentially but not always nonaqueous samples. Liquid
chromatography may be used to show the presence of unusual or foreign
substances, which is especially valuable in cases of injury to the organism
[2].

Chelating agents have been used to extract metallic ions from these
matrices with subsequent chromatographic separation and analytical identification. In addition to the analysis of impurities in the matrices, liquid
chromatography can be used to separate valuable constituents, e.g.,
enzymes and bioactive substances, from these waste materials [2].

For nonvolatile airborne waste solids and aerosol liquids, an adsorption
or reverse-phase column may be used to collect the sample from the air.
Separation of the constituents is accomplished by solvent gradient elution
followed by an appropriate detection system.

IV. EQUIPMENT

The very nature of waste analysis dictates that concentrations of any component be small, possibly less than 1 μg/liter. Contaminants in reagents or
adsorbents will, in many cases, be at a level to cause interference. Reagent
grade chemicals should not be considered adequate. Column solid phases
should be thoroughly washed with the eluant and the washing monitored by
whatever method of measurement is selected. Reagent blanks should always
be included in any analysis as the degree of contamination is most likely to
vary from lot to lot. Glassware should be thoroughly cleaned with whatever
solution proves to be effective but care must be taken that the "cleaning"
solution does not iself contribute contamination. Prior to use, glassware
should be thoroughly rinsed with the solvent they are to contain and dried

only if dilution errors are unavoidable. Etched glassware should never be
used as this type of glassware may contribute to contamination by leaching.
Also, the use of rubber or cork stoppers should be avoided. Polyethylene
stoppers and containers have also been known to adsorb species as well as
to leach contaminants, and use of equipment made of these materials should
be predetermined experimentally in the laboratory.

A. Detectors

Without detection of the separated constituents, the best separation method
is of little practical value for analysis. The liquid chromatographic detector
is a device or system that measures the quantity or composition of the
column effluent. A detector's performance for a particular separation
problem depends not only on the quantity or composition measured but also
on the ability to develop a useful readout from it. Peak detection limits,
specificity, response time, noise levels, interferences, and the linearity
of the concentration-response relationship are important performance
characteristics of detectors. The limiting effects of these characteristics
differ for each waste separation problem.

Five different types of liquid chromatographic detectors have been evaluated by Huber [4], who concludes that all five detectors can be used with
highly efficient columns but that none of them is best for all problems.

Instrumental parameters and limitations of the standard and new liquid
chromatographic detection systems have been presented in and excellent
review by Conlon [5]. A good reference when considering the use of a
refractometer as a detector for liquid chromatography is a paper by Waters
[6].

B. Columns

A constant flow of new column packing materials (e.g., Zipak, Corasil, and
Durapak) for liquid chromatography has been offered. Many of these materials [7, 8] have helped, and will continue to help, resolve waste material
constituents by adsorption and partition chromatography. Greater uniformity
of particle size, better mechanical strength, and better column packing
techniques have improved separation efficiency and column-to-column uniformity to the extent that permits high pressure, high-speed separations.

The controlled surface porosity columns developed by Kirkland [9-12]
also yield fast, efficient separations. A Kieselguhr column with chemically
bound alkyl groups has been developed by Stewart and Perry [13]. Various
silicones have been used as liquid phases on several supports by Aue et al.
[14]. A Teflon support with a liquid reverse phase is described by Conder
[15]. A single dual-packed column of silica gel and alumina was used by
Hirsch [16] to separate high-boiling petroleum residues into four fractions.

Mendel [17] describes methods using alumina column packing. With the use of a tetranitrobenzylpolystyrene resin as a stationary phase Ayres and Mann [18] have separated 26 polynuclear (polycyclic) aromatics.

C. Pumps

In high-pressure liquid chromatography, the eluting solvent is generally pumped onto the column. Pulses, generated by these pumps, can affect the detector output and have a detrimental effect on the separating efficiency of the column. These pulse effects can be minimized through the use of large surge tanks. A pulseless high-pressure pumping system is described by Jentoft and Gouw [19].

V. TECHNIQUES

Waste analyses are generally not available in the literature, partially due to lack of interest on the part of the waste generator and partially due to the fact that few wastes are qualitatively or quantitatively similar enough to be of interest to others. The lack of pure standard materials for confirmation of suspected identity also contributes to the lack of complete characterization of waste.

However, several of the following techniques, either specifically or as general guides, may be useful for waste separation and characterization.

A. Reverse-Phase Partition Chromatography

This column chromatographic technique uses an inert packing material to support a stationary liquid phase on the column. The stationary liquid is usually a nonpolar material with low solubility in the mobile polar liquid phase. The less polar materials on the column may be removed by washing with additional volumes of the mobile liquid phase or other suitable solvents.

Howard and Martin [20] separated saturated acids from lauric to stearic on silanized Kieselguhr columns. The treated Kieselguhr retains the less polar phase of the solvent system as the stationary phase, e.g., for acetone-paraffin, the paraffin is retained. Methanol-n-octane was another solvent system used successfully.

A 25% squalane stationary phase on 100/200 mesh Chromosorb P column was used by Locke [21] to separate naphthenes, paraffins, olefins, and aromatic hydrocarbons (C_4-C_{14}). As a mobile phase, acetonitrile was used over a $15°$-$35°C$ temperature range with detection by a differential refract-ometer. This study indicates that the acetonitrile-squalane has good potential as a general purpose solvent pair in hydrocarbon analysis.

To separate polar materials, a polar stationary phase, n-butanol, on 30/80 mesh polyethylene and a nonpolar mobile phase, petroleum ether, has been found useful. The sequential combination of the two techniques has resolved many polar constituents as well as several nonpolar hydrocarbons from a petroleum waste residue. UV spectrophotometric detection was used for the polar and IR detection for the nonpolar stationary-phase column eluants. If the stationary and mobile phases selected have appreciable mutual solubility, each phase should be saturated with the other phase.

Chelation of metallic ions in waste waters with subsequent reverse-phase partition chromatography has been used to insure greater reliability as well as improved sensitivity for silver, vanadium, copper, cadmium, zinc, and molybdenum. Atomic absorption detection was employed on the elution phase.

The advent of chemically bonded stationary phases has simplified or eliminated many of the problems associated with loss of the stationary phase from the column packing. For waste analyses, probably the greatest single advantage of the chemically bonded stationary phase is that a wide variety of polar to nonpolar solvents can be used. This variation in solvents, without effect on the stationary phase, makes gradient elution practical.

Helm [22] successfully used 20/40 mesh Fluoropak 80 as an inert column packing. Cyclohexane saturated with nitromethane was the stationary phase, with nitromethane saturated with cyclohexane as the mobile phase. Infrared spectra of fractions were compared against a solvent gradient elution (heptane, benzene, benzene-methanol, methanol) of the asphalt sample from a silica gel column.

Orr [23] describes a technique that uses mercuric acetate-aqueous acetic acid as the stationary phase on 100/200 mesh silica gel to separate alkyl, cycloalkyl, and aryl sulfides and disulfides from most other classes of organic compounds found in petroleum sludges. With hexane as the mobile phase, hydrocarbons and many other compounds are eluted quickly. The sulfides and disulfides are eluted according to molecular weight or carbon number through C_{18}. A 70% sample of aqueous acetic acid, saturated with mercuric acetate, showed better separation of C_{12}-C_{18} disulfides. Thiols are irreversibly adsorbed and are not recovered with the hexane solvent. However, I have recovered thiols from this column with isopropanol-acetic acid, 90:10, as the mobile phase.

A liquid membrane diffusion tower or column has been used by Li [24, 25] to separate constituents of a mixture by diffusing one or more components through a liquid membrane surrounding a tiny liquid droplet. Although this technique is somewhat of a mixture of the solvent extraction and dialysis membrane techniques, it may have analytical as well as process capabilities.

B. Adsorption Column Chromatographic Cleanup

The increased emphasis on the analysis of various pesticide residues in
the environment has required the examination of plant and animal as well as
industrial wastes. Both thin-layer and adsorption column chromatography
have proved invaluable as cleanup separation tools [26-29].

The authors (26-29] have found the adsorption column an effective means
of sample collection for trace organic materials in air or water. For many
compounds, however, the adsorption is irreversible or at least a good
desorption solvent and conditions are unknown. Hydrolysis and oxidation
reactions on the adsorbent surface alters not only the organic identity but
also the recovery procedure. Proper pH control during the adsorption and
elution steps can sometimes reduce or eliminate hydrolysis reactions.
The solvent polarity series described by Tischer and Baitsholts [30] has
been of some assistance in solvent selection or combination.

VI. APPLICATIONS

As pointed out in Sec. I, applications of liquid chromatography have been
limited due primarily to poor performance. All applications have been
directed to the area of organic analysis. Inorganic materials have been
largely determined by ion-exchange or, in some cases, thin-layer chromatography.

Albersmeyer [31] determined heterocyclic aromatic hydrocarbons and
phenols in sewage by column chromatography. Organic solvents were used
to extract the compounds first from the aqueous solutions, were then passed
through the column, and subsequently eluted with additional organic solvents.
The solvents were evaporated to dryness and the residues reconstituted in
water. Copper(II) chloride (3%) was added and the resulting solution read
in a spectrophotometer at 650 or 370 nm. Sensitivity was 0.5-2.5 mg of
hydrocarbons per sample. Acetic acid was used to clear up any turbidity.
Daniels et al. [32] determined organic contamination in wastes by adsorbing
the materials from the aqueous solution onto activated carbon. The compounds were eluted with chloroform-ethanol and then determined. Istomine
et al. [33] determined benzene contamination in sewage by placing the
water in a flask connected by a reflux condenser to a column containing
silica gel saturated with formaldehyde and sulfuric acid. One liter of waste
was equilibrated at 30°C and nitrogen flushed to remove oxygen. The flask
was then refluxed at 100°C and the benzene reacted in the column and turned
black-violet. The degree of discoloration of the column was compared to
a standard. Golovina et al. [34] determined trichloroethylene, benzene,
chlorobenzene, and methoxylol in plant circulating water by first passing
4-5 liters of waste through a 60-m long column filled with KAD-iodine grade
coal (the column was 4 cm in diameter). Ether was then passed through
the column in the opposite direction to elute the compounds. The ether was

dried using calcium chloride and evaporated to dryness at 20°C. The compounds were then assayed by GLC using a 2-m column of 20% tricresol phosphate on diamite (brick) at 100°C. With decane used as the internal standard, the sensitivity was 1-3 mg/liter. Lure et al. [35] determined paraffinic, naphthenic, and aromatic materials in surface waste water by isolating the compounds from the water on silica columns. GC analysis resulted in quantitation to within 4-5%. They also determined the same compounds in water by first extracting with hexane and then passing the solvent through activated charcoal. The compounds were then eluted with chloroform and determined by GLC on SE-30. Mueller et al. [36] determined organic acids of low volatility by adsorbing them from portions of river waters onto silica gel columns. The compounds were eluted with 1-butanol-chloroform. The eluate was then titrated with 0.02N alcoholic sodium hydroxide, using 0.1% Bromothymol Blue as the indicator. Each acid had a particularly characteristic effluent volume. Acids determined in this manner consisted of butyric, crotonic, propionic, acetic, pyruvic, formic, lactic, succinic, gallic, malonic, oxalic, aconitic, 1-malic, citric, isocitric, and tartaric. The best acid concentration for analysis was recommended to be 2-7.5 microequivalents. Recovery was greater than 90% but less than 100%. All the acids could not be separated by this method due to similarities of two or more acids in the elution pattern.

The American Journal of Public Health [37] reported adoption of a new column method for extracting hydrocarbons from water waste samples for grease determinations. The water is passed through an alumina column and washed with n-hexane.

Leibnitz et al. [38] determined monophenols in trade effluents by first passing the water through a column of silica gel containing water as the stationary phase. The samples were eluted with cyclohexane and 4-5 ml fractions collected. The fractions were then treated with p-aminodimethyl-aniline and potassium ferricyanide at pH 8.3-8.7 to give the indenophenol derivative. Phenols were then determined spectrophotometrically. Belova [39] determined phenol and three cresols in peat-gas generating plants by passing the samples through columns of silica gel, with water as the liquid phase. Isooctane was used to elute the materials and different fractions were quantitated by UV at 240-300 nm. The method was used to determine phenols present at 2% concentration and cresols at 0.5% concentration. Stradomskaya et al. [40] determined phenols by adsorption on silica columns. Interfering esters were hydrolyzed first by alkaline evaporation and extracted with ether. The column was eluted with chloroform to recover phenols and then with 50% butanol-chloroform to recover naphthenic acids. The eluates were then titrated with sodium hydroxide for analysis.

Leoni [41] separated 50 pesticides and related compounds along with polychlorobiphenyls in surface waters into four groups by silica gel micro-column chromatography. All solvents were pesticide grade and tested by GC using an electron capture detector. A 10-25 liter aqueous sample (depending on the source) was repeatedly extracted with petroleum ether and benzene after acidification (pH = 1-3). After various purification techniques, usually extraction, the solvent was evaporated to dryness and reconstituted in 1 ml of hexane. A column was prepared consisting of silica gel grade 950 such that 1 g filled a 4.2-mm diameter column to a height of 10 cm. The height was important for reproducibility. Anhydrous sodium sulfate was added to the column to remove water from the four eluants. The eluants consisted of n-hexane, 60% benzene-n-hexane, benzene, and 50% ethyl acetate in benzene. Flow rate was ca. 1 ml/min. The eluates were evaporated to dryness, reconstituted in hexane, and assayed by GC, using OV-17 as the liquid phase. Recovery was reported to be 95-100% for all cases except for disyston (55%) and dimethoate (50%). Henry et al. [42] used high-speed liquid chromatography to determine Abate (O,O,O',O'-tetramethyl-O,O'-thiodi-p-phenylene phosphorothioate), a larvacide in waters. A DuPont Model 820 liquid chromatograph was employed, using 1 m × 2.1 mm stainless steel columns packed with Zipax coated with 1% β, β'-oxydipropionitrile. The detection system was UV measurement at 254 nm. Heptane served as the mobile phase. Typical concentrations of Abate were 0.02 ppm. After acidification with 5 ml of 1N HCl, a 100-ml portion of water was extracted with one 50-ml portion of chloroform. The injector sample size was 5 μl. Recovery ranged from 96 to 104%, with an accuracy of ±2%.

Finally, Takeru [43] reported that isolation of pesticides from bodies of water by partition chromatography using a chemically bonded stationary phase was unsatisfactory.

VII. CONCLUSIONS

As can be seen, applications of liquid chromatography over a 15-year period are certainly limited. Except for the publication on pesticide analysis by high-speed liquid chromatography, recoveries were generally low and not very reproducible. It must be observed that other chromatographic techniques are faster, more selective, and easier to employ with greater accuracy, and precision. A particular example would be thin-layer chromatography, or even gas chromatography within limits.

The advent of high-speed equipment will greatly change this picture. Even here, however, equipment limitations and time delays in bringing the sample from the field to the laboratory will restrict the amount of practical applications for several years at least.

REFERENCES

1. K. H. Mancy and W. J. Weber, Analysis of Industrial Wastewaters, "Treatise on Analytical Chemistry," (Kalthoff and Elving, eds.), Part III, Section B, Wiley-Interscience, New York, 1971, p. 413.

2. C. E. Hamilton, unpublished work.

3. F. Feigl, Spot Tests, Vol. II, Organic Applications, Elsevier, New York, 1954.

4. J. F. K. Huber, J. Chromatog. Sci., 7, 172 (1969).

5. R. D. Conlon, Anal. Chem., 41(4), 107A (1969).

6. J. L. Waters, Amer. Lab., May, 61-65 (1971).

7. R. F. Hornbeck, J. Chromatog., 30, 438 (1967).

8. S. R. Himmelhoch and E. A. Peterson, Anal. Biochem., 17, 383 (1966).

9. J. J. Kirkland, Anal. Chem., 40, 391 (1968).

10. J. J. Kirkland, J. Chromatog. Sci., 7, 7(1969).

11. J. J. Kirkland, Anal. Chem., 43, 36A (1971).

12. J. J. Kirkland, J. Chromatog. Sci., 7, 361 (1969).

13. H. N. M. Stewart and S. G. Perry, J. Chromatog., 37, 97 (1968).

14. W. A. Aue and C. R. Hadtings, J. Chromatog., 42, 319 (1969).

15. J. R. Conder, Anal. Chem., 43, 367 (1971).

16. D. E. Hirsch, R. L. Hopkins, H. J. Coleman, F. O. Cotton, and C. J. Thompson, Anal. Chem., 44, 915 (1972).

17. A. Mendel, J. Chromatog., 17(2), 411(1965).

18. J. T. Ayres and C. K. Mann, Anal. Chem., 38, 861 (1966).

19. R. E. Jentoft and T. H. Gouw, Anal. Chem., 38, 949 (1966).

20. G. A. Howard and A. J. P. Martin, J. Biochem. 56, 532 (1950).

21. D. L. Locke, J. Chromatog., 35, 24 (1968).

22. R. U. Helm, Anal. Chem., 41, 342 (1969).

23. W. L. Orr, Anal. Chem., 38, 558 (1966).

24. N. N. Li, Chem. Eng. News, Oct. 5, 36 (1970).

25. N. N. Li, Chem. Eng., Nov. 2, 52 (1970).

26. Pesticide Analytical Manual, Vols. I and II, U.S. Dept. Health Education Welfare, Federal Drug Administration, Washington, D.C., 1967 and 1968.

27. F. A. Gunther (ed.), Residue Reviews, Springer-Verlag, New York, 1965.

28. P. A. Mills, J.A.O.A.C., 44, 171 (1961).

29. P. A. Mills, J.A.O.A.C., 51, 29 (1968).

30. T. N. Tischer and A. D. Baitsholts, Amer. Lab., May, 69 (1970).

31. A. Albersmeyer, Gas-U. Wassterfach, 101, 447 (1960); Chem. Abstr.,
     54, 20027g (1960).

32. S. L. Daniels, L. L. Kempe, E. S. Graham, and A. M. Beeton,
     Great Lakes Res. Inst. Sci. Tech., Univ. Mich. Publ., 10, 118
     (1963); Chem. Abstr., 60, 13013g (19).

33. K. E. Istomine, L. A. Ionova, and L. N. Andreeva, Vestn. Tekhn.
     i Ekon. Inform. Nauchn.-Issled. Inst. Tekhn.-Ekon. Essled. Gas.
     Kam. Sov. Min. SSSR po Khim., 11, 48 (1961); Chem. Abstr., 58,
     6565a (1963).

34. Z. M. Golovina and A. M. Agal'tsov, Sostoyvanii Puti Uluchsheniya
     Ochiski Slochnykh Vod. Gazov. Vybrosov Atmos., 125 (1969); Chem.
     Abstr., 75, 24990x (1971).

35. Yu, Lure, and V. A. Panova, Gidrokhim, Mater., 55, 108 (1971);
     Chem. Abstr., 75, 121164p (1971).

36. H. F. Mueller, T. E. Larson, and M. Ferritti, Anal. Chem., 32,
     687 (1960).

37. F. W. Gilcreas, Amer. J. Publ. Health, 50, 1199 (1960).

38. E. Leibnitz, U. Behrens, and A. Gabert, Wasserwirtsch-Wassertech.,
     8, 170 (1958); Chem. Abstr., 53, 6492g (1959).

39. M. Belova, Ochistka Pram. Stochnykh Vod., Akad. Stroit, i Arkitekt.
     SSSR, Vses. Nauchn.-Insled. Gidrogeol., Tr. Sovmestnoi Knof.,
     Moscow, 333 (1960); Chem. Abstr., 59, 12495c (1963).

40. A. G. Stradomskaya and I. A. Goncharova, Gidrokhim. Mater, 41,
     78 (1966); Chem. Abstr., 66, 58750y (1967).

41. V. Leoni, J. Chromatog., 62, 63 (1971).

42. R. A. Henry, J. Schmit, J. Diechman, and F. Murphey, Anal. Chem.,
     43(8), 1053 (1971).

43. Ito, Takeru, U.S. Nat. Tech. Inform. Serv., PB Rep. 1971, #204490.

Chapter 18

PAPER CHROMATOGRAPHIC ANALYSIS IN WASTE CHEMISTRY

Joseph V. Hunter

Department of Environmental Science
Cook College
Rutgers University
New Brunswick, New Jersey

I.   INTRODUCTION

Chemical analyses designed to produce information on the organic constituents of waste waters have been in progress for almost 50 years. The interest motivating this research has originated in a desire to improve or better control waste water treatment, to detect useful materials that may be present, to better assess the impact of waste waters on the aquatic environment, and perhaps also in simple curiosity.

Starting out with crude determinations designed to reveal only the total amount of organics present, and evolving into highly detailed studies of the specific compounds present, many analytical techniques have been employed in the separation, detection, and estimation of waste water organics [1]. Chromatographic techniques have played an important part in these analyses, and it is the purpose of this chapter to describe how paper chromatography has been employed in such studies.

II. SYSTEMS ANALYZED

The major problems encountered in waste water analysis are the low concentrations of the organics present, the chemical complexity of the mixture, the diurnal variations in their concentration and distribution, their biological degradability, and the physical inhomogeniety of the mixture. Many of these, such as those involving concentration and complexity, are typical analytical problems and are solved by the usual approaches. Others, such as substrate degradability and time variability, can be more or less adequately handled by suitable preservation and sampling techniques.

The problem that most influences the understanding and interpretation of waste water analyses is that of physical inhomogeniety. Analyses may be run on whole waste water, particulates separated from the waste water, or the soluble fraction remaining after the removal of such particulates. Figure 1 shows the types of physical fractionation procedures that have been used to separate waste water solids before analysis. An associated problem is encountered with treated sewages, as shown in Figure 2, also produce certain solids that are frequently analyzed. The effluent from such waste water treatment plants may be physically fractionated in a manner similar to untreated waste waters. In general, "waste water solids I" from Figure 1 may be assumed to be roughly equivalent to "primary sludge" in Figure 2. "Effluent solids I" may also be assumed to be roughly equivalent to either waste activated sludge or trickling filter humus, depending on which type of treatment is employed.

Another source of confusion lies in the various methods employed in reporting results. Analyses of waste water, effluents, and fractions isolated from them are frequently expressed as simple concentrations in units of milligrams per liter. Less frequently, the results may be reported in terms of the constituent in a given fraction as a percent of the total constituent present in all the fractions. Sludge and particulate analysis results are also frequently reported in terms of milligrams of constituent per gram of dry sludge.

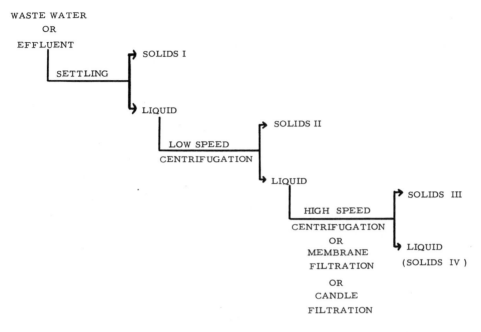

Figure 1.   Physical fractionation scheme for waste waters and effluents.

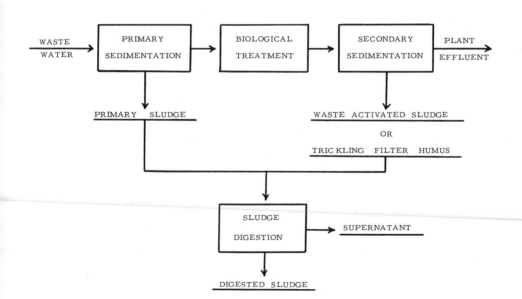

Figure 2.   Waste water treatment flow diagram.

III. GROUPS DETECTED

As may be expected, chromatographic analysis has mainly involved the separation, identification, and estimation of groups of closely related compounds. Therefore, this section concerns the detection and determination of the amino acids, sugars, higher fatty acids, and volatile acids.

A. Amino Acids

Whether they are present in soluble or particulate fractions or in sludges produced during waste water treatment, the amino acids exist mainly as polymers of various molecular weights. To determine the amino acid contents of these materials, some type of hydrolytic procedure is required. The recovery of amino acids after such hydrolytic treatment is variable, ranging from 40 to 70% of the total organic (Kjeldahl) nitrogen [2-5]. Whether this is due to the destruction of a significant part of the amino acids or to the presence of significant quantities of non-amino acid nitrogenous compounds is not known at present.

The greater part of the research in this area originated with investigations at the Indian Institute of Science [6, 7]. Using circular paper chromatography, extensive investigations were made into the amino acid contents of waste waters, effluents, primary solids, and activated sludges. To determine free amino acid, the waste water or effluent was evaporated to dryness and the amino acids were desalted by extracting them from the residue (approximately 0.4 g) with 95% ethanol. The alcoholic solution was then extracted with chloroform to eliminate lipids and then decolorized with charcoal. The extract was then evaporated to 1 ml, and 100-150 μl portions were used for spotting on the chromatograms.

To determine total amino acids, 0.3 g of dried material was hydrolyzed with 25 ml of 6N hydrochloric acid at 15 lb pressure in an autoclave. At the end of this time the suspension was filtered and evaporated to dryness. Additional water was added and the sample again evaporated to dryness, and this procedure was repeated until all hydrochloric acid was removed from the sample. The residue was then dissolved in 10 ml of water and 5-30 μl of water and 5-30 μl portions were used for spotting.

The circular chromatographic technique did not separate the following amino acid sets: lysine-histidine; serine-glycine-aspartic acid; threonine-glutamic acid; methionine-valine. Although quantitated together, the presence of lysine, histidine, serine, glycine, aspartic acid, and threonine was established by running two-dimensional chromatograms. The two solvent systems employed were phenol saturated with water and butanol-acetic acid-water. Methionine was identified by the use of platinic iodide reagent, and valine by eliminating the methionine using hydrogen peroxide. The other amino acids were detected using ninhydrin (triketohydrindine hydrate).

TABLE 1

Amino Acids Present in Waste Water Sludges[a]

Amino acid	Primary solids[b]	Activated sludge[b]	Digested sludge[b]
Cystine	4.2	14.9	0
Lysine and histidine	6.9	27.3	2.8
Arginine	6.1	18.4	4.6
Serine, glycine, and aspartic acid	13.0	38.2	7.0
Threonine and glutamic acid	13.9	35.1	3.6
Alanine	7.3	23.0	2.6
Proline	0	Trace	0
Tyrosine	5.4	13.4	Trace
Methionine and valine	11.1	32.8	4.1
Phenylalanine	14.5	28.2	1.8
Leucines	6.8	31.2	1.4

[a]From Ref. 7.

[b]As milligrams per gram of dried solids.

The quantitative determination of the amino acids was achieved by elution of the blue colored amino acid copper sulfate complexes with 75% ethanol and comparison of the color intensity with known quantities of the same amino acid run under the same conditions. Typical results obtained with this technique are shown in Table 1.

Another extensive analysis of waste water particulates was made by Heukelekian and Balmat [3]. After the solids were obtained using a scheme similar to the one described in Figure 1, the fractions were dried and 40 mg were hydrolyzed by refluxing with 150 ml of 6N hydrochloric acid for 20 hr. At the end of the hydrolysis period the sample was filtered and the filtrate was passed through an Amberlite IR-120 (H^+) cation-exchange column that removed all of the amino acids and the other cations, allowing the anions to pass through.

After the column was washed with distilled water, the amino acids were eluted using 2M ammonium hydroxide. The salt-free eluate was evaporated to dryness and dissolved in 5 ml of water, and aliquots containing 150 μg of total amino acids were used to spot Whatman No. 1 paper. The chromatogram was developed two dimensionally employing first, as the mobile phase, phenol saturated with water, and then lutidine-ethanol. After separation the sheets were dryed, sprayed with ninhydrin, and heated to expose the spots. The spots were cut out and eluted with water, and the amino acids were quantatively determined in solution by the ninhydrin reaction, using leucine as a standard. The individual amino acids were estimated using the leucine equivalent values of Moore and Stein [8]; a typical example of the results obtained by the authors is shown in Table 2.

A comparison of the amino acid contents of waste waters and effluents was made by Kahn and Wayman [9]. To determine free amino acids, a 1-liter waste water sample was evaporated to dryness at 40°C and extracted with ethanol. The remainder of the procedure was as described in

TABLE 2

Amino Acid Constituents of Domestic Waste Water Particulates[a]

Amino acid	Solids I[b]	Solids II[b]	Solids III[b]
Alanine	1.30	2.26	0.86
Cystine	0.63	1.27	—
Aspartic acid	1.21	2.01	1.07
Glutamic acid	1.88	2.50	0.79
Glycine	0.81	2.22	0.36
Histidine	1.31	—	0.72
Leucines and phenylalanine	2.23	2.32	0.87
Methionine and valine	1.68	1.74	0.79
Serine	1.16	0.64	0.03
Threonine	0.78	0.78	0.19
Tyrosine	1.40	0.47	—

[a]From Ref. 3.

[b]As milligrams per liter in the original waste water.

TABLE 3

Amino Acids in Waste Waters and Effluents[a]

Amino acid	Waste waters[b]		Effluent combined[b]
	Free	Combined	
Lysine	10	15	0
Cystine	5	10	0
Arginine	10	10	0
Glutamic acid	10	10	0
Aspartic acid	10	15	0
Serine	10	15	0
Histidine	10	15	0
Threonine	10	15	0
Glycine	10	15	0
Alanine	10	15	0
Valine	10	10	5
Methionine	5	10	0
Luecine	5	10	5

[a]From Ref. 9.

[b]As milligrams per liter in original waste water or effluent.

references [6] and [7]. To determine the combined amino acids, a 1-liter sample was evaporated to dryness at 60°C and hydrolyzed with 25 ml of 6N hydrochloric acid at reflux temperature for 12-18 hr. The hydrolyzate was evaporated to dryness, redissolved in water, decolorized with charcoal, and filtered. As tryptophane is lost during acid hydrolysis, samples to be analyzed for this amino acid are hydrolyzed with 50 ml of 14% barium hydroxide at reflux temperature for 18 hr and then filtered and decolorized as above.

The colorless solutions obtained by these procedures were evaporated to dryness, dissolved in 10 ml of a 10% isopropanol solution, and desalted with an electric desalter. The desalted samples were evaporated to dryness and dissolved in 5 ml of a 10% isopropanol solution. These solutions were spotted on Whatman No. 1 paper, and the chromatograms were developed using two-dimensional techniques. The first mobile phase was 100 ml of 88% phenol and 20 ml of distilled water containing 50 mg of 8-hydroxyquinoline, γ-collidine, distilled water, and diethylamine (100:100:100:3). The spots were detected as usual using a ninhydrin spray and approximately quantitated by comparison to standard amino acid samples. An example of the results obtained using this procedure is shown in Table 3.

B. Sugars

Painter and Viney [2] have reported an extensive investigation into the individual sugar constituents of waste water carbohydrate. These authors also employed a fractionation scheme similar to the one presented in Figure 1 to obtain three particulate fractions and one soluble waste water fraction. The carbohydrates were hydrolyzed by refluxing 1 hr with 0.25N hydrochloric acid or heating in a sealed tube for 1 hr with 1N sulfuric acid. At the end of the hydrolysis, the mixture was neutralized and evaporated to dryness.

The sugars were extracted from the dry residue by treatment with pyridine below 40°C, and the pyridine extract was evaporated to dryness in vacuo below 40°C. The solids so obtained were dissolved in a small volume of 10% isopropanol, and 5–25 µl samples were used to spot the paper.

The chromatograms were developed twice for a total of 40 hr using descending chromatography. A number of solvent systems were employed. Those proving most satisfactory were:

Proportion	Constituents	Reference
4:1:5	n-Butanol-acetic acid-water	10
2:1:2	Ethyl acetate-pyridine-water	11
45:25:40	n-Butanol-pyridine-water	12
	Phenol saturated with water	12

The sprays used to detect the individual sugars were p-anisidine hydrochloride, aniline hydrogen phthalate, naphthoresorcinol [12], and urea [13]. No attempt was made here to obtain strictly quantitative results, and the relative abundances of the sugars detected are shown in Table 4. Free sugars were detected by the same procedure except the hydrolysis was eliminated.

C. Higher Fatty Acids

In addition to their research on the amino acid contents of waste waters and sludges, the investigators at the Indian Institute of Science also studied the higher fatty acid contents of similar materials [14]. To obtain the fatty acid mixture for analysis, the sewage or effluent solids were dried and finely ground. The dry residue was then acidified with 1N hydrochloric acid

TABLE 4

Sugars Present in Domestic Waste Waters[a]

Sugar	Solids I[b]	Solids II[b]	Solids III[b]	Solids IV[b]
Glucose	+++	+++	+++	+++
Galactose	++	+	+	+
Mannose	Trace	Trace	-	-
Lactose	-	-	-	++
Sucrose	-	-	-	+++
Maltose	Trace	-	-	-
Arabinose	++	+	Trace	+
Xylose	+	+	+	+
Ribose	-	Trace	+	-
Rhaminose	+	+	++	-
Fucose	-	-	Trace	-
Fructose	-	-	-	+

[a]From Ref. 2.

[b]Relative abundance proportional to symbols.

(3 ml/g) and repeatedly extracted with cold and then hot ethanol. The ethanol extracts were then evaporated under reduced pressure and the lipids in the residue were extracted with ethyl ether.

The lipid fractions obtained by evaporation of the ethyl ether are dissolved in 1:1 ethanol-benzene and made alkaline, the solids are evaporated and dissolved in water, and the non-acid lipids are extracted with ethyl ether. The water solution of the fatty acid salts is then acidified with 0.1N hydrochloric acid, and the liberated free fatty acids are extracted with ethyl ether.

The esterified fatty acids present in the neutral lipid fraction are saponified with 0.5N alcoholic potassium hydroxide after evaporation of the ethyl ether. The alcohol is removed by evaporation under reduced pressure, and the preceeding procedure is repeated to obtain the ester fatty acids.

Due to the insolubility of the ester fatty acids in the usual aqueous polar stationary phases, reverse-phase circular paper chromatography was the technique employed. The stationary phase was 10% paraffin-impregnated Whatman No. 3 paper, and the mobile phase was 90% aqueous acetic acid. This procedure separates the saturated fatty acids from each other and the

TABLE 5

Fatty Acid Contents of Waste Water and
Effluent Particulates[a]

Fatty acid	Waste water, solids I	Effluent, solids I
Saturated		
Lauric	0.12	0.11
Myristic	0.24	0.13
Palmitic	11.7	1.3
Stearic	4.6	0.93
Unsaturated		
Oleic	17.0	1.1
Linoleic	9.96	2.1
Linolenic	0	0.06

[a]From Ref. 11.

[b]As milligrams per liter in the original
waste water as effluent.

unsaturated fatty acids from each other; however, a saturated and a non-saturated acid may have similar R_f values. The unsaturated fatty acids are detected by exposing the developed chromatogram to iodine vapors in a closed chamber. The saturated fatty acids were detected after oxidation of the unsaturated acids into fragments that did not interfere in the determination.

After the chromatogram was developed by treatment with mercuric acetate and 5-diphenylcarbazide, quantitation was achieved. Quantitation was achieved by extracting the total fatty acid mercury complex and the fatty acid mercury complex obtained from the oxidant treated fatty acids with a 1:1 methanol-toluene mixture. The quantity of fatty acid present was estimated by spectrophotometric determination at 530 nm and comparison of the observed absorption to those of standards. The unsaturated fatty acids were determined from the difference between the total and saturated fatty acids. Typical results obtained using this procedure are shown in Table 5.

D.   Volatile Acids

Volatile acids have been of interest particularly in the control of anaerobic digestion of waste water and biological treatment sludges. For this reason, there are standard procedures for the estimation of this group [15] and considerable interest as to the meaning of the actual distribution of the volatile acids.

Manganelli and Brofazi [16] adapted a method particularly for this purpose. The volatile acids are first obtained from acidified digestor sludge by steam distillation and the distillate is collected in a solution of sodium hydroxide. The alkaline solution is evaporated until the volatile acid normality is approximately 0.40. Ammonium sulfate is added in equivalent concentrations to the sodium hydroxide, and the sample is acidified with sulfuric acid to pH 3.0. The pH is adjusted to 8-9 with ethylamine and the sample is diluted to a final volatile acid concentration of 0.2N.

Five microliters of this solution is used to spot Whatman No. 1 filter paper equilibrated with water saturated with n-butanol (from an initial 1:1 n-butanol-water mixture). The mobile phase is butanol saturated with water (from an initial 1:1 n-butanol-water mixture) and equilibrated with 2% by volume of a 33.3% aqueous ethylamine solution. The chromatogram is run using the descending mode, and after evaporation of the solvent the spots are developed by dipping in a Chlorophenol Red solution (200 mg per 100 ml ethanol).

This procedure separates acetic, proprionic, butyric, and valeric acids, which are identified by their R_f values. Quantitation is achieved by measuring spot areas and comparing them to spot areas of standard solutions run under comparable conditions. Here, concentration was a linear function of spot area in the range of 15-40 μg per 5 μl, with an average error of estimate ranging from 0.6716 μg per 5 μl for acetic acid to 2.9190 μg/μl for valeric acid.

A later study by Buswell and co-workers [17] also used chromatographic procedure based on the separation of volatile acid ethylamine salts. This procedure, for purposes of simplicity, eliminated the initial separation and concentration steps. Five to 10 μl samples were used to spot Whatman No. 1 paper, and the chromatograms were developed in the descending mode employing the butanol (upper) phase of a mixture of 1:2 2N ethylamine solution and n-butanol as the mobile phase. After development, the spots are made visible with a bromcresol Green indicator (0.4% in ethanol). Again, acetic, proprionic, butyric, and valeric acids may be separated and identified by their R_f values. For spots to be easily detected, the volatile acid concentration in solution must be at least 500 mg/liter. No quantitation was achieved as no simple relationship was observed between the spot area and concentration. As these studies involve techniques rather than detailed investigations, no examples of the use of these procedures are reported here.

Painter and Viney [2] detected acetic, proprionic, butyric, and valeric acids in domestic waste waters after steam distillation and concentration of the total volatile acids present. The method employed was that of Duncan and Porteous [18], which involves a separation of the ammonium salts employing n-butanol-1.5N ammonia as a mobile phase. No quantitation was reported using this technique, and the actual concentrations of the individual volatile acids reported were obtained by silica gel column chromatography.

E. Miscellaneous

There has been some use of paper chromatography abroad in waste water and effluent analyses. However, as details are lacking in the abstracts and as the journals are not readily available, such investigations are beyond of the scope of this presentation. However, brief mention is made of these methods as many of them do not overlap with those already described. Thus, o-, m-, and p-cresols have been separated and detected using paper chromatography [19, 20], as have naphthenic acid [21] and bitumens [22].

IV. COMPOUND IDENTIFICATION

In addition to its use in the separation, identification, and estimation of closely related organic species, paper chromatography has been employed in studies involving the chemical nature of an unknown material. This material, labeled "hestianic acid," was isolated from activated sludge effluent by precipitation from an acidified activated sludge effluent solution, followed by certain purification procedures [23].

The major use of paper chromatography in this research was the study of the degradation products of hestianic acid to obtain some insight into its chemical nature. For this purpose, two solvent (mobile-phase) systems were employed. These were: solvent A, isopropanol 65%, 2N hydrochloric acid; solvent B, n-butanol, acetic acid, and water (80:20:100)

1. Alkaline permanganate degradation. One of the degradation products separated by paper chromatography using solvent A gave a strong ureide reaction (yellow color with dimethylamino benzaldehyde), amines or amino acids (purple color with ninhydrin), a tertiary amine (Dragendorff reagent), and slight indications of phenols (color with diazotized sulfanilic acid).

2. Dilute sulfuric acid degradation. Using solvent system B, paper chromatography revealed at least five different amino acids and/or amines and two substances reacting with diazotized sulfanilic acid. Use of solvent systems A and B in two-dimensional chromatography indicated that the positive substance were amines rather than amino acids. No ureides were detected.

3. Perchloric acid hydrolyses. Using solvent system A no tertiary amines or ureides were detected and only one spot was detected.
4. Dichromate oxidation. Using solvent system B, no amines, amino acids, phenols, or ureides were detected.
5. Reduction with zinc and hydrochloric acid. Solvent system B gave no indications of amines, ureides, tertiary amines, or amino acids as phenols.

The authors concluded from this that no proteins or polysaccharides were part of the hestianic acid "molecule," but the complex picture provided by the degradation products detected did not allow them to propose even an approximate structure.

V. SUMMARY

Although paper chromatography provided considerable information on the nature and concentration of waste water effluent and sludge organics, its period of major use in this field was from 1957 to 1965. Since then, there has been relatively little use of this technique in waste water analysis by investigators in this country. It seems reasonably evident that the rapid development of both gas and thin-layer chromatography during the last decade has considerably reduced investigator interest in this technique for waste water and effluent analysis.

REFERENCES

1. J. V. Hunter and D. A. Rickert, "Organic Analytical Chemistry in Aqueous Systems," in Water Pollution Handbook, (L. Ciaccio, ed.), Marcel Dekker, New York, 1972.

2. H. Painter and M. Viney, J. Biochem. Microbiol. Tech. Eng., 1, 143 (1959).

3. H. Heukelekian and J. Balmat, Sewage Ind. Wastes, 31, 413 (1959).

4. J. V. Hunter and H. Heukelekian, J. Water Poll. Contr. Fed., 37, 1142 (1965).

5. B. Bai, C. Viswanathan, and S. Pillai, J. Sci. Ind. Res., 21C, 72 (1962).

6. C. Sastry, P. Subrahanyam, and S. Pillai, Sewage Ind. Wastes, 30, 1241 (1958).

7. P. Subrahanyam, C. Sastry, A. Rao, and S. Pillai, J. Water Poll. Contr. Fed., 32, 344 (1960).

8. S. Moore and W. Stein, J. Biol. Chem., 176, 367 (1948).

9. L. Kahn and C. Wayman, J. Water Poll. Contr. Fed., 36, 1368 (1964).

10. A. Gottschalk, Biochem. J., 61, 298 (1955).

11. M. Jermyn and F. Isherwood, Biochem. J., 44, 402 (1949).

12. R. Block, E. Durrum, and G. Zweig, A Manual of Paper Chromatography and Paper Electrophoresis, Academic, New York, 1955.

13. F. Isherwood, Brit. Med. Bull., 10, 202 (1954).

14. C. Viswanathan, B. Bai, and S. Pillai, J. Water Poll. Contr. Fed., 34, 189 (1962).

15. Standard Methods for the Analysis of Water and Waste Water, 12th ed., American Public Health Assoc., New York, 1965.

16. R. Manganelli and F. Brofazi, Anal. Chem., 29, 1441 (1957).

17. A. Buswell, J. Boring III, and J. Milam, J. Water Poll. Contr. Fed., 32, 721 (1960).

18. R. Duncan and J. Porteous, Analyst, 78, 641 (1953).

19. Y. Lure and Z. Nicolaeva, Zavodsk Lab., 30, 937 (1964); Chem. Abstr., 61, 13035 (1964).

20. K. Koch, Gas u. Wasserfach, 106, 1390 (1965); Water Poll. Abstr., 41, 207 (1968).

21. E. Bykova, Byul. Nauchn-Tekhn. Inform., Gos. Geol. Kom. SSSR, Otd. Nauchn.-Tekhn, Inform., Vses, Nauchn. Issled. Inst. Mineral'n Syr'ya, 87, 937 (1964); Chem. Abstr., 61, 13035d (1964).

22. A. Sinel'nikov, Kazansk Med. Zh., 3, 83 (1968); Chem. Abstr., 69, 7995 (1968).

23. A. Fredericks and D. Hood, Atomic Energy Commission Publ. AD618932, Washington, D.C, 1965.

Chapter 19

THIN-LAYER CHROMATOGRAPHIC ANALYSIS IN WASTE CHEMISTRY

Eugene J. McGonigle

Merck Sharp and Dohme Research Laboratories
Pharmaceutical Research and Development
West Point, Pennsylvania

I. INTRODUCTION

"Springs should be tested and proved in advance .... if they run free and
open, inspect and observe the physique of the people who dwell in the vicinity
before beginning to conduct the water, and if their frames are strong, their
complexions fresh, legs sound and eyes clear, the springs deserve complete
approval. If a spring is just dug out, its water can be sprinkled into a

Corinthian vase or any other sort of good bronze without leaving a spot on
it .... if boiled in a bronze cauldron and left for a time, then poured off
without leaving sand or mud being found at the bottom of the cauldron, that
water also will have proved its excellence." The preceeding are criteria for
certification of drinking water in the era of Julius Caesar [1]. Today, our
definitions of purity and our standards of certification have improved
slightly, although concern about contamination has remained. In 1960, about
51% of the water withdrawn from surface and ground resources in the United
States was consumed by industry or about 138 billion gal/day. At the same
time, nearly 98% of industrially used water was returned to these same
sources [2]. Public consumption and subsequent waste have only added to
the problem. Carelessness in applying pesticides and herbicides is an old
concern with a relatively new awareness for the emphasis on routine monitoring occurring in the last 5 years or so.

Thin-layer chromatography (TLC) is an excellent analytical technique
being employed more and more in recent years for such analyses. Before
beginning any treatment of specific applications of TLC in analyses of waters
contaminated by various wastes, some basics must be reviewed regarding
the technique.

A. Sensitivity and Selectivity

Sensitivity and selectivity (specificity) are distinct parameters, although
closely related in thin-layer chromatography. Both depend on purity of
sample and reagents, technique and the means of measurement or detection.
Sensitivity may be regarded as the lowest concentration of a substance
measurable in the presence of some sample and reagent blank. The sensitivity of a TLC method of analysis may very likely depend on how specific
the system is for the component of interest. In general, sensitivity and
selectivity must be determined according to the application intended by the
analyst. In the applications reviewed, there are three general classifications
for use of thin-layer chromatography: (1) screening of samples for component identification; (2) cleanup procedures, in which the sample is purified
by TLC and components are eluted and quantitated individually by various
methods of measurement; and, (3) quantitative or semiquantitative analysis,
in which components are determined directly on the plate.

1. Screening of Samples

In screening analyses of waste-contaminated water, the analyst is concerned
primarily with determining what is present rather than how much. The
sample is applied to a TLC plate as a spot (or streak) and developed with an
appropriate solvent. After development, resolved spots (or bands) may be
eluted and identified or respotted on different plates and developed again to
improve resolution.

(a) Adsorbent.   An adsorbent is selected according to degree of polarity. Many are available, the most common being alumina, silica, cellulose, or Kieselguhr. If the nature of the sample is completely unknown, more than one adsorbent may have to be evaluated. The plate is spread at a thickness (usually 0.25 mm) depending on desired sample capacity, activated by heating at temperatures between 90° and 110°C in an oven free of contaminating vapors, and stored over a desiccant. Samples should be applied as soon as possible after the plate is removed from storage to avoid variations in plate activity.

(b) Sample Application.   The unknown may be applied to the plate as a spot or streak using a syringe or pipet. For best resolution, a general recommendation is that the spot be kept as small as possible but this is very difficult when sample solvents are low in volatility. The problem is compounded in screening analyses as the presence of more than one component makes overloading almost unavoidable. Putting more than one concentration of the sample on the plate and keeping the spot area constant will aid in selecting the most suitable sample size.

(c) Development.   The plate is placed in an appropriate chamber containing a developing solvent or solvent mixture. If the developing solvent is more volatile than, say, methanol, the chamber should be lined with solvent-washed filter paper to aid in maintaining solvent-vapor equilibrium. More than one developing solvent should be tried, varying in degree of polarity. The components move from the origin and separate on the basis of differences in affinity for the solvent and the adsorbent. The affinities depend on solubility, polarity, and molecular weight of the components. When the solvent has traversed the plate, the plate is dried and examined by some means of detection. No sample residue should remain at the plate origin as this may represent unresolved mixtures. Components may be scraped off, eluted, and applied to another plate and developed with a different developing solvent for further resolution. This process is repeated until the analyst is satisfied that all components have been resolved. At this point, such a decision is purely arbitrary and based primarily on experience.

(d) Elution Solvent.   The selection of the elution solvent is very important and difficult without prior knowledge of the desired component(s). A good technique is to rotate the developed plate 90° and develop it again using the proposed elution solvent. For best recovery, the elution solvent should move the entire spot at least 80% of the length of the solvent front.

(e) Spot Analysis.   After elution, the components should be individually examined by independent techniques (GC, IR, UV, etc.) for identification and purity confirmation. Depending on these studies, the analyst may modify the TLC system(s) by additional changes in the adsorbent and/or developing solvents to improve resolution.

2. Cleanup Procedures

In cleanup procedures, the primary concern of the analyst is to separate
one or more specific components from a matrix that is known or suspected to interfere with quantitative (or qualitative) analysis. TLC may be
employed to resolve components the properties of which are too similar
to effect separation in other ways. The component of interest is eluted from
the plate and determined by some means of measurement. However, in
many waste samples, the component is present at concentrations of parts
per million or less, while interferences may be thousands of times greater.
If such grossly contaminated samples are concentrated and applied directly
to TLC plates, the analyst may expect difficulties. The component may be
leached by the developing solvent from sample residue at the origin, resulting in tailing, or a variation in HETP (see Chapter 1, p. 11) may be observed
due to overloading. To avoid such difficulties, most authors have recommended a preliminary extraction of the component (or class of components) from
the aqueous sample. The most popular approach has been to buffer the sample
to a pH that minimizes aqueous solubility of the component and then extract
with some immiscible solvent selected according to degree of polarity. The
number of extractions required to completely recover the components should
be experimentally determined whenever possible using an actual sample
"spiked" with standard. One disadvantage of this approach is that large
volumes of sample and/or extracting solvent are sometimes required.
Another technique is to collect components by passing the aqueous sample
through a column filled with some adsorbing medium, such as activated
carbon or an ion-exchange resin. The adsorbing medium should completely
remove the components from the aqueous phase since a large volume of
sample may be required to collect a measurable quantity of the component.
  Whatever technique is employed for preliminary extraction, care should
be taken to avoid additional contamination by anything that has been brought
into contact with the sample. Reagents and adsorbents should be prewashed
and blanks determined.
  Once the analyst has eliminated the problem of gross contamination, he
or she must determine the behavior of some reference standard in the TLC
system. The standard should represent the purest available form of the
component to be analyzed. Because the component is eluted, recovery and
sensitivity limitations using the selected method of measurement must be
established first. Recovery of the standard is measured over a concentration
range less than to greater than the expected concentration of the component
in the sample. Sufficient data should be collected to provide a statistical
evaluation of reproducibility, particularly if recovery is less than 100%. In
this way, the analyst has established accuracy and precision limitations
under ideal conditions — in the absence of the sample matrix. A standard
calibration curve should be plotted and compared to a similar plot of sample
"spiked" with standard over the same concentration range. As the sample
will also contain the specific component to be analyzed, the curves will not

be superimposable. However, the shapes of the curves should be identical, and any differences represent sample interference and necessitate a re-evaluation of the screening study.

TLC adsorbents vary in "activity" from plate to plate depending on preparation and on laboratory environment. In addition, the partial pressures of the solvents in the developing chamber may not always be reproducible. As a result, R_f values are a poor indication of component identification and in any subsequent analyses standards should always be incorporated in more than one area of the TLC plate along with samples.

3. Quantitative or Semiquantitative Analysis

Components may be quantitated directly on the TLC plate. Generally, the plate is sprayed with a reagent that, after incubation, yields a chromagen or fluorophore, which may be quantitated using a densitometer. In some cases, areas of standard spots have been measured and plotted vs concentration. Component assay may then be determined by comparing sample areas to the standards. Such methods are highly dependent on technique and plate surface variability. As a result, while some authors regard them as reliable, others consider them semiquantitative at best. In any event, these methods have been used and are reviewed in this chapter. However, I feel that these methods should be employed only when elution analysis is precluded for some reason, and then only after careful evaluation.

B. Sampling

In most analyses of water contaminated by wastes, the sample must be random and representative. There may be instances when randomness is precluded by the specific objective of the analysis. One example would be an analysis to determine the uniformity of contamination by a specific component in some drainage effluent. For best results, Mancy and Weber [2] recommended a sampling system based on (1) method of sampling, (2) size of samples, (3) total number of samples, and (4) site of sampling.

The "grab" sampling technique was used by most authors. In the "grab" technique, an isolated sample is taken from a specific location. Several "grab" samples may be combined from several sources to form a composite sample. This minimizes source variation if some average evaluation is the objective but may require large storage facilities. It is more advisable to analyze a large number of "grab" samples from the same site at the same time and then to repeat the process at several sites. The statistical interpretation of the collected data determines whether the sampling technique is valid. Automatic sampling systems have been employed but it has been reported that manual techniques are more desirable [3].

The size of samples is partly a practical consideration depending on storage and/or transportation requirements. However, the minimum size of the sample will always be determined by sensitivity limitations of the measurement system for the component to be analyzed.

Wastes are not uniformly added to waters but vary at a specific site with time. The number of samples will depend primarily on the variation of sample composition.

Selection of a site is very important. If the analyst is sampling a river, he or she must be aware of any incoming streams that may dilute the sample supply. If the analyst is sampling an effluent pipe, he or she must be sure to adequately sample a cross-section to assure a representative sample.

Once the sample is collected, the analysis should be conducted as quickly as possible to avoid possible changes in composition. If the sample must be transported some distance before analysis, there is risk of component interaction or other chemical changes, such as biodegradation. If such time lapse is unavoidable, it is advisable to store the sample in a light shielded container at low temperature but above the freezing point. It is not advisable to add preservatives to suppress biological or chemical activity, but if they are added they should be accounted for separately in the analysis [4]. Field analysis is more desirable and TLC is particularly suited for this purpose. A mobile laboratory can be easily and economically designed provided sophisticated means of detection or measurement are not required and/or eluted samples are stable in the appropriate solvents.

C. Reagents and Equipment

In TLC analysis of water wastes, the analyst is working with components present at concentrations in the area of parts per million or less. Bevenue et al. [5] report pesticide residues in grab samples of 1 gallon or less to be in the parts per trillion range. At these levels, precautions must be taken to avoid contamination. In the applications to be treated, recommended purification steps are included where advised. However, there are certain general considerations common to most analyses encountered.

Virtually all adsorbents and binders, regardless of source, are too contaminated for analyses at low concentrations. Whether or not this is critical depends on the specific application. However, a good purification technique, applicable in most instances, is to prewash the prepared plate with the selected developing solvent. The surface should be examined between washes and the selected method of measurement (GC, UV, etc.) used to determine treatment efficiency. There may be instances when the developing solvent does not elute the contaminant. When this occurs, another more effective solvent must be used. Whatever the case, the "cleaning solvent" should be volatile and should not modify the activity of the plate.

Reagent grade solvents should not be regarded as contaminant free, particularly if such measurement techniques as gas chromatography or fluorescence are to be employed. If purer grades are unavailable, then the reagents should be examined individually using the method of measurement and, if necessary, purified in the laboratory by distillation or some other technique experimentally determined to be effective.

Some authors recommend cleaning glassware with dichromate-sulfuric acid cleaning solution followed by thorough rinsing with distilled water. However, this solution may itself contribute contamination depending on the analysis. Others have recommended heating at temperatures 300°C or more. Whether this is effective remains to be seen, particularly if equipment cannot withstand these temperatures. Prior to use, glassware should be thoroughly rinsed with the purified solvents they are to contain and dried only if necessary to avoid dilution errors. Etched glassware should never be used as contaminants may be trapped in grooves and leached out during the analysis. Ground glass fittings present similar difficulties and should be avoided if possible. If ground glass fittings must be used, special attention should be given to cleaning. Bevenue et al. [5] suggested that solution transfers using this type of stopper be made with clean disposable pipets. Rubber or cork stoppers should never be used. In the laboratory, I have experienced contamination, presumably by leaching, using polyethylene containers and stoppers and these, too, should be avoided unless predetermined experimentally to cause no interference.

The analyst may have need to separate a multiphase system. Centrifugation is preferred but if filtering is necessary, the medium should be copiously cleaned with any solvent with which it is to come into contact. This applies also to filter paper, which may be used to line the developing chamber, saturated with developing solvent, and used to aid in maintaining solvent-vapor equilibrium.

Sample applicators usually consist of syringes or microliter spotting pipets. The choice will depend on the nature of the sample. If possible, applicators should be soaked in the sample solvent overnight prior to use.

D. Sample Application

Sample application is critical to the success of any TLC analysis. The sample may be applied as a spot or streak, depending on availability of sample and sensitivity limitations. Plate capacity should be predetermined as overloading can cause tailing and/or poor resolution. Kovacs [6] recommended microliter spotting pipets because they were easier to manipulate and less likely to cause error. He further recommended spotting volumes be kept at a minimum, 10 μl or less, to avoid excessive origin diffusion. A plate developed vertically should be scored on both sides prior to use to minimize horizontal diffusion. Samples should be applied at a

distance from the score(s) such that the developed spot lies at least 1 cm
from the score. This is important for two reasons. First, and most
obvious, sample may be lost should it develop too close to the edge.
Second, plate surface uniformity is worst near the edges. Quantitation
using a densitometer or spot area determination are extremely dependent
on plate surface uniformity. As these techniques are more than just
occasionally recommended, the analyst should remember when employing
them that the developed spot is a three-dimensional figure. Developed
spots are not always circular and surface dimensions (length and width)
depend on sample penetration (depth) as well as concentration.

Standards should be spotted alternately across the plate with samples
for best results. Whenever possible, the analyst should always spot or
streak identical volumes of sample and standard solutions.

II. APPLICATIONS

A. Nonionic Detergents

Synthetic detergents are used in many households as well as in industry.
While detergents are generally classified as anionic, nonionic, and cationic
[2], anionic detergents are the most common [7] and many methods are
available for their determination. Nonionic detergents are not readily biodegradable, cause considerable foaming problems, and are difficult, at
best, to analyze in small concentrations in effluents [8, 9]. Patterson et
al. [9, 10], described a procedure for determining polyoxyethylene-type
nonionic detergents. This type of detergent consists of condensed ethylene
oxide groups ranging from about three to 15 groups, with an average of
about nine. The procedure described was an application for the determination
of alkylphenol-9-ethylene oxide. A sensitivity of 0.01 mg/liter was reported
with an accuracy of ±10%, in analyses of effluents and river waters.
Dragendorf reagent was used as the means of detection. This reagent was
made by suspending 1.7 g of bismuth oxynitrate in 220 ml of glacial acetic
acid, adding 100 ml of 40%, w/v, potassium iodide (in water), and diluting
to 1 liter with water (solution A); to 10 ml of solution A was added 1 ml of
orthophosphoric acid and 5 ml of 20% w/v barium chloride (in water). Using
this reagent spray, a sensitivity of 1 μg per spot was observed but because
the spots faded with time in air, the authors covered the TLC plates with
plain glass plates prior to measurement. Using this technique, spots over
a range of 1-5 μg could be discerned after several days. A 250-ml portion
effluent was treated with magnesium sulfate (60% w/v) to salt out the detergent and was then extracted with four 50-ml portions of chloroform.
Emulsions were eliminated by filtering the organic phase through a cotton
pledget. (If this was unsuccessful or impractical, the effluent was reduced
in volume by boiling from 250 ml to 100 ml and then acidified with 25 ml of
1N hydrochloric acid and brought to 250 ml with water). The chloroform

was washed with acid or base solutions depending on the quality of the effluent. The chloroform was evaporated to dryness and the residue dissolved in 0.5 ml of chloroform. Silica gel plates were prepared at a thickness of 0.25 mm on 20 × 20 cm plates, air dried, and activated by heating at 110°C for 1 hr. Plates should be activated on the day to be used and stored over dried silica gel. One developing tank (22 × 22 × 9 cm) was lined with filter paper and 100 ml of ethyl acetate-water-glacial acetic acid (4:3:3) was added. The solvent was allowed to migrate up the filter paper (lining). A 1-5 μg sample of standard was applied to the plate (2-10 μl of chloroform solution) and 10 and 20 μl of sample solution was spotted adjacent to the standards using a 5-μl pipet. Developed plates were air dried and heated at 100°C for 10 min and sprayed with the modified Dragendorf reagent. The plates were covered immediately with plain glass plates and the samples were quantitated by comparing unknowns with standards, presumably by eye. A second developing solvent may be used to confirm the detergent as a polyoxyethylene type, consisting of 100 ml of ethyl acetate-glacial acetic acid-water (7:1.5:1.6). This tank was unlined. In cases of excessive background, a third developing solvent may be used consisting of ethyl acetate-water-glacial acetic acid (80:10:11.5). Plates should be developed in an area where the temperature does not exceed 22°C, as higher temperatures caused spot diffusion. Thicker plates (0.3-0.35 mm) tolerated higher temperatures. Recovery data were reported using standards over a concentration range of 0.2-10.0 mg/liter in detergent-free effluent. Included as part of the recovery studies were the results of a collaborative study employing eight analysts. As described, the procedure was capable of estimating levels of detergent over a range of 0.1-1.0 mg/liter. Assay values as low as 0.01 mg/liter were achieved by using larger volumes of aqueous effluent and spotting volumes up to 40 μl. Anionic detergents were carried through the extraction procedure but migrated ahead of the nonionic detergents; also, color reactions with Dragendorf reagent were different. Polyethylene glycols remained at or near the spotting origin. Analytical reagent grade chemicals were used throughout.

In another publication, Patterson et al. [11] described the application of the above procedure for studying the biodegradation of alcohol polyethoxylates. Various degradates containing three or more ethylene oxide groups were investigated. The extraction procedure was modified slightly to include an additional chloroform extraction of the effluent made 2N with sulfuric acid. This allowed recovery of acidic degradates remaining in the effluent after the preliminary chloroform extractions. The polyglycol degradates were separated by molecular weight using the ethyl acetate-glacial acetic acid-water (7:1.6:1.5) solvent system (higher molecular weight components migrated least). The TLC system was used effectively to corroborate foaming studies.

In two additional reports, Patterson et al. [12, 13] used the same TLC system to study the degradation properties of alkylphenol polyethoxylates in sewage effluents and rivers. They were able to correlate foaming properties and rate of degradation with such factors as pH of the effluent and bacterial degradation. Anionic detergents could also be detected in small quantities. During these studies, they found it necessary to prewash the TLC plates with the developing and extracting solvents. Plates so treated required reactivation. Silica plates as thick as 0.5 mm were used. After TLC cleanup of the effluent, IR and NMR were used to confirm identification of unknowns.

Ellerker et al. [8] followed the general TLC procedure used by Patterson in their studies of nonionic detergents in sewage but made certain modifications. They experienced difficulties in preparing samples (emulsions, etc.) and found it more practical to prepare samples one day and run the TLC analysis on the following day. Using 250 ml of effluent, the extraction required 3 hr in the absence of any emulsions, allowing them to complete four analyses daily. They recommended that all reagents be redistilled. Redistilled chloroform contributed a blank equivalent to 10 mg/liter of detergent. They changed the ratio of components in the developing solvent to ethyl acetate–glacial acetic acid–distilled water (3:4:3). Samples (presumably grab samples) were pooled over a 1-month period using mercuric chloride as a preservative. The authors caution that patience and much skill in interpreting spot intensities are required but generally regard the TLC procedure as applicable for simple, routine analyses of nonionic detergents.

B. Greases and Oils

Greases and oils have received recent notoriety primarily as a result of the damage these sediments cause to the environment due to gross spillages. The materials cause additional problems by their tendency to clog normal drainage conduits and screens.

In the analysis of greases and oils in wastes, the acquisition of a representative sample is most difficult as is the case in most multiphase systems. The difficulty is worse in instances of wide contamination or gross spillage due to variations in sediment thickness [7]. The simultaneous presence of detergents or other chemicals may cause these to emulsify or react in some other way to solubilize the materials [14].

Generally, liquid-liquid extraction of an aqueous grab sample using such solvents as petroleum ether or chloroform have been most popular. Jenkins et al. [15] discussed some techniques for extraction and fractionation of oils and greases in sewage and wastes. In extracting these materials, care must be taken to assure no loss of volatiles, particularly if heating or evaporation under reduced pressure are employed in any concentration steps.

Lambert [16] described a technique for identifying oils and greases from treatment plants, ponds, rivers, and streams by TLC. After adjusting the pH to 4-4.5, the sample was extracted with a mixture of chloroform

and methanol. Addition of water gave a biphasic system that enabled the
analyst to isolate the chloroform layer. The solvent was evaporated to dryness and the residue redissolved in 1 ml of chloroform. A working concentration of 1 mg/ml was advised. Silica gel plates were activated at 110°C for
2 hr. No thickness was reported in the author's reference. Hexane-ethyl
acetate-glacial acetic acid (90:10:2), hexane-pentane-ethyl acetate (50:40:10),
and hexane-chloroform-glacial acetic acid (90:10:1) were recommended as
developing solvents. Plates were sprayed with either 0.5% formaldehyde
in sulfuric acid or 4:1 carbon tetrachloride-antimony pentachloride and
components were identified by color difference.

Matthews [17] used TLC for extensive identification of heavy lubricating
oils, greases, engineering oils and fuels, coal tar and products, residual
and crude petroleum products, and various special materials, such as oil
of wintergreen, liquid paraffin, and different vegetable oils. These materials
were identified in all types of sewage effluents. The technique was not
recommended for separations of mixtures of components, such as greases
and lubricating oils, as each is a mixture in its own right. It is of considerable aid therefore if the analyst is familiar with the types of components that
may be present. Aqueous samples were extracted with petroleum ether or
methylene chloride. Any fats were eliminated by saponification. TLC plates
(5 × 2 cm and 10 × 20 cm) consisted of Kieselguhr G, Alumina T, and Silica
Gel T prepared at a thickness of 0.25 mm. The plates were air dried overnight and activated at 105°C for 30 min prior to use. A drop of extracted oil
was applied to a TLC plate using a melting point tube. In case of high
viscosity, the oil or grease was heated or prepared as a chloroform concentrate. Samples (as 10% toluene solution, w/v) may also be applied as a band
consisting of five spots spread over a distance of 1.5 cm. The solvent must
be spotted carefully to avoid diffusion and dried completely prior to development. A variety of developing solvents were employed consisting of petroleum
ether, acetone, and ethanol used individually or in combinations. Ascending
development was used throughout. Due to the complex nature of the samples
studied, the data and methods of interpretation were many and are not discussed here. These are treated extensively by the author in five tables included
in the original reference. Suffice it to say that all plates were examined under
UV light at 254 and 350 nm. Color (fluorescence) and R_f values are used primarily for identification and, as a result, standards are recommended when
possible.

C. Metals

The presence of trace metals in waters can have various effects on the
environment [7]. Water hardness is the reason for such problems as resis-
tence to lathering and can be caused not only by calcium and magnesium but
also by strontium, iron, manganese, aluminum, and zinc. Variations in
water taste have been attributed to such metals as iron, manganese, sodium,
potassium, copper, and zinc. Water coloring indicates the presence of

certain transition metals. High concentrations of metals may prove toxic. A summary of the more common metals found in trace quantities in U.S. waters is reported by Allen and Mancy [7] and is shown in Table 1.

TABLE 1

Summary of Trace Elements in Waters of the United States[a,b]

Element	Number of positive occurrences	Frequency of detection %	Observed positive values (μg/1)		
			Minimum	Maximum	Mean
Zinc	1207	76.5	2	1183	64
Cadmium	40	2.5	1	120	9.5
Arsenic	87	5.5	5	336	64
Boron	1546	98.0	1	5000	101
Phosphorus	747	47.4	2	5040	120
Iron	1192	75.6	1	4600	52
Molybdenum	516	32.7	2	1500	68
Manganese	810	51.4	0.3	3230	58
Aluminum[c]	456	31.2	1	2760	74
Beryllium	85	5.4	0.01	1.22	0.19
Copper	1173	74.4	1	280	15
Silver	104	6.6	0.1	38	2.6
Nickel	256	16.2	1	130	19
Cobalt	44	2.8	1	48	17
Lead	305	19.3	2	140	23
Chromium	386	24.5	1	112	9.7
Vanadium	54	3.4	2	300	40
Barium	1568	99.4	2	340	43
Strontium	1571	99.6	3	5000	217

[a]1577 samples (October 1, 1962 to September 30, 1967.

[b]Reprinted from Ref. 7, p. 979, through the courtesy of Marcel Dekker Publishing Co., New York, New York.

[c]1464 aluminum analyses.

Baily [18] recommended TLC for screening studies and quantitative
estimates of metals in effluents. An effluent sample (presumably a grab
sample) of 100 ml was placed in a 250-ml flat-bottomed boiling flask, acidified with nitric acid, and evaporated to dryness. The residue was redissolved
in and brought to 10.0 ml with 50% hydrochloric acid. Standards were
prepared using the purest form of the metal available. For most metals,
the standard was dissolved in hydrochloric acid-nitric acid (10:1), heated
until solubilized, and then diluted to a concentration of 0.5 $\mu g/\mu l$. There
were slight variations in the preparation of chromium and lead. Standard
concentrates of individual metals could be kept up to a month while mixtures,
prepared by various combinations during dilution steps, were stable for
about a week. An appropriate volume of the standard mixture was evaporated
and reconstituted in the same manner as the sample. TLC plates (10 × 20
cm) were prepared using various grades of cellulose that were prewashed
overnight with Decon 75 (to remove grease) and distilled water. The plates
were spread at a thickness of 0.25 mm, dried overnight (presumably air
dried), and activated in an aluminum oven for 2 hr at 105°C prior to use.
Samples should be spotted while the plates are still hot. For zinc analyses,
the plates should also be washed with the suggested developing solvent.
While constant spotting volumes were not emphasized, an origin spot of 7
mm was recommended. The size of the spot, Baily felt, could be controlled by careful spotting technique. The plates were placed in a developing
tank lined with filter paper, saturated with developing solvent, and developed
for 13 cm (chromium required 15 cm). The tank should be equilibrated for
1 hr prior to use. Four developing solvent systems were advised depending
on the analysis:

1. Methylethylketone-hydrochloric acid-distilled water (60:12:8) for
   cobalt, lead, and manganese. Solvent is stable for about 5 days.
2. Industrial methylated spirits 98%-2N acetic acid-2,5-hexanedione
   (120:20:2) for zinc. Solvent should be prepared fresh.
3. n-Butanol-1.4N hydrochloric acid-2,5-hexanedione (100:20:0.5) for
   cadmium. Solvent is stable for about 5 days.
4. Hydrochloric acid-acetone (45:55) for copper, nickel, manganese,
   and chromium. This system may require two solvent passes, with
   the plate dried between passes. A plate developed using this system
   may also require some special treatment, including spraying with
   alkaline bromine solution.

Developed plates were air dried and "neutralized" for 1 hr in a separate
tank saturated with ammonia vapor. For spot visualization, three color-
producing reagent sprays were recommended:

1. 0.1% alcoholic dithiooxamide for cobalt (yellow), copper (green),
   nickel (blue). Reagent may be regarded as stable.

2. 0.1 g of sym. diphenyl-thiocarbazone, 0.05 g of dimethylglyoxime in 35 ml of methylated spirits, 15 ml of ammonia, 150 ml of alcohol for lead (pink), cadmium (pink). Reagent should be prepared fresh daily.

3. 0.25% sym. diphenylcarbazide in methylate spirits-glacial acetic acid (9:1) for chromium (magenta). Reagent may be regarded as stable.

Manganese (pink) and zinc (red) were visualized by spraying first with reagent spray 1 followed by reagent spray 2. For qualitative studies (screening), reagent spray 2 was recommended.

Baily reported that with his scheme, organic materials generally did not interfere. Sensitivities varied depending on the metal and the means of detection. For copper, nickel, cobalt, and chromium, an application of 3.5–12.5 μg was recommended. For zinc, cadmium, lead, and manganese, the range was 2.5–5.0 μg. Components could be quantitated in a number of ways. Baily used a pantograph and cut out and weighted the tracings. He also gave reference to a computation that related the square root of the spot area with the logarithm of the weight of the component. However, reasonable estimates can be obtained by visual comparison spot intensities between samples and standards. While no prior knowledge of anticipated amounts is required, observing such phenomena as color intensity of effluents or residues can be an aid to the analyst. Tables were presented that compared quantitation by TLC with atomic absorption analysis.

Other TLC systems have been described for trace metal determination. While applications are not specifically directed to waste analysis, with suitable sample preparation and precautions against contamination, they could be suitable and some are briefly discussed.

Oguma [19] described the separation of arsenic(III) and arsenic(V) on silica gel. Samples were prepared in dilute sodium hydroxide solution. TLC plates (2.5 × 20 cm) were prepared at a thickness of 0.25 mm and air dried overnight. No activation conditions are given and are presumed unnecessary. Acetone-phosphoric acid (50:1) was reported as the best developing solvent with arsenic(V) having the greatest mobility. Resolution could be modified by varying the phosphoric acid concentration. Ten to 20 μl) of either arsenic(III) or arsenic(V) could be visualized by spraying the plate first with saturated tin(II) chloride followed by spraying with 5% sodium thiosulfate. Both metals appeared as yellow spots. Other developing solvents were discussed.

Frei et al. [20] achieved a similar separation of arsenic(III) or arsenic(V) using silica and cellulose on plates or chromatographic sheets. "Compact" spots could be obtained on cellulose sheets requiring no pretreatment of any kind. Cellulose plates gave spots good for quantitation using any one of five developing solvents:

1. Acetonitrile-nitric acid-water-2,4-pentanedione (78:2:20:1).
2. Isopropanol-ethyl acetoacetate-acetonitrile-25% nitric acid (4:4:2:2).
3. Benzyl alcohol-acetone-acetonotrile-25% nitric acid (4:4:2:2).
4. n-Butanol-acetone-25% nitric acid (7:3:3).
5. Acetone-benzene-25% nitric acid (4:1:1).

Spots were visualized by spraying successively with 1% nitric acid in ethanol followed by 5% glycerol in ethanol and finally 1% silver nitrate in 1N ammonia. Arsenic(V) appeared as a brown spot, while arsenic(III) appeared yellow. Sensitivity was about $2\mu g$ and spots were stable for about 20 min.

Miketukova and Frei [21, 22] reported the separation of cobalt, nickel, copper, bismuth, lead, manganese, cadmium, vanadium(V), and uranium(VI) on cellulose and silica plates, 0.25 mm thick, and on cellulose and silica chromatographic sheets as well. Developing solvents consisted of various combinations of acetone, isopropanol, acetic acid, hydrochloric acid, and water. Tailing of lead could be avoided by using cellulose layers developed with methanol-25% nitric acid-water (8:1:1). Chromatographic sheets were preferable because of better surface uniformity. Metals could be detected at concentrations of 0.01-0.1 μg per spot using 4-(2-thiazolylazo)resorcinol, 1-(2-thiazolylazo)-2-naphthol, 4-(2-pyridylazo)resorcinol, or 1-(2-pyridylazo)resorcinol, or 1-(2-pyridylazo)-2-naphthol; all were prepared as 0.1% solutions in 95% ethanol. Plates and sheets were prewashed using the developing solvents and stored carefully to avoid contamination by iron (dust). Glassware was cleaned with nitric acid and distilled water.

Zetlmeisl and Haworth [23] separated 21 cations on microcrystalline cellulose using various developing solvents and reagent sprays. A tabulation of their data may be found in Table 2. TLC plates (5 × 20 cm) were prepared at a thickness of 0.75 mm, dried overnight, and stored in a desiccator in the presence of sodium hydroxide. Samples were applied by syringe from aqueous solutions of their nitrate or chloride salts. All reagents were reagent grade or redistilled. Variations in plate uniformity caused, variations in R_f's. An advantage of using microcrystalline cellulose was its ability to withstand a variety of solvents.

Hashmi et al. [24] identified 40 cations and 19 anions using circular TLC. Cations were separated on aluminum oxide D-5 (with 5% calcium sulfate) and on Silica D-O (without binder). Anions were separated on Aluminum Oxide S and Silica Gel S. Development of the chromatogram was reported to take 2 min, making the method suited to rapid, routine analysis. Various developing solvents and reagent sprays for detection were discussed.

D. Sterols

Sterols in water wastes are associated primarily with fecal contamination and, as a result, are usually found close to shore, with the highest concentrations reported to be within a mile of a sewage outlet. A secondary source

TABLE 2

Inorganic Cation Mixtures, Solvent Systems, and R_f Values by TLC
Using Microcrystalline Cellulose[a]

No.	Cation mixture	Solvent system	Detection	R_f
1	Hg(I)	85% n-butyl alcohol–	aq. K_2CrO_4	0.13
	Ag	15% H_2O		0.11
	Pb	pH 3.0 (HOAc)		0.05
2	Ag	85% butyl alcohol–	aq. K_2CrO_4	0.14
	Hg(I)	15% H_2O		0.12
	Pb	pH 2.1 (HOAc)		0.10
3	Hg(II)	90% ethyl alcohol–	H_2S	0.97
	Cd	10% 5M HCl		0.93
	Bi			0.89
	Cu			0.44
4	Hg(II)	22% ethyl alcohol–	H_2S	1.00
	Cd	10% t-butyl alcohol–		0.89
	Cu	11% acetone–		0.82
	Bi	28% H_2O–		0.00
		15% pyridine–		
		1% HNO_3		
		1% HCl–		
		12% n-butyl alcohol		
5	Sn	45% chloroform–	H_2S	0.84
	Sb	45% t-butyl alcohol–		0.73
	As	10% 8M HCl		0.65
6	Sn	47.5% n-butyl alcohol–	H_2S	0.83
	Sb	47.5% 1M HNO_3–		0.77
	As	5% acetoacetic ester		0.66
7	Sn	49.5% n-butyl alcohol–	H_2S	0.94
	Sb	49.5% 2M HNO_3–		0.82
	As	1% acetylacetone		0.74
8	Zn	87% acetone–	Dithizone	0.91
	Fe	8% HCl–	Self-indicating	0.77
	Co	5% H_2O	β-Nitrosonaphthol	0.37
	Ni		Dimethylglyoxime	0.10
9	Zn	90% ethyl alcohol–	Dithizone	0.93
	Fe	10% 5M HCl	Self-indicating	0.80
	Co		β-Nitrosonaphthol	0.33
	Ni		Dimethylglyoxime	0.33

TABLE 2. — Continued

No.	Cation mixture	Solvent system	Detection	R_f
10	Zn Al	70% t-butyl alcohol- 26% H_2O- 4% 6M HCl	1-(2-Pyridylazo)-2- naphthol + 6M NH_3	0.85 0.28
11	Cd Zn	96.5% dioxane- 2.5% H_2O- 1% conc. HNO_3- 1 g antipyrine	N-8-Quinolyl-p-toluene- sulfonamide (1% in $CHCl_3$)	0.22 0.10
12	Ca Sr Ba	40% isopropyl alcohol- 40% H_2O- 20% 1M HCl	Alcoholic alizarin Potassium rhodizonate Potassium rhodizonate	0.73 0.66 0.55
13	Ca Sr Ba	50% collidine- 50% 0.4M HNO_3	Alcoholic alizarin Potassium rhodizonate Potassium rhodizonate	0.64 0.56 0.46
14	Li Na K	70% ethyl alcohol- 30% H_2O	Fluorescein + silver nitrate	0.87 0.73 0.61

[a]Reprinted from Ref. 23 through the courtesy of the Elsevier Publishing Co., Amsterdam, The Netherlands.

is rivers that serve as runoffs from pastures and the like. That industrial wastes contribute to sterol contamination has been implied by a few authors but not established.

Murtaugh and Bunch [25] used TLC as a cleanup technique in studying sterols as a measure of fecal pollution in both domestic wastes and natural runoffs. Quantitation was by GLC using the trimethylsily ether derivatives of components eluted from TLC plates.

Effluent samples as small as 50 ml were acidified and treated with sodium chloride (depending on the sample source) to salt out the sterol. The resulting acid salt solution was then extracted with hexane, which was evaporated to dryness. (When emulsions occurred they were eliminated by washing the hexane with 70% ethanol.) The residue was refluxed in alcoholic potassium hydroxide for 3 hr to saponify the sterol esters and diluted, and the free sterols were extracted twice with hexane. The extract was evaporated to dryness and the residue reconstituted in 20-50 μl of acetone. Standards were prepared at a concentration of 1 μg/μl in acetone and stored at 5°C (fresh solutions were prepared weekly). All glassware was rinsed

with the solvents they would contain and reagents were pre-extracted with
hexane. TLC plates (20 × 20 cm) were prepared using Silica Gel G spread
at a thickness of 0.25 mm. The plates were dried and activated for 1 hr at
110°C prior to use. The entire sample (20-50 µl) was spotted on the plate
and standards were applied adjacent to the samples at various intervals
across the plate. The concentration of the standards ranged from 2 to 4
µg per spot. The plate was developed in chloroform-ether (9:1). As samples
were interspersed with standards, sample components were located by
covering the sample sections of the dried plate and spraying the standards
with phosphomolybdic acid in 95% ethanol. Sample component areas corresponding to the R_f's of the standards were scraped off and eluted with
chloroform-ethanol (4:1). The elution solvent was evaporated and the residue
used to make the trimethylsilyl ether derivative and determined by GLC.

TLC cleanup was essential for GLC analysis of low concentrations of
sterols depending on the sample source. Linearity was observed by GLC
over a 1-4 µg range using a FID. Sensitivity was reported to be 0.02 µg/
liter using 2-liter samples. The procedure was suited to mobile field
analyses with two people completing ten assays per day.

Matthews and Smith [26] also used TLC and GLC to identify cholesterol,
stigmasterol, and β-sitosterol in large bodies of water. Samples (17-45
liters) were collected in lipid-free glass bottles over a 6-month period.
Each sample was filtered using Celite 545 diatomaceous earth. The pH of
the sample was adjusted to 2.5-3.5 and the sample was extracted with
redistilled hexane. Extractions were continued until no further sterol material could be detected in the extract by TLC. The hexane was evaporated
under reduced pressure at temperatures less than 35°C and preserved by
storing under vacuum or by freezing. TLC plates (20 × 20 cm) were
prepared using Silica Gel HF_{254} spread at a thickness of 1 or 2 mm. Sample
residues were reconstituted in methylene chloride and spotted on the plates.
Plates were developed using ethyl acetate-heptane (1:1) or acetone-heptane
(1:1). In some analyses, depending on the source, more than one solvent
pass was required. Two-dimensional chromatography was used to detect
autoxidation. Autoxidation could be avoided by storing the samples in the
dark but was catalyzed by aeration. I suggest storing the samples in amber
bottles (or bottles covered with aluminum foil if amber bottles are impractical); samples can then be flushed with solvent-saturated nitrogen or
treated with bisulfite if these present no contamination (interference) problems. Components were located by comparison with standard R_f's spotted
on the edge of the plate. These standards could be sprayed with 50% sulfuric
acid reagent while the rest of the plate containing the sample components
was shielded. Components were eluted with 5-10% diethyl ether in hexane
and identified by GLC using the trimethylsily ether or acetate derivatives.
Studies indicated that at least 83% of the sterols could be recovered. GLC
was necessary to effect separations in many cases with TLC used as a
cleanup procedure, particularly in the presence of cholesterol.

Smith and Gouron [27] described a procedure for determining 5-β-cholestan-3β-ol in effluents. Samples (8 liters) were collected in lipid-free containers and extracted with 4 liters of hexane. Repeated extractions were used to confirm complete recovery. The extracts were evaporated to dryness under reduced pressure. The residue was dissolved in chloroform-methanol (7:1) and applied to a 20 × 20 cm Silica Gel HF$_{254}$ TLC plate spread at a thickness of 0.25 mm. The plate was developed in benzene-ethyl acetate (3:2). Component spots were visualized under UV light or by spraying with 50% sulfuric acid. Components were eluted with methanol. The methanol was evaporated to dryness and the residue was reconstituted in methylene chloride and assayed by gas chromatography. Other developing solvents were suggested, including those used by Matthews and Smith [26]. Preparatory work was also described. Sensitivity is about 1 μg/liter using TLC in conjunction with GLC.

In the absence of recommended internal standards and because of questionable interferences, I feel that quantitation using these techniques [26, 27] should be experimentally evaluated by the analyst prior to use.

E. Phenolic Compounds

Phenols are found in the effluents of such industries as coal tar, gasoline, plastics, rubber proofing, disinfectants, pharmaceuticals, and steel (as coke oven effluents) [2, 7]. A primary concern about the presence of phenols in waste waters is their toxic effect on aquatic life. Chlorination of phenol-containing water contributes to poor taste and coloring due to the formation of chlorophenols (also found in some pesticides).

Wallwork et al. [28] determined phenol and o-, m-, and p-cresol content in industrially contaminated river waters by TLC. Samples (2 liters) were acidified and extracted with chloroform (four 25-ml portions were adequate). The chloroform was then extracted with two 10-ml portions of 10% sodium hydroxide and the aqueous phase was acidified and then extracted with two 10-ml portions of ether. The ether was evaporated to dryness leaving an 8 mg of residue. TLC plates (20 × 20 cm) were prepared using silica gel spread at a thickness of 0.25 mm. Plates were air dried for 15 min, activated at 105°C for 1 hr, and stored in a desiccator over silica gel. Standard solutions of phenol and o-, m-, and p-cresol were prepared in acetone at a concentration of 1 μg/μl. Standard (2 μl) and sample (presumably 10 μl of an acetone solution) were spotted on the plate, dried, and "overspotted" with 2 μl of diazotizing solution. Diazotizing solution was prepared from two solutions: solution I consisted of 1.9 g of p-nitroaniline dissolved in 45 ml of concentrated hydrochloric acid and diluted to 1 liter with water; solution II was 5% sodium nitrite. The diazotizing solution was a 1:5 mixture of solutions I and II. The plate was developed with methylethylketone for 17 cm at 20°-22°C. The developing chamber was not lined with filter paper.

Developed spots were highly colored (each different) but faded as the developing solvent evaporated, so measurements were made immediately after the
plate was removed from the chamber (I recommend covering the TLC
plates with a plain glass plate immediately after removing them from the
developing chamber to slow the fading process). Total phenol content as
low as 10 mg/liter was reported but no sensitivity was given. However, it
was implied that GLC was more sensitive.

Smith and Lichtenberg [29] measured alkyl-, chloro-, amino-, and
nitrophenols at a concentration of less than 1 µg/liter in surface waters
using TLC. They reported their method was more sensitive than UV and
IR methods and more sensitive and selective than the 4-aminoantipyrine
(colorimetric) method. GLC was more sensitive than TLC but GLC was
not regarded as suited for analyses of raw surface waters at that time.
Samples could be collected either as 1 liter grab samples or carbon adsorption representing the phenol content of 1000-20,000 liters of effluent. Glass
sampling containers were lined with Teflon and transported to the laboratory
in polystyrene cartons. The 1-liter "grab" samples were adjusted to pH
2 and extracted with ether or chloroform (three 100-ml portions). The
extract was evaporated in stages to 0.1 ml, the last stage occurring in a
15-ml centrifuge tube. Samples collected by carbon adsorption were oven
dried at 40°C for 2 days and extracted with 2500 ml of chloroform in a
soxhlet extractor for 35 hr. The extract was filtered through chloroform-
washed filter paper and evaporated to dryness in stages, the last stage
occurring in a tared beaker. The weight of the residue was then determined.
TLC plates (20 × 20 cm) were prepared using Silica Gel G spread at a thickness of 0.25 mm. Plates were air dried for 5 min and activated in an oven
at 100°C. (I caution that plates air dried for only 5 min using aqueous
slurries of silica gel should be spread out during activation, as stacking
the plates in the oven causes condensation from the surface of the bottom
plate to build up on the glass back of the top plate. Drippings will cause
distortions in plate surface.) Smith and Lichtenberg prepared a weak-acid
fraction of the concentrated chloroform extracts as described by a Shriner
and Fuson [30] solubility class extraction technique. This weak-acid fraction
was concentrated to a spotting volume of 0.5-1.0 ml. A 10-20% sample of
the total fraction was spotted along with standards in varying concentrations
(1-50 µg) adjacent to samples. Two plates were spotted. The first plate
was developed in benzene-cyclohexane-diethylamine (5:4:1). This solvent
must age 1 day and be replaced after every second development. The plate
was air dried and nitrophenols were determined without further treatment
as these appeared as brightly colored (yellow) spots. The plate was then
sprayed with freshly prepared p-nitrobenzenediazonium fluoroborate (p-
NBDF) reagent (1% in acetone), dried, and sprayed again with 1N potassium
hydroxide solution. The resulting colors may be compared with standards
and the components quantitated. The second plate was developed in chloroform using two solvent passes (to remove interferences). The plate was
dried and sprayed with Gibbs' reagent (2,6-dibromoquinone chlorimide, 1%

in chloroform). Gibbs' reagent is stable for several weeks if stored in the dark. The first developing solvent was selective for the alkylphenols. Chloro-, nitro-, and aminophenols were only slightly mobile. The second solvent system was a more general solvent and compounds with functional groups ortho to the phenolic hydroxyl group migrated farthest. Gibbs' reagent was more selective (producing more colors) than p-NBDF reagent and about five to ten times as sensitive; Gibbs' reagent was sensitive to about 0.1-1 μg except for para-substituted phenols, where the limit was about 0.5-10 μg. Rhodamine B, 4-aminoantipyrine, or silver nitrate may be more selective (as visualization sprays) but are limited in sensitivity.

Aly [31] separated and identified mixtures of phenol; o-, m-, p-chloro-phenols; and 2,4- and 2,6-dichlorophenols in river waters and sewage by TLC. Samples (100 ml) containing about 100 μg of phenol were treated with either (1) 4-aminoantipyrine reagent (2% in water) and potassium ferricyan-ide solution (1% in water) in alkaline medium (pH 8) or (2) p-nitrobenzene-diazonium reagent (0.2% in methanol) in ca. 0.04N sodium hydroxide solution. The resulting dyes were extracted with chloroform (antipyrine derivative) or ether (azo derivative, after acidification). The spotting volumes were 0.5 ml in chloroform. TLC plates (20 × 20 cm) were prepared using Silica Gel G spread at a thickness of 0.25 mm. Plates were air dried for 15 min and activated for 30 min at 110°C. The antipyrine derivatives were spotted on plates prepared using silica-water slurries, while the azo derivatives were spotted on plates prepared using silica-0.1N sodium hydroxide slurries. Five to 10 μl (1-2 μg) were spotted at 1-cm intervals and developed at room temperature using:

Antipyrine derivatives (use any)
1. Methylene chloride-acetylacetone (70:30)
2. Chloroform-ethyl acetate (80:20)
3. Ethyl acetate-hexane-acetic acid (70:20:10)

Azo derivatives
1. Chloroform-acetone (90:10)

Sensitivity using the 4-aminoantipyrine derivative was 0.05-0.1 μg of phenol, whereas use of the p-NBDF derivatives was limited to 0.1-0.5 μg of phenol per spot.

Zigler and Phillips [32] used two-dimensional TLC to identify and estimate chlorophenols in water samples before and after treatment in a purification plant. A 1-liter sample was acidified to pH 1.5 and extracted with petroleum ether (pesticide quality). The extract was evaporated in stages to 0.1 ml and spotted on TLC plates prepared using aluminum oxide activated at 120°C for 1 hr [33]. The plates were developed first with benzene for a distance of 3-4 in. in a "sandwich" developing system. This development removes undesirable impurities and may require more than one solvent pass. Chlorophenol standards (0.05-0.5 μg) were spotted above the benzene solvent front in the same line as the samples. The plate was then rotated 90° and developed using 6% 1N sodium hydroxide in acetone. Spots were visualized (light to dark brown spots) using silver nitrate solution, which consisted of 0.5 g of silver nitrate in 5 ml of distilled water, 100 ml of 2-phenoxyethanol, three drops of 30% hydrogen peroxide, diluted to 1 liter with acetone; 4-aminoantipyrine spray reagent may also be used but is less sensitive. Recoveries on raw water samples were reported to be 75-100%, with a sensitivity of less than 0.1 μg of chlorophenol per liter of sample.

Heier [34] reported using TLC to identify and measure phenols in water wastes using Kieselguhr G made alkaline with sodium carbonate. The developing solvent was toluene-ethanol-acetone (60:6:0.5). Spots were visualized using freshly diazotized p-nitroaniline, presumably as a spray. Components were quantitated using a TLC densitometer. Sensitivity was reported to be 0.04-0.3 μg per spot with a precision of ±6-18% depending on the phenol and the amount present.

Dyatlovitskaya and Maktaz [35] determined as low as 2 μg of phenol and o-, m-, and p-cresol by TLC in sewage. Samples were acidified to pH 5 and extracted with ether. The ether was then extracted with a 1% alkali solution. An aliquot was treated with 1 ml of 20% sulfuric acid, 25 ml of 2N sodium carbonate, and 2.5 ml of diazotized 0.1N p-nitroaniline solution. Chlorobenzene extracts were applied to TLC plates consisting of alumina spread at a thickness of 1-2 mm. Chlorobenzene also served as a developing solvent. The colored spots were scraped off and eluted with isopropanol-water-1N sodium hydroxide (75:25:1) and determined spectrophotometrically.

Seeboth [36] reported the separation and quantitation of phenol, catechol, resorcinol, and hydroquinone on Silica Gel A or Supergel or a mixture of both with acidic alumina. Developing solvents were chloroform-acetic acid (5:1), chloroform-acetone-acetic acid (10:2:1), benzene-acetic acid (5:1), light petroleum (b.p. 60°-80° C)-carbon tetrachloride-acetic acid (4:6:1), or chloroform-acetone-diethylamine (20:10:1). Sample spots were sprayed using p-NBDF reagent, eluted with methanol, and determined spectrophotometrically. For reliable quantitative analyses of waste water, phenolic content greater than 10 μg was recommended. Sensitivity was reported to be about 1 μg.

Chambon [37] also used TLC to determine small quantities of phenols in water to supplement the nonspecific 4-aminoantipyrine colorimetric method.

F. Miscellaneous Hydrocarbons

As effluents are very complex, the analyst should have a specific organic compound or class of compounds in mind before attempting an analysis of any kind. Also, he or she should consider environmental influences on the component of interest, e.g., oxidation, bacterial effects, and temperature variations. Some of the more common classes of organic compounds have already been discussed. In this section, some not

so common but equally important compounds are treated as their presence
in effluents affect the environment.

Lambert [38] and Berthold [39] described the use of TLC to identify
tract hydrocarbon pollutants in waters. Ershova and Mints [40] determined
1,2-benzanthracene in petroleum industry effluents by liquid-liquid extraction and TLC on alumina. The component was located by fluorescence under
UV light and eluted from the plate with ether. The ether was evaporated to
dryness and the residue dissolved in n-octane; 1,2-benzanthracene was
quantitated by fluorescence at liquid nitrogen temperatures.

Svishchuk et al. [41] determined hexamethylenediamine in waste waters
by first reacting the components in the sample matrix with 1-(dimethylamino)-
naphthalene-5-sulfonyl chloride. After the solution was extracted with
chloroform, the colored materials were applied to TLC plates, developed,
and eluted with ethanol. Hexamethylene-diamine was quantitated by
fluorescence.

Dyatlovitshaya and Maktaz [42] determined caprolactam in sewage using
TLC. Portions of effluents (25 ml) containing caprolactam at a concentration
of 0.25-10 mg/liter were hydrolyzed in acid to form 6-aminohexanoic acid.
The solution was evaporated to dryness and the residue dissolved in hot
water. The sample was then chromatographed on alumina plates and
developed with butanol-acetic acid-water. Spots could be visualized by
spraying with ninhydrin or copper-ninhydrin reagent and heating at $110^\circ C$.
The colored area was scraped off, eluted with methanol, and quantitated
spectrophotometrically.

Scholz and Altmann [43] reported measuring benzo[a]pyrene in ground
waters by TLC after extracting with cyclohexane. The extract was concentrated and spotted on Silica Gel H containing Polyoxyethylene Glycol 1000
and developed with benzene-hexane (1:3). Benzo[a]pyrene was located by
fluorescence under UV light, scraped off, and eluted with cyclohexane.
The elution solvent was evaporated to dryness and the residue reconstituted
in dioxane. Benzo[a]pyrene was quantitated by fluorescence (excitation
maximum = 365 nm; emission maximum = 429). The method was sensitive
to 0.1 ng/liter with a relative error of ±15% using samples containing
benzo[a]pyrene at a concentration of 1-10 ng/liter.

Borneff and Kunte [44] separated and determined some carcinogenic
substances in ground water contaminated by sewage. They were able to
measure fluoranthene, 3,4-benzofluoranthene, 11,12-benzofluoranthene,
3,4-benzopyrene, 1,12-benzoperylene, and indeno[1,2,3-cd]pyrene.
Samples (10 liters) of ground water containing about 50 μg of these components were collected in aluminum barrels and extracted overnight with 600
ml of fluorescent grade benzene that had been redistilled three times (each
distillate was examined chromatographically for polycyclic contaminants).
Carbon adsorption was not adequate to remove contaminants. The residue
from extraction was dissolved in 0.1 ml of solvent and spotted on TLC plates
along with standards. Plates consisted of Kieselguhr, alumina, or cellulose
spread at a thickness of 0.3 mm. Plates were air dried and activated at

110°C for 1 hr. Three developing solvents were employed: (1) isooctane; (2) n-hexane-benzene, 9:1; and (3) methanol-ether-water, 4:4:1. The plates were developed with solvent 1 for about 40 min, followed by solvent 2 for about 60 min. Solvent 3 (optional) may require two solvent passes and required 60 min for development. Components were quantitated using a TLC fluorodensitometer. The measurements were made quickly as samples were light sensitive and subject to oxidation and sublimation. Spot sensitivity was reported to be 0.01 μg per spot. Glassware should be scrupulously cleaned using acetone, detergent, and distilled water. Aluminum sampling bottles should stand overnight in demineralized water containing about 1 g of potassium permanganate. Care should be taken to avoid contamination of the water samples with grease or oils (use of pumps should be avoided).

G. Pesticides, Herbicides, and Fungicides

Hundreds of compounds are available commercially for use in protecting crops and waters from various flora, fungi, insects, etc., which prove harmful to man directly or through his nourishment. However, these useful compounds have, in themselves, a potential for harm or annoyance to man. In water, they have been known to contribute to foul odor and taste and in sufficient quantities have even proved toxic. Some have a very low rate of degradation, while others degrade rapidly to more toxic substances. Their treatment in this chapter is based on their presence in waters as wastes and in this capacity the concern is primarily with ground leaching, manu-facturing wastes, and spillages. TLC has proved very useful as a technique for class screening, cleanup procedures, and quantitative determinations. Clarke [45] and Smith et al. [46] reported extensive use of TLC as a cleanup procedure in conjunction with GC analyses of pesticides. Kovacs [6] and Bevenue et al. [5] published excellent reviews on TLC techniques as applied to pesticide residue analysis with discussions on sample preparation, spotting techniques, sources of contamination, etc.

Shmigidina and Klisenko [47] reported the use of TLC to determine Acrex, a fungicide, in drinking water. The fungicide was extracted from a 200-ml sample of water using ca. 50 ml of chloroform. The chloroform was evap-orated and the residue dissolved in 0.2 ml of ether. The sample was applied to a silica gel TLC plate containing powdered zinc and developed using hexane–acetone (4:1). Standards (100 μg) were spotted adjacent to the sample. Spots were visualized using alcoholic ninhydrin spray. Sensitivity was reported to 3 μg per spot with a precision of ±6.8%.

Tamus [48] separated and identified malathion, dimethoate, and ethion in waters on silica gel. Samples were extracted with chloroform or ethylene chloride and spotted on 0.25-mm plates activated at 130°C. Best developing solvents were malathion/hexane-ether (9:2), dimethoate/hexane-ether (5:1), or benzene-ether (5:1); ethion/benzene; or hexane. Maximum sensitivity (0.2 μg per spot) was achieved by exposing the plates to bromine vapor

followed by spraying with 0.5%, 3,5-dibromo-p-benzoquinone chlorimine
in DMF. Dimethyl phosphorodithioate or malaoxon did not interfere.

Samosvat [49] was able to determine trace quantities of monuron on
alumina plates using chloroform as the developing solvent. Monuron was
extracted from water with chloroform. After development, the plate was
air dried and heated for 1 hr at 160°C and the monuron was visualized by
spraying with sodium nitrite in hydrochloric acid followed by naphthol in
alkali. Semiquantitative analysis was possible by comparing the intensity
of the red spots of samples with known standards.

Katz [50] estimated some substituted urea herbicides in surface waters
by TLC on Eastman Chromatographic Sheets. Linuron, monuron, diuron,
fenuron, and neburon were extracted from a 100-ml portion of water with
four 50-ml portions of chloroform. The organic phase was evaporated to
dryness and the residue was dissolved in 0.2 ml of acetone. Ten microliters were spotted on an Eastman Sheet No. K301R2. The sheet was
developed vertically for 15 cm using methanol-methylchloroform-2,2,4-
trimethylpentane (1:7:12). The sheet was dried and sprayed with ninhydrin
reagent and heated at 140°C for 10 min. As little as 0.2 μg of the herbicides
was detected and estimates were made by comparing sample and standard
spot intensities.

Abbott et al. [51] identified and estimated carbamates, urea, and uracil-
type herbicides in water by TLC on silica and alumina. Samples were
extracted from neutral solutions with methylene chloride. The organic
phase was evaporated to dryness and reconstituted in 40 μl of carbon tetrachloride. Samples (1-10 μg) were applied to silica or alumina TLC plates
spread at a thickness of 0.25 mm. The plate was developed using hexane-
acetone (7:3). On silica, spots were visualized by spraying with Brilliant
Green solution followed by bromination; on alumina, spots were visualized
by spraying with dichlorofluoroscein or ethanolic sulfuric acid. Spots were
scraped off and a sheet of blocked graph paper was applied to the plate
surface. A sheet of light sensitive diazo paper was placed on top of the
graph paper, compressed with a plain glass plate, and exposed to light for
20 sec. After exposure to ammonia in a suitable chamber, the diazo paper
showed the image of the plate with spot "holes" superimposed on the blocked
graph paper. The logarithm of the weight of standard was plotted vs the
square root of the standard spot area. Unknowns were quantitated from
their spot areas and a standard curve by interpolation. Using the extraction
procedure described, sensitivity was reported to be 0.1 ppm.

Cohen and Wheals [52] quantitated some substituted ureas and carbamates
by GC after TLC cleanup. Water samples (1 liter) were extracted using
50-ml portions of chloroform. The solvent was evaporated to a volume
such that the entire residue could be spotted. Standards (5-10 μg) were
spotted along with samples. TLC plates (20 × 20 cm) were prepared using
Silica Gel G spread at a thickness of 0.25 mm. The plates were developed
in unlined chambers using chloroform or hexane-acetone (5:1). Spots were
visualized by treating as follows: the plate was sprayed with 7 ml of 5%

hydrochloric acid and sprayed again using 7 ml of 1-fluoro-2,4-dinitroben-
zene (4% w/v in acetone); a glass plate was clamped over the TLC plate
tightly and it was heated in a vertical position at 100°C for 40 min. The
resulting yellow spots may be eluted in acetone and assayed by GC. Recovery
data were included for "spiked" water controls at a concentration of 0.01
ppm but at this level there were some extraction losses.

El-Dib [53] also determined carbamate and phenylurea residues in
waters by TLC. Samples (500 ml) were acidified and extracted with chloroform or methylene chloride. Plates (20 × 20 cm) were prepared using silica
gel spread at a thickness of 0.25 mm, air dried, and activated at 110°C.
Samples (2-5 μg) were spotted on the plate along with standards (prepared
in acetone at a concentration of 100 μg/ml). Various solvent systems
were suggested, including (1) benzene, (2) benzene-acetone (95:5), (3)
benzene-acetone (85-15), (4) cyclohexane-ethanol (85:15), (5) hexane-
acetone-toluene (6:2:2), and (6) hexane-acetone (7:3). Carbamates and
phenylureas could be detected by spraying with p-dimethylaminobenzaldehyde
reagent and heating at 150°C. Phenyl carbamates required additional treatment with sodium nitrite and 1-naphthol (sprays). The sample portion of
the plate could be shielded from the sprays using a glass plate. R_f's of
standards were then noted and the corresponding sample areas were scraped
off, eluted in chloroform (5 ml), and identified by UV in 2-cm cells.

Ellerker et al. [8] studied the degradation of 2,4-dichlorophenoxyacetic
acid (2,4-D) and 2,4,5-trichlorophenoxyacetic acid (2,4,5-T) by TLC.
These herbicides were used to control weeds that were blocking a sewage
conduit. The method was a modification of the TLC system developed by
Abbott et al. [54], who determined these materials in waters contaminated
by runoffs. Ellerker found that recovery was best using a single (2:1)
preliminary ether extraction of the aqueous phase. The ether was dried
using anhydrous sodium sulfate and evaporated to dryness, and the residue
was dissolved in 40 μl of ethyl acetate. TLC plates (20 × 20 cm), similar
to those described by Abbott et al., consisted of a 60:40 mixture of Kieselguhr G-Silica Gel G spread at a thickness of 0.25 mm, air dried, and
activated at 120°C for 2 hr. Best results were obtained when plates were
used within 36 hr of preparation. The recommended sample spotting
volume was 1 μl. The developing tank was equilibrated with 200 ml of
liquid paraffin-benzene-glacial acetic acid-cyclohexane (1:3:2:14). The
plate was placed in the tank at an angle of 135° (to give the fastest developing time with the least spot diffusion) and allowed to develop for 15 cm.
After development, the plate was air dried, heated for 10 min at 120°C,
and sprayed with 0.5% alcoholic silver nitrate; after an additional 10 min
of heating at 120°C, sample and standards appeared as black spots. The
author reported the analysis effective for herbicide concentrations of 10
mg/liter effluent.

Askew et al. [55] reported TLC as the only successful way to characterize phosphate pesticides and their metabolites in waters contaminated by
accidental or intentional spillages. No specific preliminary cleanup

procedure was recommended as this was highly dependent on the source of
the sample. However, adsorption chromatography using 1 g of Nuchar
carbon in a column gave good results. Analyses of 40 different organophosphorus pesticides were analyzed. Samples (1 liter) were extracted
with three 50-ml portions of chloroform. The chloroform was dried with
anhydrous sodium sulfate and evaporated to an appropriately small
volume (depending on concentration of the pesticide). TLC plates (20 × 20
cm) were prepared using Silica Gel G spread at a thickness of 0.25 mm.
Plates were activated at 120°C for at least 2 hr. Sample and standards
were spotted and the plate was developed using any of three developing
solvents: (1) hexane-acetone (5:1), (2) chloroform-acetone (9:1), or
(3) chloroform-acetic acid (9:1). Solvent 1 was the most generally useful.
Solvents 2 or 3 were recommended when R_f's were low using solvent 1.
All 40 pesticides were visualized by first spraying with hydriodic acid:glacial
acetic acid:water (1:1:2); this reagent was stable for several weeks. After
spraying, the plates were covered with a plain glass plate and heated at
180°C for 30 min. After cooling, the protective glass plate was removed
(in a fume hood as toxic fumes were emitted) and the plate sprayed with
ammonium molybdate solution [2 g of ammonium molybdate + 20 ml of
hydrochloric acid-water (1:1), dissolved and 80 ml of water added]; this
reagent was also stable for several weeks. The plate was heated again at
180°C for 5 min. Finally, the plate is cooled and sprayed with tin(II)
chloride solution [1 g tin(II) chloride in 10 ml of concentrated hydrochloric
acid + 40 ml of water + 50 ml of acetone); this reagent must be prepared
fresh daily. The plate was placed in a tank containing two beakers of
ammonia and the pesticides appeared as blue spots. Samples were estimated
by comparing spot intensities with standards. Sensitivity was reported to
1 μg per spot.

Howe and Petty [56] determined $0,0,0',0'$-tetramethyl-$0,0'$thiodi-p-
phenylene phosphorothioate (Abate), a pesticide effective in killing mosquito
larvae, in water by TLC using silica gel. The analysis was applied to
controlled samples where Abate stock solution (2 g/liter in acetone) was
diluted 1:200 with distilled water and added in the desired amounts to (1)
distilled water; (2) surface water; or (3) a 4:1 mixture of surface water-
settled sewage. Samples (1500 ml) were acidified and extracted with 500
ml of redistilled chloroform. The chloroform was dried with anhydrous
sodium sulfate and evaporated to dryness. Residues were reconstituted in
acetone at a concentration of ca. 1 μg/μl of anticipated Abate. TLC plates
(20 × 20 cm) were prepared using Silica Gel G spread at a thickness of
0.25 mm. Plates were air dried and activated at 90°C for 1 hr. Fifteen
microliters of sample and standard solutions were spotted on the plate and
developed using hexane-acetone (10:1). The plate was dried, exposed to
bromine vapor for 1 min, and aerated for 3 min (in a hood). Spots were
visualized by spraying with 1% N,N-dimethyl-p-phenylazoaniline in 95%
2-propanol; spots appeared red on a yellow background. Spot areas were
calculated and a straight line was obtained for a plot of standard areas over

a concentration range of 9-24 μg with deviations no greater than 0.5 μg.
Best results were obtained using fresh plates. Care must be taken to be
sure all the bromine vapor has been removed. Average recovery was 75%
with an average deviation of 6%. No background interference from sewage
samples was observed.

Kawahara et al. [57] used TLC as a cleanup procedure for GC analysis
of parathion and methyl parathion in the presence of chlorinated hydrocarbons. Samples were collected by either the grab technique or by carbon
adsorption; the latter required much time and was reported to be less
accurate. Grab samples (1-4 liters) were extracted with either ether-
hexane (1:1) or benzene-hexane (1:1). The solvents were dried (anhydrous
sodium sulfate) and concentrated to a volume of 0.1-1.0 ml. TLC plates
were prepared using Silica Gel G spread at a thickness of 0.25 mm. Up to
80% of the extracted sample residues were spotted on the plate. Plates
were developed using carbon tetrachloride. After development, the plate
was divided into zones that contained known classes of components; these
classes were then scraped off and eluted with 5 ml of ether-petroleum
ether (1:1) for determination by GC. This would be a poor technique for
use with an unfamiliar sample.

Gebott [1] determined chlorophenols in various pesticides and herbicides
by TLC and fluorescence. Plates were prepared using silica gel, alumina,
and Gelman Instant TLC media type SG. Plates were spread at a thickness
of 0.3 mm, air dried, and activated at 150°F for 60-90 min. Standards
were prepared at chlorophenol concentrations of 1 μg/ml in acetone. Several
liters of aqueous sample were extracted with petroleum ether (five to ten
100-ml extractions). The solvent was concentrated to 0.1 ml and samples
were spotted (2 μl) along with standards on silica plates. After development
in benzene (15 cm), the plate was dried and viewed under short-wavelength
UV light (254 nm); sensitivity was 0.5 μg per spot. On alumina, the
developing solvent was petroleum ether-benzene (1:1), and detection was
also by UV. Plates made using Gelman Instant TLC medium were developed
using benzene-petroleum ether-acetic acid (20:10:1). Components in this
system were visualized by spraying with Rhodamine B solution (0.05% in
ethanol). Spots appeared orange-pink on a purple background when exposed
to long-wavelength UV light (350 nm). Sensitivity was 0.5 μg per spot.

Taylor and Bogacha [58] determined DDT, methoxychlor, and γ-BHC
in water and effluents by TLC. Samples (1 liter) were extracted with 100-
200 ml of petroleum ether. The extract was filtered, dried (anhydrous
sodium sulfate), and evaporated to dryness. The residue was dissolved
in 0.1 ml of ethyl ether and applied to a Silica Gel G plate. Plates were
prepared at a thickness of 0.25 mm and activated at 110°C for 30 min.
After development with carbon tetrachloride, the plate was dried and
sprayed with ethanolamine, heated at 120°C, and sprayed with 0.1N silver
nitrate-nitric acid solution (3:1). Under UV light, the area of the spots was
determined using a planimeter. Sample components could be estimated by
comparison with standards. Sensitivity could be enhanced by using such

modifications as column adsorption during sample preparation. Recoveries were considerably less than 100% but correction factors were used and an accuracy of ±20% was reported.

Dyatlovitskaya [59] also determined DDT, plus hexachloran, aldrin, and dieldrin, in sewage using TLC. Components were extracted from water with benzene and were spotted on alumina plates. After development in hexane, spots were visualized by spraying with an ammoniacal solution of silver nitrate in acetone and exposing the plate to UV light (Hg lamp) for up to 1 hr. There was no reference to accuracy, sensitivity, or reproducibility in this work.

Kawahara [60] determined dieldrin, endrin, lindane, heptachlor epoxide, heptachlor, DDT, DDE, and aldrin in waste waters. The extraction technique was not discussed in this paper but grab samples of effluents were used. TLC plates (20 × 20 cm) were prepared using alumina spread at a thickness of 0.25 mm, air dried for 16 hr at room temperature, and activated immediately prior to use at 75°C for 10 min. Plates were developed in carbon tetrachloride-hexane (1:4). While the samples were shielded, standards were sprayed with Rhodamine B solution for identification. The remaining unsprayed samples were then located by reference with the sprayed standard(s) and removed from the plate by suction through a plugged inverted medicine dropper. These were eluted with 4 ml of ether-hexane-acetone (5:4:1). The eluents could then be determined by GC. Recoveries varied depending on the compound but a standard deviation)based on 13 determinations) of ±9% was reported. Losses were attributed primarily to volatility but some TLC retention was observed resulting in low recoveries.

Schutzmann and Borthel [61] used two-dimensional TLC to identify submicrogram quantities of thionophosphate compounds (P=S) and their oxygen analogs (P=O). Water samples (5 gallons) were decanted into a tared extractor. The samples were extracted with 1 liter of hexane-ether (3:1). The organic phase was dried using anhydrous sodium sulfate and concentrated to 100 ml. TLC plates (20 × 20 cm) were prepared using Adsorbosil-1 (silica gel) according to the procedure outlined by Stahl [62]. An amount of sample containing ca. 0.1-1 μg of pesticide was spotted on the plate. A standard was spotted 11 cm to the right of the sample and another spotted 11 cm above the sample. Preparations were given for preparing oxidative products of the standards. Spot areas were kept as small as possible. Plates were developed using either (1) benzene-tetrahydrofuran (8:2) or (2) benzene-methanol (9:1), for a distance of 10 cm. After drying, the plate was rotated 90° and developed for another 10 cm using either (1) hexane-tetrahydrofuran (8:2) or (2) hexane-methanol (9:1). The plate was dried and sprayed with cholinesterase serum until the plate was saturated. Cholinesterase serum spray was prepared by diluting 10 ml of Bacto horse serum (available from Difco Laboratories, Detroit, Mich.) with 35 ml of distilled water. Saturating the plate with spray or moistening with steam prior to spraying helped to avoid flaking. After 25 min the plate was sprayed with indoxyl acetate solution (0.5% in distilled acetone). The plate was exposed

to UV light (253 nm) for 15 min. Components were visualized as white
spots on a blue background. The white spots indicated the area of enzyme
inhibition by the pesticide. No reference was made to quantitation.

Leoni and Puccetti [63] also studied organic phosphorus pesticides in
Italian surface waters. Sample extracts were prepared and purified according to standard methods [64] or by a semiautomatic extraction of 10-20
liters using nonpolar solvents such as hexane. Standards were prepared in
n-hexane at a concentration of 0.5, 5, and 50 μg/ml. TLC plates (20 × 20
cm) were prepared using alumina spread at a thickness of 0.25 mm, activated for 1 hr at 120°C, and stored in a dry area (presumably over silica
gel). Adsorbents could be prewashed but this was not found essential.
Samples (10-40 μl, equivalent to 500 ml of effluent sample) were spotted on
the plate along with standards. The developing chamber was lined with
filter paper and saturated with developing solvent consisting of 7% acetone
in n-hexane. Two plates were prepared and each was developed for a
distance of 10 cm. Components were visualized by enzymatic and chroma-
genic agents. In enzymatic detection, reagent A consisted of 10 ml of
unhemolyzed horse serum diluted to 25 ml with distilled water; reagent B
consisted 1% indoxyl acetate in acetone (prepared fresh). The plate was
exposed to bromine vapor to convert phosphate esters to oxide analogs and
then air dried. The plate was sprayed with reagent A (10 ml) and heated at
37°C for 10 min; then the plate was sprayed with reagent B (10 ml) and
immediately resprayed with reagent A (10 ml). Components appeared as
a blue-black spot on a violet background under UV light and could be estimated by comparing intensities with standards. Heating the plate at 37°C
for 30 min made the spots appear as white spots on a blue background. On
the average, sensitivity was 0.05 μg per spot. If bromine exposure was
omitted only the oxide analogs were visible. In chemical detection, reagent
A consisted of tetrabromophenolphthalein ethyl ester (0.2% in acetone);
reagent B consisted of 500 mg of silver nitrate in 25 ml of distilled water,
brought to 100 ml with acetone; reagent C consisted of 2.5 g citric acid in
25 ml of distilled water, brought to 50 ml with acetone. The plate was
dried and sprayed with 10 ml of reagent A and then 10 ml of reagent B.
After 2 min, the plate was sprayed with reagent C. This spray was used
again after another 10 min. (P=S) components appeared as blue spots on a
yellow background. Sensitivity varied depending on the pesticide but, in
general, enzymatic detection was about five to ten times more sensitive
(ca. 0.5 μg per spot). In samples not sensitive to enzymatic detection but
sensitive to chemical detection other techniques, e.g., GC, were used to
quantitate components.

In Table 3 are listed all the spray reagents for visualization of the spots
on TLC plates that have been discussed in this chapter.

TABLE 3

Spray Reagents for Visualization of Spots on TLC Plates Referred
to Throughout this Chapter

Reagent	Page
4-Aminoantipyrine — for phenols	659,660
Antimony pentachloride — for grease and oils	649
Brillian Green solution — for pesticides	663
Bromine vapor — for pesticides	662,663,668
Cholinesterase serum (from Bacto horse serum) — for pesticides	667
Copper-ninhydrin — for caprolactam	661
Diazotized p-nitroaniline — for phenols	660
3,5-Dibromo-p-benzoquinone chlorimine — for malathion, dimethoate, and ethion	663
Dichlorofluoroscein — for carbamates, urea, and uracil type herbicides	663
p-Dimethylaminobenzaldehyde — for phenyl urea and carbamates	664
N,N-Dimethyl-p-phenylazoaniline — for pesticides	665
Diphenylcarbazide — for chromium	652
Diphenyl-thiocarbazone/dimethylglyoxime — for lead and cadmium	652
Dithiooxamide — for cobalt, copper, and nickel	651
Dithiooxamide — for manganese and zinc	652
Dragendorf reagent — for detergents	646,647
Ethanolamine/silver nitrate — for pesticides	666
1-Fluoro-2,4-dinitrobenzene — for ureas and carbamates	663,664
Formaldehyde/sulfuric acid — for greases and oils	649
Gibb's reagent — for phenols	658,659
Hydriodic acid, ammonium molybdate, and tin(II) chloride — for pesticides	665
Indoxyl acetate — for pesticides	667,668
Ninhydrin — for caprolactam	661
Ninhydrin — for pesticides	662,663

TABLE 3. — Continued

Reagent	Page
p-Nitrobenzenediazonium fluoroborate — for phenols	658,660
Phosphomolybdic acid — for sterols	656
1-(2-Pyridylazo)-2-naphthol — for metals	653
4-(2-Pyridylazo) resorcinol — for metals	653
Rhodamine B — for phenols	659
Rhodamine B — for pesticides	666
Silver Nitrate — for herbicides	664,666,668
Silver nitrate — for phenols	659,660
Sodium nitrite/napthol — for herbicides	663,664
Sulfuric acid (50%) — for sterols	656
Sulfuric acid-ethanol — for pesticides	663
1-(2-Thiazolyl)-2-naphthol — for metals	653
4-(2-Thiazolylazo)resorcinol — for metals	653
Tin chloride-sodium thiosulfate — for arsenic	652
Tetrabromophenolphthalein ethyl ester, silver nitrate, and citric acid — for pesticides	668

ACKNOWLEDGMENTS

I wish to express appreciation to the Merck, Sharp, and Dohme Research Library staff (West Point) for their prompt and patient assistance, and also to Dr. G. V. Downing and Mr. F. A. Bacher for their constructive comments on the introduction to this chapter.

REFERENCES

1. M. Gebott, Solutions, 6(1), 8 (1967).

2. K. Mancy and W. Weber, Jr., in Treatise on Analytical Chemistry, Part III, Vol. 2, Wiley, New York, 1971, p. 415.

3. J. G. Rabosky and D. L. Koraido, Chem. Eng., 80(1), 111 (1973).

4. R. Bunch, in Treatise on Analytical Chemistry, Part III, Vol. 2, Wiley, New York, 1971, p. 563.

5. A. Bevenue, T. Kelly, and J. Hylin, J. Chromatog., 54, 71 (1971).

6. M. Kovacs, J. Assoc. Off. Agr. Chem., 48(5), 1018 (1965).

7. H. Allen and K. Mancy, in Water and Water Pollution Control Handbook Marcel Dekker, New York, 1972, p. 979.

8. R. Ellerker, H. Dee, F. Lax, and D. Sargent, Water Poll. Contr. (London), 67, 542 (1968).

9. S. Patterson, E. Hunt, and K. Tucker, J. Proc. Inst. Sewage Purif., 2, 190 (1966).

10. S. Patterson, E. Hunt, and K. Tucker, Proc. Soc. Anal. Chem. Conf., Nottingham, Eng., 339 (1965); Chem. Abstr., 65, 19826g (1966).

11. S. Patterson, C. Scott, and K. Tucker, J. Amer. Oil Chem. Soc., 44(7), 407 (1967).

12. S. Patterson, C. Scott, and K. Tucker, J. Amer. Oil Chem. Soc., 45(7), 528 (1968).

13. S. Patterson, C. Scott, and K. Tucker, J. Amer. Oil Chem. Soc., 47(2), 37 (1970).

14. M. Skougstad, Encyclopedia of Chemical Technology, Vol. 21, 2nd ed., Interscience, New York, 1970, p. 688.

15. S. Jenkins, N. Harkness, P. Hewitt, M. Snaddon, R. Ellerker, B. Divito, and H. Dee, J. Inst. Sewage Purif., 6, 633 (1965).

16. G. Lambert, Tribune CEBEDEAU, 20(271-2), 279 (1966); Chem. Abstr., 66, 21993d (1966).

17. P. Matthews, J. Appl. Chem., 20, 87 (1970).

18. A. Bailey, Water Poll. Contr., 68, 449 (1969).

19. K. Oguma, Talanta, 14, 685 (1967).

20. R. Frei, V. Miketuhova, M. Kinnon, Chromatographia, 3, 519 (1970).

21. V. Miketuhova and R. Frei, J. Chromatog., 47, 435 (1970).

22. V. Miketuhova and R. Frei, J. Chromatog., 47, 427 (1970).

23. S. Zetlmeisl and D. Haworth, J. Chromatog., 30, 637 (1967).

24. S. Hashmi, Ayaz, A. F. Chughtai, and N. Hassan, Anal. Chem., 38, 1554 (1966).

25. J. Murtaugh and R. Bunch, Water Poll. Contr. Fed., 39, 404 (1967).

26. W. Matthews and L. Smith, Lipids, 3, 239 (1968).

27. L. Smith and R. Gouron, Water Res., 3, 141 (1969).

28. J. Wallwork, M. Bentley, and D. Symonds, Water Treat. Exam., 18, 203 (1969).

29. D. Smith and J. Lichtenberg, Microorganic Matter in Water, A.S.T.M., Spec. Tech. Pub. No. 448, 78 (1969).

30. R. L. Shriner and R. C. Fuson, Systematic Identification of Chemical Compounds, 5th ed., Wiley, New York, 1964.

31. O. Aly, Water Res., 2, 587 (1968).

32. M. Zigler and W. Phillips, Env. Sci. Tech., 1(1), 65 (1967).

33. M. Kovacs, J. Assoc. Off. Agr. Chem., 46, 884 (1963).

34. H. Heier, Fortschr. Wasserchem. Ihrer. Grenzgeb., 12, 20 (1970); Chem. Abstr., 73, 59094u (1970).

35. F. Dyatlovitskaya and E. Maktaz, Gig. Sanit., 6, 60 (1965); Anal. Abstr., 14, 1738 (1967).

36. H. Seeboth, Monatsber. Dtsch. Akad. Wiss. (Berlin), 5(11-12), 693 (1963); Anal. Abstr., 12, 3064 (1965).

37. P. Chambon, R. Chambon-Mougenot, J. Bringiuer, Rev. Inst. Pasteur Lyon, 3(4), 395 (1970); Chem. Abstr., 76, 49603g (1972).

38. G. Lambert, Tribune CEBEDEAU, 18(266), 21 (1966); Anal. Abstr., 14, 2880 (1967).

39. I. Berthold, Z. Analyt. Chem., 240(5), 320 (1968); Anal. Abstr., 18, 656 (1970).

40. K. Ershova and I. Mints, Gig. Sanit., 33(9), 52 (1968); Chem. Abstr., 69, 109670 (1968).

41. A. Svishchuk, M. Smygun, and L. Seredguk, Otkrytiya, Izobret., Prom. Obraztsy, Tovarnye Znaki, 48(28), 183 (1971); Chem. Abstr., 76, 27839m (1972).

42. F. Dyatlovitskaya and E. Maktaz, Referat. Zh. Khim., 19GD(7), (1969); Khim. Volokna, 6, 69 (1968); Anal. Abstr., 17, 3892 (1969).

43. L. Scholz and H. Altmann, Z. Analyt. Chem., 240(2), 81 (1968); Anal. Chem., 18, 633 (1970).

44. J. Borneff and H. Kunte, Arch. Hyg. Bakterial, 153(3), 220 (1969).

45. H. Clarke, Proc. 14th Ont. Ind. Waste Conf., June 1967, p. 23-37. Chem. Abstr., 68, 812052 (1968).

46. D. Smith, J. Eichelberger, J. Water Poll. Contr. Fed., 37(1), 77 (1965).

47. A. Shmigidina and M. Klisenko, Vop. Pitan., 29(5), 91 (1970); Chem. Abstr., 74, 11938r (1971).

48. R. Tamus, Revta Chim., 20(4), 259 (1969); Anal. Abstr., 19, 2721 (1970).

49. L. Samosvat, Gidrolicol Zh., 4(4), 78 (1968); Anal. Abstr., 17, 3876 (1969).

50. S. Katz, J. Assoc. Off. Agr. Chem., 49(2), 452 (1966).

51. D. Abbott, K. Blake, K. Tarrant, and J. Thomson, J. Chromatog., 30, 136 (1967).

52. I. Cohen and B. Wheals, J. Chromatog., 43, 233 (1969).

53. M. El-Dib, J. Assoc. Off. Agr. Chem., 53(4), 756 (1970).

54. D. Abbott, H. Egan, E. Hammond, and J. Thomson, Analyst, 89, 480 (1964).

55. J. Askew, J. Ruzicha, and B. Wheals, Analyst, 94, 275 (1969).

56. L. Howe and C. Petty, J. Agr. Food Chem., 17(2), 401 (1969).

57. F. Kawahara, J. Lichtenberg, J. Eichelberger, Water Poll. Contr. Fed., 39, 446 (1967).

58. R. Taylor and T. Bogacha, Chem. Anal., 13(2), 227 (1968); Anal. Abstr., 17, 1206 (1969).

59. F. Dyatlovitskaya and E. Gladenko, Gig. Sanit., 33(11), 53 (1968); Chem. Abstr., 70, 60625n (1969).

60. F. Kawahara, J. Gas Chromatog., 6, 24 (1968).

61. R. Schutzmann and W. Borthel, J. Assoc. Off. Agr. Chem., 52, 151 (1969).

62. E. Stahl, Chemiker Ztg., 82, 323 (1958).

63. V. Leoni and G. Puccetti, Il Farmaco (Ed. Pr.), 26(7), 383 (1971).

64. Pesticide Analytical Manual, Vol. 1, Food and Drug Administration, U.S. Dept. of Health, Education, and Welfare, Washington, D.C., 1968.

ADDENDUM FOR CHAPTER 3

The column of Vivilecchia et al. [see Ref. 253 on p. 177] derived its selectivity from the interaction of the silver atom with the lone pair of electrons on the heterocyclic nitrogen atom. Last year, Frei et al. [1'] applied the silver absorbent to the actual quantitative and qualitative analysis of aromatic nitrogen heterocycles in environmental air samples. With a uv monitor (254 nm) linear calibration curves were obtained and detection limits were 0.2-25 ng per injection (1-m column) for the compounds studied. The analysis of spiked air filter extracts gave results that were within 1-10% of known values. Unfortunately the air samples tested did not contain significant amounts of the aromatic nitrogen heterocycles under investigation, a fact attributed to sample age. It was conjectured that in fresh air samples, interferences from less volatile pollutants would not preclude efficient and conclusive monitoring of aromatic nitrogen heterocycles. However the authors caution that before adopting high-speed liquid chromatography with silver absorbent as a method, considerable work would have to be done on improving air sampling procedures.

Last year, Wolkoff and Larose [2'] also reported the detection of phenols at concentrations as low as 0.4 ppm using a high-speed liquid chromatography system in which cerium (IV) sulfate is allowed to react with phenols in the column effluent. The resulting cerium (III) is detected by fluorescent spectroscopy. Good linearity was shown in the region 10-230 ppb of phenol. A Waters Associates model ALC 202 liquid chromatograph equipped with a modified Milton Roy pump (pressures 3000 p.s.i.g.) was used. Water samples from Lake Ontario at Hamilton were spiked with a mixture of phenols for the tests of the method. The authors emphasized that high speed LC with the sensitive cerium sulfate-fluorescence detection system filled an important role in phenol detection which could not be accomplished as effectively by existing gas chromatographic methods. In particular respect to phenols as water pollutants, it was noted that 2-chlorophenol when present in concentrations as low as 2 ppb caused disagreeable taste and odor [3'], hence the importance of the highly sensitive LC system.

REFERENCES

1'. R. W. Frei, K. Beall, and R. M. Cassidy, <u>Mikrochimica Acta</u>, 859 (1974).

2'. A. W. Wolkoff and R. H. Larose, <u>J. Chromatography</u>, <u>99</u>, 731 (1974).

3'. R. H. Burttschell, A. A. Rosen, F. M. Middleton, and M. B. Ettinger, <u>J. Amer. Water Works Assoc.</u>, <u>51</u>, 205 (1959).

AUTHOR INDEX

Numbers in parentheses are reference numbers and indicate that an author's work is referred to although his name is not cited in the text. Underlined numbers give the page on which the complete reference is listed.

A

Abbott, D.C., 384, <u>389</u>, 505, 506 (10), <u>513</u>, 538, 541(22), 542(23), 545(23), 558(22,23), <u>564</u>, 663, 664, <u>673</u>

Abdel-Akher, M., 264(203), <u>286</u>

Abel, K., 101(17), <u>119</u>

Abrahams, L., 164(281), <u>178</u>, 468 (146), <u>486</u>

Abrams, J., 453(197), <u>484</u>

Acher, R., 347(237), <u>372</u>

Ackman, R.G., 402(21), 416(67, 69), 417(21), <u>423</u>, <u>425</u>

Adams, A.P., <u>321</u>, 340(140), <u>368</u>

Adams, D.F., 113, <u>121</u>

Adams, I.M., 604, <u>608</u>

Addison, R.F., 416, <u>425</u>

Adlard, E.R., 421, <u>426</u>

Agal'tsov, A.M., 620(34), <u>624</u>

Ahling, B., 433(23), 435(23), 437 (23), <u>480</u>

Ahlmann, J., 131(132), <u>172</u>, 181(14), <u>190</u>

Ahren, E.H., Jr., 440(58), 442(58), 451(58), <u>482</u>

Aizawa, A., 329(12), 333(12), <u>362</u>

Akasaki, K., 260(156), <u>283</u>

Akiya, T., 343(188), <u>370</u>

Albersmeyer, A., 620, <u>624</u>

Albon, N., 336(98), <u>366</u>

Albrecht, P., 269(240), 271(261), <u>287</u>, <u>288</u>

Alexander, E., 604, <u>608</u>

Aliprandi, B., 251(72), <u>279</u>

Allegrini, I., 114(54), <u>121</u>

Allen, C.F.H., 186(52), <u>192</u>

Allen, H., 646(7), 648-650(7), <u>671</u>

Allie, W., 572(36), <u>577</u>

Alt, K.O., 382(12), <u>389</u>

Altgelt, K.H., 431, <u>480</u>

Altmann, H., 661, <u>672</u>

Altshuller, A.P., 85, 117(61), <u>119</u>, <u>121</u>, 180(6), <u>190</u>

Alvord, E.T., 131(124), 136(124), <u>171</u>

Alvsaker, E., 304(56), <u>321</u>, 335(82), 344(204), <u>366</u>, <u>370</u>

Aly, O.M., 526, 528(17), 530(17), 532(17), <u>563</u>, <u>564</u>, 600, <u>606</u>, 659, <u>672</u>

Amato, M., 266(341), <u>292</u>

Anders, E., 267(237), <u>287</u>

Andersen, W.R., 573(53), <u>578</u>

Anderso, C.A., 262(170), <u>284</u>

Anderson, A.B., 334(114), <u>367</u>

Anderson, C.A., 302(52), <u>321</u>

Anderson, D.W., 252(90), <u>280</u>

Anderson, G., 340(141-143,150), 341(150), 342, 361(310), <u>368</u>, <u>369</u>, <u>375</u>

Andreeva, L.N., 620(33), <u>624</u>